GRAPHS & FUNCTIONS

for JEE Main & Advanced

Conceptual Graphs • Practice Problems • Detailed Explanations

Dr. S.K. Goyal

S Chand And Company Limited

(ISO 9001 Certified Company)

S Chand And Company Limited

(ISO 9001 Certified Company)

Head Office: D-92, Sector–2, Noida 201301, U.P. (India), Ph. 91-120-4682700

Registered Office: A-27, 2nd Floor, Mohan Co-operative Industrial Estate, New Delhi – 110 044
Phone: 011-49731800

www.**schandpublishing.com**; e-mail: **info@schandpublishing.com**

Marketing Offices:

Chennai	:	Ph: 044-2363 2120, chennai@schandpublishing.com
Guwahati	:	Ph: 0361-4066 369, guwahati@schandpublishing.com
Hyderabad	:	Ph: 040-4018 6018, hyderabad@schandpublishing.com
Jalandhar	:	Ph: 0181-4645 630, jalandhar@schandpublishing.com
Kolkata	:	Ph: 033-2335 7458, 2335 3914, kolkata@schandpublishing.com
Lucknow	:	Ph: 0522-4003 633, lucknow@schandpublishing.com
Mumbai	:	Ph: 022-2500 0297, mumbai@schandpublishing.com
Patna	:	Ph: 0612-2260 011, patna@schandpublishing.com

First Edition 2024

ISBN : 978-93-5501-976-9

Product Code: H3GRF54MATH40ENAA24O

PRINTED IN INDIA

By Vikas Publishing House Private Limited, Plot 20/4, Site-IV, Industrial Area Sahibabad, Ghaziabad – 201 010 and
Published by S Chand And Company Limited, A-27, 2nd Floor, Mohan Co-operative Industrial Estate, New Delhi – 110 044

PREFACE

NTA's Joint Entrance Examination (JEE) is an engineering entrance test conducted for admission to various engineering colleges in India including 23 Indian Institutes of Technology (IITs), 31 National Institutes of Technology (NITs), 25 Indian Institutes of Information Technology (IIITs) campuses and other Government Funded Technical Intitutes (GFTIs). It it is conducted in 2 levels: JEE Main and JEE Advanced. This book will guide you through **Graphs and Functions** which are vital in mathematics, helping to visualize concepts, analyze data, solve equations, and understand function behaviour.

This book is designed to build your understanding from the basics to advanced levels, preparing you for the JEE Main and Advanced exams. With clear explanations, practical examples, and comprehensive exercises, this book equips you with the tools needed to solve any graph or function-related questions with confidence. It is tailored to foster problem-solving skills and critical thinking, ensuring that you are well-prepared for the challenges of both the examination and academic pursuits that follow.

This book additionally serves as a bridge to mastering higher-level mathematical skills that are crucial for success not only in JEE but also in your future engineering coursework.

Dr. S.K. Goyal

kqmath@gmail.com

CONTENTS

Chapter

1

Graphs

Concavity and Points of Inflection

In the plane, we consider a curve $y = f(x)$, which is the graph of a single valued differentiable function $f(x)$.

(a) Concave up (or Convex down)

If the graph of the function $y = f(x)$ lies above all of its tangents on an interval I, then it is said to be concave up on I.

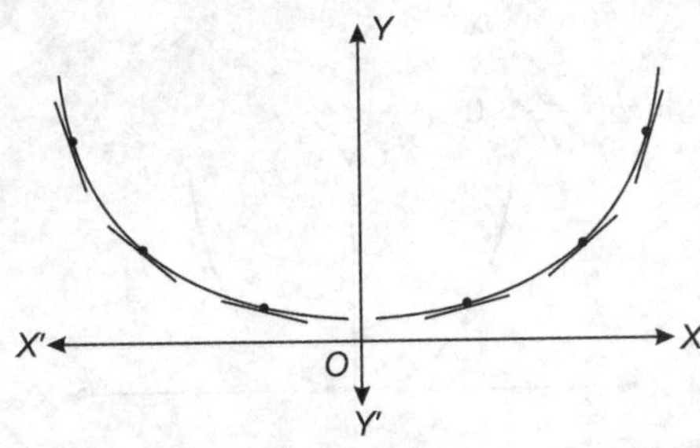

Conversely, a graph of the function $y = f(x)$ will be concave up on an interval I where the slope of the tangent lines increases as x increases *i.e.*,

$$\frac{dy}{dx} \text{ increases as } x \text{ increases} \Rightarrow \frac{d^2y}{dx^2} > 0$$

Note : If shape of the graph of $y = f(x)$ has a "cup" symbol ($\cup$) *i.e.*, "opens" up on an interval I, then f is concave up on an interval I.

(b) Concave down (or Convex up)

If the graph of the function $y = f(x)$ lies below all of its tangents on an interval I, then it is said to be concave down on I.

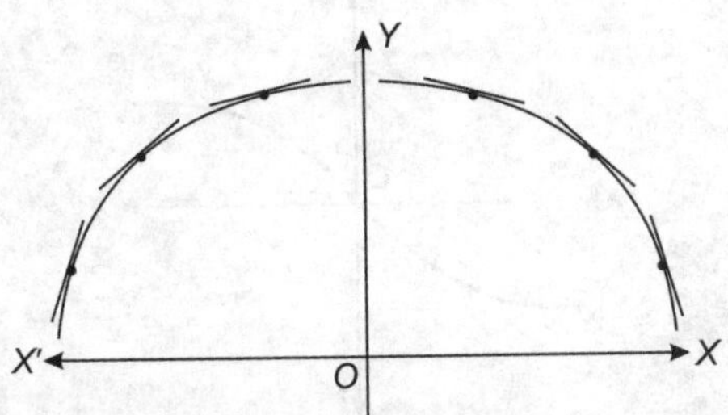

Conversely, a graph of the function $y = f(x)$ will be concave down on an interval I where the slope of the tangent lines decreases as x increases *i.e.*,

$$\frac{dy}{dx} \text{ decreases as } x \text{ increases} \Rightarrow \frac{d^2y}{dx^2} < 0$$

Note : If shape of the graph of $y = f(x)$ has a "cap" symbol ($\cap$) *i.e.*, "opens" down on an interval I, then f is concave down on an interval I.

Illustration 1.

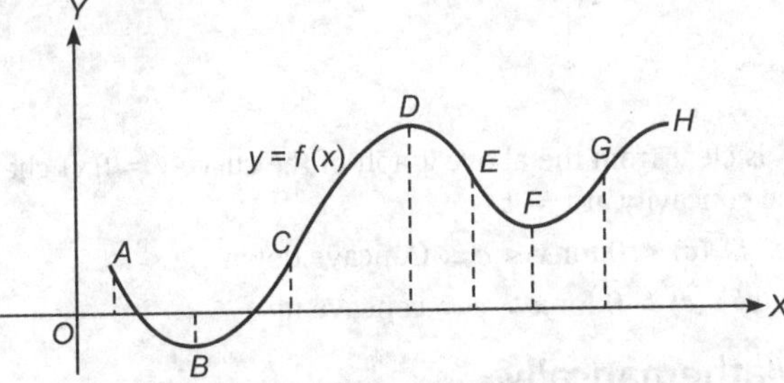

In the above graph the arc ABC of the curve $y = f(x)$ is concave up, CDE is concave down again EFG is concave up and again GH is concave down.

Illustration 2.

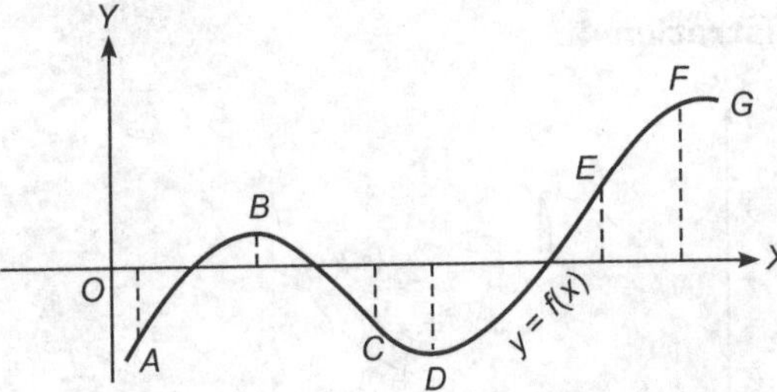

In the above graph the arc ABC of the curve $y = f(x)$ is concave down, CDE is concave up again EFG is concave down.

Illustration 3. Find where the graph of $y = x^3 + 4x + 2$ is concave up and where it is concave down.

Sol. We have,

$$y = x^3 + 4x + 2 \text{ then,}$$

$$\frac{dy}{dx} = 3x^2 + 4 \text{ and } \frac{d^2y}{dx^2} = 6x$$

Therefore, $\frac{d^2y}{dx^2} > 0$ if $x > 0$

and $\frac{d^2y}{dx^2} < 0$ if $x < 0$

So, the graph of $y = x^3 + 4x + 2$ is concave up for $x > 0$ and concave down for $x < 0$.

(c) Point of Inflection

A point P $(c, f(c))$ on a curve $y = f(x)$ is called a point of inflection if the graph of $y = f(x)$ is concave up on one side of P and concave down on the other side.

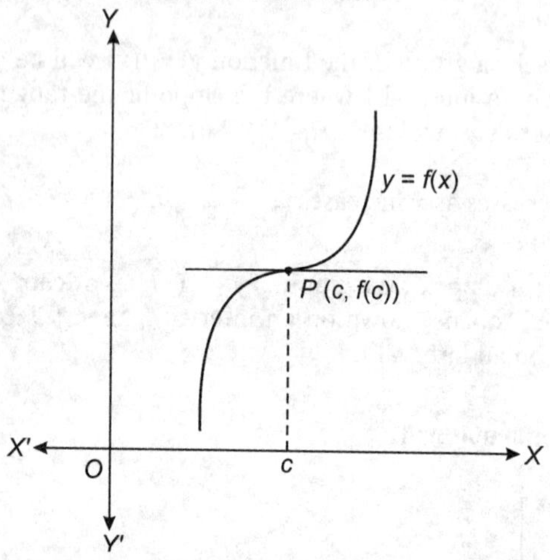

It is clear from the above graph of the curve $y = f(x)$ changes the concavity at $x = c$.

i.e., $f''(c) < 0$ for $x < c \Rightarrow$ Concave down

and $f''(c) > 0$ for $x > c \Rightarrow$ concave up

Mathematically

(i) $\frac{d^2y}{dx^2} = 0$ or does not exist at the point.

(ii) $\frac{d^2y}{dx^2}$ changes its sign about the point.

Illustration 4.

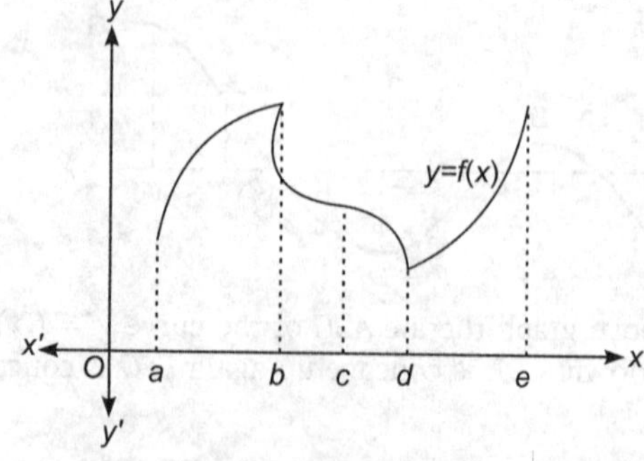

In the above graph of the curve $y = f(x)$ is concave up on $b < x < c$ and $d < x < e$ and concave down on $a < x < b$ and $c < x < d$.

The points of inflection at $x = b$, $x = c$ and $x = d$.

Illustration 5. The curve $y = x^3$ has a point of inflection at $x = 0$, where $\frac{d^2y}{dx^2} = 6x$

Sol. Since $\frac{d^2y}{dx^2} > 0$ at $x > 0$

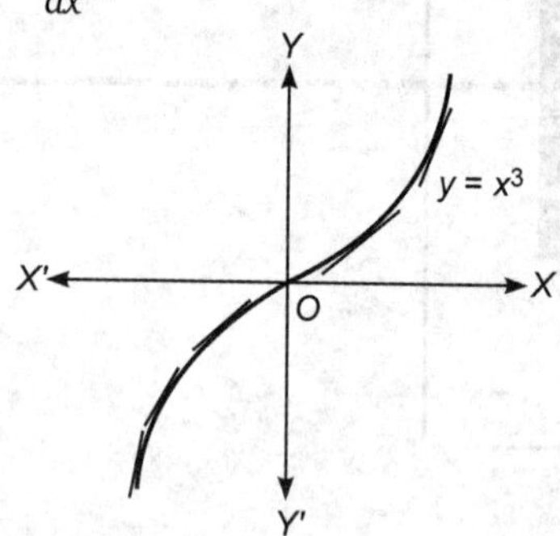

and $\frac{d^2y}{dx^2} < 0$ at $x < 0$

i.e., sign changes $\frac{d^2y}{dx^2}$ at $x = 0$

Hence, the curve $y = x^3$ has a point of inflection at $x = 0$.

Illustration 6. The curve $y = x^4$ has no inflection point at $x = 0$, where $\frac{d^2y}{dx^2} = 12x^2$

Sol. Since $\frac{d^2y}{dx^2} > 0$ at $x > 0$

and $\frac{d^2y}{dx^2} > 0$ at $x < 0$

i.e., no sign changes of $\frac{d^2y}{dx^2}$ at $x = 0$.

Hence, at $x = 0$, the curve $y = x^4$ has no inflection point.

Illustration 7. The curve $y = x^{1/5}$ has a point of inflection at $x = 0$.

Sol. We calculate $\frac{d^2y}{dx^2}$ at $x = 0$

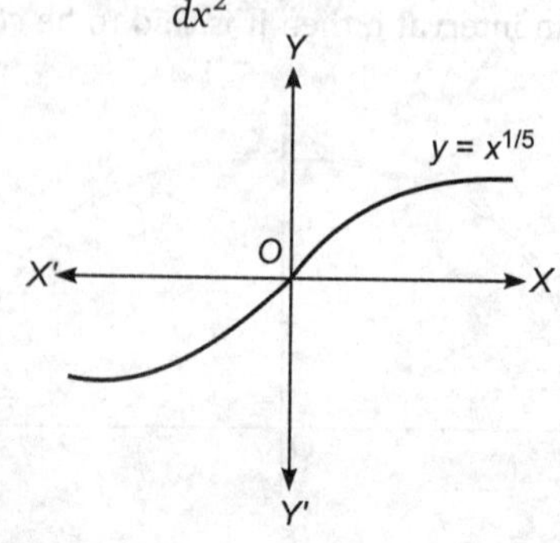

$\therefore \quad \frac{dy}{dx} = \frac{1}{5}x^{-4/5} \quad \Rightarrow \quad \frac{d^2y}{dx^2} = -\frac{4}{25}x^{-9/5}$

As $x \to 0$, $\frac{d^2y}{dx^2} \to -\infty$ and curve is concave up for $x < 0$, $\frac{d^2y}{dx^2} > 0$, then $\frac{dy}{dx}$ is increase and concave down for $x > 0$, $\frac{d^2y}{dx^2} < 0$, then $\frac{dy}{dx}$ is decrease.

Illustration 8. Find the points of inflection of the curve $y = x^3 - 6x^2 + 12x - 3$.

Sol. We have,

$$y = x^3 - 6x^2 + 12x - 3$$

$$\therefore \quad \frac{dy}{dx} = 3x^2 - 12x + 12$$

and $$\frac{d^2y}{dx^2} = 6x - 12$$

For $\frac{d^2y}{dx^2} = 0$, then $x = 2$ (for $x = 2, y = 5$)

and $\left(\frac{d^2y}{dx^2}\right)_{x<2}$ is negative

and $\left(\frac{d^2y}{dx^2}\right)_{x>2}$ is positive

Since the sign of $\frac{d^2y}{dx^2}$ changes about $x = 2$.

Hence, (2, 5) is a point of inflection.

Symmetry

Let $F(x, y) = 0$ be an equation in x and y determines a graph in the plane *i.e.*, the set of all points (x, y) whose co-ordinates satisfy the equation.

Certain kinds of symmetries in the graph of the equation $F(x, y) = 0$.

(i) Symmetry about the x-axis : The graph is symmetrical about the x-axis if the equation of curve unaltered when y is replaced by $-y$

i.e., if $F(x, y) = F(x, -y)$ and if $F(x, |y|) = 0$

Or

Only even powers of y occur in the equation of curve *i.e.*, $F(x, y^{2n}) = 0, n \in N$

Graph of $y^2 = x$:

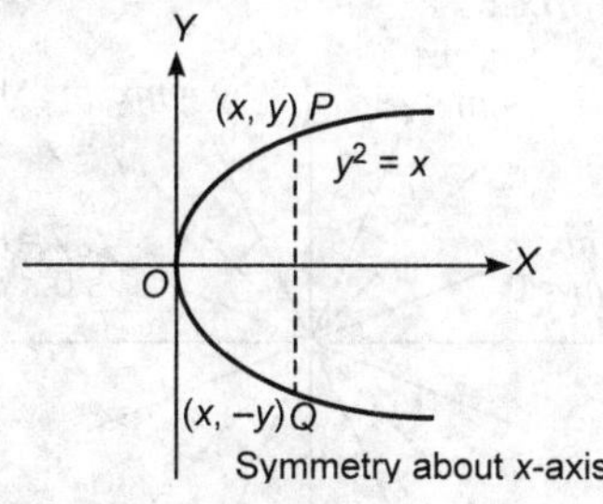

Symmetry about x-axis

Graph of $y = \sqrt{x}$:

It means that y is positive $\quad (\because \sqrt{4} = 2 \neq \pm 2)$

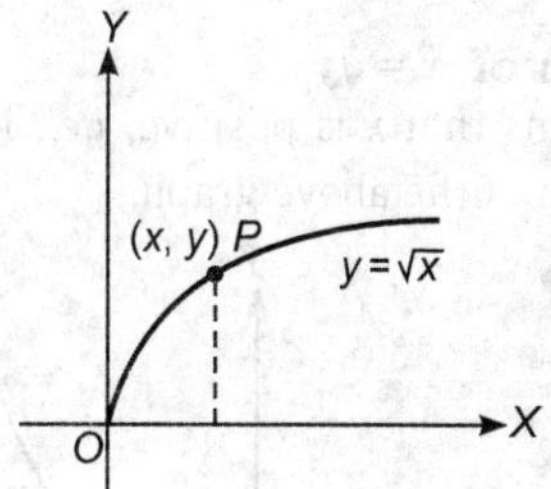

wrong process $\Rightarrow y = \sqrt{x} \Rightarrow y^2 = x$

i.e., ignore the negative portion of the above graph.

Graph of $|y| = x$:

But in the equation of curve $|y| = x$

$|-y| = x \quad \Rightarrow |y| = x$

Curve is symmetrical about x-axis and x always positive.

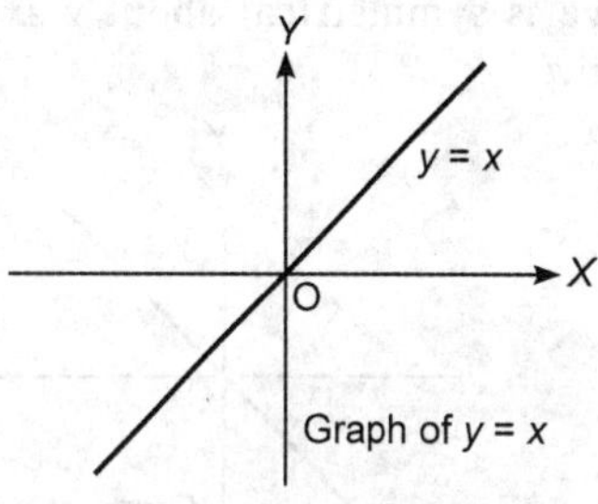

Graph of $y = x$

Or

Negative portion of above graph (about y-axis) convert in positive side of x.

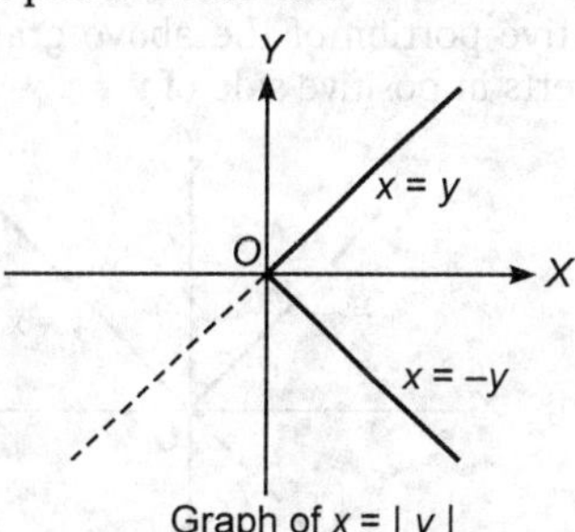

Graph of $x = |y|$

(ii) Symmetry about the y-axis : The graph is symmetrical about the y-axis if the equation of curve unaltered when x is replaced by $-x$

i.e., if $F(x, y) = F(-x, y)$

and if $F(|x|, y) = 0$

Or

Only even powers of x occur in the equation of curve

i.e., $F(x^{2n}, y) = 0, n \in N$.

Graph of $x^2 = y$:

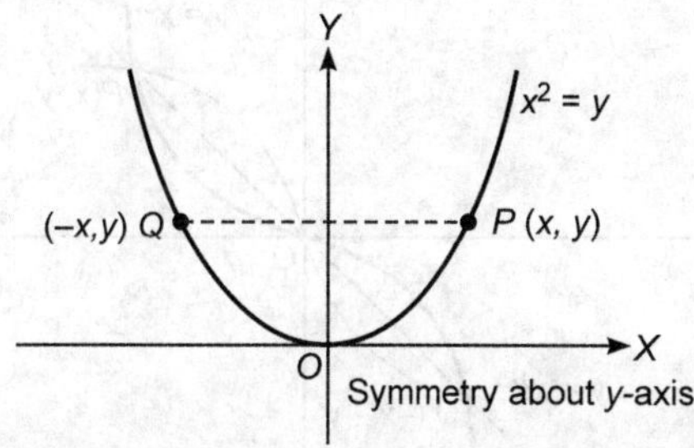

Symmetry about y-axis

Graph of $x = \sqrt{y}$

It means that x is positive, *i.e.*, ignore the negative portion of the above graph.

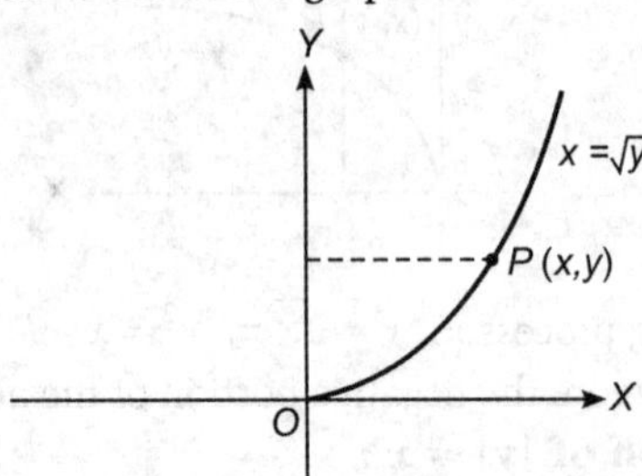

Graph of $y = |x|$: But in the equation of the curve

$$y = |x|$$
$$y = |-x| \quad \Rightarrow \quad y = |x|$$

Curve is symmetrical about y-axis and y always positive.

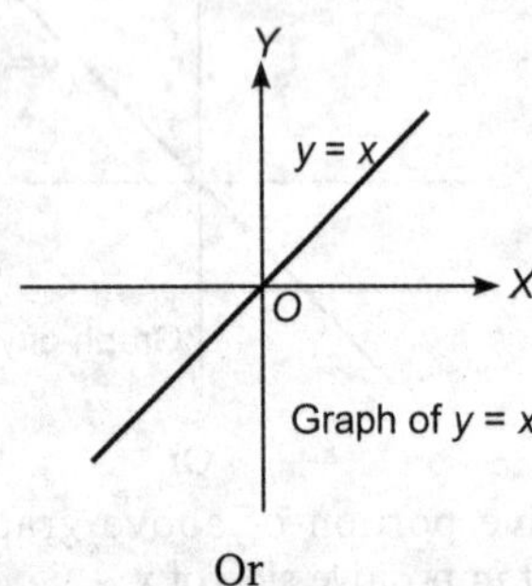

Graph of $y = x$

Or

negative portion of the above graph (about x-axis) converts in positive side of y.

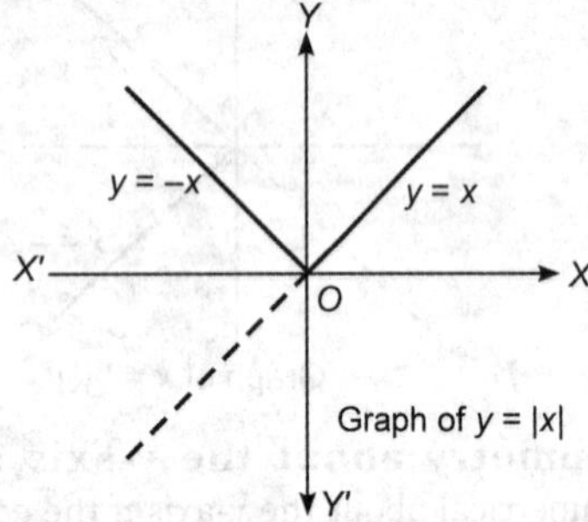

Graph of $y = |x|$

(iii) Symmetry about the Origin : The graph is symmetrical about the origin if the equation of curve remains unaltered when x and y are replaced by $-x$ and $-y$ *i.e.*, $F(x, y) = F(-x, -y)$

Graph of $y = x^3$ and $y = x^{1/3}$:

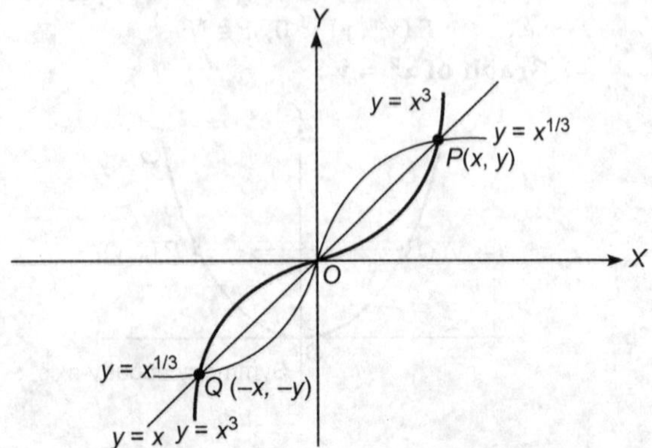

Note : Graphs of $y = x^3$ and $y = x^{1/3}$ are symmetrical about the line $y = x$. Hence $y = x^3$ is the inverse function of $y = x^{1/3}$ and vice-versa.

(iv) Symmetry about the line $y = x$: The graph is symmetrical about the line $y = x$ if the equation of curve remains unaltered when x and y are interchanged.

i.e., $F(x, y) = F(y, x)$

Graph of $\sqrt{x} + \sqrt{y} = 1$: Here $x \geq 0$ and $y \geq 0$.

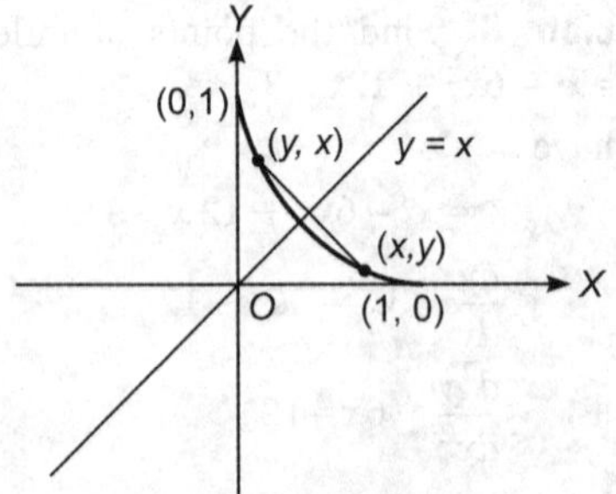

Graph of $x^3 + y^3 = 3axy$:

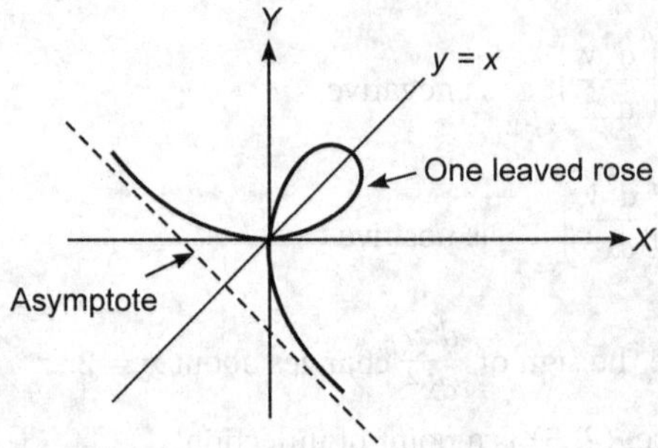

Graph of Basic Functions

Graph of basic functions is essential for plotting typical graphs.

Functions are of two types:

(i) Algebraic Functions

(ii) Transcendental Functions

(1) Algebraic Functions

Functions of the type :

$f(x) = ax + b, f(x) = ax^2 + bx + c,$

$f(x) = x^n, f(x) = \dfrac{1}{ax + b}, f(x) = \dfrac{1}{ax^2 + bx + c},$

$f(x) = \dfrac{1}{x^n}, f(x) = \sqrt{x}$ etc. come under this type.

Graph of $y = mx$:

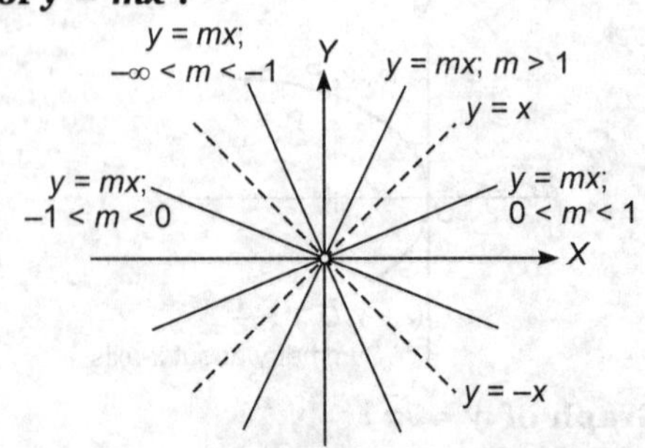

$\therefore$ $y = mx$ is a line passing through origin.

Graph of straight line $ax + by + c = 0$:

First write $ax + by + c = 0$ in intercept form

i.e., $$\frac{x}{\left(-\frac{c}{a}\right)} + \frac{y}{\left(-\frac{c}{b}\right)} = 1$$

and then draw :

Case I. If $a > 0, b > 0, c < 0$.

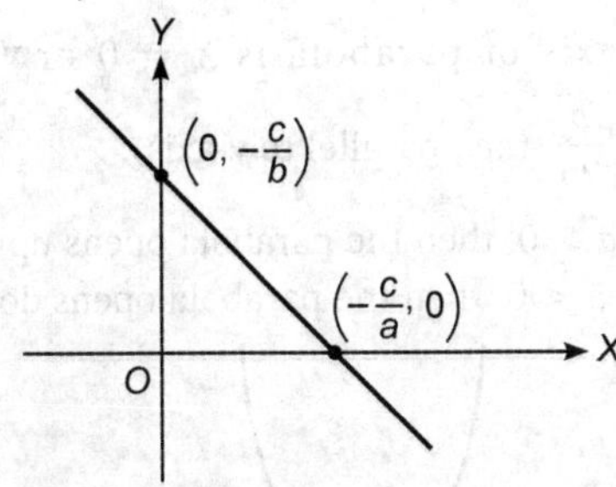

Case II. If $a < 0, b > 0, c < 0$.

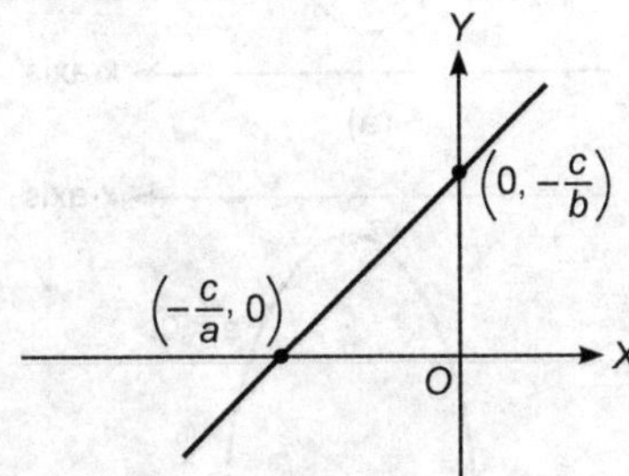

Case III. If $a < 0, b < 0, c < 0$.

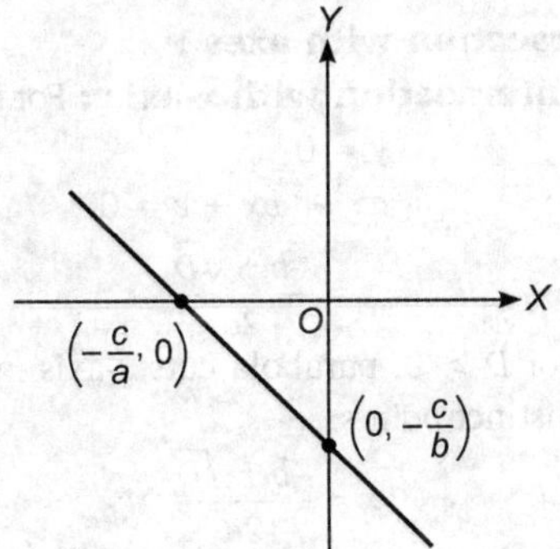

Case IV. If $a > 0, b < 0, c < 0$.

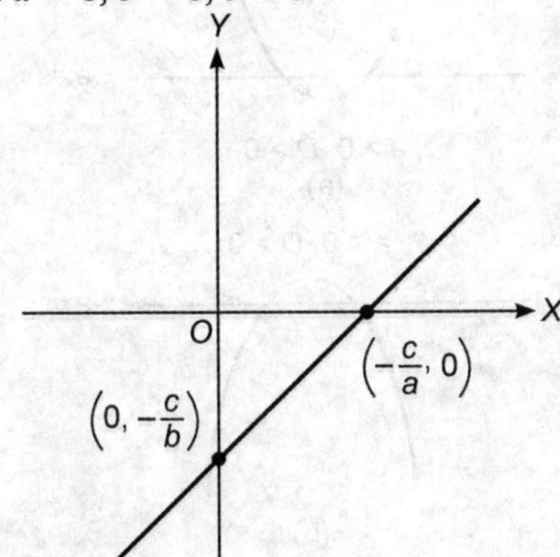

Graph of Circle:

Circle $x^2 + y^2 = a^2$ with centre at (0, 0) and radius a.

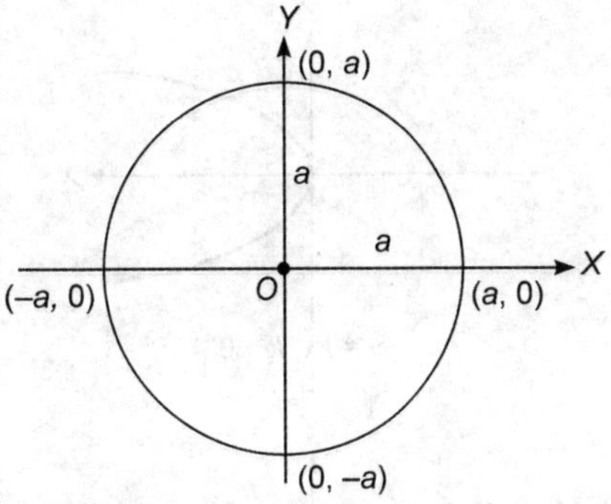

Graph of $x = \sqrt{(a^2 - y^2)}$:

Here $x \geq 0$ and $-a \leq y \leq a$ *i.e.*, a part of circle.

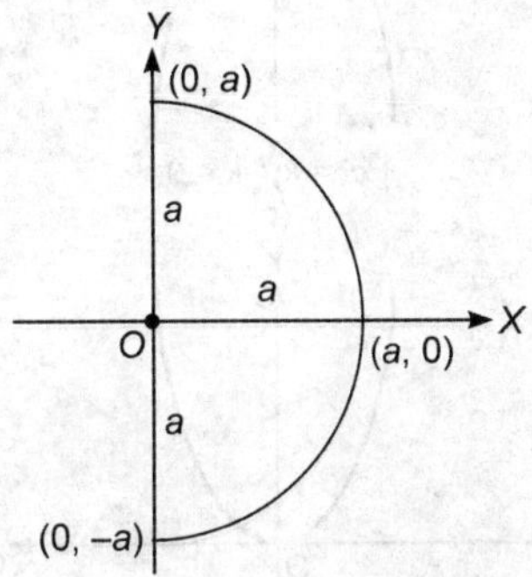

Graph of $y = \sqrt{a^2 - x^2}$:

Here $y \geq 0$ and $-a \leq x \leq a$ *i.e.*, a part of circle.

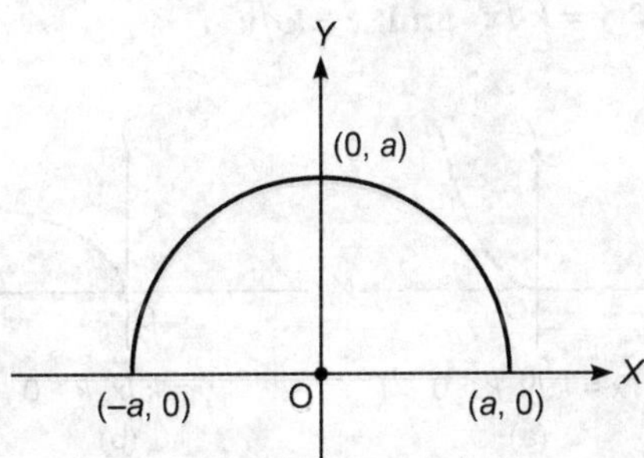

Note: We can write $x^2 + y^2 = a^2 \Rightarrow y = \pm\sqrt{a^2 - x^2}$ but we can not write $y = \sqrt{a^2 - x^2} \Rightarrow x^2 + y^2 = a^2$. Since $y = \sqrt{a^2 - x^2}$ is symmetrical about y-axes but $x^2 + y^2 = a^2$ is symmetrical about both axes.

Graph of Parabolas

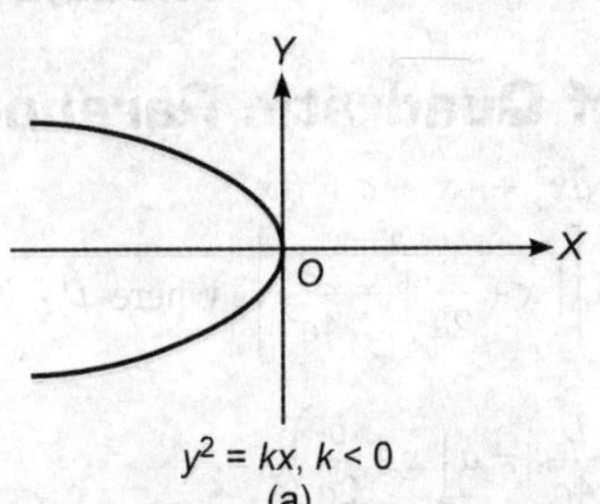

$y^2 = kx, k < 0$

(a)

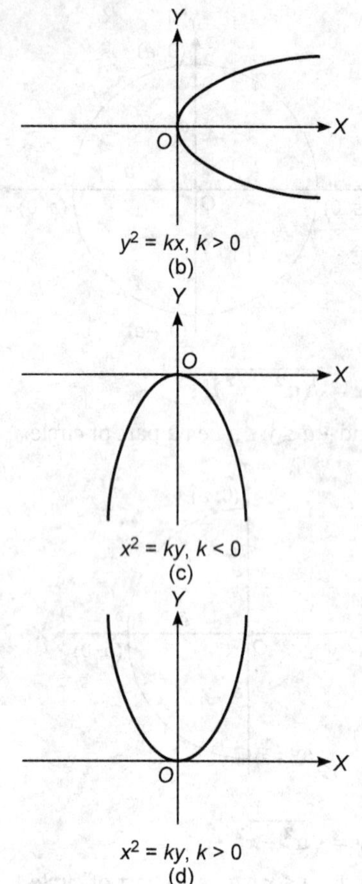

$y^2 = kx, k > 0$
(b)

$x^2 = ky, k < 0$
(c)

$x^2 = ky, k > 0$
(d)

Graph of $y = k\sqrt{x}$ and $x = k\sqrt{y}$:

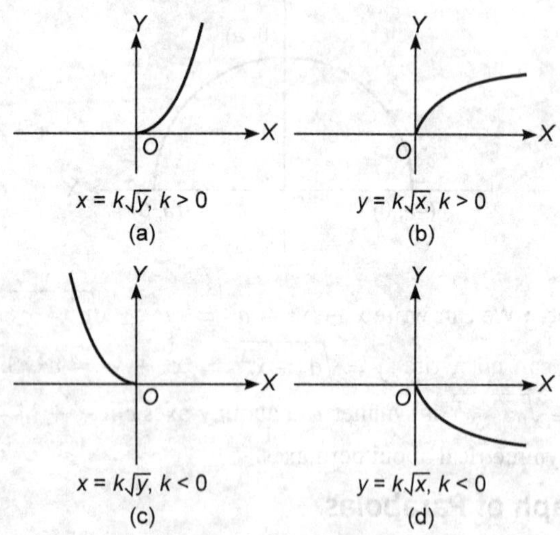

$x = k\sqrt{y}, k > 0$
(a)

$y = k\sqrt{x}, k > 0$
(b)

$x = k\sqrt{y}, k < 0$
(c)

$y = k\sqrt{x}, k < 0$
(d)

Graph of Quadratic Parabola

We have $y = ax^2 + bx + c = f(x)$

$\Rightarrow \quad y = a\left[\left(x+\dfrac{b}{2a}\right)^2 - \dfrac{D}{4a^2}\right]$, where $D = b^2 - 4ac$

or $\quad \left(y+\dfrac{D}{4a}\right) = a\left(x+\dfrac{b}{2a}\right)^2$

Now let $\quad y + \dfrac{D}{4a} = Y$

and $\quad x + \dfrac{b}{2a} = X$

$\therefore \quad Y = aX^2 \Rightarrow X^2 = \dfrac{Y}{a}$

1. The shape of the curve $y = f(x)$ is parabolic.

2. The axis of parabola is $X = 0$ or $x + \dfrac{b}{2a} = 0$ or $x = -\dfrac{b}{2a}$ (*i.e.*, parallel to y-axis).

3. (i) If $a > 0$, then the parabola opens upwards.
(ii) If $a < 0$, then the parabola opens downwards.

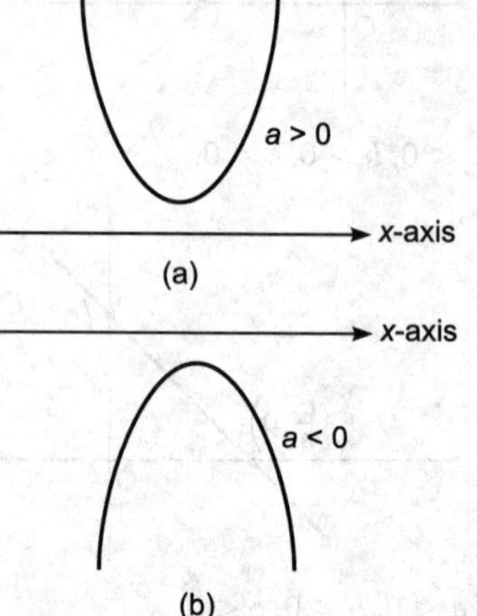

(a)

(b)

4. Intersection with axes :

(i) Intersection with x-axis : For x-axis,

$$y = 0$$

$$\Rightarrow \quad ax^2 + bx + c = 0$$

$$\Rightarrow \quad x = \frac{-b \pm \sqrt{D}}{2a}$$

For $D > 0$, parabola cuts x-axis in two real and distinct points.

i.e., $\quad x = \dfrac{-b \pm \sqrt{D}}{2a}$

x-axis

$a > 0, D > 0$
(a)

$a < 0, D > 0$

x-axis

(b)

For $D = 0$, parabola touches x-axis in one point

i.e., $\quad x = -\dfrac{b}{2a}$

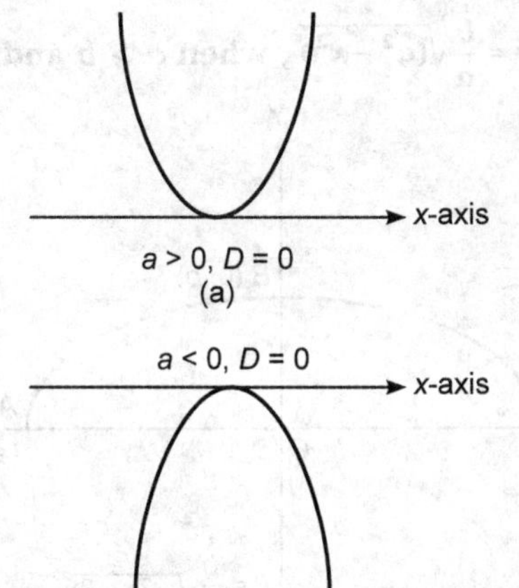

For $D < 0$, parabola does not cut x-axis (*i.e.*, imaginary values of x)

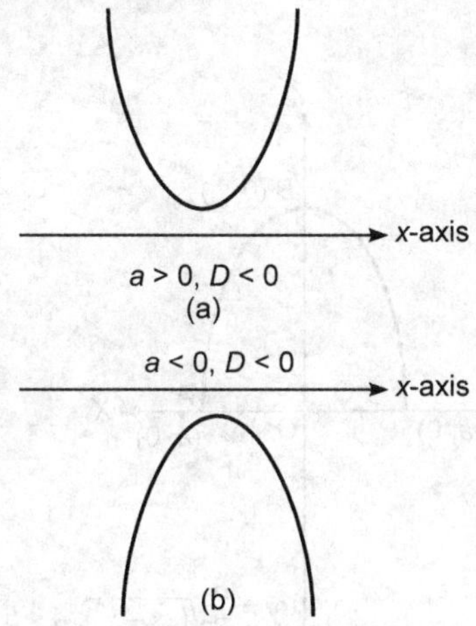

(ii) Intersection with y-axis:

For y-axis, $x = 0$

$\Rightarrow y = c$ (*i.e.*, parallel to x-axis)

5. Greatest and least value of $f(x)$:

Vertex of the parabola $Y = aX^2$ and $X = 0, Y = 0$

$$\therefore \quad x + \frac{b}{2a} = 0 \text{ and } y + \frac{D}{4a} = 0$$

Hence, vertex of $y = ax^2 + bx + c$ is $\left(-\frac{b}{2a}, -\frac{D}{4a}\right)$

- For $a > 0$, $f(x)$ has least value at $x = -\frac{b}{2a}$. This least value is given by

$$f\left(-\frac{b}{2a}\right) = -\frac{D}{4a}$$

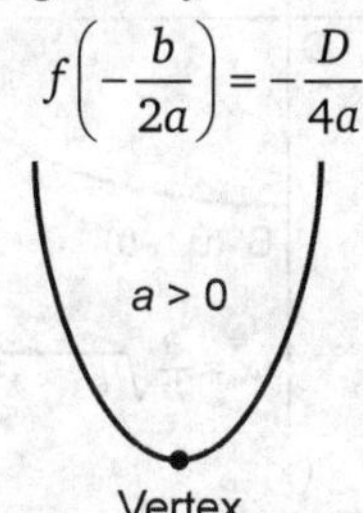

- For $a < 0$, $f(x)$ has greatest value at $x = -\frac{b}{2a}$. This greatest value is given by

$$f\left(-\frac{b}{2a}\right) = -\frac{D}{4a}$$

Vertex

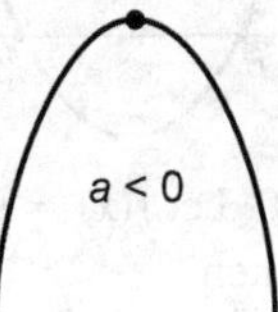

Sign of Quadratic Expression

Let $\quad f(x) = ax^2 + bx + c$

or $\quad y = ax^2 + bx + c$

when $a, b, c \in R$ and $a \neq 0$, for some values of x, $f(x)$ may be positive, negative or zero. This gives the following cases :

(1) $a > 0$ and $D < 0$ so, $f(x) > 0$ for all $x \in R$ *i.e.*, $f(x)$ is positive for all $x \in R$.

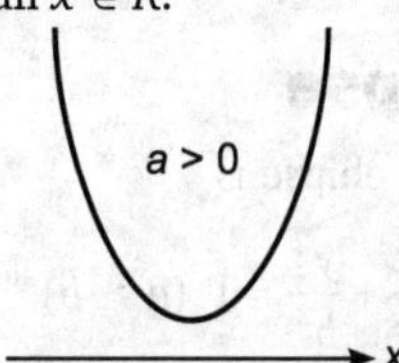

(2) $a < 0$ and $D < 0$ so, $f(x) < 0$ for all $x \in R$ *i.e.*, $f(x)$ is negative for all $x \in R$.

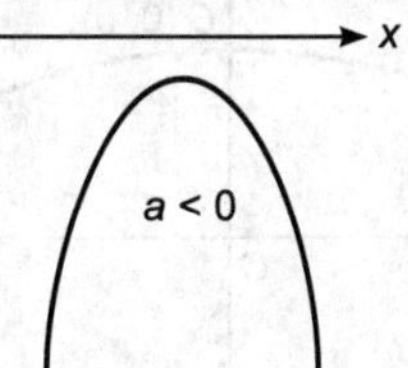

(3) $a > 0$ and $D = 0$ so, $f(x) \geq 0$ for all $x \in R$ *i.e.*, $f(x)$ is positive for all real values of x except at vertex when $f(x) = 0$.

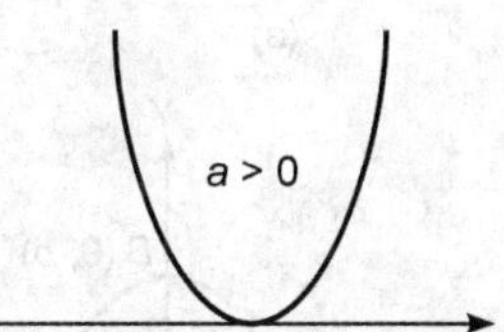

(4) $a < 0$ and $D = 0$ so, $f(x) \leq 0$ for all $x \in R$ *i.e.*, $f(x)$ is negative for all real values of x except at vertex when $f(x) = 0$.

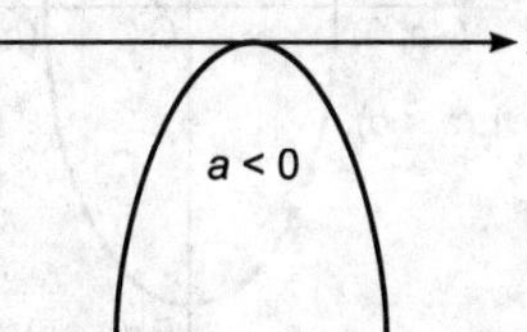

(5) $a > 0$ and $D > 0$. Let $f(x) = 0$ has two real roots α and β $(\alpha < \beta)$, then

$$f(x) > 0 \ \forall \ x \in (-\infty, \alpha) \cup (\beta, \infty)$$

and $\quad f(x) < 0 \ \forall \ x \in (\alpha, \beta)$

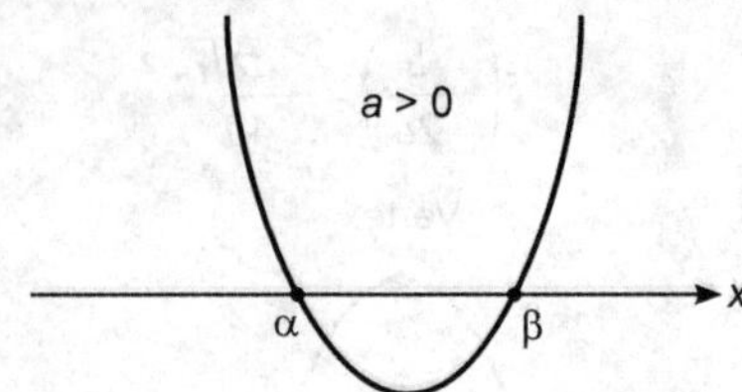

(6) $a < 0$ and $D > 0$. Let $f(x) = 0$ have two real roots α and β ($\alpha < \beta$), then

$$f(x) < 0 \ \forall \ x \in (-\infty, \alpha) \cup (\beta, \infty)$$

and $$f(x) > 0 \ \forall \ x \in (\alpha, \beta)$$

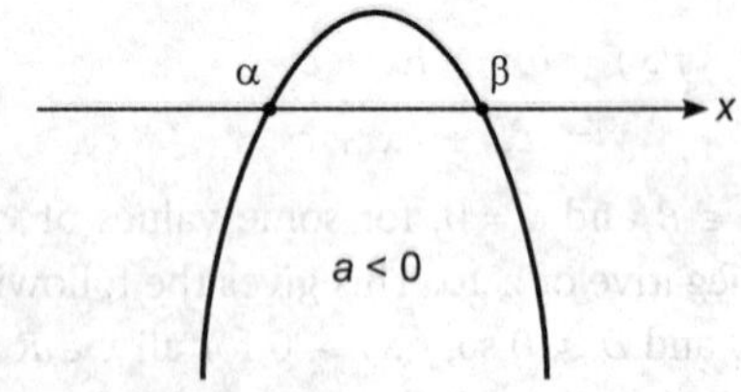

Graph of Ellipse

Standard equation of ellipse is

$$\frac{x^2}{a^2} + \frac{y^2}{b^2} = 1 \quad (a > b)$$

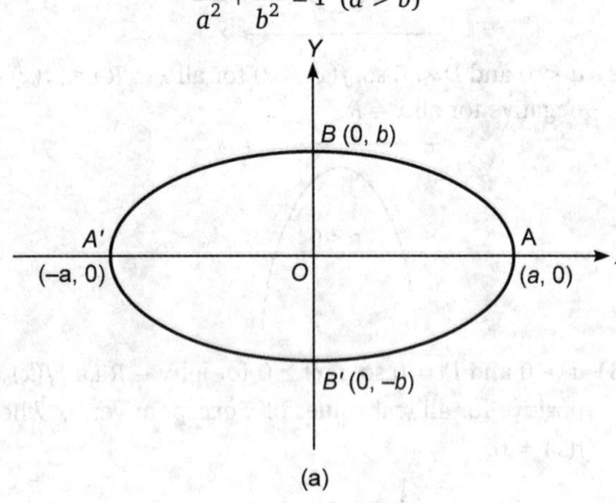

(a)

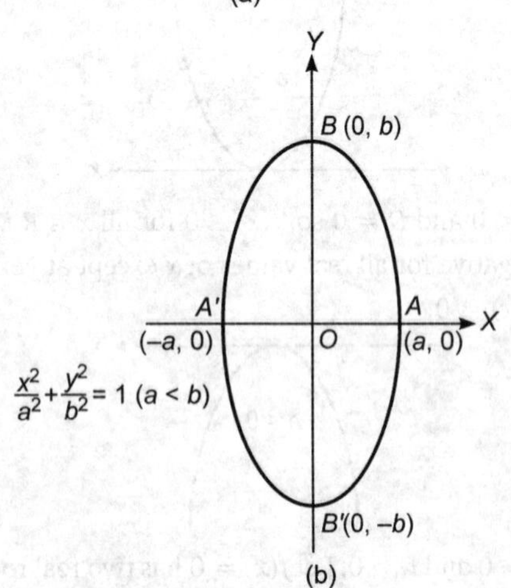

(b)

Graph of $y = \frac{b}{a}\sqrt{(a^2 - x^2)}$**, when** $a > b$ **and** $a < b$

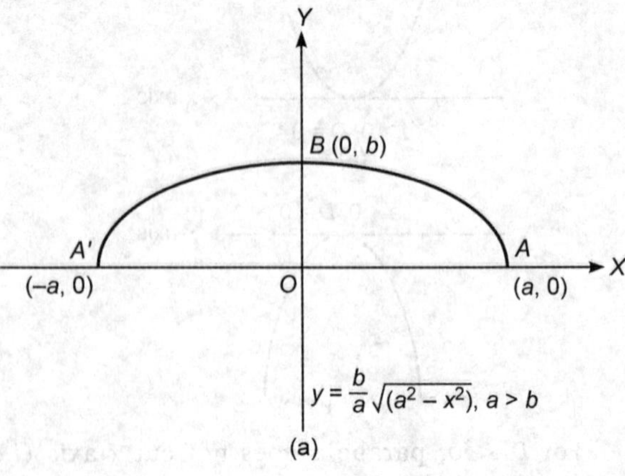

(a)

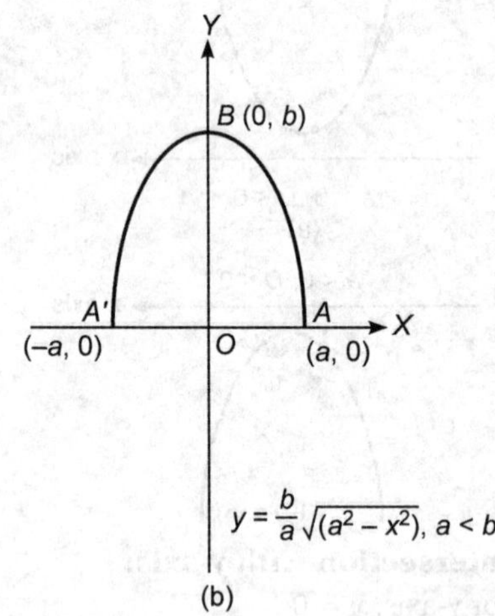

(b)

Graph of $x = \frac{a}{b}\sqrt{(b^2 - y^2)}$**, when** $a > b$ **and** $a < b$

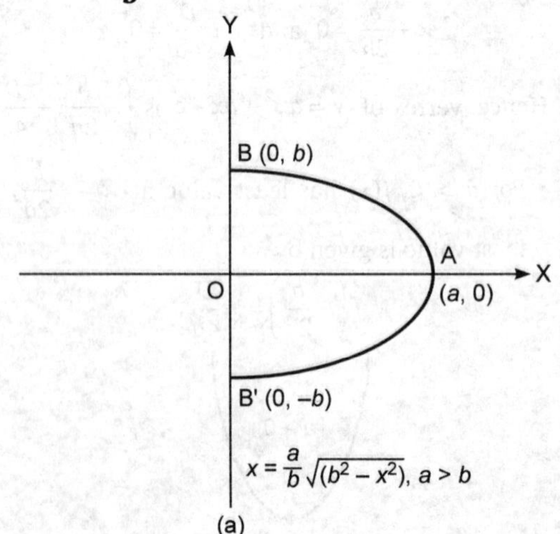

(a)

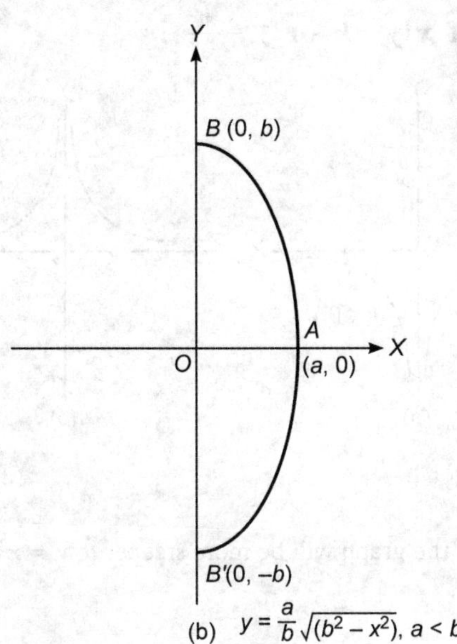

Graph of Hyperbola

Standard equation of hyperbola is

$$\frac{x^2}{a^2} - \frac{y^2}{b^2} = 1 \quad \text{(hyperbola)}$$

or

$$-\frac{x^2}{a^2} + \frac{y^2}{b^2} = 1 \quad \text{(conjugate hyperbola)}$$

Graphs of hyperbolas will be shown in the following figures :

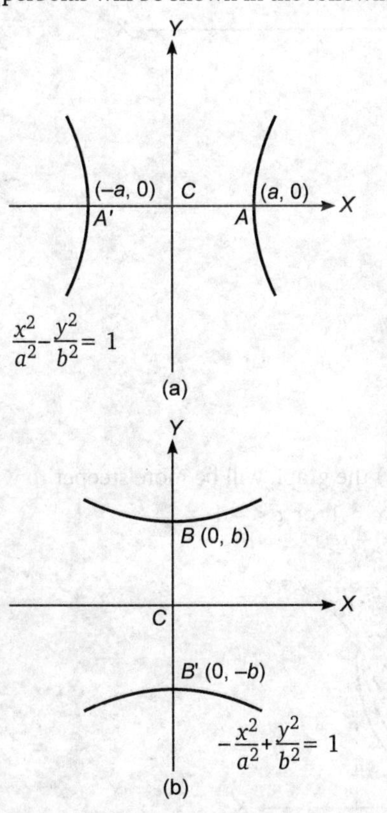

Graphs of $y = \frac{b}{a}\sqrt{(x^2 - a^2)}$, $y = \frac{b}{a}\sqrt{(x^2 + a^2)}$

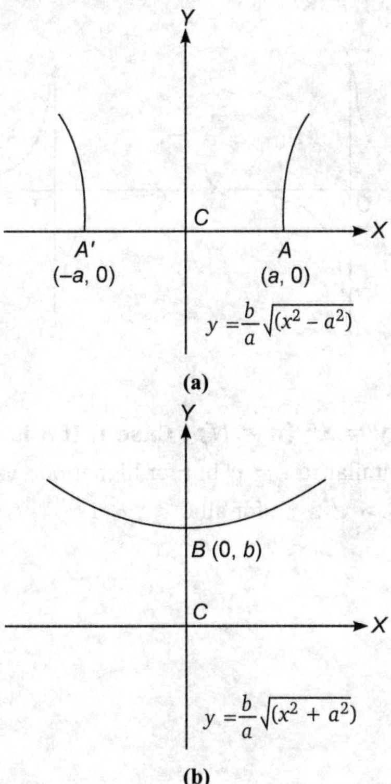

Graphs of $x = \frac{a}{b}\sqrt{(y^2 + b^2)}$, $x = \frac{a}{b}\sqrt{(y^2 - b^2)}$:

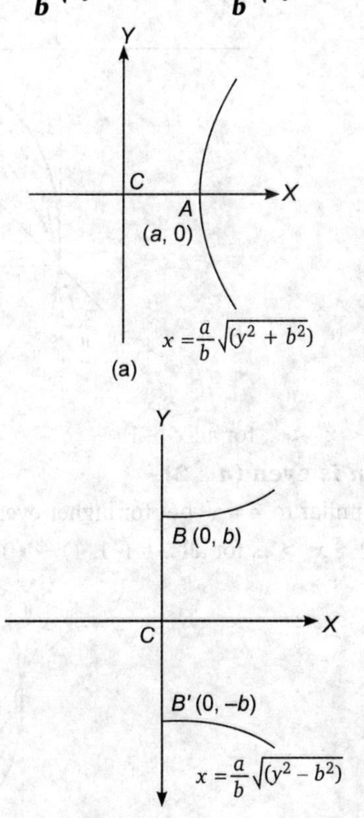

Graph of Rectangular Hyperbola

Graph of $xy = k$:

(a) (b)

Graph of $x^2y = k$ or $y = \dfrac{k}{x^2}$:

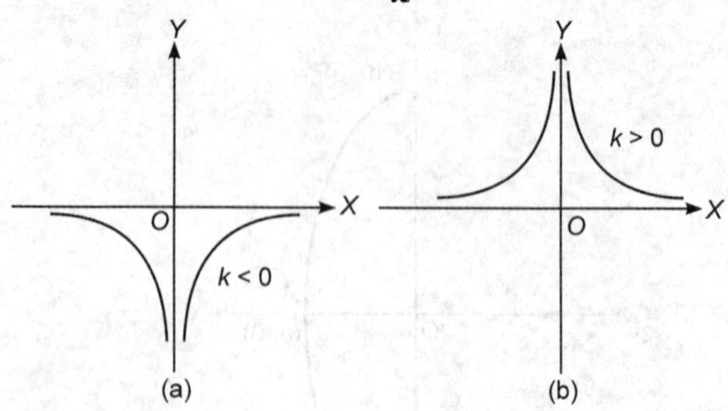

Graph of $y = x^n$ ($n \in N$) : Case I. If n is odd ($n \geq 3$)

It will look similar to $y = x^3$ but for higher odd values of n (*i.e.*, 5, 7, 9, ...) the graph will be more steeper to $y = x^3$. Also, $x^3 > x^5 > x^7 > \dots$ for all $0 < x < 1$

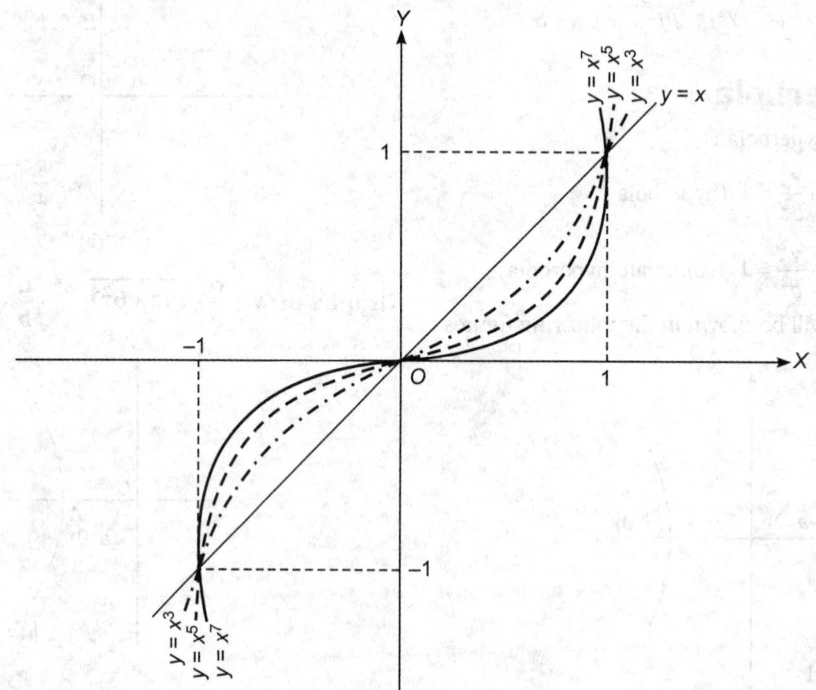

and $x^3 < x^5 < x^7 < \dots$ for all $x > 1$.

Case II. If n is even ($n \geq 2$)

It will look similar to $y = x^2$ but for higher even values of n (*i.e.*, 4, 6, 8, ...) the graph will be more steeper to $y = x^2$. Also, $x^2 > x^4 > x^6 > \dots$ for all $x \in (-1, 1) \sim \{0\}$ and $x^2 < x^4 < x^6 < \dots$ for $x > 1$ or $x < -1$

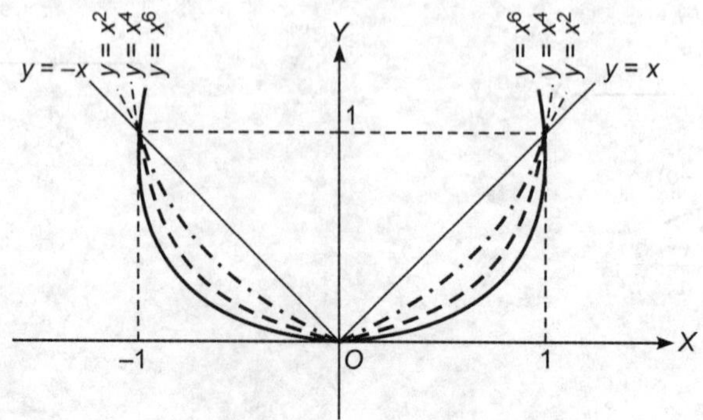

Graph of $y = \frac{1}{x^n} (n \in N)$:

Case I. If n is odd ($n \geq 1$)

It will look similar to $y = 1/x$ but for higher odd values of n (*i.e.*, 3, 5, 7, ...) the graph will be more steeper to $y = 1/x$.

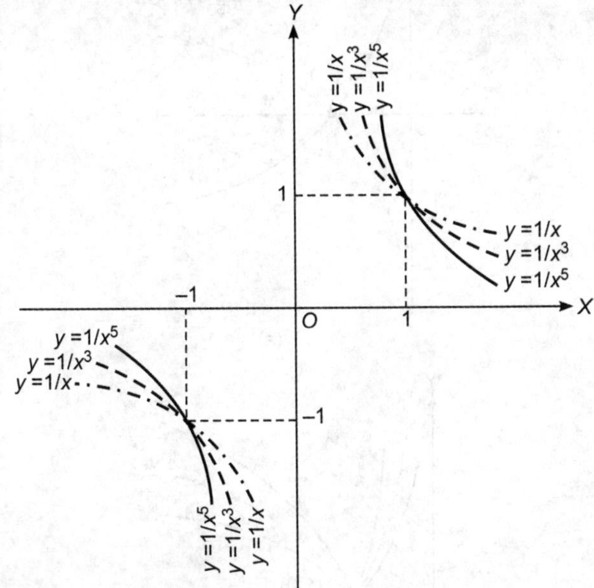

Also, $\frac{1}{x} < \frac{1}{x^3} < \frac{1}{x^5} < \frac{1}{x^7} < \ldots$ for all $0 < x < 1$ and

$\frac{1}{x} > \frac{1}{x^3} > \frac{1}{x^5} > \frac{1}{x^7} \ldots$ for all $x > 1$.

Case II. If n is even ($n \geq 2$)

It will look similar to $y = \frac{1}{x^2}$ but for higher even values of n (*i.e.*, 4, 6, 8, ...) the graph will be more steeper to $y = \frac{1}{x^2}$.

Also $\frac{1}{x^2} < \frac{1}{x^4} < \frac{1}{x^6} < \ldots$. for all $x \in (-1, 1) \sim \{0\}$

and $\frac{1}{x^2} > \frac{1}{x^4} > \frac{1}{x^6} > \ldots$ for all $x > 1$ or $x < -1$

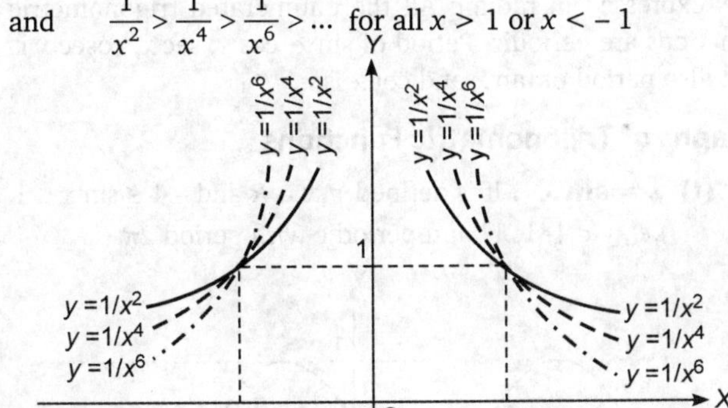

Graph of $y = x^{1/n} (n \in N)$:

Case I. If n is odd ($n \geq 3$)

Since $y = x^{1/n}$ is inverse of $y = x^n$ but inverse function is symmetrical about the line $y = x$.

Also, $x^3 > x^5 > x^7 > \ldots$. for all $0 < x < 1$

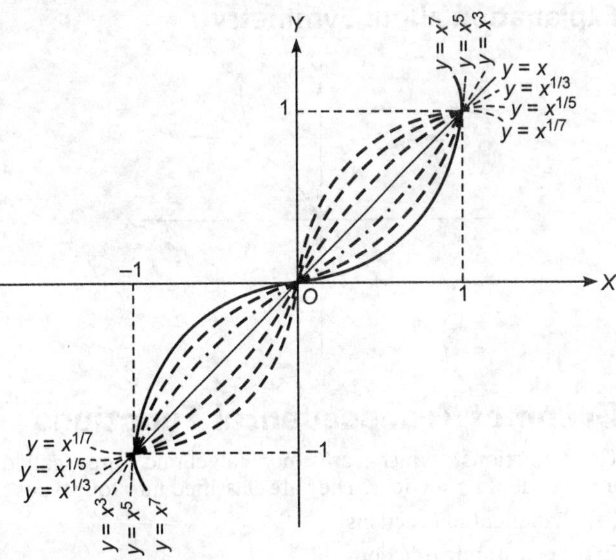

then $x^{1/3} < x^{1/5} < x^{1/7} < \ldots$ for all $0 < x < 1$

and $x^3 < x^5 < x^7 < \ldots$ for all $x > 1$

then $x^{1/3} > x^{1/5} > x^{1/7} >$ for all $x > 1$

Case II. If n is even ($n \geq 2$)

Since $y = x^{1/n}$ is inverse of $y = x^n$ but $y = x^{1/n} \geq 0$ (right hand side of y-axis) but inverse function is symmetrical about the line $y = x$.

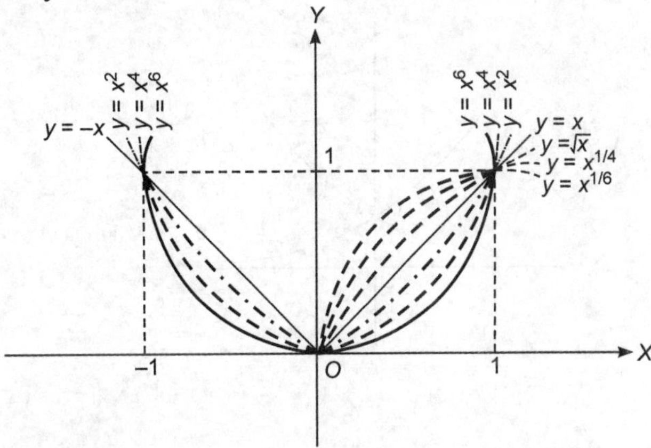

Also, $x^2 > x^4 > x^6 > \ldots$ for all $x \in (-1, 1) \sim \{0\}$

then $\sqrt{x} < \sqrt[4]{x} < \sqrt[6]{x} < \ldots$ for all $0 < x < 1$

and $x^2 < x^4 < x^6 < \ldots$ for all $x > 1$ or $x < -1$

then $\sqrt{x} > \sqrt[4]{x} > \sqrt[6]{x} > \ldots$ for all $x > 1$

Graph of $y = x^{2/3}$ and $y = x^{3/2}$:

Since $y = x^{2/3}$ and $y = x^{3/2}$ are inverse functions to each other which are symmetrical about the line $y = x$.

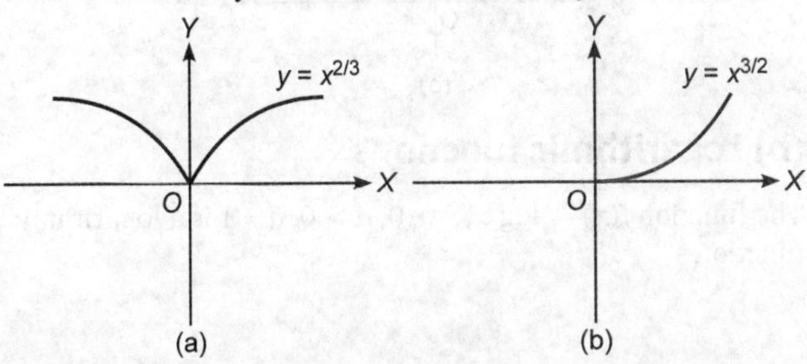

Explanation about Symmetry

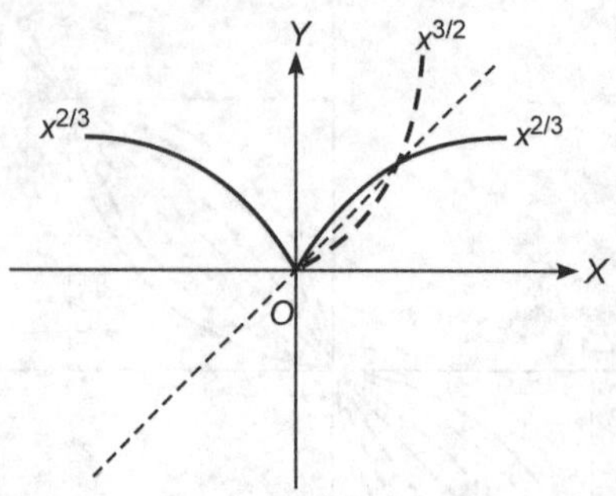

Graph of Transcedental Functions

The functions which are not algebraic are called transcendental functions. They are classified into four types :

(a) Exponential functions
(b) Logarithmic functions
(c) Trigonometric functions
(d) Inverse trigonometric functions

(a) Exponential Functions :

The function $f(x) = a^x$, $a > 0$, $a \neq 1$ is called an exponential function, where variable x is the exponent of a constant value

or $\quad \exp(A) = e^A \qquad (e = 2.718281828 \ldots.)$

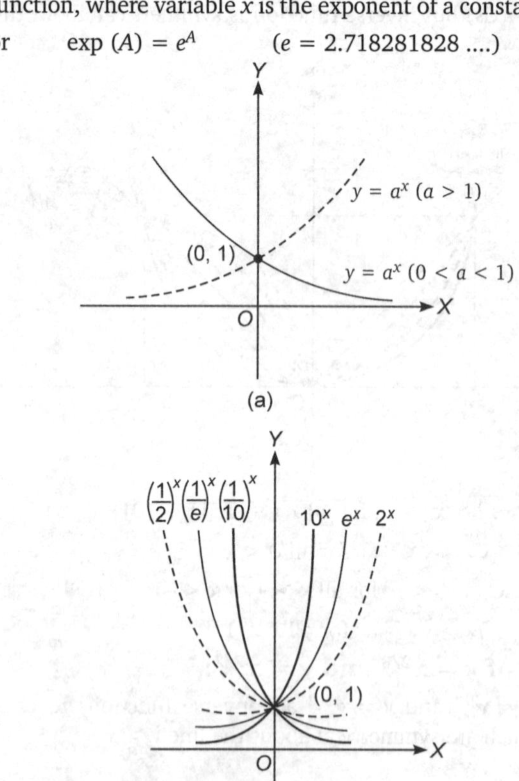

(b) Logarithmic functions

The function $f(x) = \log_a x$, $x > 0$, $a > 0$, $a \neq 1$ is a logarithmic function

or $\quad \ln x = \log_e x$

and $\quad \log x = \log_{10} x$.

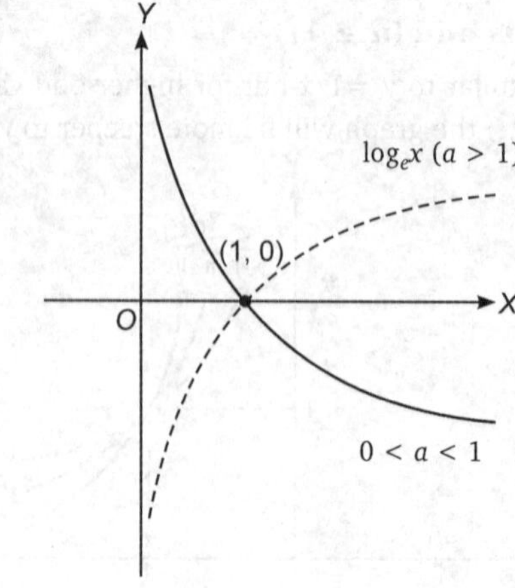

(a)

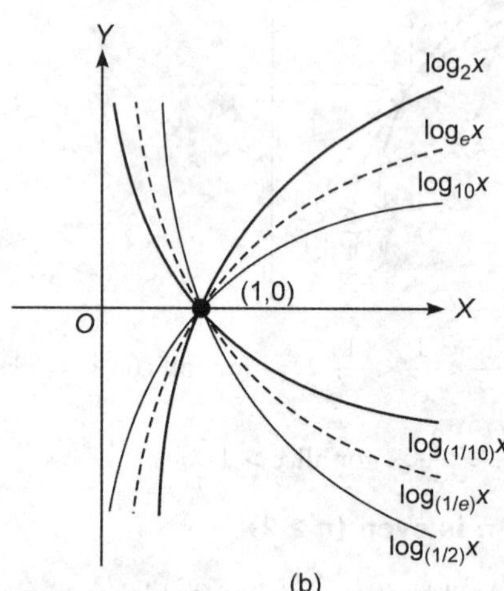

(b)

(c) Trigonometric Functions:

There are six basic trigonometric functions, they are sin x, cos x, than x, cot x, sec x and cosec x. The independent variable x is expressed in radian. All the enumerated trigonometric functions are periodic. Period of sin x, cos x, sec x, cosec x is 2π; also period of tan x and cot x is π.

Graphs of Trigonometric Functions

(i) $\mathbf{y = \sin x}$ **:** It is defined $\forall\ x \in R$ and $-1 \leq \sin x \leq 1$ *i.e.*, $y \in [-1, 1]$ and periodic with period 2π.

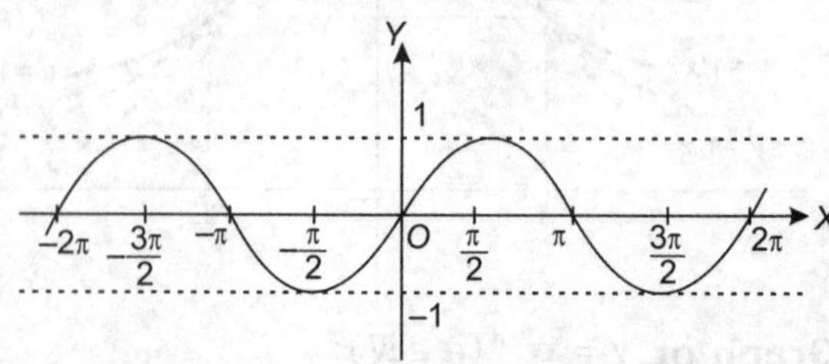

(ii) $\mathbf{y = \cos x}$ **:** It is defined $\forall\ x \in R$ and $-1 \leq \cos x \leq 1$ *i.e.*, $y \in [-1, 1]$ and periodic with period 2π.

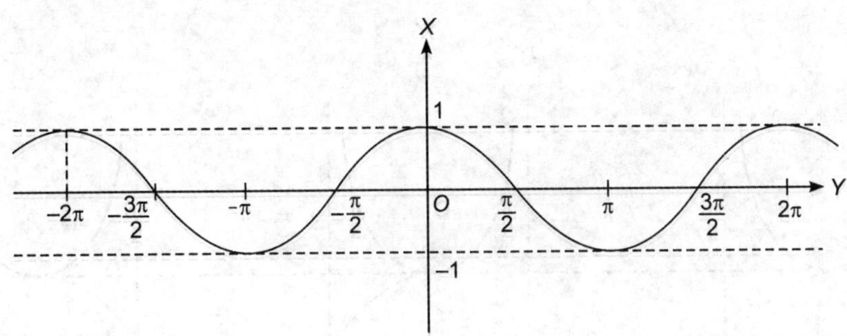

(iii) $y = \tan x$: It is defined for $x \in R - \left\{(2n+1)\frac{\pi}{2}, n \in I\right\}$ because tan x is not defined if $\cos x = 0$

i.e., $x = n\pi + \frac{\pi}{2} = (2n+1)\frac{\pi}{2}, n \in I$ and $-\infty < \tan x < \infty$

i.e., $y \in R$ and periodic with period π.

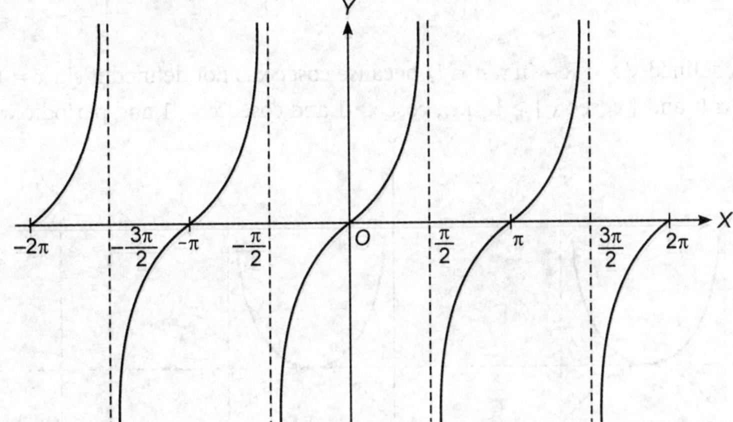

(iv) $y = \cot x$: It is defined for $x \in R - \{n\pi, n \in I\}$ for because cot x is not defined if $\sin x = 0$

i.e., $x = n\pi, n \in I$ and $-\infty < \cot x < \infty$

i.e., $y \in R$ and periodic with period π

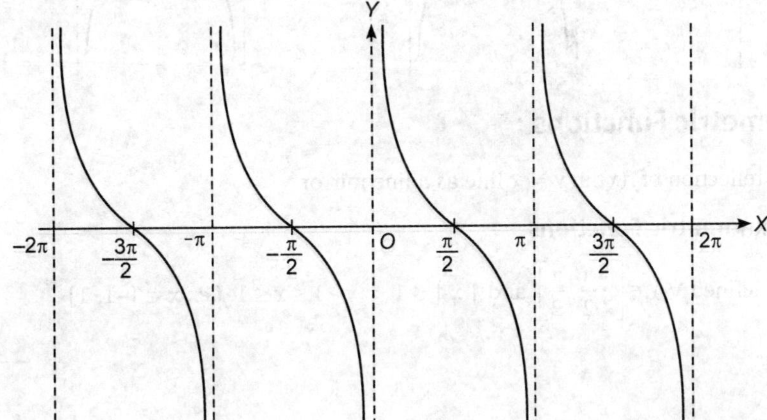

(v) $y = \sec x$: It is defined $\forall\ x \in R \left\{(2n+1)\frac{\pi}{2}, n \in I\right\}$ because sec x is not defined if $\cos x = 0$

i.e., $x = n\pi + \frac{\pi}{2} = (2n+1)\frac{\pi}{2}$ and $|\sec x| \geq 1$

i.e., $\sec x \geq 1$ and $\sec x \leq -1$ and periodic with period 2π.

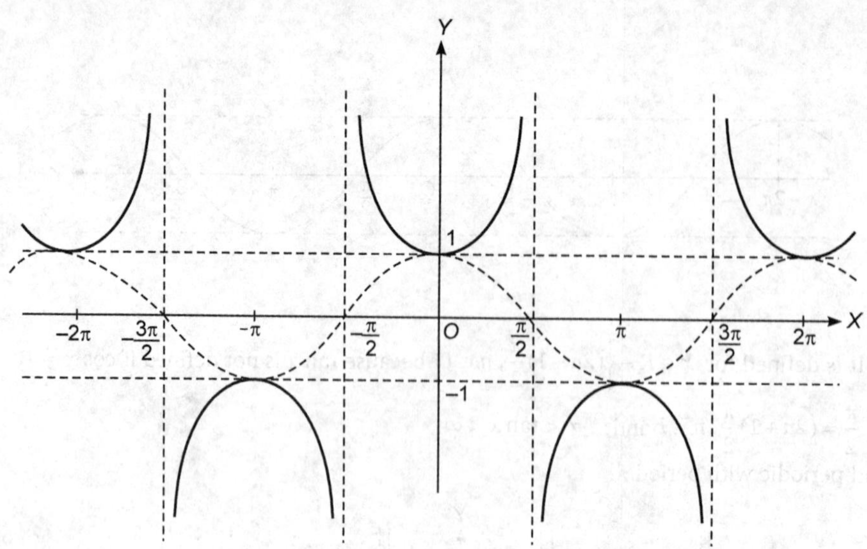

(vi) y = cosec x : It is defined $\forall\, x \in R - \{n\pi, n \in I\}$ because cosec x is not defined if $\sin x = 0$.

i.e., $x = n\pi, n \in$ I and $|\operatorname{cosec} x| \geq 1$ *i.e.,* cosec ≥ 1 and cosec $x \leq -1$ and periodic with period 2π.

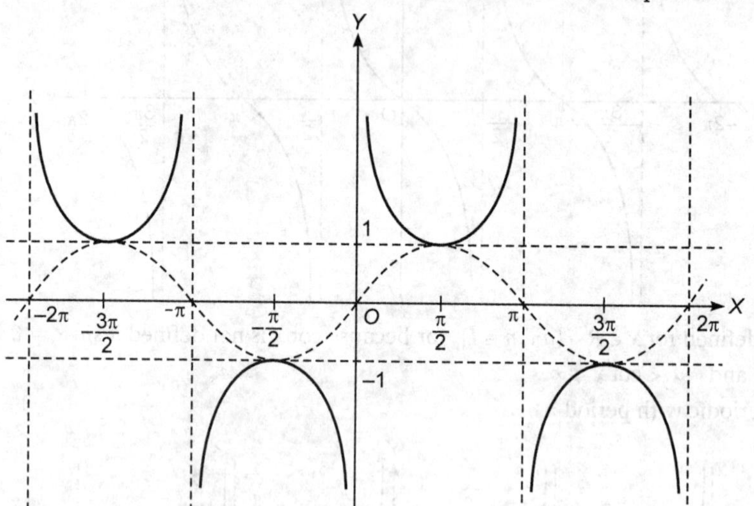

(d) Inverse Trigonometric Functions :

The graph of $f^{-1}(x)$ is the reflection of $f(x)$ in $y = x$ line as a line mirror.

Graphs of Inverse Trigonometric Functions

(i) $y = \sin^{-1} x$: It is defined $\forall\, y \in \left[-\frac{\pi}{2}, \frac{\pi}{2}\right]$ and $|x| \leq 1 \Rightarrow -1 \leq x \leq 1$ *i.e.,* $x \in [-1, 1]$

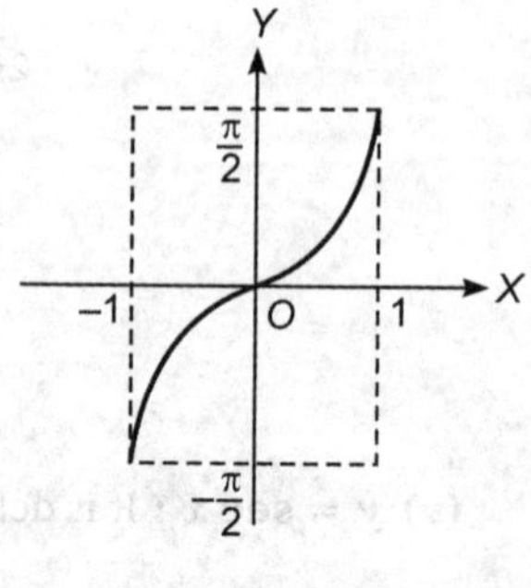

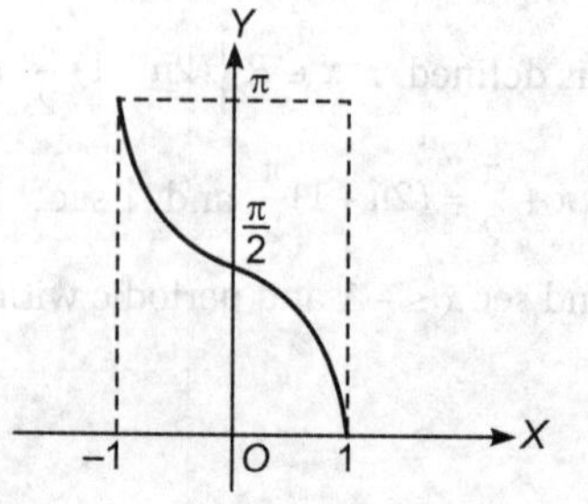

(ii) $y = \cos^{-1} x$: It is defined $\forall\, y \in [0, \pi]$ and $|x| \leq 1$

$\Rightarrow$ $-1 \leq x \leq 1$

i.e., $x \in [-1, 1]$

(iii) $y = \tan^{-1}x$: It is defined $\forall\ y \in \left(-\dfrac{\pi}{2}, \dfrac{\pi}{2}\right)$ and $x \in R$

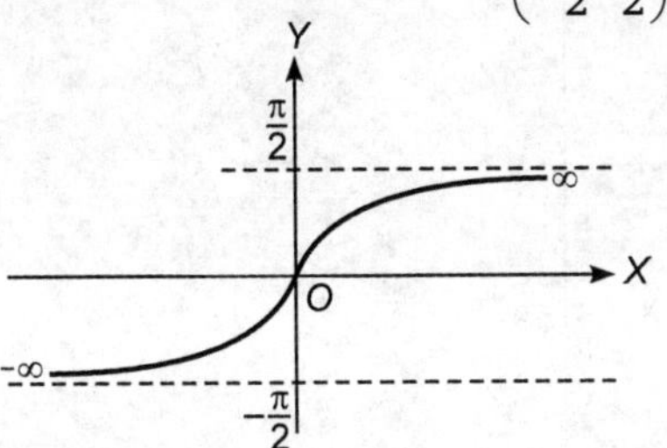

(iv) $y = \cot^{-1}x$: It is defined $\forall\ y \in (0, \pi)$ and $x \in R$

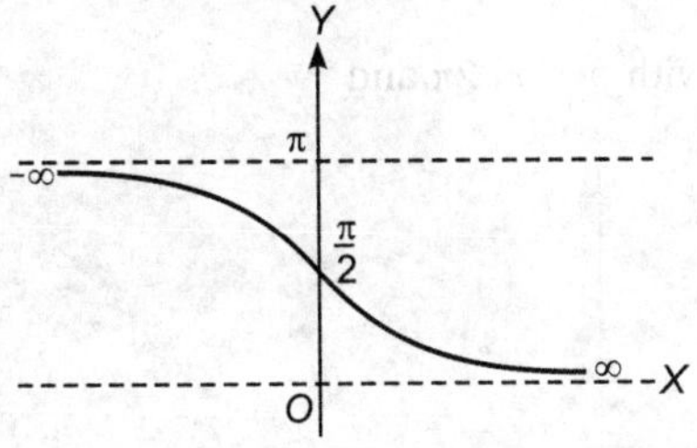

(v) $y = \sec^{-1}x$: It is defined $\forall\ y \in \left[0, \dfrac{\pi}{2}\right) \cup \left(\dfrac{\pi}{2}, \pi\right]$

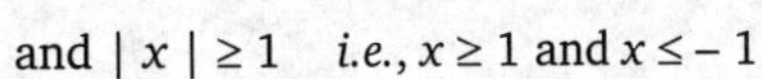

and $|x| \geq 1$ *i.e.*, $x \geq 1$ and $x \leq -1$

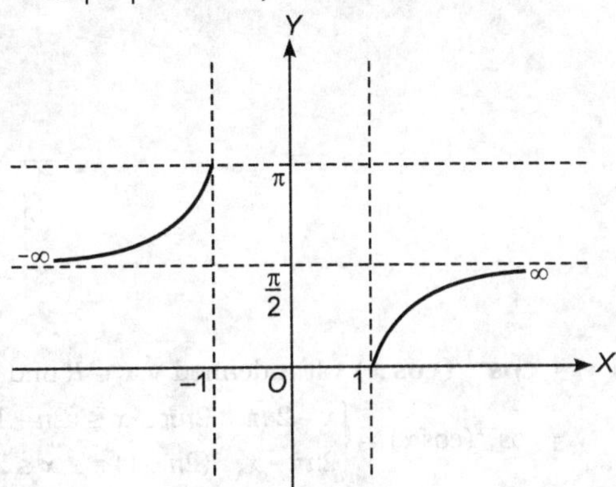

(vi) $y = \text{cosec}^{-1}x$: It is defined $\forall\ y \in \left[-\dfrac{\pi}{2}, 0\right) \cup \left(0, \dfrac{\pi}{2}\right]$ and $|x| \geq 1$ *i.e.*, $x \geq 1$ and $x \leq -1$

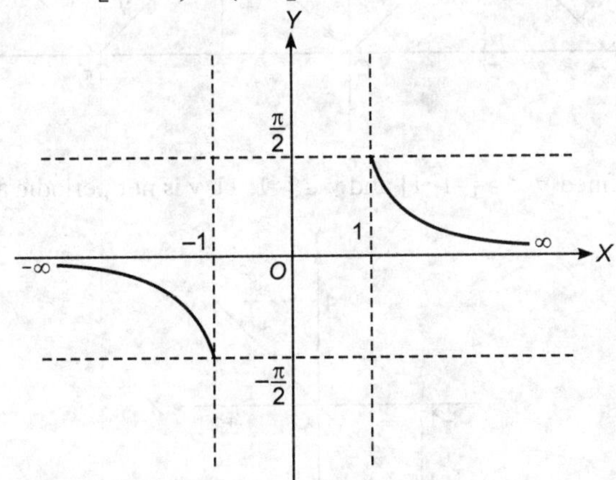

(vii) $y = \sin^{-1}(\sin x)$: It is defined $\forall\ x \in R$ and $y \in \left[-\dfrac{\pi}{2}, \dfrac{\pi}{2}\right]$, periodic with period 2π and

$$y = \sin^{-1}(\sin x) = \begin{cases} x - 2n\pi, & (4n-1)\dfrac{\pi}{2} \leq x \leq (4n+1)\dfrac{\pi}{2}\ \forall\ n \in I \\ (2n-1)\pi - x, & (4n-3)\dfrac{\pi}{2} \leq x \leq (4n-1)\dfrac{\pi}{2}\ \forall\ n \in I \end{cases}$$

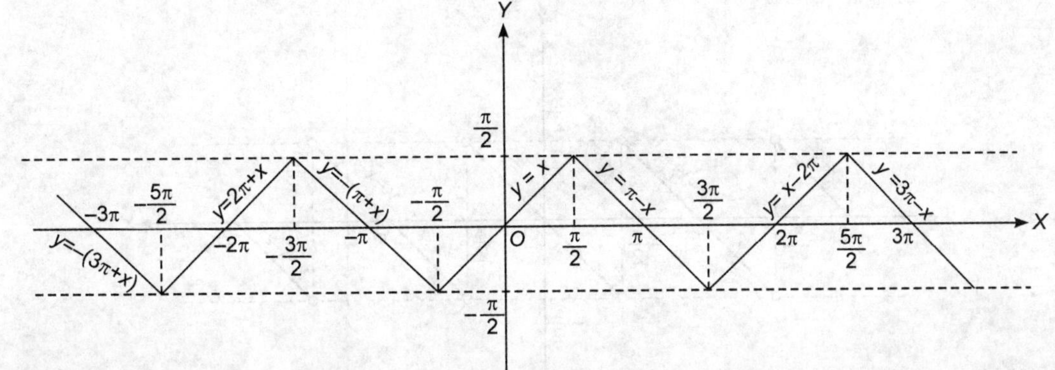

(viii) $y = \sin(\sin^{-1}x)$: It is defined $\forall\ x \in [-1, 1]$ and $y \in [-1, 1]$, y is not periodic and $y = \sin(\sin^{-1}x) = x$.

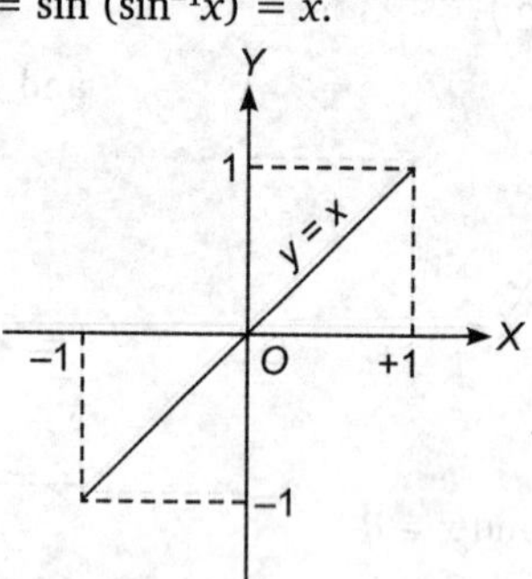

(ix) $y = \cos^{-1}(\cos x)$: It is defined $\forall\ x \in R$ and $y \in [0, \pi]$, periodic with period 2π and

$$y = \cos^{-1}(\cos x) = \begin{cases} x - 2n\pi, & 2n\pi \le x \le (2n+1)\pi\ \forall\, n \in I \\ 2n\pi - x, & (2n-1)\pi \le x \le 2n\pi\ \forall\, n \in I \end{cases}$$

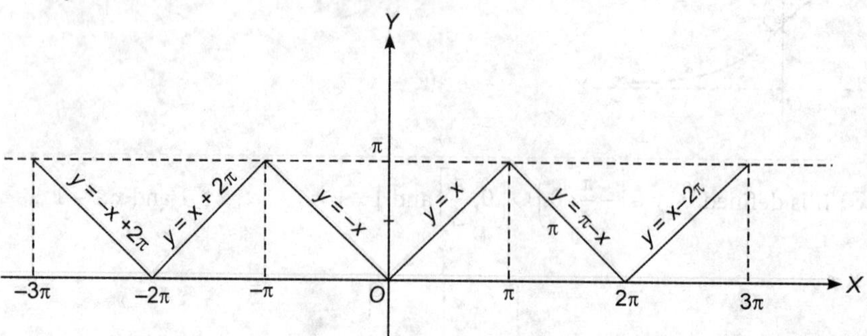

(x) $y = \cos(\cos^{-1}x)$: It is defined $\forall\ x \in [-1, 1]$ and $y \in [-1, 1]$, y is not periodic and $y = \cos(\cos^{-1}x) = x$.

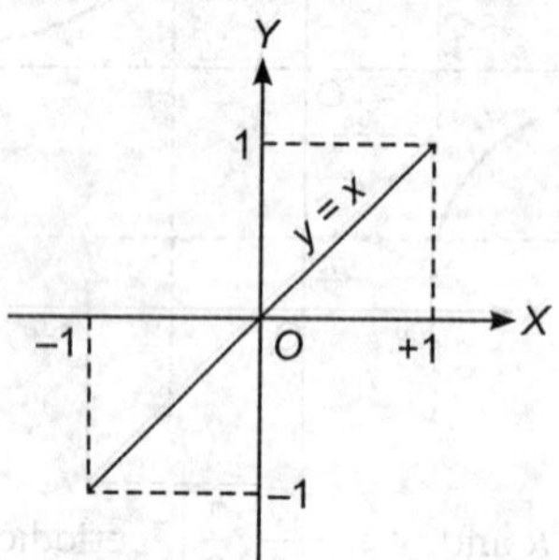

(xi) $y = \tan^{-1}(\tan x)$: It is defined $\forall\ x \in R - \left\{(2n+1)\frac{\pi}{2}, n \in I\right\}$ and $y \in \left(-\frac{\pi}{2}, \frac{\pi}{2}\right)$, periodic with period π and $y = \tan^{-1}(\tan x) = x - n\pi,\ n\pi - \frac{\pi}{2} < x < n\pi + \frac{\pi}{2}\ \forall\, n \in I$.

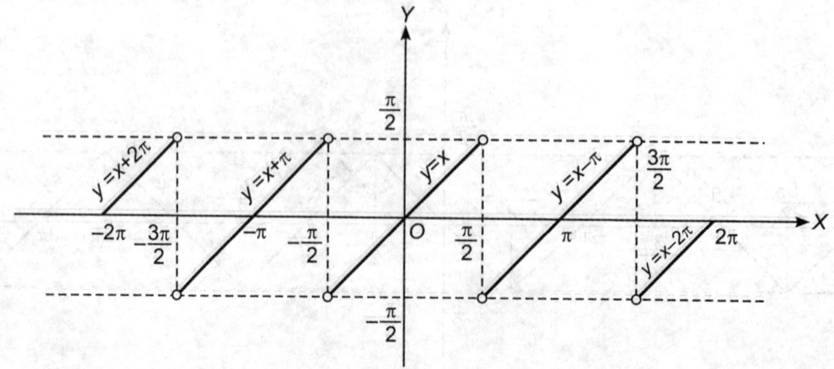

(xii) $y = \tan(\tan^{-1}x)$: It is defined $\forall\, x \in R$ and $y \in R$, y is not periodic and $y = \tan(\tan^{-1} x) = x$.

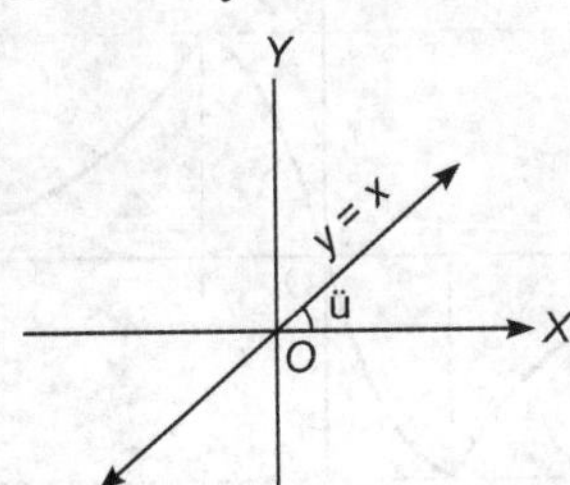

(xiii) $y = \cot^{-1}(\cot x)$: It is defined $\forall\, x \in R - \{n\pi, n \in I\}$ and $y \in (0, \pi)$, periodic with period π and $y = \cot^{-1}(\cot x) = x - n\pi$, $n\pi < x < (n+1)\pi\ \forall\, n \in I$.

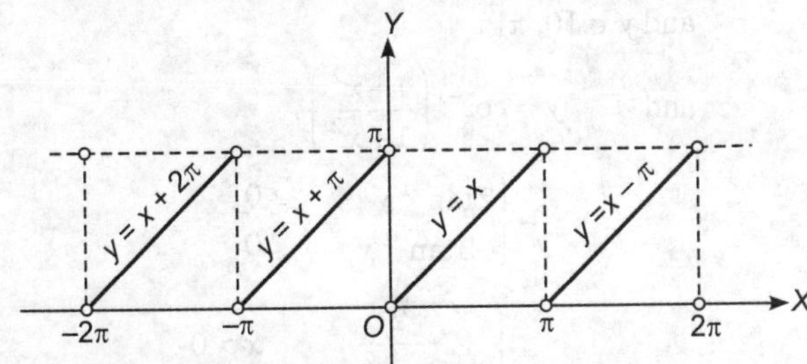

(xiv) $y = \cot(\cot^{-1}x)$: It is defined $\forall\, x \in R$ and $y \in R$, y is not periodic and $y = \cot(\cot^{-1}x) = x - n\pi$, $nx < x < (n+1)\pi\ \forall\, n \in I$.

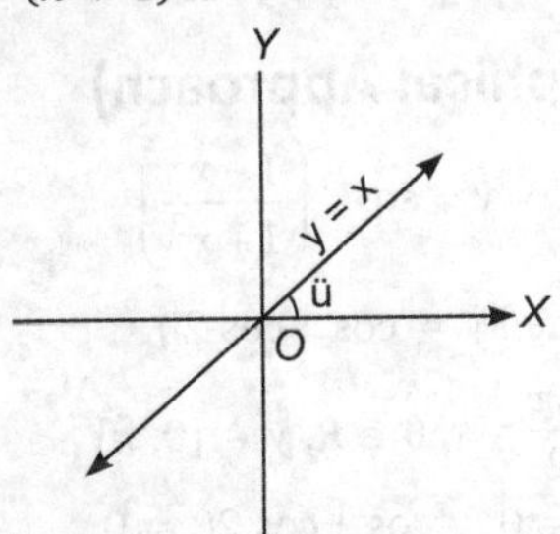

(xv) $y = \sec^{-1}(\sec x)$: If is defined $\forall\, x \in R - \left\{(2n+1)\frac{\pi}{2}, n \in I\right\}$ and $y \in \left[0, \frac{\pi}{2}\right) \cup \left(\frac{\pi}{2}, \pi\right]$,

Periodic with period 2π and

$$y = \sec^{-1}(\sec x) = \begin{cases} x - 2n\pi, & 2n\pi \le x \le (2n+1)\pi \\ & \text{Except } x = (4n+1)\frac{\pi}{2}\ \forall\, n \in I \\ 2n\pi - x, & (2n-1)\pi \le x \le 2n\pi \\ & \text{Except } x = (4n-1)\frac{\pi}{2}\ \forall\, n \in I \end{cases}$$

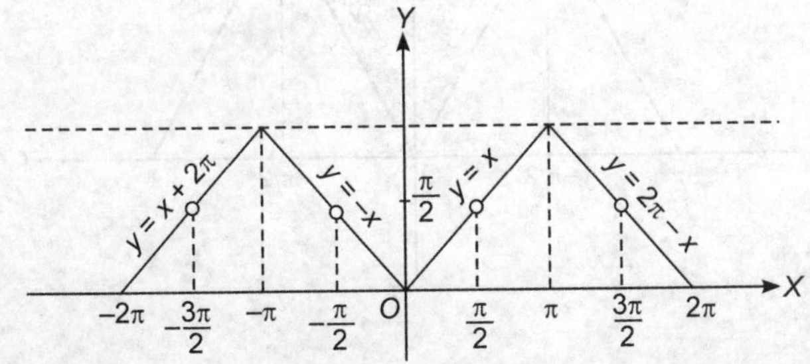

(xvi) $y = \sec(\sec^{-1}x)$: It is defined $\forall\ |x| \ge 1$ and $|y| \ge 1$, y is not periodic and $y = \sec(\sec^{-1}x) = x$.

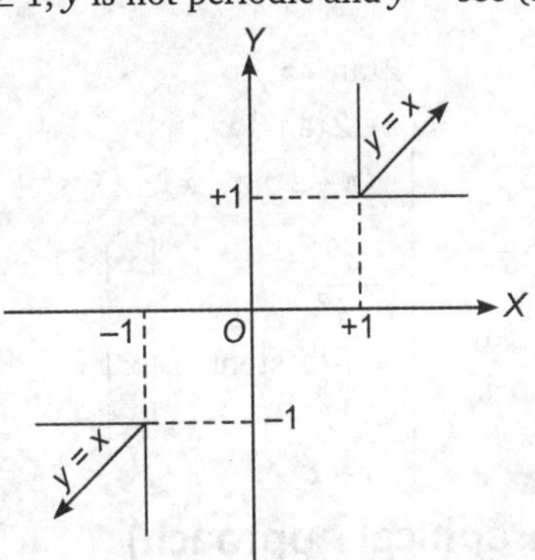

(xvii) $y = \operatorname{cosec}^{-1}(\operatorname{cosec} x)$: It is defined $\forall\, x \in R - \{n\pi, n \in I\}$, $y \in \left[-\frac{\pi}{2}, 0\right) \cup \left(0, \frac{\pi}{2}\right]$, periodic with period 2π and $y = \operatorname{cosec}^{-1}(\operatorname{cosec} x)$

$$= \begin{cases} x - 2n\pi, & (4n-1)\frac{\pi}{2} \le x \le (4n+1)\frac{\pi}{2} \\ & \text{Except } x = 2n\pi\ \forall\, n \in I \\ (2n-1)\pi - x, & (4n-3)\frac{\pi}{2} \le x \le (4n-1)\frac{\pi}{2} \\ & \text{Except } x = (2n-1)\pi\ \forall\, n \in I \end{cases}$$

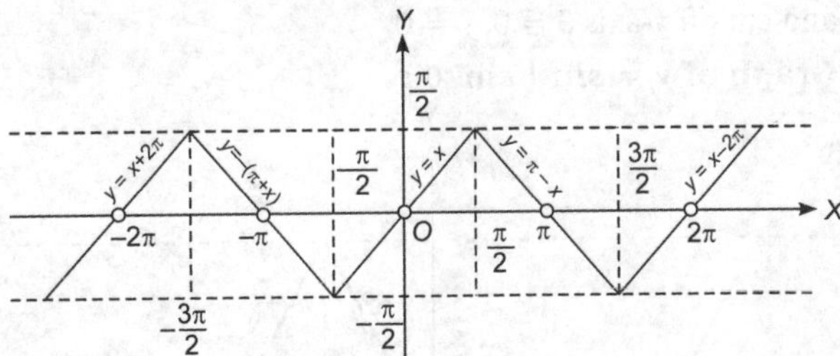

(xviii) $y = \operatorname{cosec}(\operatorname{cosec}^{-1}x)$: It is defined $\forall\ |x| \ge 1$, $|y| \ge 1$, y is not periodic and $y = \operatorname{cosec}(\operatorname{cosec}^{-1} x) = x$.

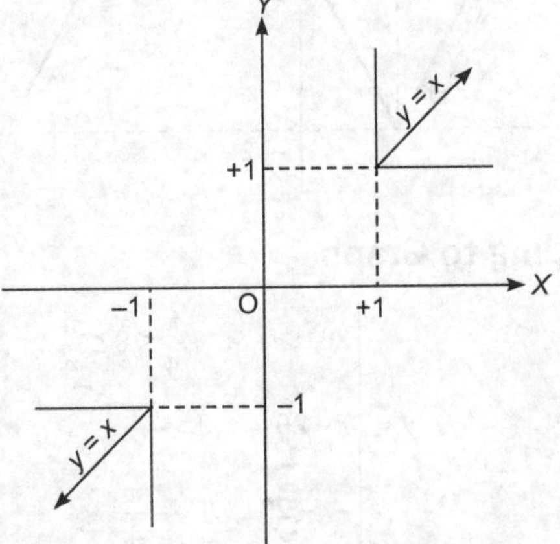

(xix) $y = \sin^{-1}\left(\frac{2x}{1+x^2}\right)$: It is defined $\forall\, x \in R$ and

$$y \in \left[-\frac{\pi}{2}, \frac{\pi}{2}\right]$$

and $\quad y = \sin^{-1}\left(\frac{2x}{1+x^2}\right)$

$$= \begin{cases} 2\tan^{-1} x & , |x| \le 1 \\ \pi - 2\tan^{-1} x & , x > 1 \\ -(\pi + 2\tan^{-1} x) & , x < -1 \end{cases}$$

and $\quad \frac{dy}{dx} = \begin{cases} \frac{2}{1+x^2} & , |x| < 1 \\ \text{non-existent} & , |x| = 1 \\ -\frac{2}{1+x^2} & , |x| > 1 \end{cases}$

Proof : (Graphical Approach)

$\because \quad y = \sin^{-1}\left(\frac{2x}{1+x^2}\right)$

Put $x = \tan\theta$, then $y = \sin^{-1}\sin 2\theta$

Now period $= \frac{2\pi}{2} = \pi$

$$\theta \in R, y \in \left[-\frac{\pi}{2}, \frac{\pi}{2}\right]$$

cut off θ-axis $y = 0$, $\sin 2\theta = 0$

$$\theta = \frac{n\pi}{2}, n \in I$$

and cut off y-axis $\theta = 0, y = 0$

Graph of $y = \sin^{-1}\sin 2\theta$:

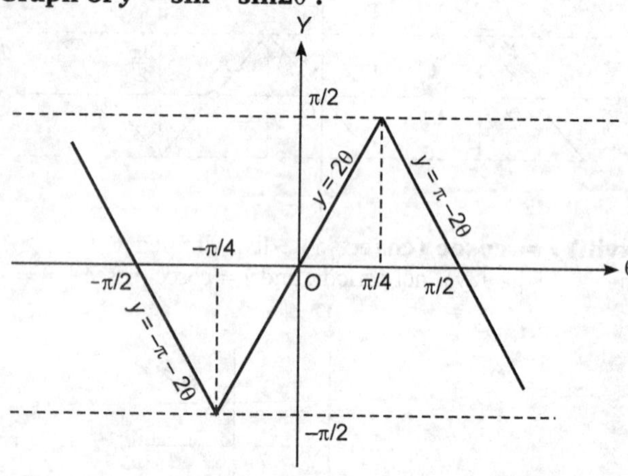

According to Graph

$$y = \begin{cases} 2\theta & , -\frac{\pi}{4} \le \theta \le \frac{\pi}{4} \\ \pi - 2\theta & , \theta > \frac{\pi}{4} \\ -\pi - 2\theta & , \theta < -\frac{\pi}{4} \end{cases}$$

or $\quad y = \begin{cases} 2\tan^{-1} x & , -1 \le x \le 1 \\ \pi - 2\tan^{-1} x & , x > 1 \\ -\pi - 2\tan^{-1} x & , x < -1 \end{cases}$

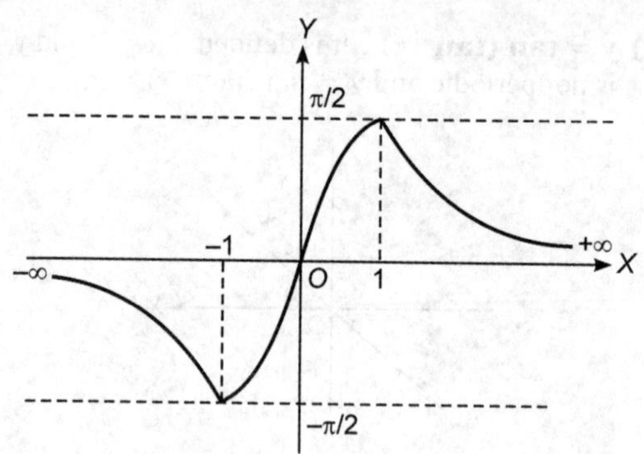

(xx) $y = \cos^{-1}\left(\frac{1-x^2}{1+x^2}\right)$: It is defined $\forall x \in R$ and $y \in [0, \pi]$

and $\quad y = \cos^{-1}\left(\frac{1-x^2}{1+x^2}\right)$

$$= \begin{cases} 2\tan^{-1} x, & x \ge 0 \\ -2\tan^{-1} x, & x < 0 \end{cases}$$

and $\quad \frac{dy}{dx} = \begin{cases} \frac{2}{1+x^2} & , x > 0 \\ \text{non-existent} & , x = 0 \\ -\frac{2}{1+x^2} & , x < 0 \end{cases}$

Proof : (Graphical Approach)

$\because \quad y = \cos^{-1}\left(\frac{1-x^2}{1+x^2}\right)$

Put $x = \tan\theta$, then $y = \cos^{-1}\cos 2\theta$

Now period $= \frac{2\pi}{2} = \pi$, $\theta \in R$, $y \in [0, \pi]$

cut off θ-axis $y = 0 \therefore \cos^{-1}\cos 2\theta = 0$

$\Rightarrow \quad \cos 2\theta = 1 \Rightarrow \quad 2\theta = 2n\pi$

$\Rightarrow \theta = n\pi, n \in I$ and cut off y-axis $\theta = 0$

$$y = \cos^{-1}\cos 0$$
$$= \cos^{-1} 1$$
$$= 0$$

Graph of $y = \cos^{-1}\cos 2\theta$:

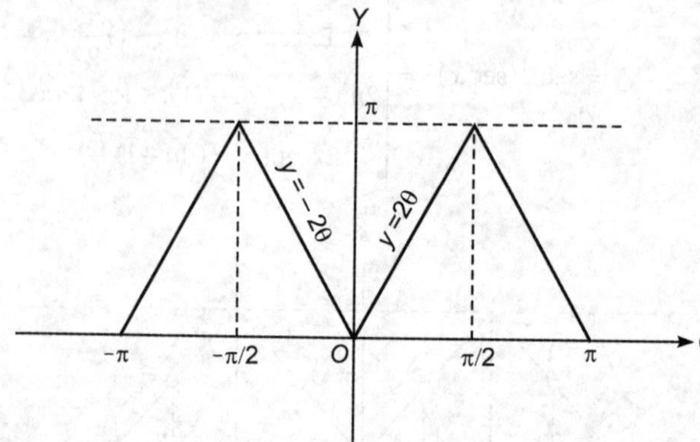

According to graph

$$y=\begin{cases}2\theta & ,\theta\ge 0\\ -2\theta & ,\theta<0\end{cases}\quad \text{or } y=\begin{cases}2\tan^{-1}x & ,x\ge 0\\ -2\tan^{-1}x & ,x<0\end{cases}$$

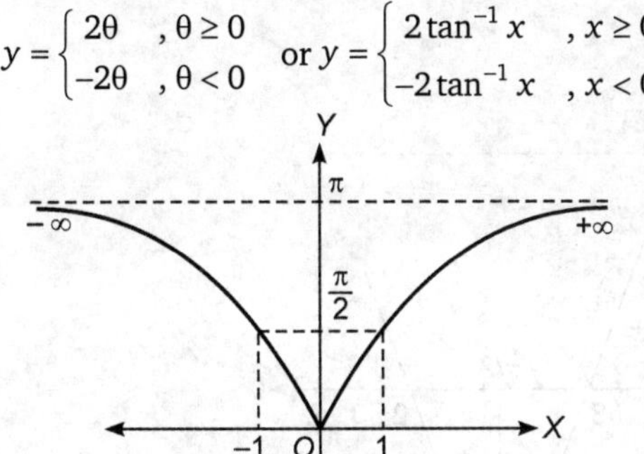

(xxi) $\mathbf{y=\tan^{-1}\left(\dfrac{2x}{1-x^2}\right)}$ **:** It is defined $\forall\, x\in R-\{-1,1\}$

and $\quad y\in\left(-\dfrac{\pi}{2},\dfrac{\pi}{2}\right)$

and $\quad y=\tan^{-1}\left(\dfrac{2x}{1-x^2}\right)$

$$=\begin{cases}2\tan^{-1}x & ,|x|<1\\ \pi+2\tan^{-1}x & ,x<-1\\ -(\pi-2\tan^{-1}x) & ,x>1\end{cases}$$

and $\quad \dfrac{dy}{dx}=\begin{cases}\dfrac{2}{1+x^2} & ,|x|\ne 1\\ \text{non-existent} & ,|x|=1\end{cases}$

Proof : (Graphical Approach)

$\because \quad y=\tan^{-1}\left(\dfrac{2x}{1-x^2}\right)$ put $x=\tan\theta$

$\therefore \quad y=\tan^{-1}\tan 2\theta$

Now period $=\dfrac{\pi}{2}$ and $\theta\in R\sim\left\{\dfrac{n\pi}{2}+\dfrac{\pi}{4},n\in I\right\}$

and $\quad y\in\left(-\dfrac{\pi}{2},\dfrac{\pi}{2}\right)$ cut off θ-axis $y=0\Rightarrow\tan 2\theta=0$

$\therefore \quad \theta=\dfrac{n\pi}{2},n\in I$ and cut off y-axis $\theta=0, y=0$

Graph of $y=\tan^{-1}\tan 2\theta$:

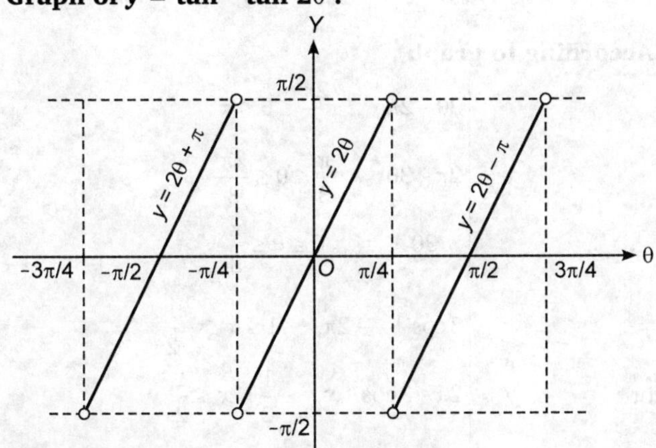

According to graph :

$$y=\begin{cases}2\theta & ,-\dfrac{\pi}{4}<\theta<\dfrac{\pi}{4}\\ \pi+2\theta & ,\theta<-\dfrac{\pi}{4}\\ -\pi+2\theta & ,\theta>\dfrac{\pi}{4}\end{cases}$$

or

$$y=\begin{cases}2\tan^{-1}x & ,\ -1<x<1\\ \pi+2\tan^{-1}x & ,\ x<-1\\ -\pi+2\tan^{-1}x & ,\ x>1\end{cases}$$

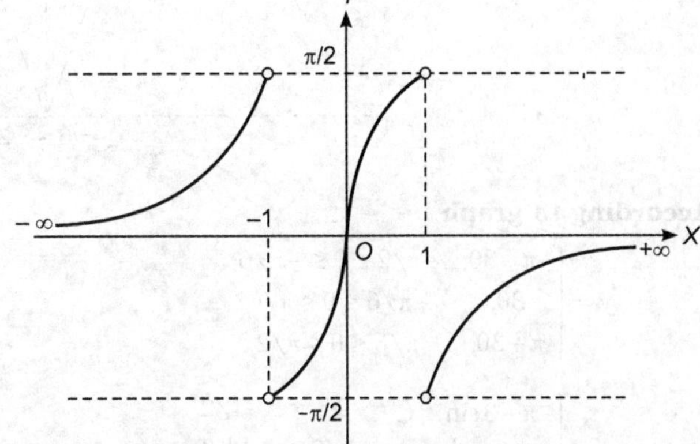

(xxii) $\mathbf{y=\sin^{-1}(3x-4x^3)}$ **:** It is defined $\forall\, x\in[-1,1]$ and $y\in\left[-\dfrac{\pi}{2},\dfrac{\pi}{2}\right]$

and $\quad y=\sin^{-1}(3x-4x^3)$

$$=\begin{cases}-(\pi+3\sin^{-1}x) & ,\ -1\le x\le-\dfrac{1}{2}\\ 3\sin^{-1}x & ,\ -\dfrac{1}{2}\le x\le\dfrac{1}{2}\\ \pi-3\sin^{-1}x & ,\ \dfrac{1}{2}\le x\le 1\end{cases}$$

and $\quad \dfrac{dy}{dx}=\begin{cases}\dfrac{3}{\sqrt{1-x^2}} & ,\ x\in\left(-\dfrac{1}{2},\dfrac{1}{2}\right)\\ -\dfrac{3}{\sqrt{1-x^2}} & ,\ x\in\left(-1,-\dfrac{1}{2}\right)\cup\left(\dfrac{1}{2},1\right)\end{cases}$

Proof : (Graphical Approach)

$\because \quad y=\sin^{-1}(3x-4x^3)$ put $x=\sin\theta$

$\therefore \quad y=\sin^{-1}\sin 3\theta$

Now period $=\dfrac{2\pi}{3}$ and $\theta\in R,\ y\in\left[-\dfrac{\pi}{2},\dfrac{\pi}{2}\right]$ cut off θ-axis $y=0$

$\therefore \quad \sin 3\theta=0\Rightarrow\theta=\dfrac{n\pi}{3},\ n\in I$

and cut off y-axis $\theta=0$

$\therefore \quad y=0$

Graph of $y = \sin^{-1}\sin 3\theta$

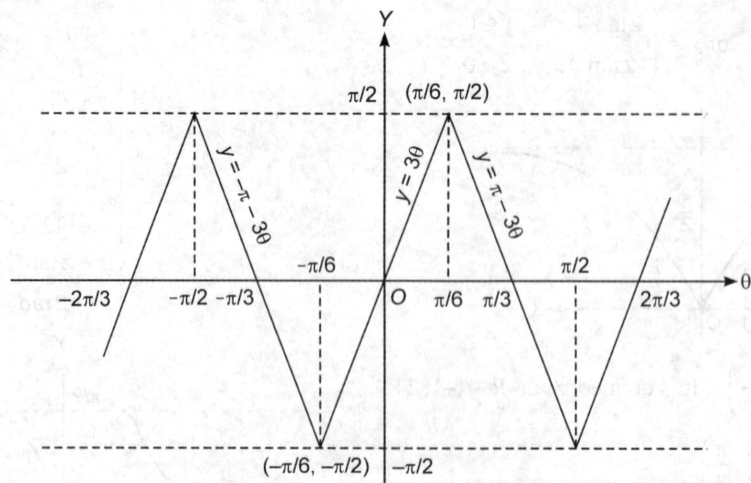

According to graph

$$y = \begin{cases} -\pi - 3\theta, & -\pi/2 \le \theta \le -\pi/6 \\ 3\theta, & -\pi/6 \le \theta \le \pi/6 \\ \pi - 3\theta, & \pi/6 \le \theta \le \pi/2 \end{cases}$$

or
$$y = \begin{cases} -\pi - 3\sin^{-1} x, & -1 \le x \le -1/2 \\ 3\sin^{-1} x, & -1/2 \le x \le 1/2 \\ \pi - 3\sin^{-1} x, & 1/2 \le x \le 1 \end{cases}$$

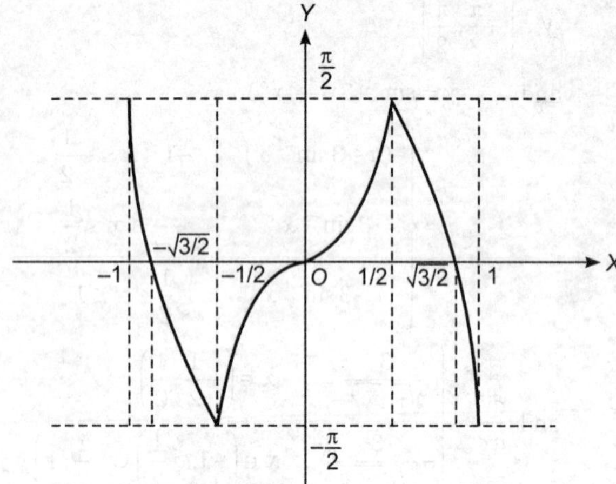

(xxiii) $y = \cos^{-1}(4x^3 - 3x)$: It is defined $\forall\, x \in [-1, 1]$ and $y \in [0, \pi]$

and $\quad y = \cos^{-1}(4x^3 - 3x)$

$$= \begin{cases} 3\cos^{-1} x - 2\pi, & -1 \le x \le -\frac{1}{2} \\ 2\pi - 3\cos^{-1} x, & -\frac{1}{2} \le x \le \frac{1}{2} \\ 3\cos^{-1} x, & \frac{1}{2} \le x \le 1 \end{cases}$$

and
$$\frac{dy}{dx} = \begin{cases} \frac{3}{\sqrt{1-x^2}}, & x \in \left(-\frac{1}{2}, \frac{1}{2}\right) \\ -\frac{3}{\sqrt{1-x^2}}, & x \in \left(-1, -\frac{1}{2}\right) \cup \left(\frac{1}{2}, 1\right) \end{cases}$$

Proof : (Graphical Approach)

$\because \quad y = \cos^{-1}(4x^3 - 3x)$

Put $x = \cos\theta$

$\therefore \quad y = \cos^{-1}\cos 3\theta$

Now period $= 2\pi/3$ and $\theta \in R$, $y \in [0, \pi]$ and cut off θ-axis

$y = 0$

$\Rightarrow \quad \cos 3\theta = 1$

$\therefore \quad 3\theta = 2n\pi$

$\theta = \frac{2n\pi}{3}$, $n \in I$ and cut off y-axis $\theta = 0$

$\therefore \quad y = \cos^{-1}\cos 0$

$= \cos^{-1} 1 = 0$

Graph of $y = \cos^{-1}\cos 3\theta$

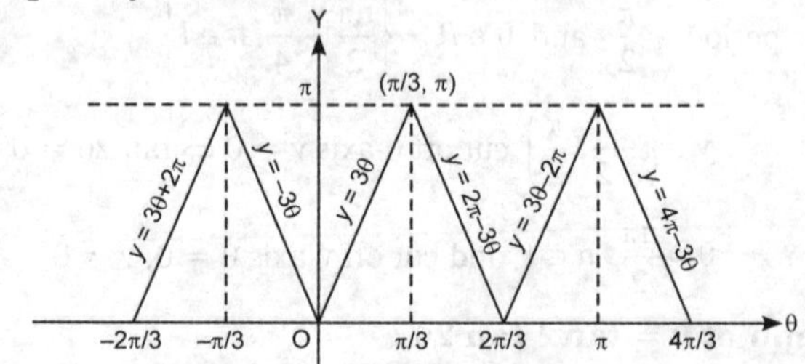

According to graph

$$y = \begin{cases} 3\theta - 2\pi, & \frac{2\pi}{3} \le \theta \le \pi \\ 2\pi - 3\theta, & \frac{\pi}{3} \le \theta \le \frac{2\pi}{3} \\ 3\theta, & 0 \le \theta \le \frac{\pi}{3} \end{cases}$$

or
$$y = \begin{cases} 3\cos^{-1} x - 2\pi, & -1 \le x \le -\frac{1}{2} \\ 2\pi - 3\cos^{-1} x, & -\frac{1}{2} \le x \le \frac{1}{2} \\ 3\cos^{-1} x, & \frac{1}{2} \le x \le 1 \end{cases}$$

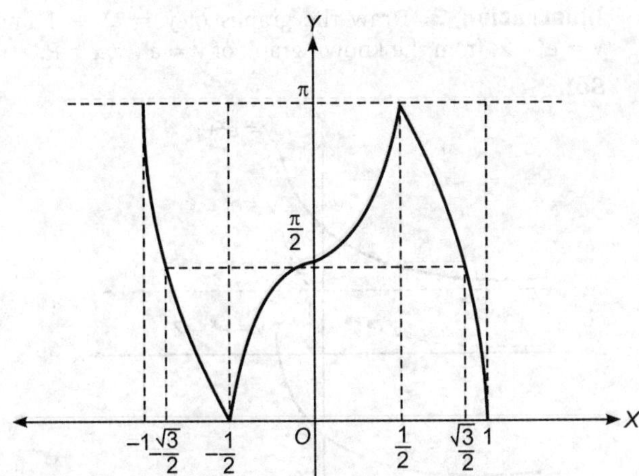

(xxiv) $y = \tan^{-1}\left(\frac{3x - x^3}{1 - 3x^2}\right)$ **: It is defined**

$$\forall\ x \in R - \left\{-\frac{1}{\sqrt{3}}, \frac{1}{\sqrt{3}}\right\}$$

and $$y \in \left(-\frac{\pi}{2}, \frac{\pi}{2}\right)$$

and $$y = \begin{cases} 3\tan^{-1} x & , |x| < \frac{1}{\sqrt{3}} \\ \pi + 3\tan^{-1} x & , x < -\frac{1}{\sqrt{3}} \\ -\pi + 3\tan^{-1} x & , x > \frac{1}{\sqrt{3}} \end{cases}$$

and $$\frac{dy}{dx} = \begin{cases} \frac{3}{1+x^2} & , |x| \neq \frac{1}{\sqrt{3}} \\ \text{non-existent} & , |x| = \frac{1}{\sqrt{3}} \end{cases}$$

Proof : (Graphical Approach)

$\because$ $$y = \tan^{-1}\left(\frac{3x - x^3}{1 - 3x^2}\right)$$

cut off x-axis $y = 0$ then $x = 0, \pm\sqrt{3}$

Put $x = \tan\theta$

$\therefore$ $$y = \tan^{-1}\tan 3\theta$$

Now period $= \frac{\pi}{3}$, $\theta \in R \sim \left\{\frac{n\pi}{3} + \frac{\pi}{6}, n \in I\right\}$

and $$y \in \left(-\frac{\pi}{2}, \frac{\pi}{2}\right)$$

cut off θ-axis $y = 0$

$\therefore$ $\tan 3\theta = 0 \Rightarrow \theta = \frac{n\pi}{3}, n \in I$

and cut off y-axis $\theta = 0$

$\therefore$ $$y = \tan^{-1}\tan 0$$
$$= \tan^{-1} 0$$
$$= 0$$

Graph of $y = \tan^{-1}\tan 3\theta$:

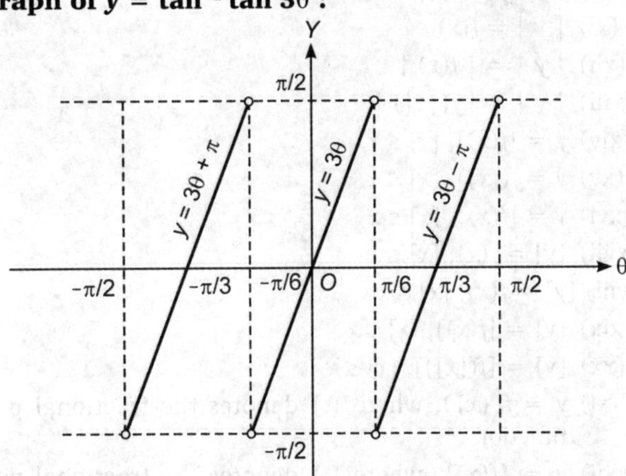

According to graph

$$y = \begin{cases} 3\theta & , -\frac{\pi}{6} < \theta < \frac{\pi}{6} \\ \pi + 3\theta & , \theta < -\frac{\pi}{6} \\ -\pi + 3\theta & , \pi > \frac{\pi}{6} \end{cases}$$

or $$y = \begin{cases} 3\tan^{-1} x & , -\frac{1}{\sqrt{3}} < x < \frac{1}{\sqrt{3}} \\ \pi + 3\tan^{-1} x & , x < -\frac{1}{\sqrt{3}} \\ -\pi + 3\tan^{-1} x & , x > \frac{1}{\sqrt{3}} \end{cases}$$

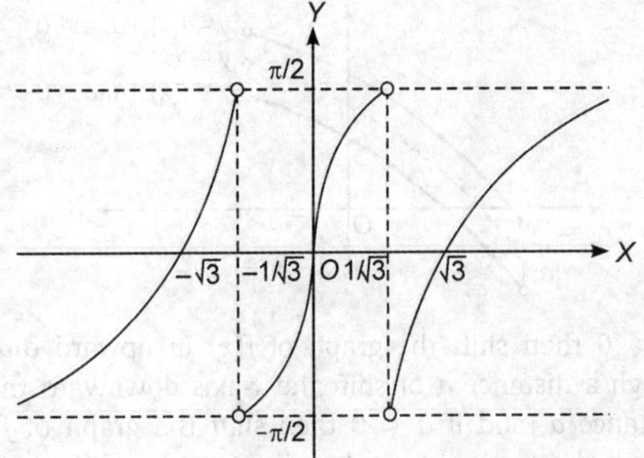

Transformation of Graphs

(I) For the given graph of $y = f(x)$, to draw the following graphs :

(i) $y = f(x) + a, a \in R$

(ii) $y = f(x + a), a \in R$

(iii) $y = af(x), a \in R$

(iv) $y = f(ax), a \in R$

(v) $y = f(-x)$

(vi) $y = -f(x)$

(vii) $y = -f(-x)$

(viii) $y = |f(x)|$

(ix) $y = f(|x|)$

(x) $y = |f(|x|)|$
(xi) $|y| = f(x)$
(xii) $|y| = |f(x)|$
(xiii) $|y| = |f(|x|)|$
(xiv) $y = [f(x)]$, $[x] \le x$
(xv) $y = f([x])$, $[x] \le x$
(xvi) $y = [f(x)]$, $[x] \le x$
(xvii) $[y] = f(x)$, $[x] \le x$
(xviii) $[y] = f(x)$, $[x] \le x$
(xix) $[y] = [f(x)]$, $[x] \le x$
(xx) $[y] = [f([x])]$, $[x] \le x$
(xxi) $y = f(\{x\})$, where $\{.\}$ denotes the fractional part function
(xxii) $y = \{f(x)\}$, where $\{.\}$ denotes the fractional part function
(xxiii) $y = \{f(\{x\})\}$, where $\{.\}$ denotes the fractional part function
(xxiv) $\{y\} = f(x)$, where $\{.\}$ denotes the fractional part function
(xxv) $\{y\} = \{f(x)\}$, where $\{.\}$ denotes the fractional part function
(xxvi) $\{y\} = f(\{x\})$ and $\{y\} = \{f\{x\}\}$, where $\{.\}$ denotes the fractional part function
(xxvii) $y = f^{-1}(x)$

(II) If $f(x)$ and $g(x)$ are given below

$y = \max\{f(x), g(x)\}$ or $y = \min\{f(x), g(x)\}$

(i) Drawing the graph of $y = f(x) + a$, $a \in R$ from the known graph of $y = f(x)$:

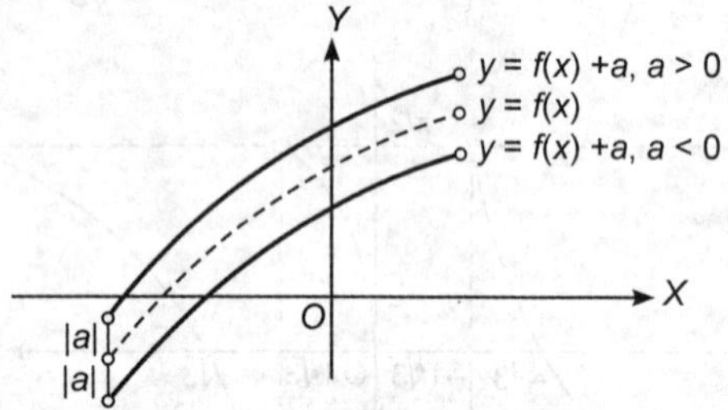

If $a > 0$ then shift the graph of $f(x)$ in upward direction through a distance a or shift the x-axis downward through a distance a, and if $a < 0$ then shift the graph of $f(x)$ in downward direction through a distance a or shift the x-axis upward through a distance a.

or

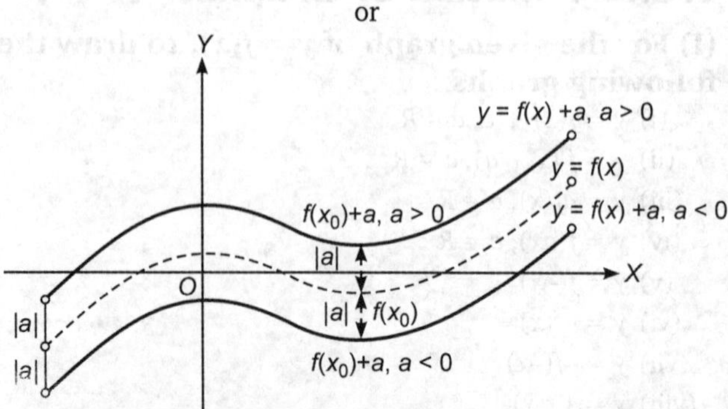

Illustration 9. Draw the graphs of $y = e^x + 1$ and $y = e^x - 2$, from the known graph of $y = e^x$, $\forall\, x \in R$.

Sol.

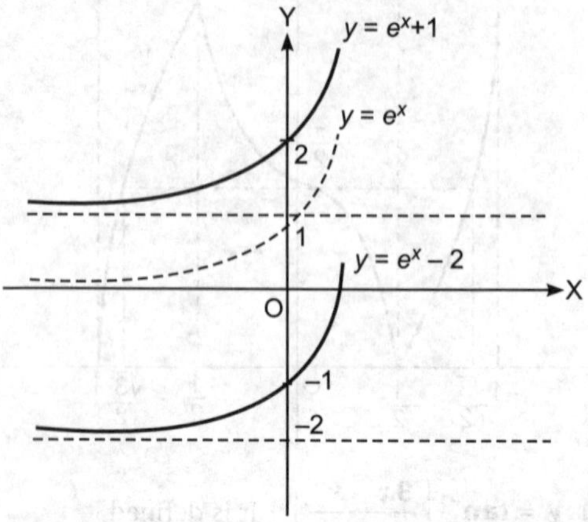

Illustration 10. Draw the graph of $y = \sin x - 1$, from the known graph of $y = \sin x$, $\forall\, x \in R$.

Sol. Since $-1 \le \sin x \le 1$

$\therefore \quad -2 \le \sin x - 1 \le 0$

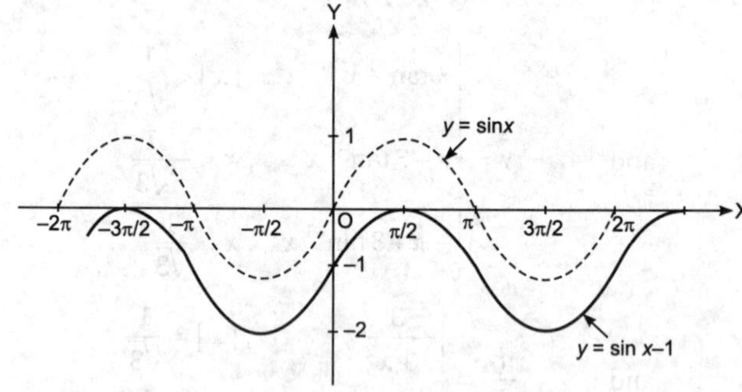

Illustration 11. Draw the graph of $y = |x| - 2$, from the known graph of $y = |x|$, $\forall\, x \in R$.

Sol.

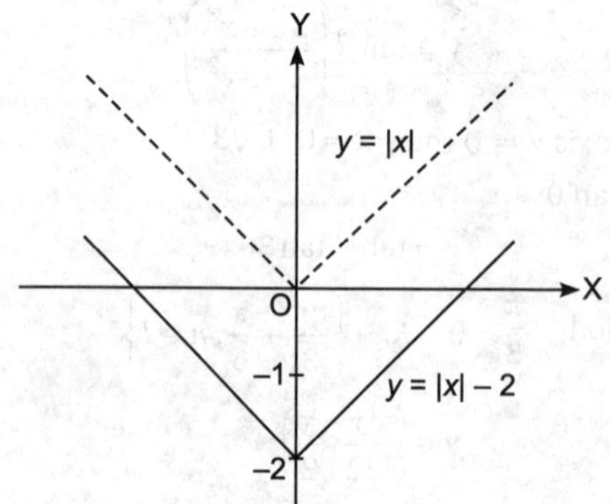

(ii) Drawing the graph of $y = f(x + a)$, $a \in R$, from the known graph of $y = f(x)$:

If $a > 0$ then shift the graph of $f(x)$ in the left hand direction through a distance a or shift the y-axis to the right hand direction by distance a, and if $a < 0$ then shift the graph of $f(x)$ in the right hand direction through a distance a or shift the y-axis to the left hand direction by distance a.

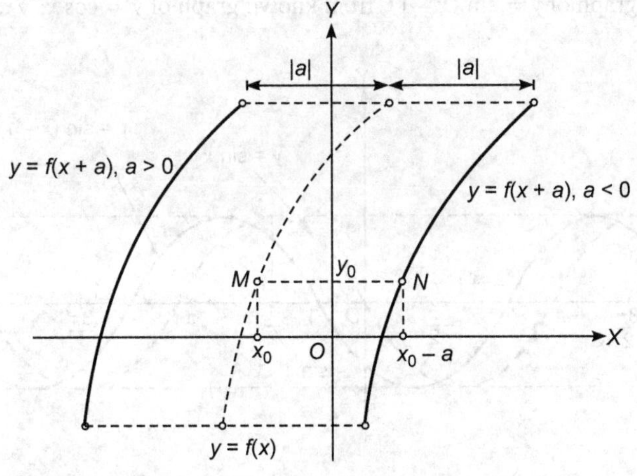

Or

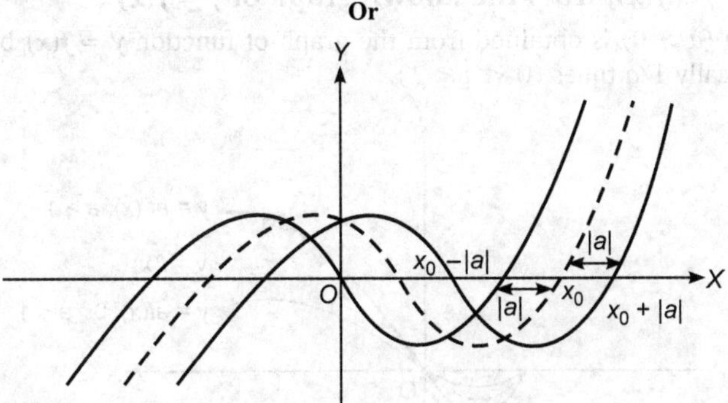

Illustration 12. Draw the graphs of $y = |x - 1|$ and $y = |x + 1|$, from the known graph of $y = |x|$, $\forall x \in R$.

Sol.

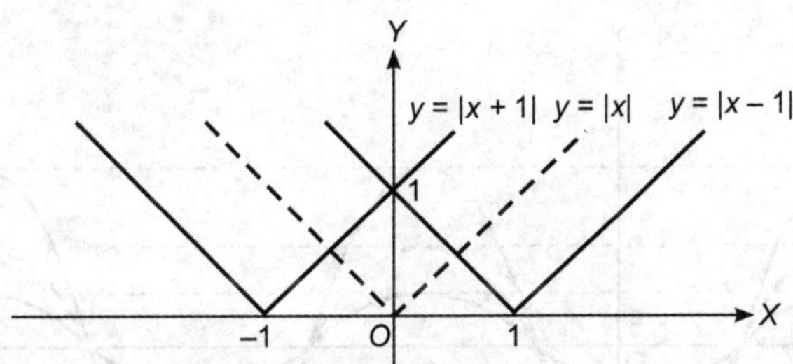

Illustration 13. Draw the graphs of $y = \sin\left(x - \frac{\pi}{4}\right)$ and $y = \sin\left(x + \frac{\pi}{4}\right)$ from known graph of $y = \sin x$, $\forall x \in R$.

Sol.

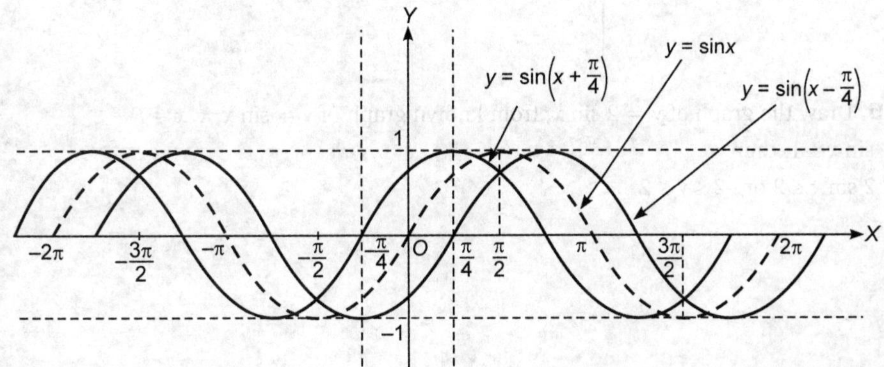

Illustration 14. Draw the graph of $y = \sin(x - 1)$, from known graph of $y = \cos x$, $\forall x \in R$.

Sol.

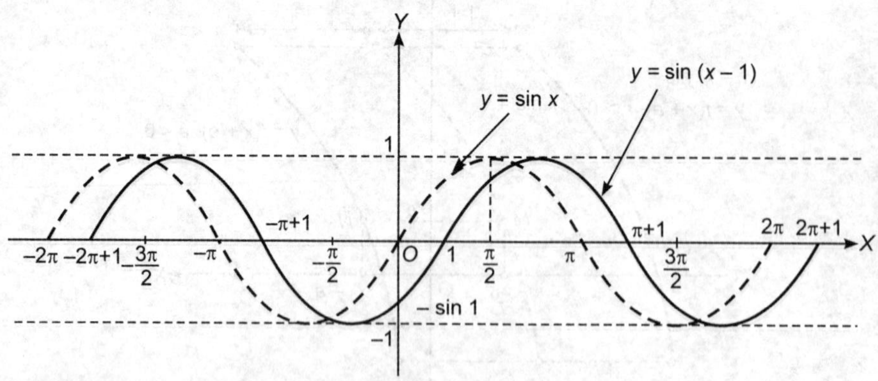

(iii) Drawing the graph of $y = af(x)$, from the known graph of $y = f(x)$:

The graph of function $y = af(x)$ $(a > 0)$ is obtained from the graph of function $y = f(x)$ by "stretching" vertically a times $(a > 1)$ and "compressing" vertically $1/a$ times $(0 < a < 1)$.

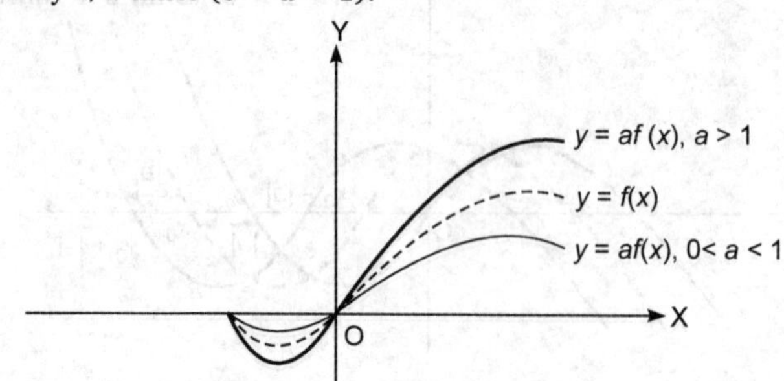

OR

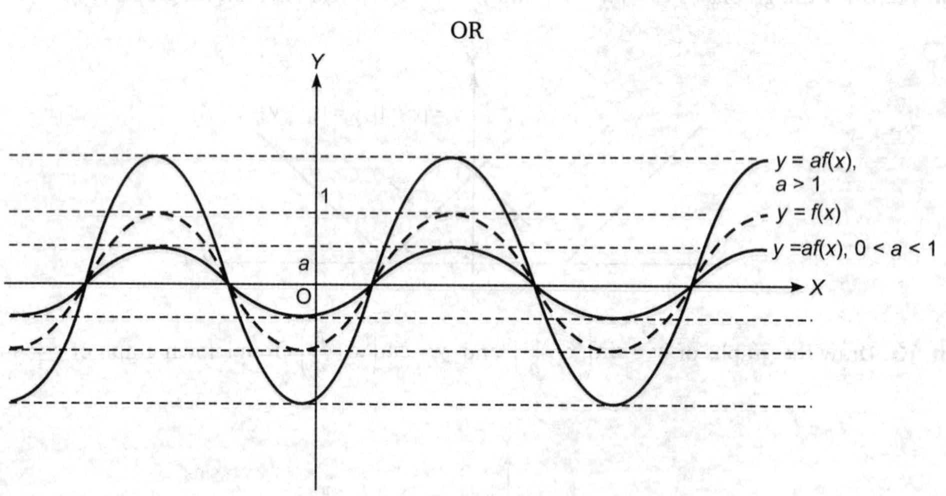

Illustration 15. Draw the graph of $y = 2 \sin x$, from known graph of $y = \sin x$, $\forall x \in R$.

Sol. Since $-1 \le \sin x \le 1$, then

$$-2 \le 2 \sin x \le 2 \text{ or } -2 \le y \le 2$$

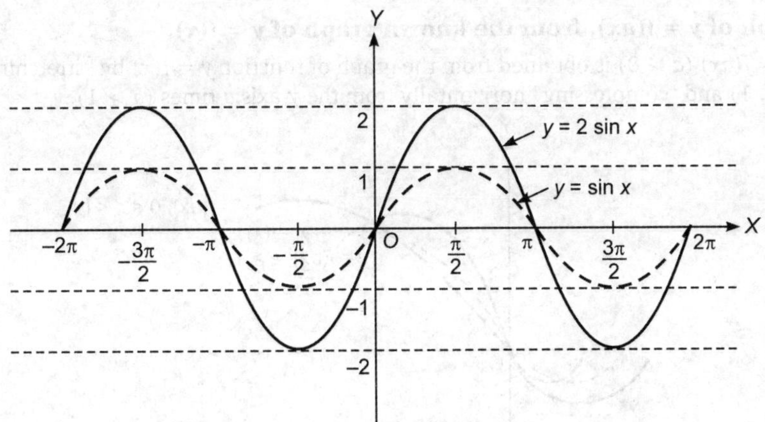

Illustration 16. Draw the graphs of $y = 4 \cos x$ and $y = \frac{1}{4} \cos x$, from the known graph of $y = \cos x$, $\forall\, x \in R$.

Sol. Since $-1 \le \cos x \le 1$

$\therefore$ $\quad -4 \le 4 \cos x \le 4 \quad$ and $\quad -\frac{1}{4} \le \frac{1}{4} \cos x \le \frac{1}{4}$

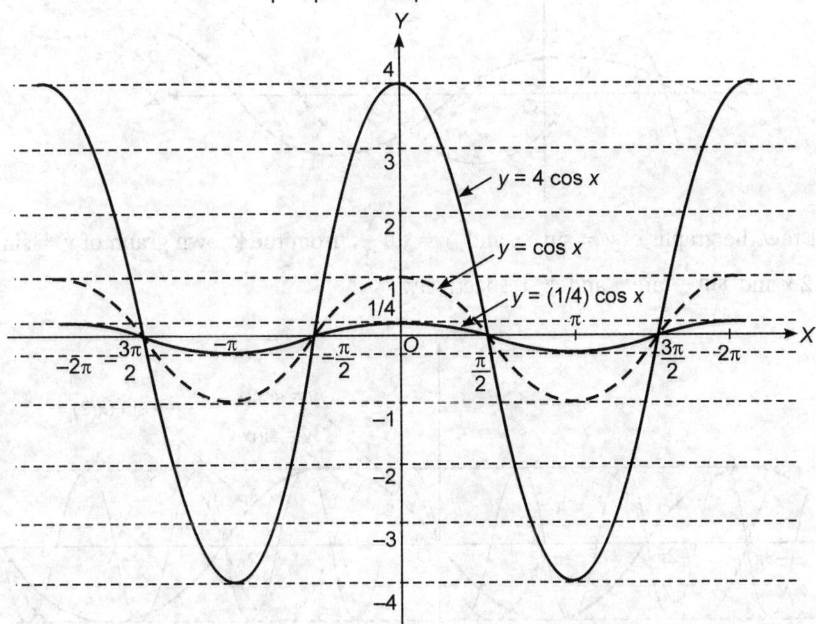

Illustration 17. Draw the graph of $y = \cos x + |\cos x|$, from the known graph of $y = \cos x$, $\forall\, x \in R$.

Sol. Since $y = \cos x + |\cos x| = \begin{cases} 2\cos x, & \cos x \ge 0 \\ 0, & \cos x < 0 \end{cases}$

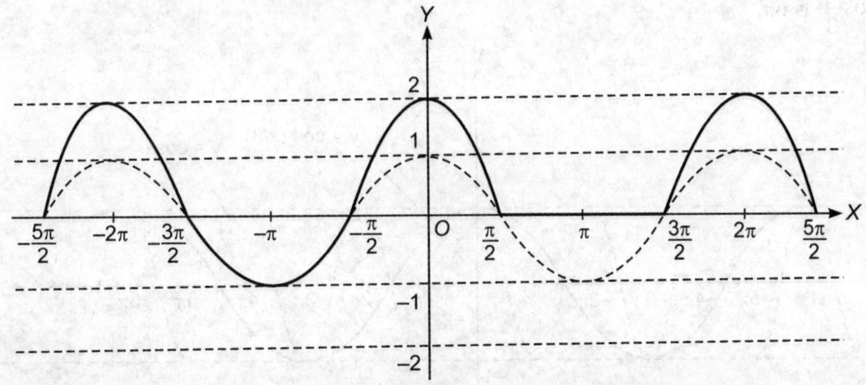

(iv) Drawing the graph of $y = f(ax)$, from the known graph of $y = f(x)$.

The graph of function $y = f(ax)$ $(a > 0)$ is obtained from the graph of function $y = f(x)$ by "stretching" horizontally from the y-axis $1/a$ times $(0 < a < 1)$ and "compressing" horizontally from the y-axis a times $(a > 1)$.

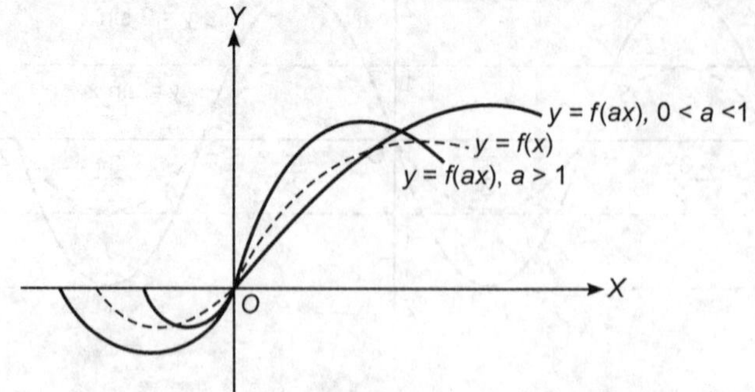

OR

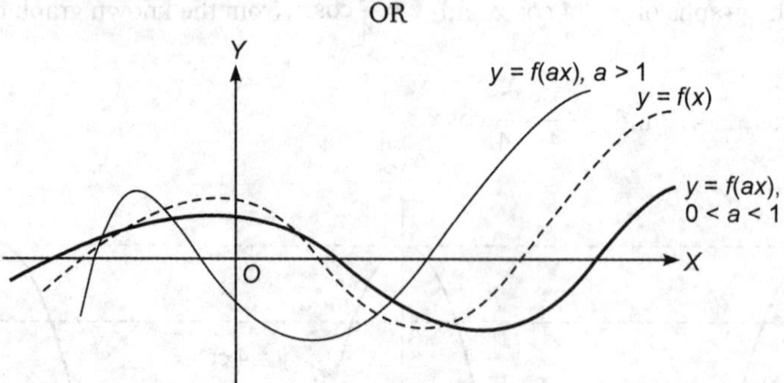

Illustration 18. Draw the graphs of $y = \sin 2x$ and $y = \sin\dfrac{x}{2}$, from the known graph of $y = \sin x$, $\forall\, x \in R$.

Sol. Periods of $\sin 2x$ and $\sin\dfrac{x}{2}$ are π and 4π respectively.

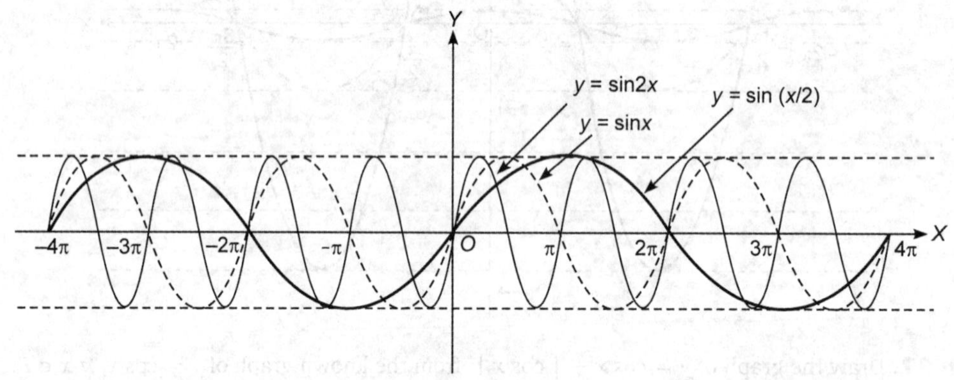

Illustration 19. Draw the graph of $y = \cos\dfrac{x}{3}$, from the known graph of $y = \cos x$, $\forall\, x \in R$.

Sol. Period of $\cos\dfrac{x}{3}$ is 6π.

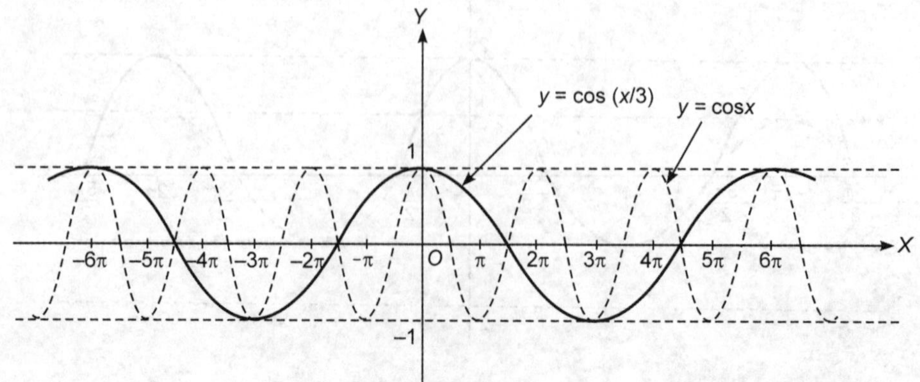

Illustration 20. Draw the graph of $y = \cos(2x + |x|)$, from the known graph $y = \cos x$, $\forall x \in R$.

Sol. $y = \cos(2x+|x|) = \begin{cases} \cos 3x, & x \geq 0 \\ \cos x, & x < 0 \end{cases}$

Periods of $\cos 3x$ and $\cos x$ are $\frac{2\pi}{3}$ and 2π respectively.

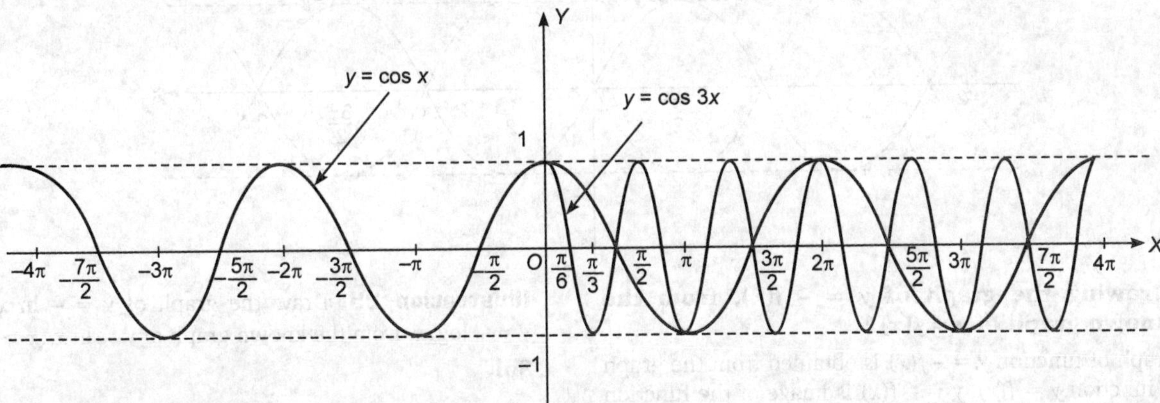

(v) Drawing the graph $y = f(-x)$, from the known graph of $y = f(x)$:

The graph of function $y = f(-x)$ is obtained from the graph of the function $y = f(x)$, $y = f(-x)$ is image of the function $y = f(x)$ about the y-axis or rotate the graph of $f(x)$ through 180° about y-axis.

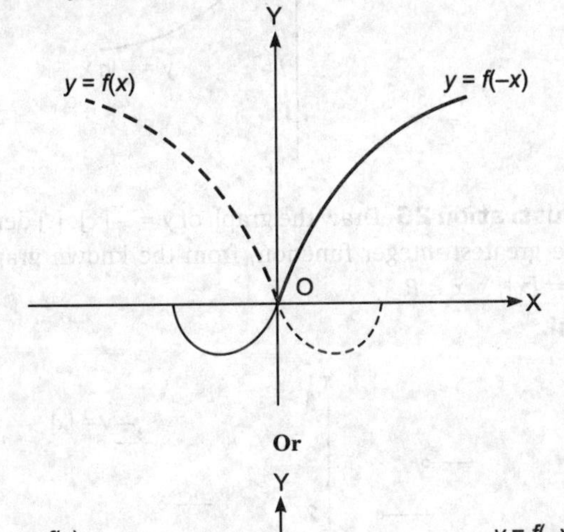

Or

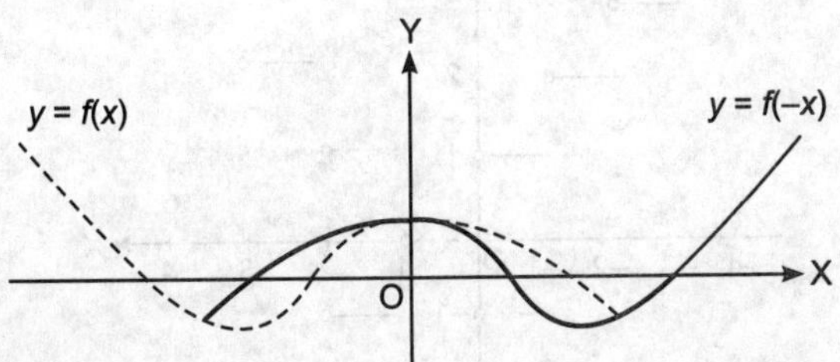

Note : Take reflection of $f(x)$ considering y-axis as plane mirror.

Illustration 21. Draw the graph of $y = e^{-x}$, from the known graph of $y = e^x$, $\forall x \in R$.

Sol.

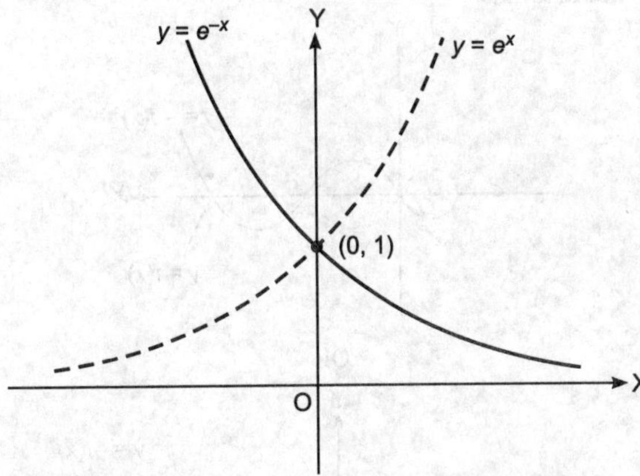

Illustration 22. Draw the graph of $y = \ln(-x)$ *i.e.*, $y = \log_e(-x)$, from the known graph of $y = \ln x$ $(x > 0)$.

Sol.

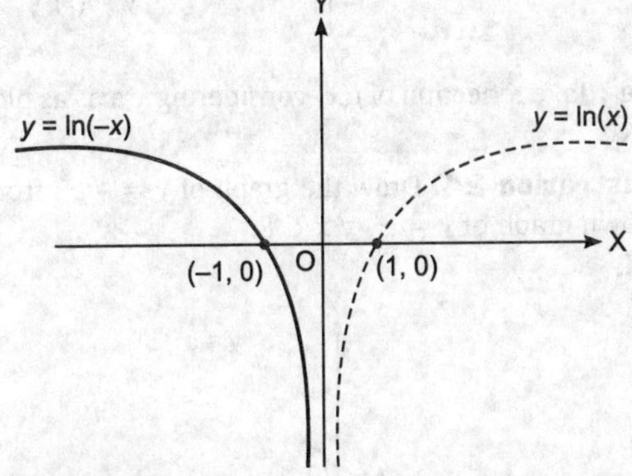

Illustration 23. Draw the graph of $y = \sin(-x)$, from the known graph of $y = \sin x$, $\forall x \in R$.

Sol.

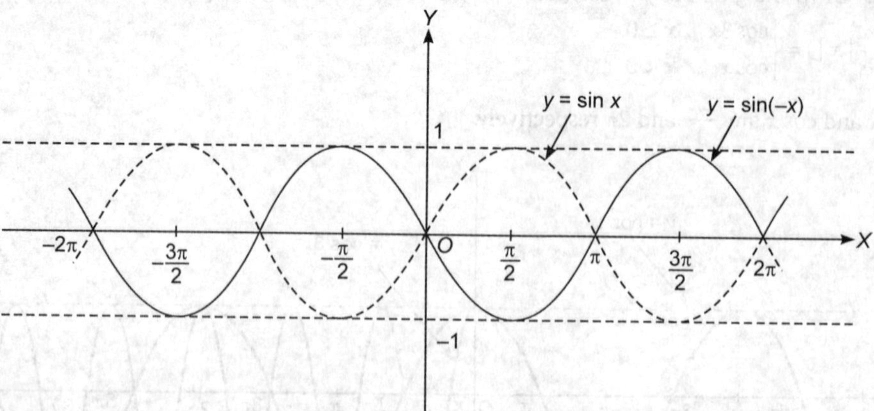

(vi) Drawing the graph of $y = -f(x)$, from the known graph of $y = f(x)$:

The graph of function $y = -f(x)$ is obtained from the graph of the function $y = f(x)$, $y = -f(x)$ is image of the function $y = f(x)$ about the x-axis or rotate the graph of $f(x)$ through 180° about x-axis.

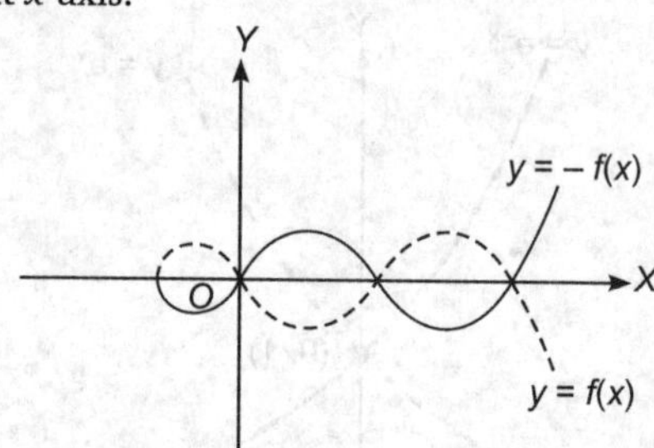

OR

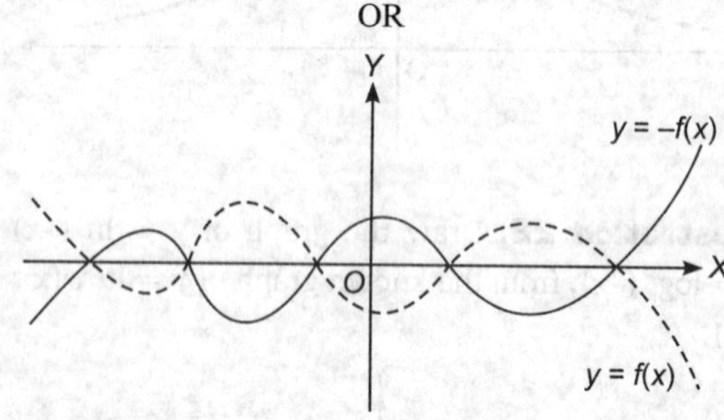

Note : Take reflection of $f(x)$ considering x-axis as plane mirror.

Illustration 24. Draw the graph of $y = -e^x$, from the known graph of $y = e^x$, $\forall x \in R$.

Sol.

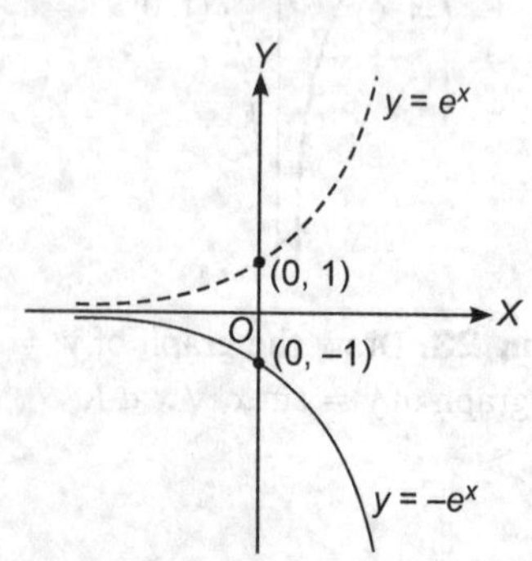

Illustration 25. Draw the graph of $y = -\ln x$ *i.e.*, $y = -\log_e x$, from the known graph of $y = \ln x$ $(x > 0)$.

Sol.

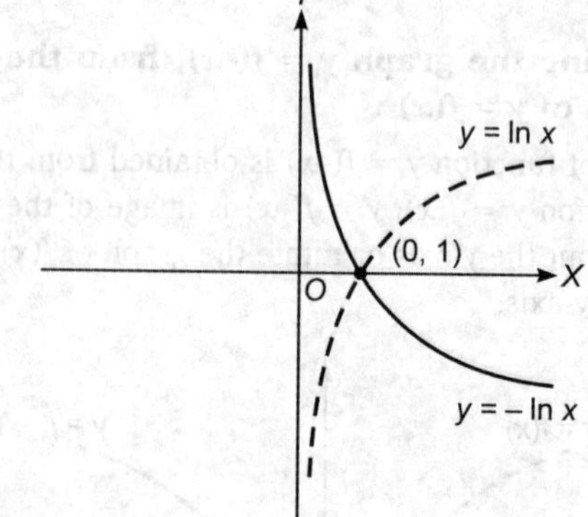

Illustration 26. Draw the graph of $y = -[x]$, [.] denotes the greatest integer function, from the known graph of $y = [x]$, $\forall x \in R$.

Sol.

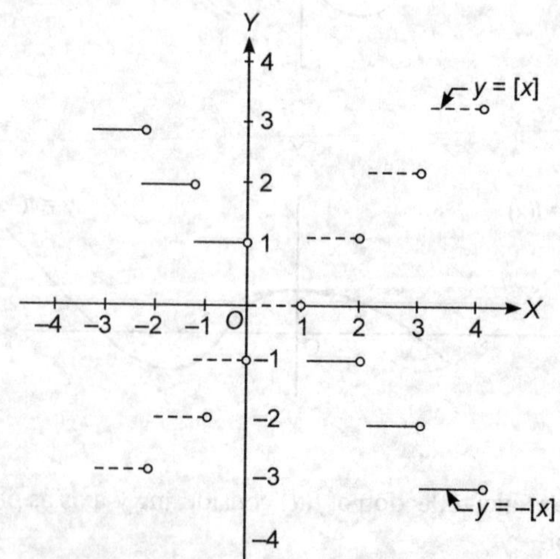

(vii) Drawing the graph of $y = -f(-x)$, from the known graph of $y = f(x)$:

The graph of function $y = -f(-x)$ is obtained from the graph of the function $y = f(x)$.

Since $f(-x)$ is image of the function $y = f(x)$ about the y-axis and $-f(-x)$ is also image of the function $y = f(-x)$ about the x-axis.

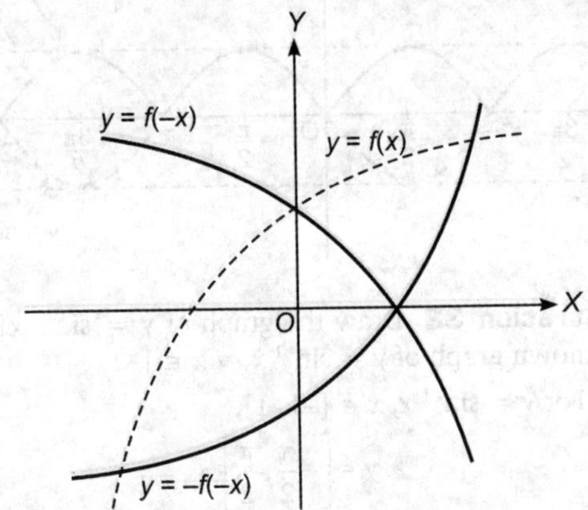

OR

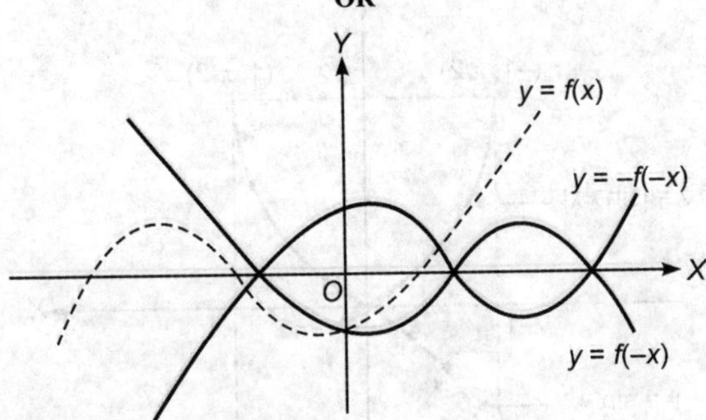

Illustration 27. Draw the graph of $y = -e^{-x}$, from the known graph of $y = e^x$, $\forall\, x \in R$.

Sol.

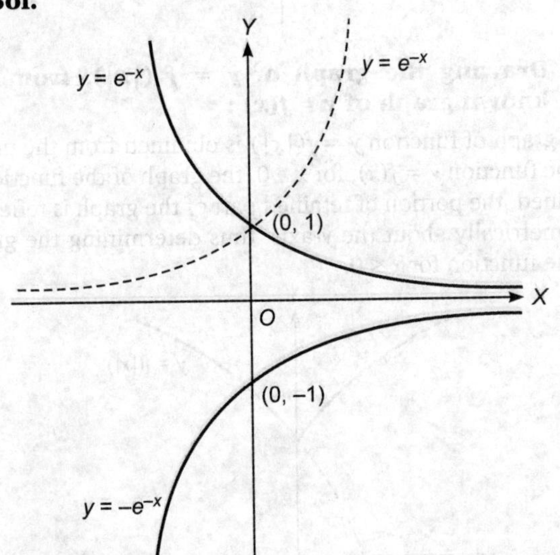

Illustration 28. Draw the graph of $y = -\ln(-x)$ *i.e.*, $y = -\log_e(-x)$, from the known graph of $y = \ln(x)$, $(x > 0)$.

Sol.

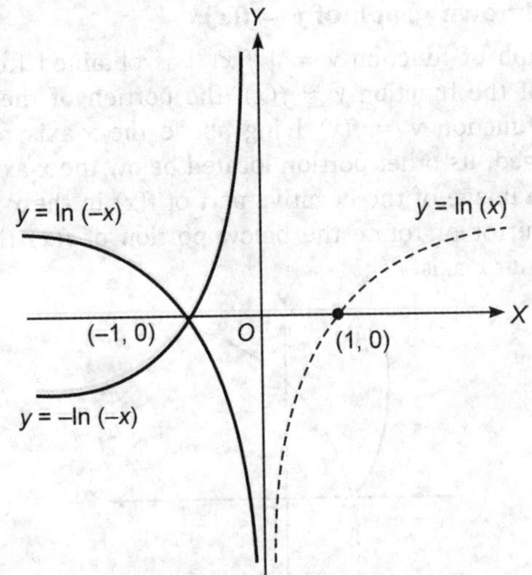

Illustration 29. Draw the graph of the function $y = -[-x]$, [.] denotes the greatest integer function, from the known graph of $y = [x]$, $\forall\, x \in R$.

Sol. Graph of $y = [-x]$:

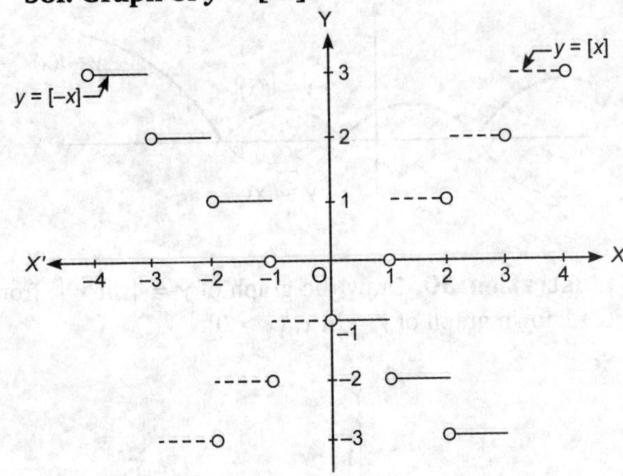

Graph of $y = -[-x]$:

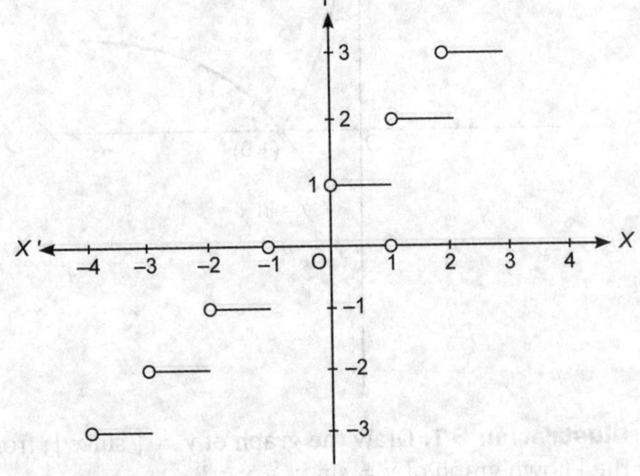

(viii) Drawing the graph of $y = | f(x) |$, from the known graph of $y = f(x)$:

The graph of function $y = | f(x) |$ is obtained from the graph of the function $y = f(x)$, the portion of the graph of the function $y = f(x)$ lying above the x-axis remains unchanged. Its other portion located below the x-axis then take the image of the negative part of $f(x)$ in the x-axis as plane mirror or rotate the below portion of $f(x)$ through 180° about x-axis.

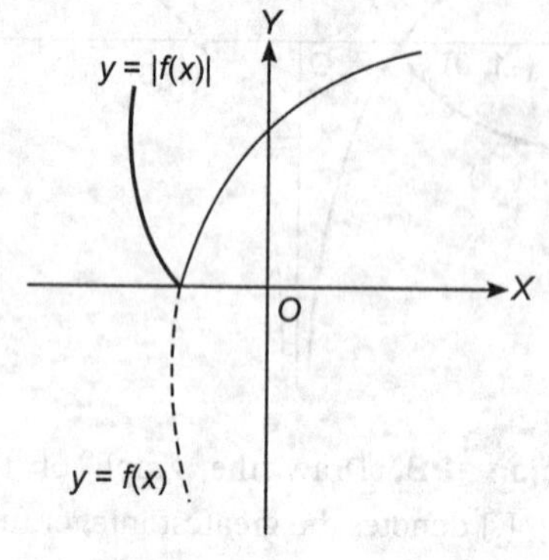

OR

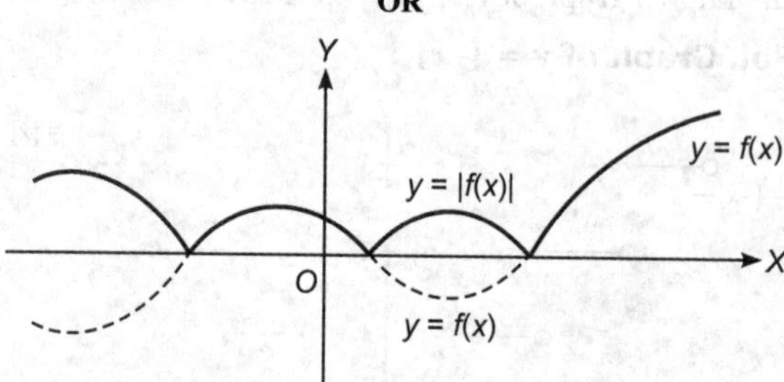

Illustration 30. Draw the graph of $y = | \ln x |$, from the known graph of $y = \ln x$, $(x > 0)$.

Sol.

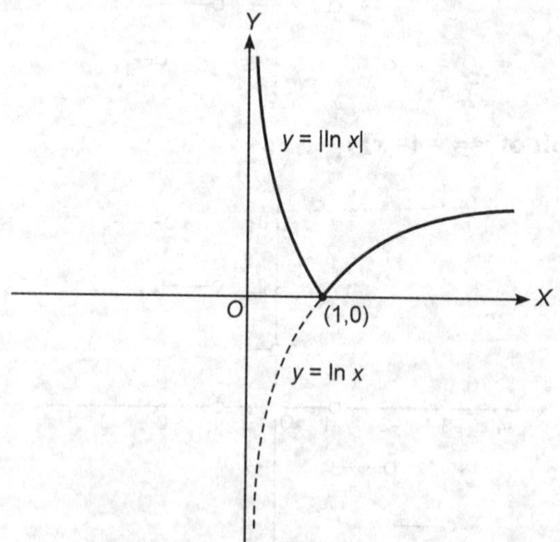

Illustration 31. Draw the graph of $y = | \sin x |$, from the known graph of $y = \sin x$, $\forall x \in R$.

Sol.

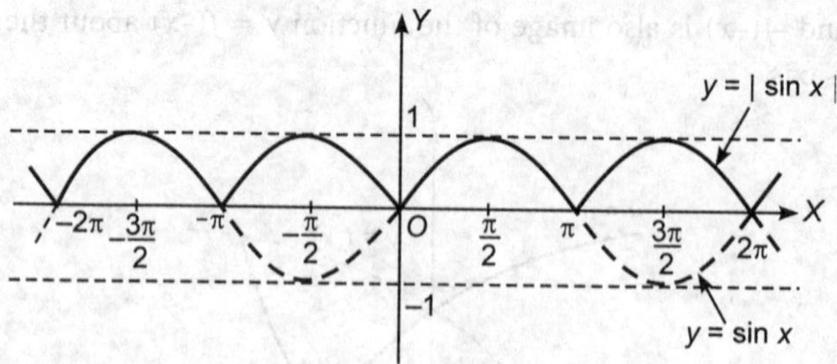

Illustration 32. Draw the graph of $y = |\sin^{-1} x|$, from the known graph of $y = \sin^{-1} x$, $\forall x \in [-1, 1]$.

Sol. For $y = \sin^{-1} x$, $x \in [-1, 1]$.

and $$y \in \left[-\frac{\pi}{2}, \frac{\pi}{2}\right]$$

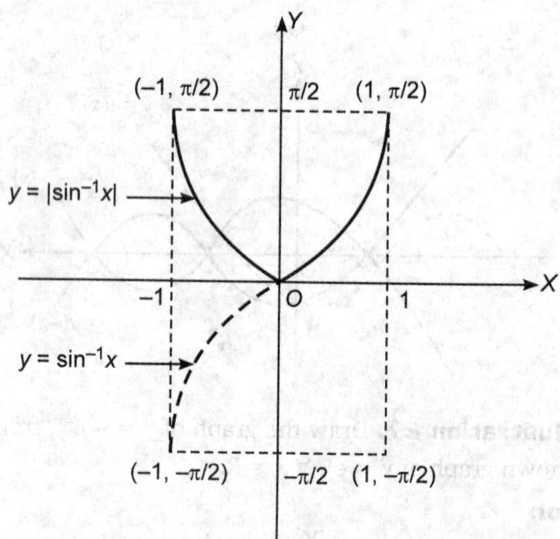

(ix) Drawing the graph of $y = f(|x|)$, from the known graph of $y = f(x)$:

The graph of function $y = f(|x|)$ is obtained from the graph of the function $y = f(x)$, for $x \geq 0$, the graph of the function is retained, the portion of retained part of the graph is reflected symmetrically about the y-axis, thus determining the graph of the function for $x \leq 0$.

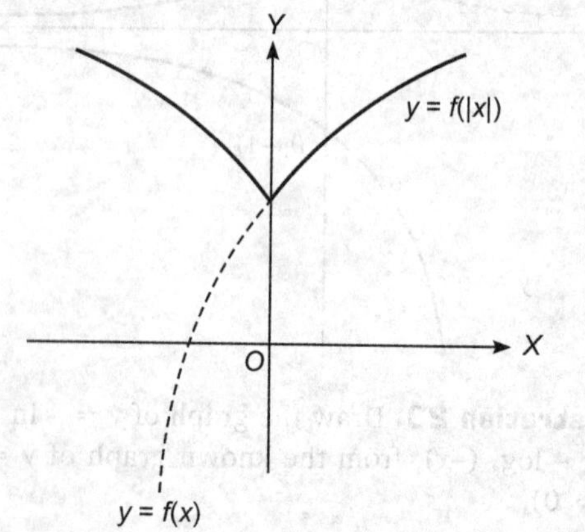

OR

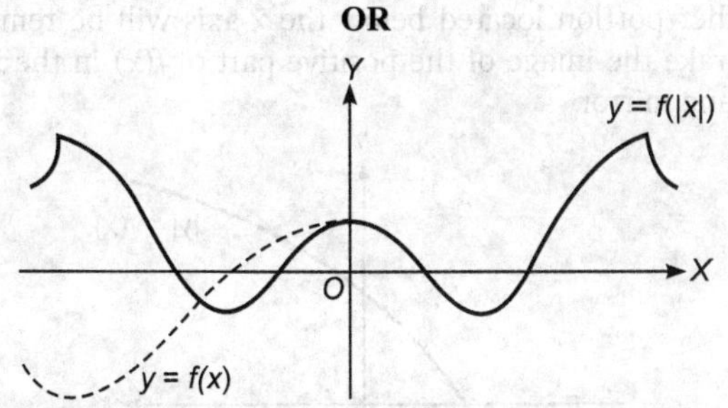

Notes : (i) The part of $f(x)$ lying leftward of the y-axis (if it exits) is omitted. (ii) $f(|x|)$ is always an even function.

Illustration 33. Draw the graph of the function $y = e^{|x|}$, from the known graph of $y = e^x$, $\forall\ x \in R$.

Sol.

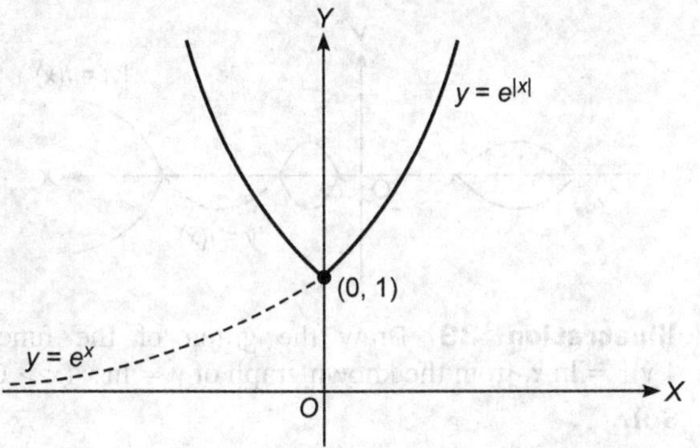

Illustration 34. Draw the graph of the function $y = \sin |x|$, from the known graph of $y = \sin x$, $\forall\ x \in R$.

Sol.

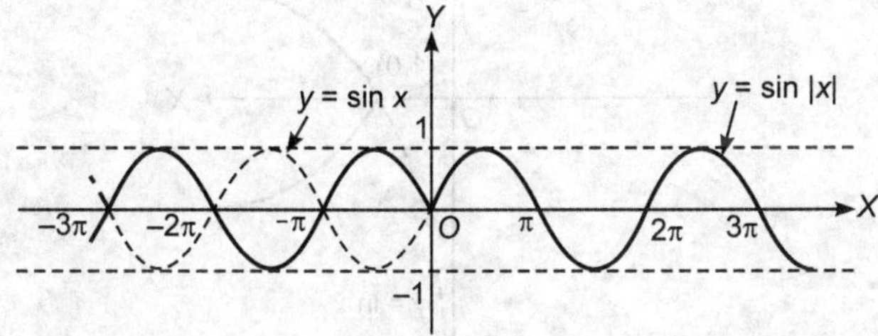

Illustration 35. Draw the graph of the function $y = \ln |x|$, from the known graph $y = \ln x$, $(x > 0)$.

Sol.

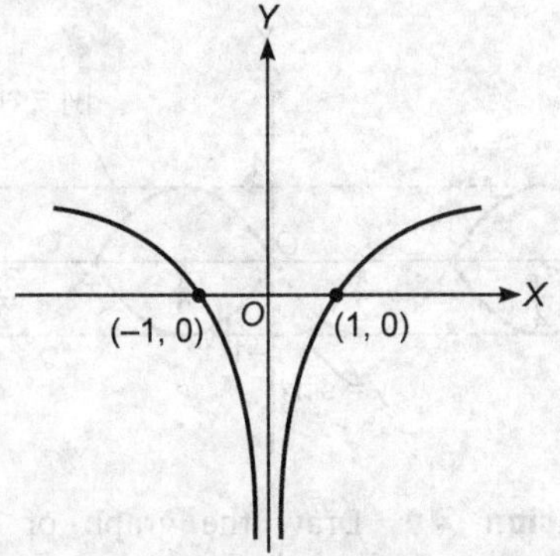

(x) Drawing the graph of $y = |f(|x|)|$, from the known graph of $y = f(x)$:

The graph of function $y = |f(|x|)$ is obtained from the graph of the function $y = f(x)$, first draw the graph of $y = f(|x|)$ from method (ix) then graph of $y = |f(|x|)|$ is the portion of the graph of the function $y = f(|x|)$ lying above the x-axis remains unchanged. Its other portion located below the x-axis then take the image of the negative part of $y = f(|x|)$ in the x-axis as mirror image.

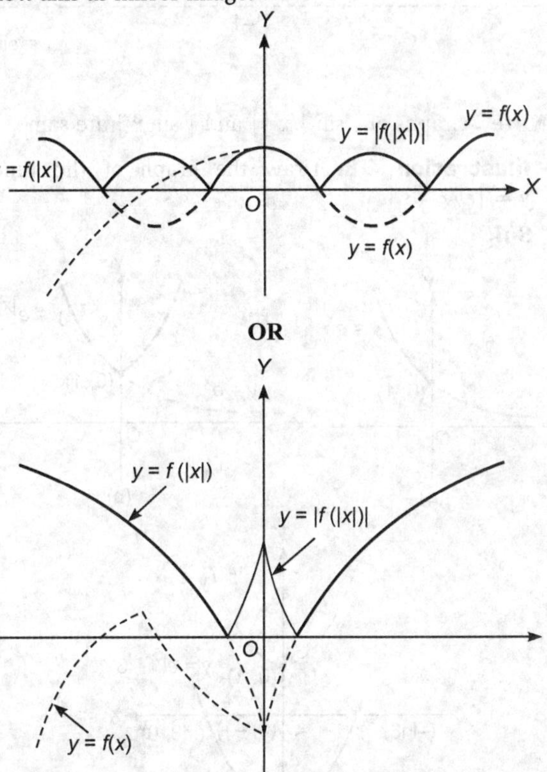

Illustration 36. Draw the graph of function $y = |\ln|x||$, from the known graph of $y = \ln x$, $(x > 0)$.

Sol.

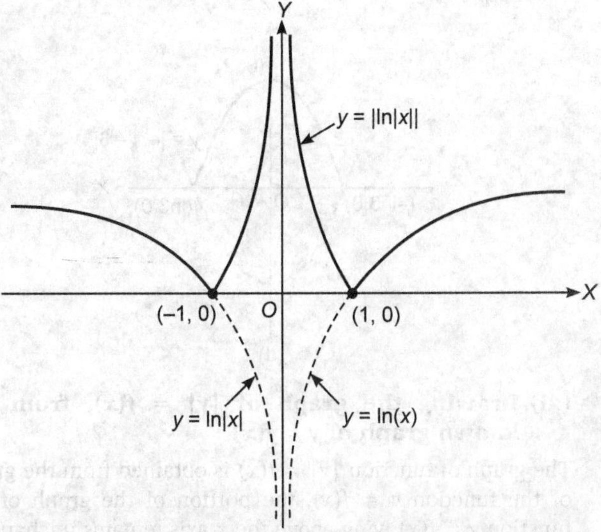

Illustration 37. Draw the graph of the function $y = |\sin|x||$, from the known graph of $y = \sin x$, $\forall\, x \in R$.

Sol.

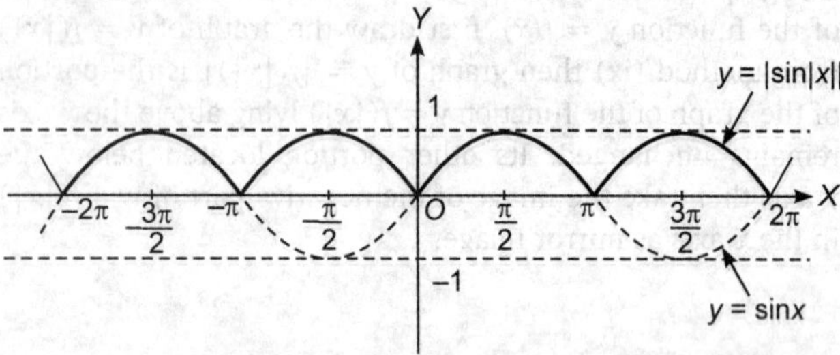

Note : Graphs of $|\sin|x||$ and $|\sin x|$ are same.

Illustration 38. Draw the graph of the function $y = |e^{|x|} - 3|$.

Sol.

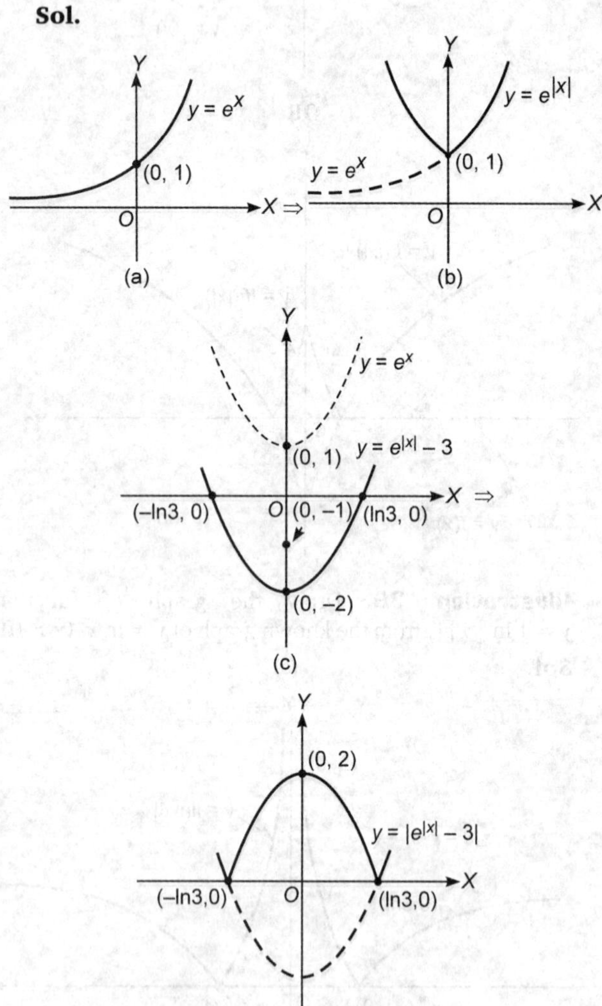

(xi) Drawing the graph of $|y| = f(x)$, from the known graph of $y = f(x)$:

The graph of function $|y| = f(x)$ is obtained from the graph of the function $y = f(x)$, the portion of the graph of the function $y = f(x)$ lying above the x-axis remains unchanged, its other portion located below the x-axis will be removed, then take the image of the positive part of $f(x)$ in the x-axis as plane mirror.

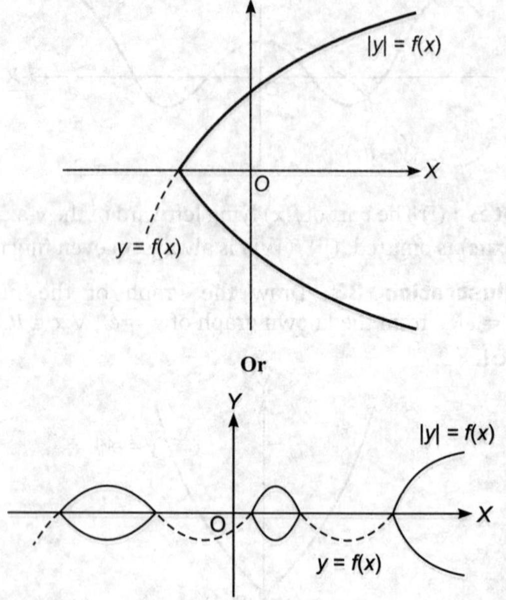

Illustration 39. Draw the graph of the function $|y| = \ln x$, from the known graph of $y = \ln x$ $(x > 0)$.

Sol.

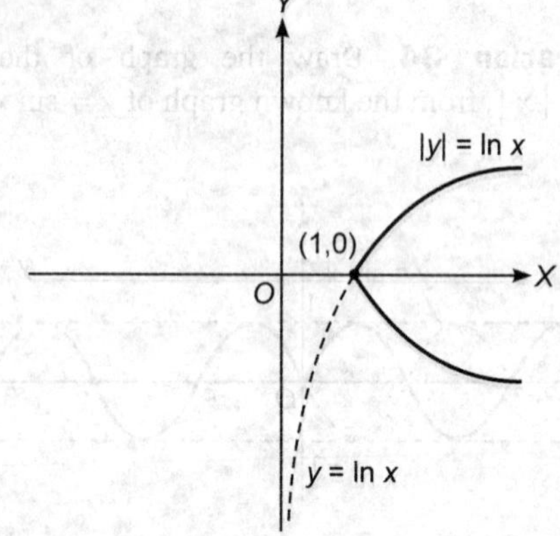

Illustration 40. Draw the graph of the function $|y| = \sin x$, from the known graph of $y = \sin x$, $\forall\, x \in R$.

Sol.

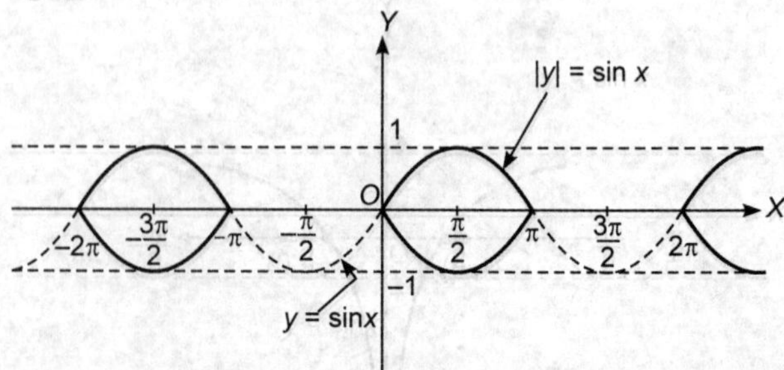

Illustration 41. Draw the graph of the function $|x| + |y| = 1$, from the known graph of $x + y = 1$, $\forall\, x, y \in R$.

Sol. Graph of $x + y = 1$:

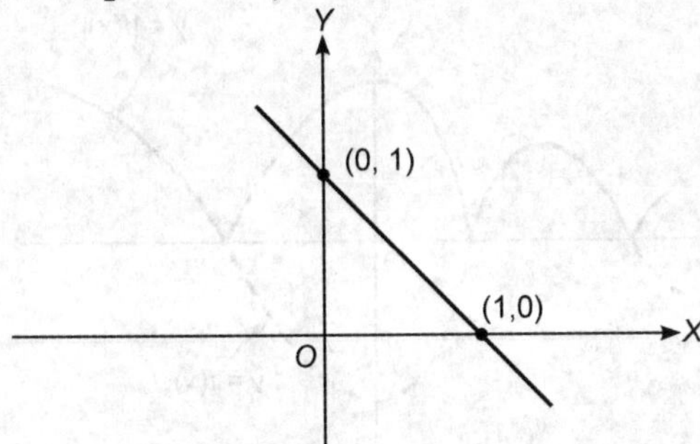

Graph of $y = 1 - |x|$:

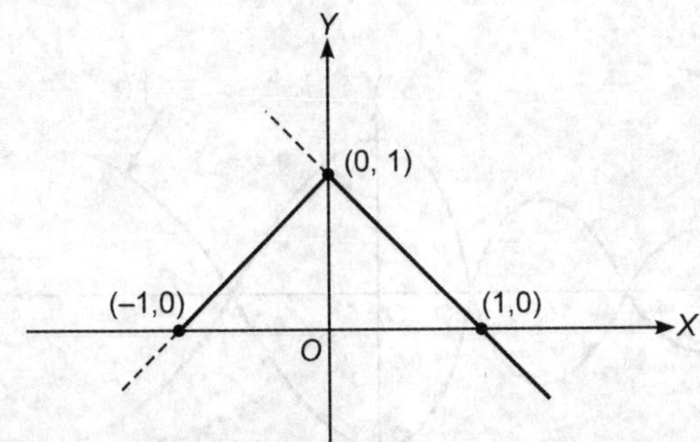

Now graph of $|y| = 1 - |x|$ or $|x| + |y| = 1$:

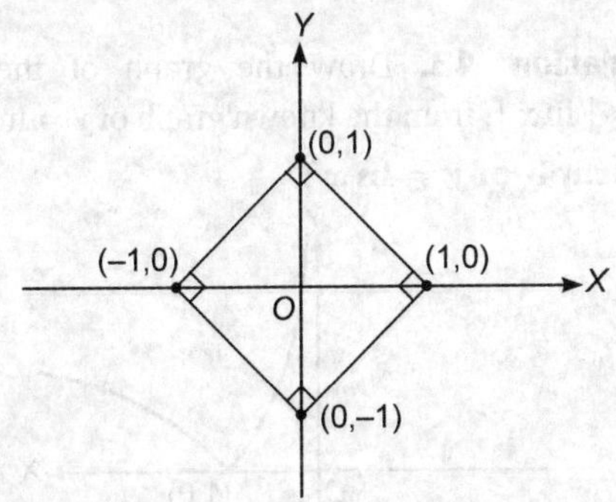

Note : Area of curve bounded by $|x| + |y| = 1$ is $(\sqrt{2})^2 = 2$ sq. units.

Illustration 42. Draw the graph of the function $2|x| + 3|y| \le 6$, from the known graph of $2x + 3y = 6, \forall x, y \in R$ and also find the bounded area.

Sol. Graph of $2x + 3y = 6$:

i.e., $\frac{x}{3} + \frac{y}{2} = 1$

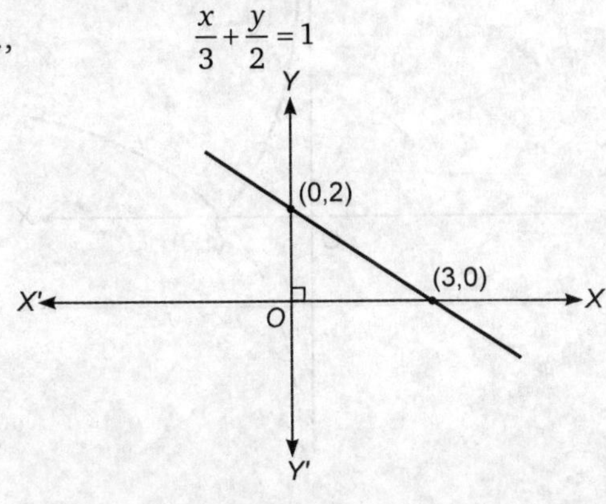

Graph of $2|x| + 3y = 6$:

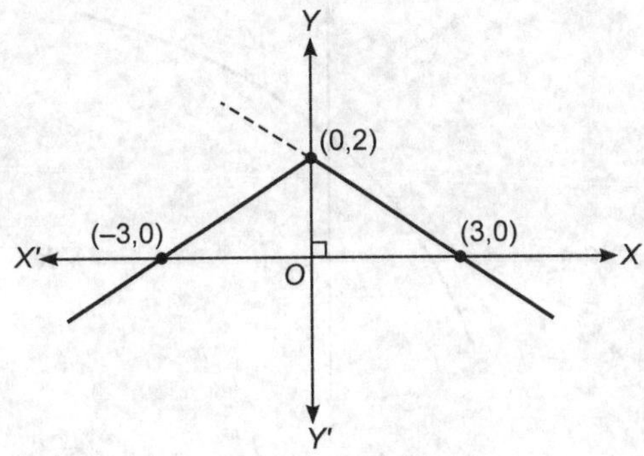

Graph of $2|x| + 3|y| = 6$:

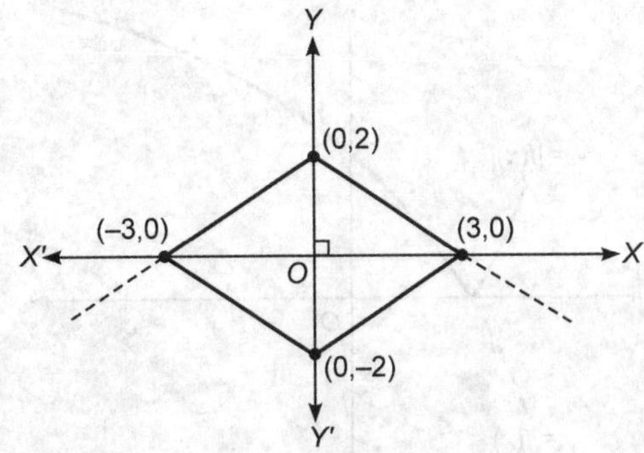

Graph of $2|x| + 3|y| \le 6$:

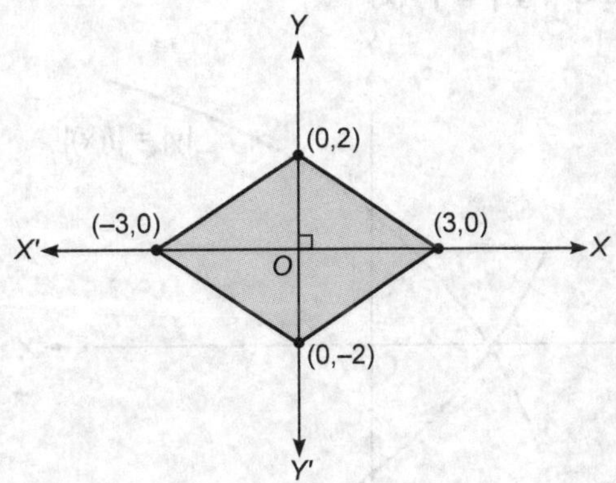

$\therefore$ Required area $= 4 \times$ Area of ΔAOB

$$= 4 \times \frac{1}{2} \times 3 \times 2$$

$$= 12 \text{ sq. units}$$

(xii) Drawing the graph of $|y| = |f(x)|$, from the known graph of $y = f(x)$:

The graph of function $|y| = |f(x)|$ is obtained from the graph of the function $y = f(x)$, first draw the graph of $y = |f(x)|$, then graph of $|y| = |f(x)|$ is the portion of the graph $y = |f(x)|$ lying above the x-axis remains unchanged. Its other portion located below the x-axis then take the image of positive part of $y = |f(x)|$ in the x-axis as plane mirror.

Graph of $y = f(x)$:

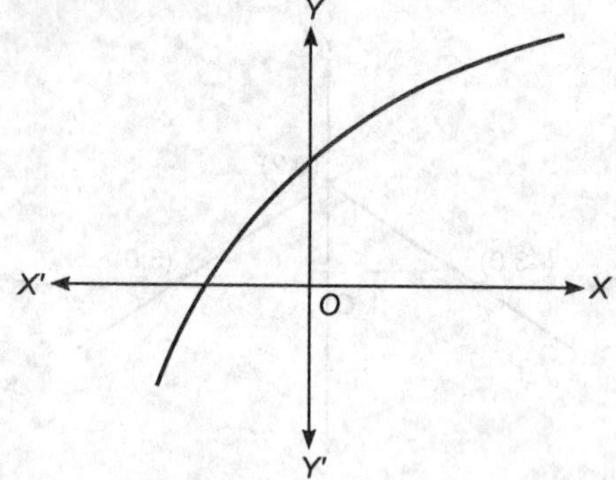

Graph of $y = | f(x) |$:

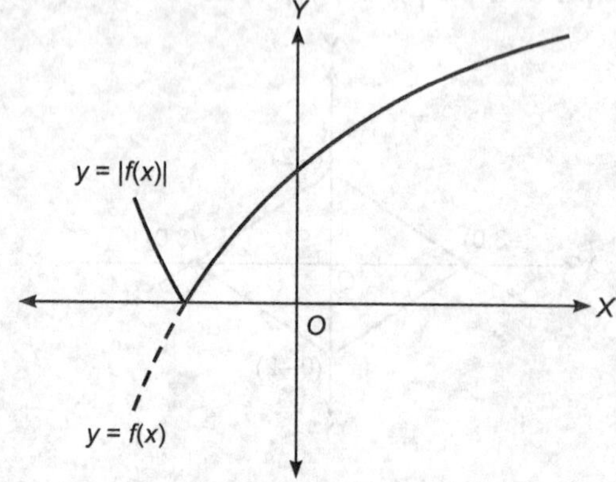

Graph of $| y | = | f(x) |$:

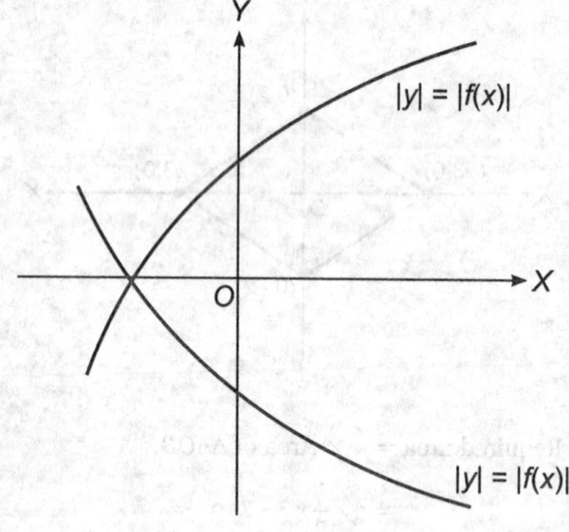

OR

Graph of $y = f(x)$:

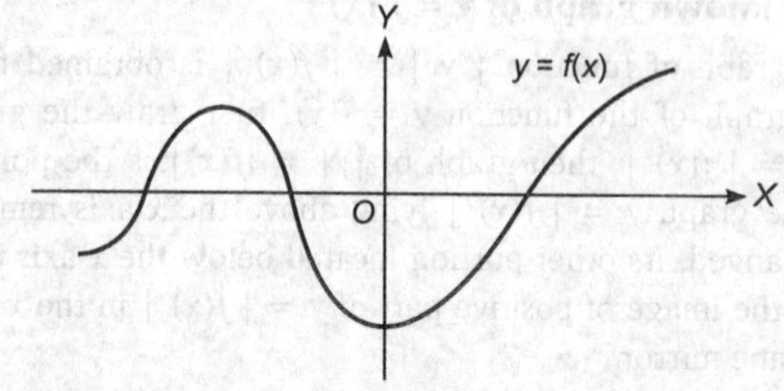

Graph of $y = | f(x) |$ with the help of above graph :

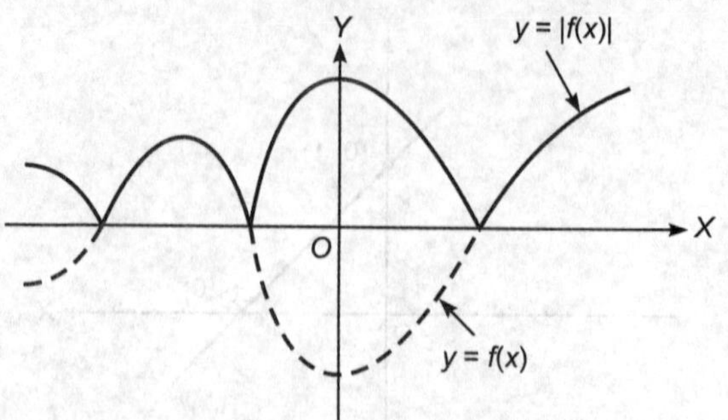

Graph of $| y | = | f(x) |$ with the help of above graph :

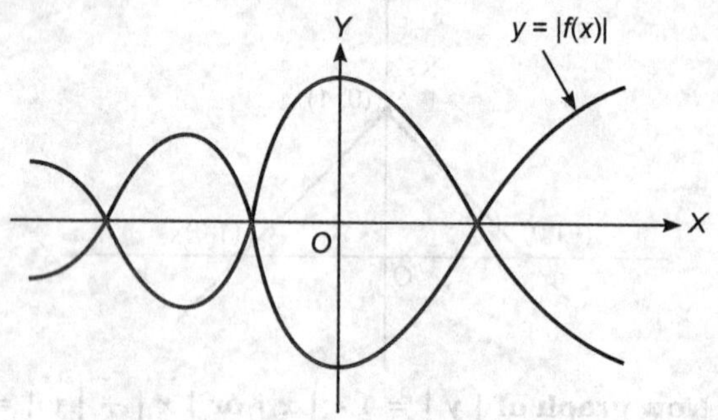

Illustration 43. Draw the graph of the function $| y | = | \ln x |$, from the known graph of $y = \ln x$ $(x > 0)$.

Sol. Graph of $y = \ln x$:

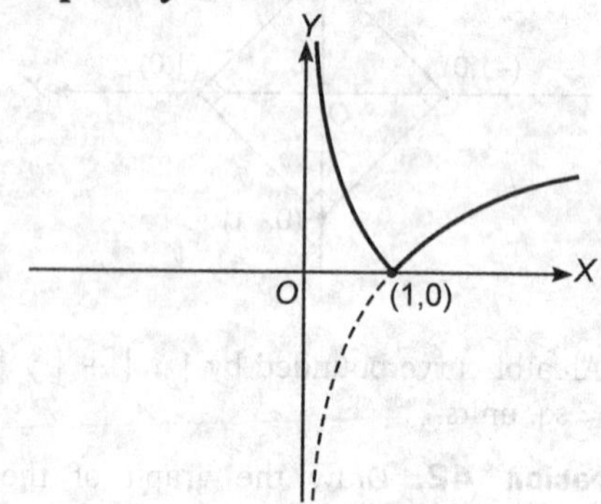

Graph of $y = | \ln x |$:

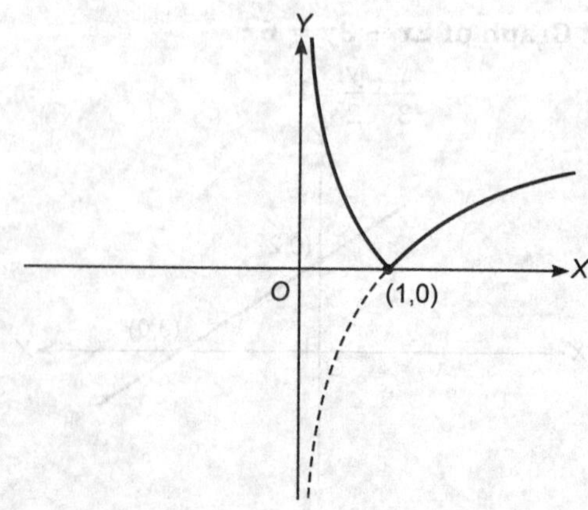

Graph of $|y| = |\ln x|$:

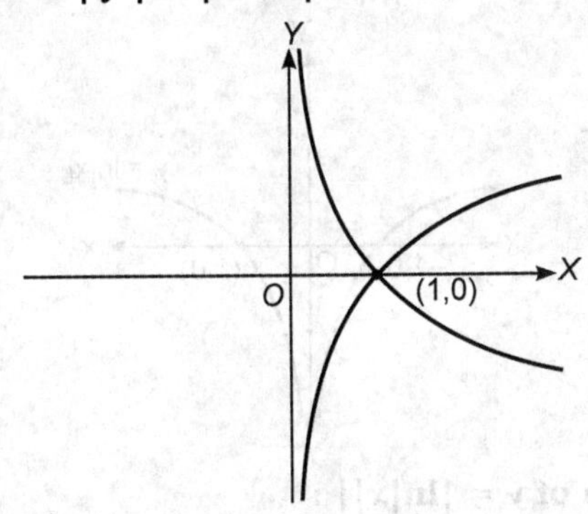

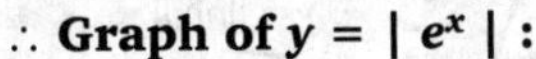

Illustration 44. Draw the graph of the function $|y| = |e^x|$, from the known graph of $y = e^x$, $\forall x \in R$.

Sol. Since $|e^x| = e^x$

$\therefore$ **Graph of $y = |e^x|$:**

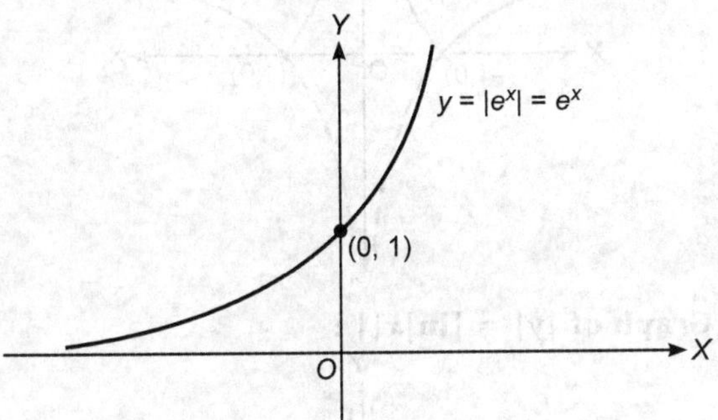

Graph of $|y| = |e^x|$:

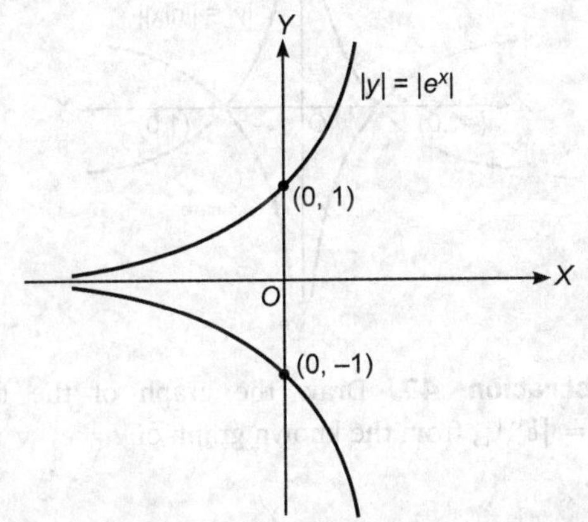

Illustration 45. Draw the graph of the function $|y| = |\cos x|$, from the known graph of $y = \cos x$, $\forall x \in R$.

Sol. Graph of $y = \cos x$:

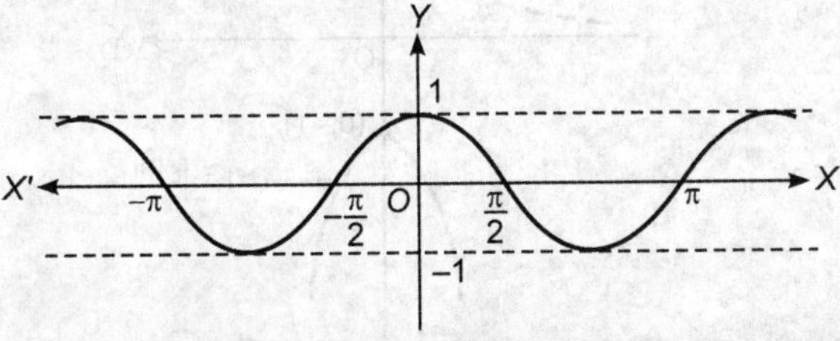

Graph of $y = |\cos x|$:

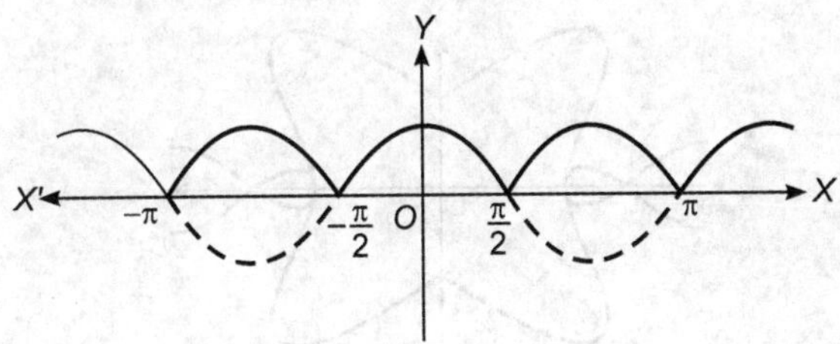

Graph of $|y| = |\cos x|$:

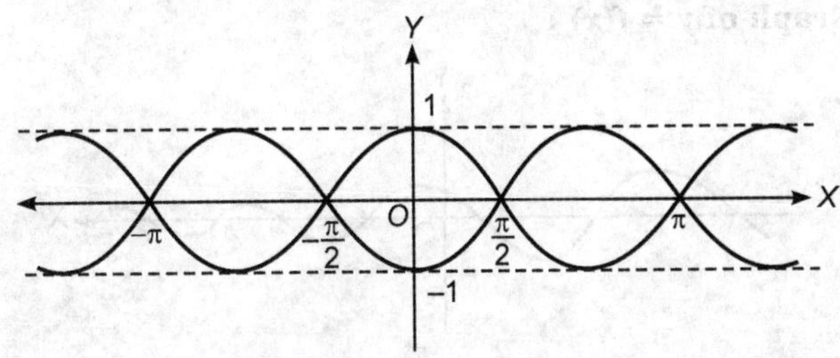

Note : Graphs of $|y| = |\cos x|$ and $|y| = |\cos |x||$ are same.

(xiii) Drawing the graph of $|y| = |f(|x|)|$ from the known graph of $y = f(x)$:

The graph of the function $|y| = |f(|x|)|$ is obtained from the graph of $y = f(x)$. First draw the graph of $y = |f(|x|)|$ from method (x). Hence the graph of $|y| = |f(|x|)|$ is image with x-axis as plane mirror.

Graph of $y = f(x)$:

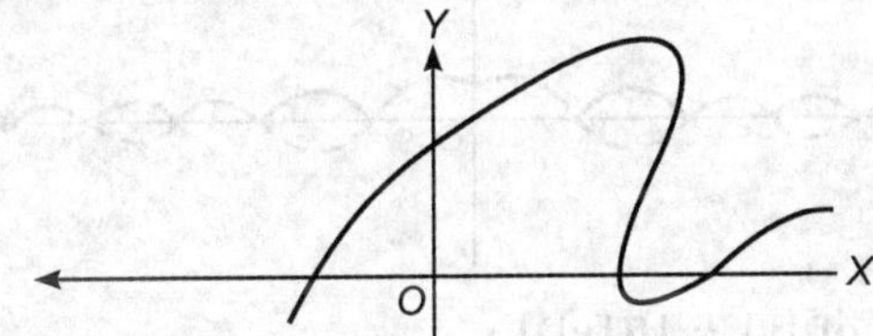

Graph of $y = f(|x|)$:

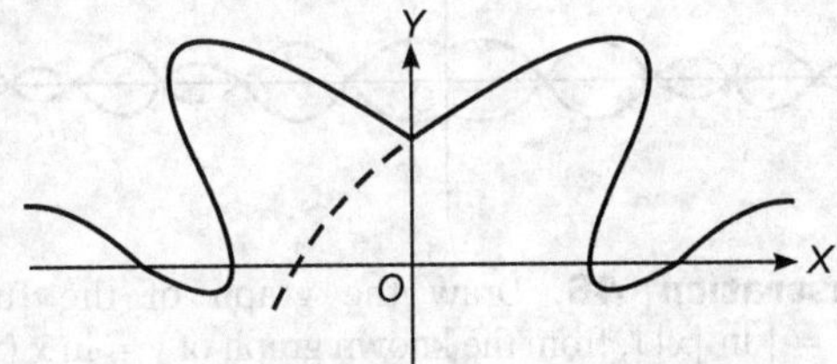

Graph of $y = |f(|x|)|$:

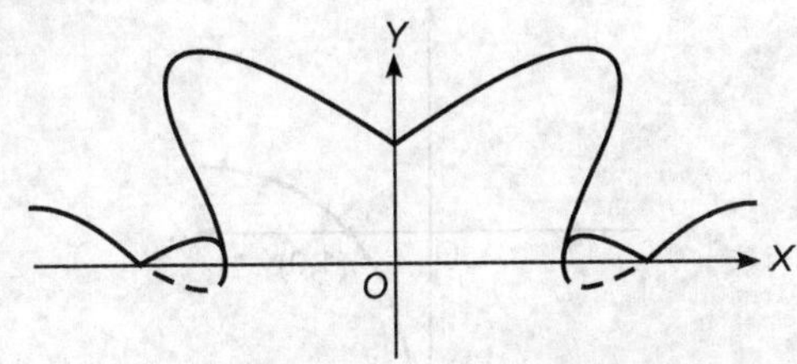

Graph of $|y| = |f(|x|)|$:

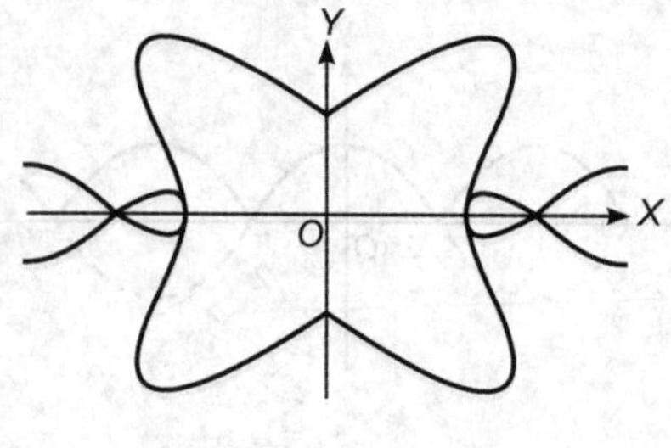

Or

Graph of $y = f(x)$:

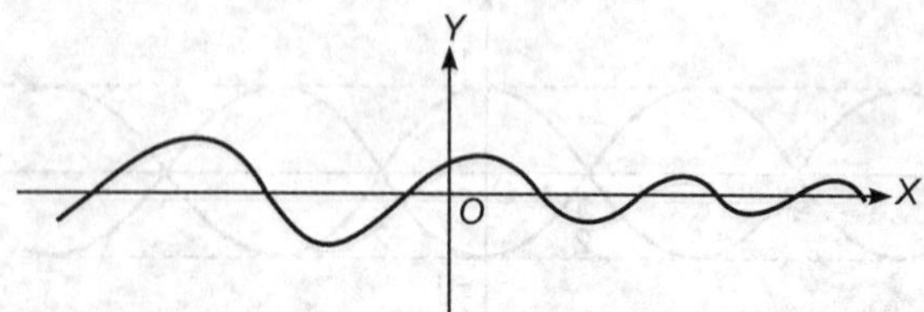

Graph of $y = f(|x|)$:

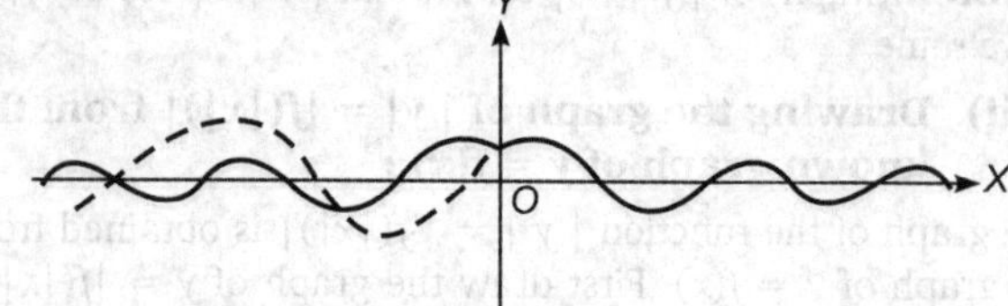

Graph of $y = |f(|x|)|$:

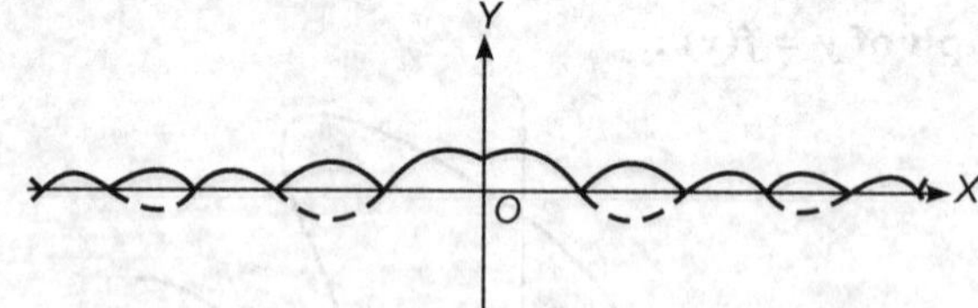

Graph of $|y| = |f(|x|)|$:

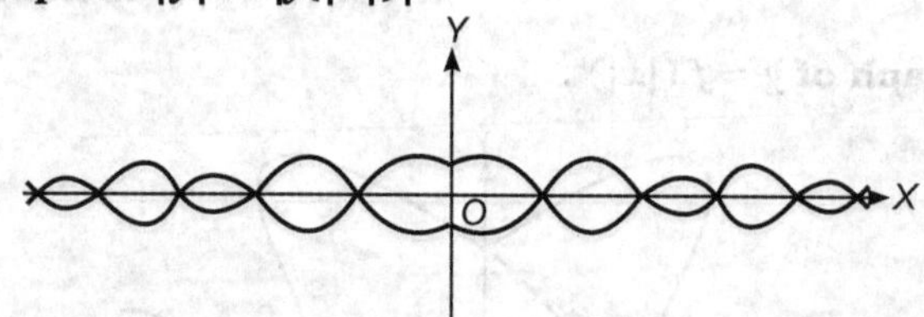

Illustration 46. Draw the graph of the function $|y| = |\ln|x||$, from the known graph of $y = \ln x$ $(x > 0)$.

Sol. Graph of $y = \ln x$:

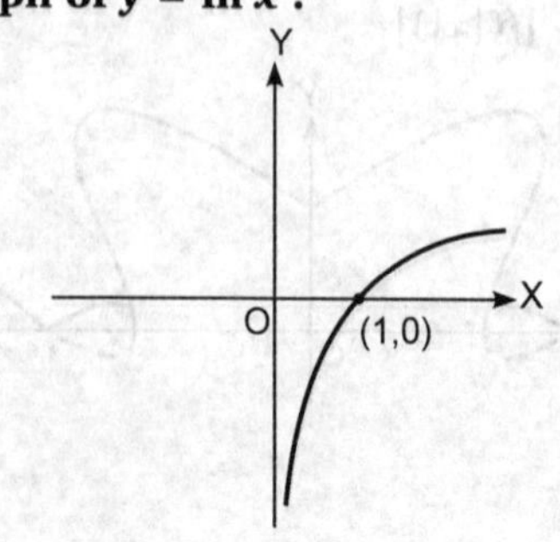

Graph of $y = \ln|x|$:

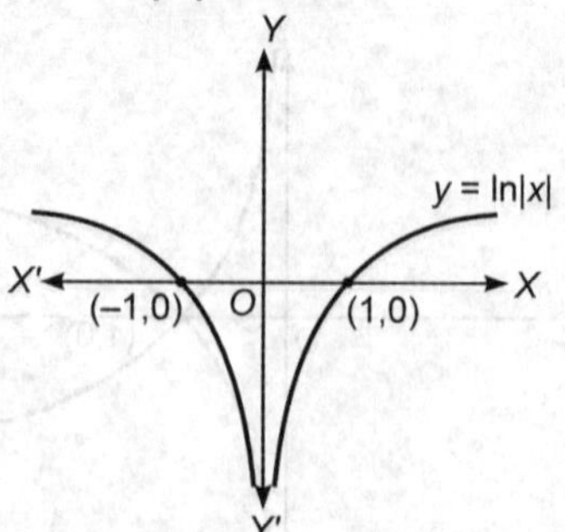

Graph of $y = |\ln|x||$:

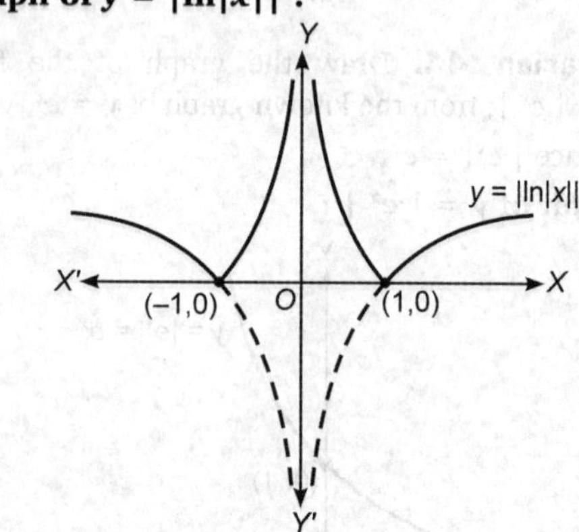

Graph of $|y| = |\ln|x||$:

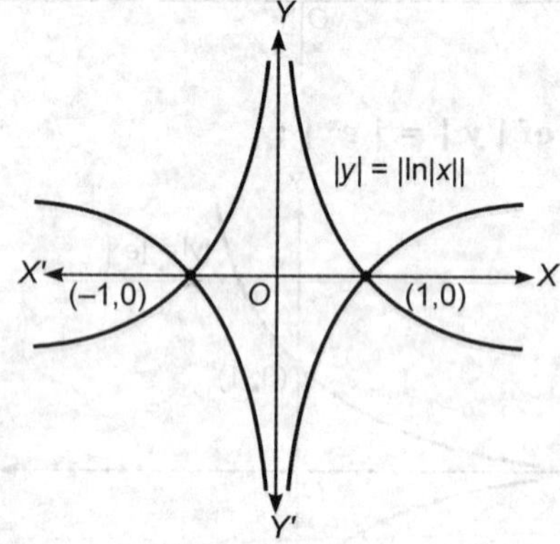

Illustration 47. Draw the graph of the function $|y| = |e^{|x|}|$, from the known graph of $y = e^x$, $\forall\ x \in R$.

Sol.

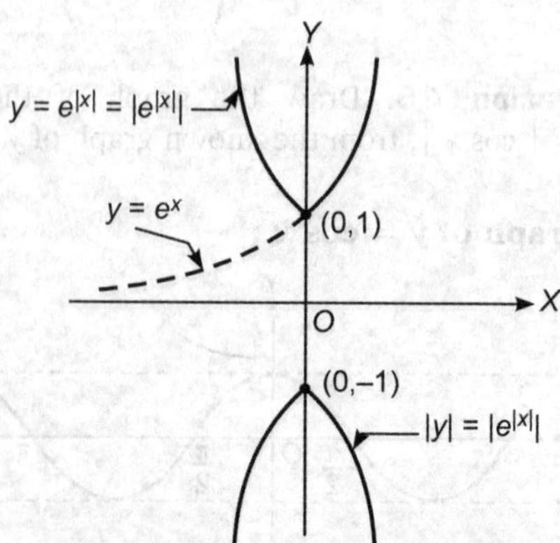

Note : Graphs of $|y| = |e^{|x|}|$ and $|y| = e^{|x|}$ are same.

Illustration 48. Draw the graph of the function $|y| = |\sin|x||$, from the known graph of $y = \sin x$, $\forall x \in R$.

Sol. Graph of $y = |\sin|x||$:

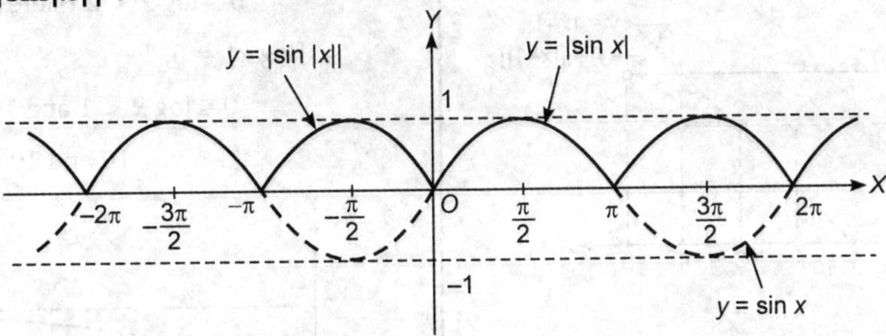

Note : Graph of $y = |\sin x|$ and $y = |\sin|x||$ are same.

Graph of $|y| = |\sin|x||$:

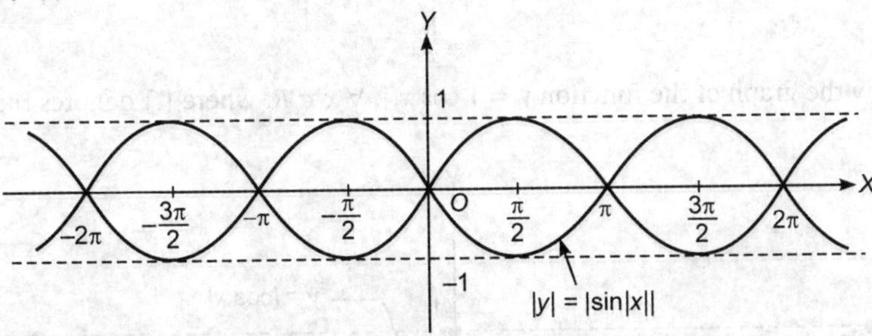

Note : Graphs of $|y| = |\sin x|$ and $|y| = |\sin|x||$ are same.

(xiv) Drawing the graph of $y = [f(x)]$ from the known graph of $y = f(x)$ ([.] denotes the greatest integer function) :

The graph of the function $y = [f(x)]$ is obtained from the graph of $y = f(x)$. Since $y = [f(x)] \in I$, mark the integers on y-axis of unit length.

If $f(x) \in [-2, -1), [-1, 0), [0, 1), [1, 2)$ etc. then $y = -2, -1, 0, 1$ etc. Now draw the horizontal dotted lines through integers till they intersect the graph. Draw vertical dotted lines from these intersection points. In last draw horizontal lines parallel to x-axis from any intersection point to the nearest vertical dotted line with blank dot at right end or at left end according as $f(x)$ increase or decrease.

Now if $f(x) = \in [n, n+1)$, $\therefore y = [f(x)] = n$

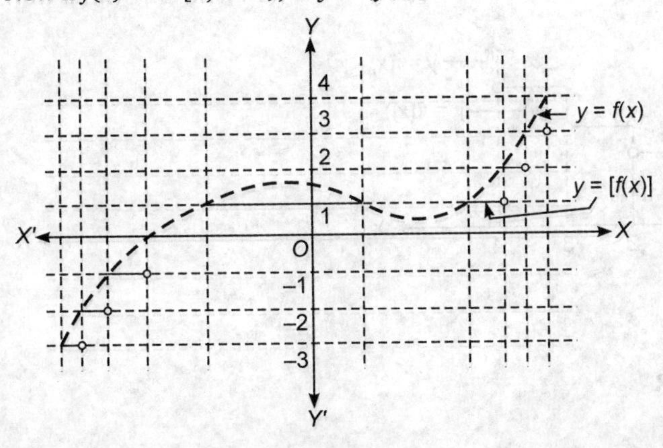

OR

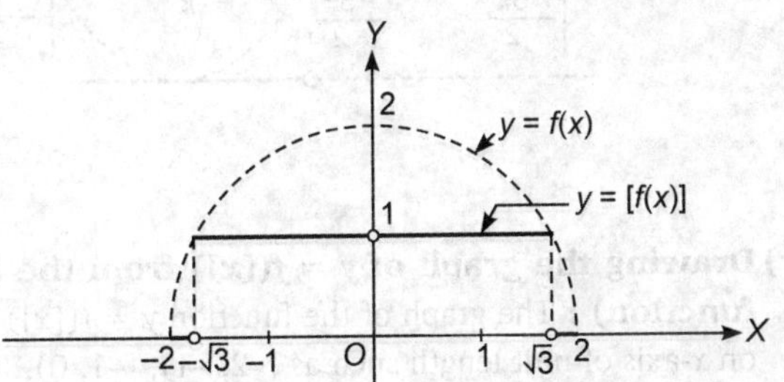

Illustration 49. Draw the graph of the function $y = [x^2]$, $x \in R$, where [.] denotes the greatest integer function.

Sol.

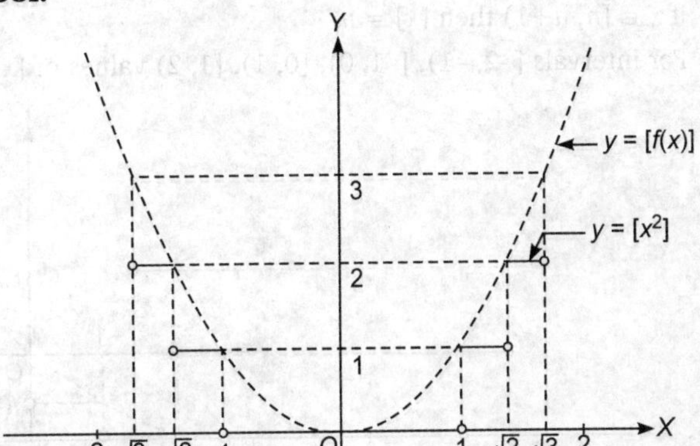

Illustration 50. Draw the graph of the function $y = [|x - 1|]$, $\forall x \in R$, where [.] denotes the greatest integer function.

Sol.

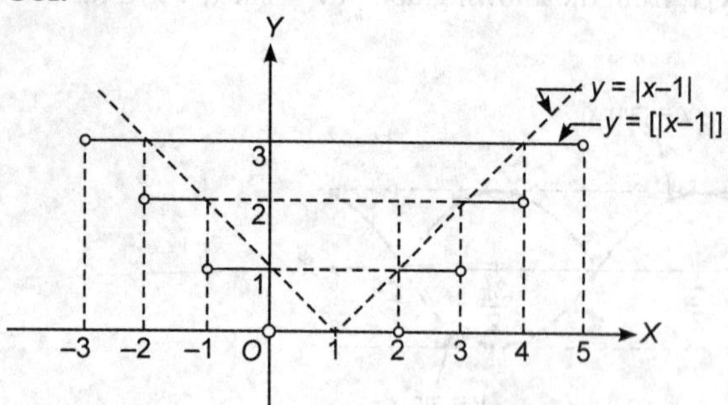

Illustration 51. Draw the graph of the function $y = [\log x]$, $1 \le x \le 26$, where [.] denotes the greatest integer function.

Sol. Since $1 \le x \le 26$

or $1 \le x < 100$ $(\because \log 10 = 1)$

$0 \le \log x < 2$ (Here 26 > 10),

$\therefore$ $[\log x] = 0, 1$

$\therefore$ $0 \le \log x < 1$ and $1 \le \log x < 2$

$\therefore$ $1 \le x < 10$ and $10 \le x < 100$

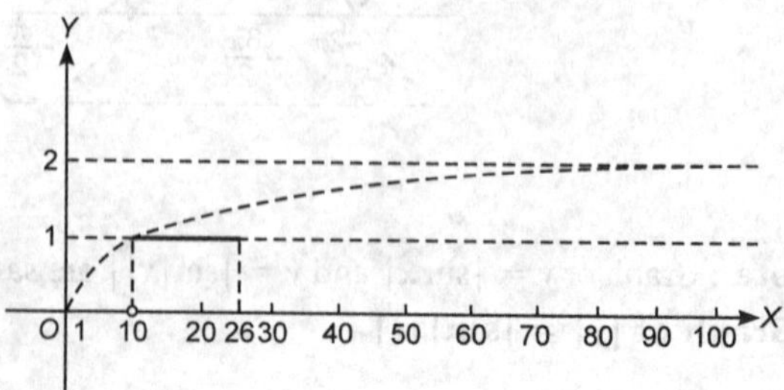

Illustration 52. Draw the graph of the function $y = [\cos x]$, $\forall x \in R$, where [.] denotes the greatest integer function.

Sol.

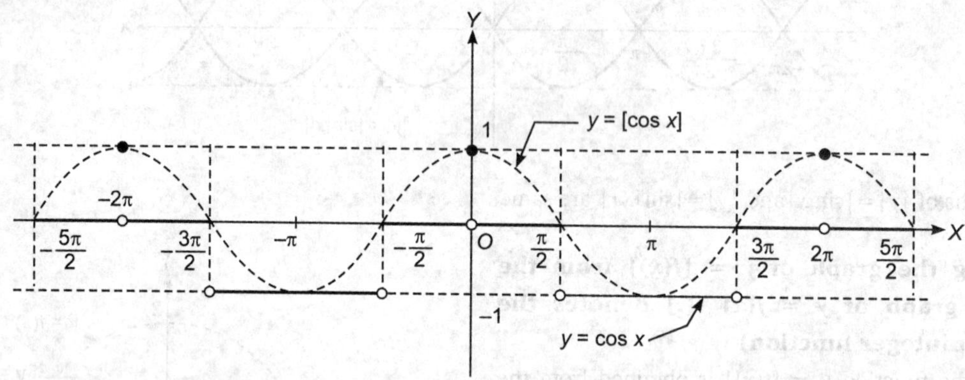

(xv) Drawing the graph of $y = f([x])$ from the known graph of $y = f(x)$ ([.] denotes the greatest integer function) : The graph of the function $y = f([x])$ is obtained from the graph $y = f(x)$, since $[x] \in I$ mark the integers on x-axis of unit length such as $[-2, -1)$, $[-1, 0)$, $[0, 1)$, $[1, 2)$ etc. Now draw vertical lines till these lines intersect the graph $y = f(x)$. From these intersection points draw a horizontal line which is parallel to x-axis to meet the nearest right vertical line with a blank dot (o).

n ●———○ $n + 1$

if $x \in [n, n+1)$ then $[x] = n$,

For intervals $[-2, -1)$, $[-1, 0)$, $[0, 1)$, $[1, 2)$ values of $[x]$ are $-2, -1, 0, 1$ respectively.

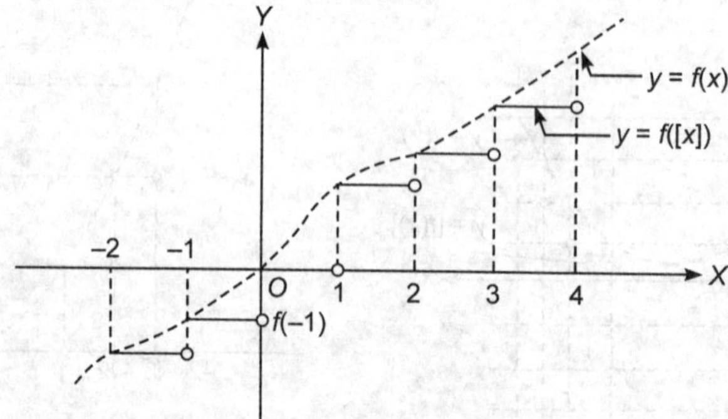

OR

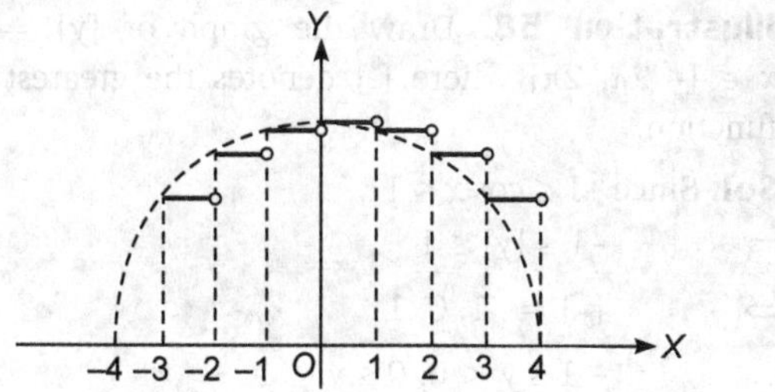

Illustration 53. Draw the graph of the function $y = e^{[x]}$, $\forall x \in R$, where [.] denotes the greatest integer function.

Sol.

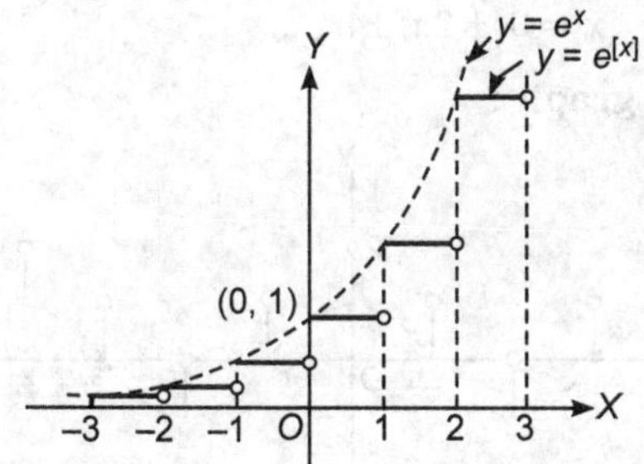

Illustration 54. Draw the graph of the function $y = [x]^2$, $\forall x \in R$, where [.] denotes the greatest integer function.

Sol.

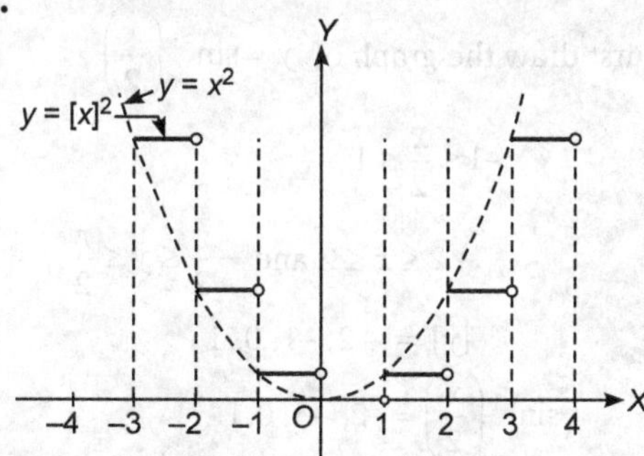

Illustration 55. Draw the graph of the function $y = \log [x]$, $x \geq 1$, where [.] denotes the greatest integer function.

Sol.

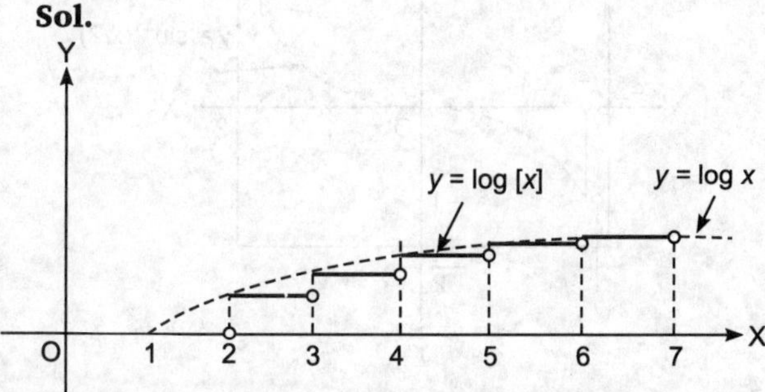

(xvi) Drawing the Graph of $y = [f([x])]$, from the known graph of $y = f(x)$ ([.] denoted the greatest integer function) :

The graph of function $y = [f([x])]$ is obtained from the graph of the function $y = f(x)$, first draw the graph $y = f([x])$ from method (xv) then graph of $y = [f([x])]$ is the integral part of $f([x])$.

If $f([x]) \in [-1, 0), [0, 1), [1, 2), [2, 3)$ etc. then $[f([x])] = -1, 0, 1, 2$ etc.

$\therefore \quad y \in I$

and also $f([x]) \in (0,1), (-2, -1), (-3, -2), (-3, -2), (-4, -3)$ where $x \in [-1, 0), [-2, -1), [-3, -2), [-4, -3), [-5, -4)$

$\therefore \quad y = [f([x])]$

then $y = 1, 0, 1, 1, 2, 3, 0, -2, -3, -3, -4$

where $x \in [0, 1), [1, 2), [2, 3), [3, 4), [4, 5), [5, 6), [-1, 0), [-2, -1), [-3, -2), [-4, -3), [-5, -4)$

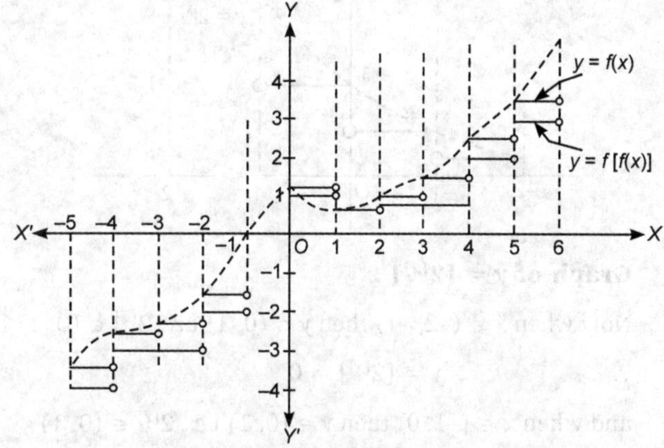

According to figure it is clear that

$f([x]) \in (1, 2), (0, 1), (1, 2), (1, 2), (2, 3), (3, 4)$

where $\quad x \in [0, 1), [1, 2), [2, 3), [3, 4), [4, 5)$ and $[5, 6)$

Illustration 56. Draw the graph of the function $y = [\sin [x]]$, where [.] denotes the greatest integer function $\forall x \in [0, \pi]$.

Sol. Graph of $y = \sin [x]$:

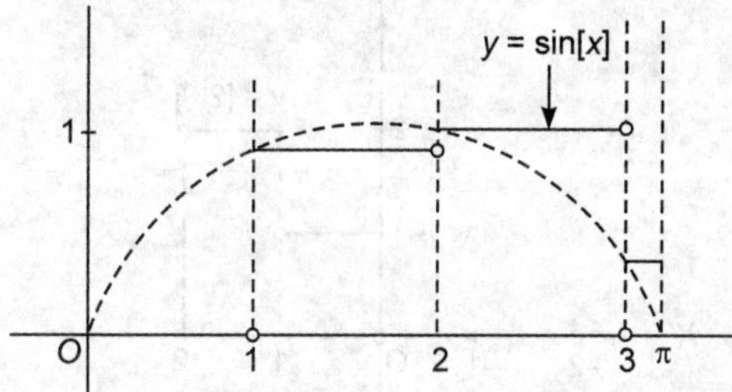

$\because \quad 0 \leq \sin [x] < 1 \ \forall x \in [0, \pi]$

$\therefore \quad y = [\sin [x]] = 0 \ \forall x \in [0, \pi]$

Graph of $y = [\sin[x]]$:

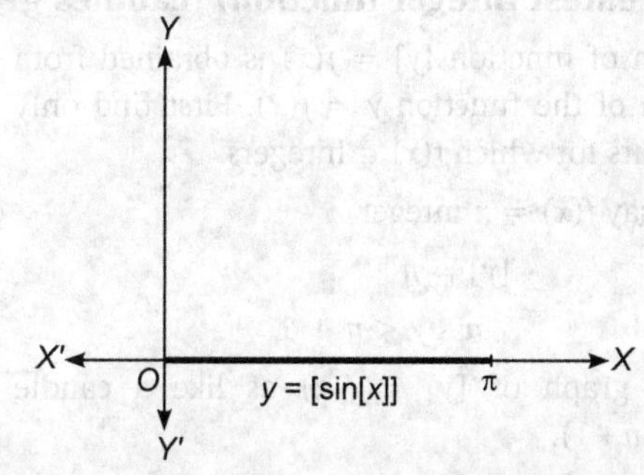

Illustration 57. Draw the graph of $y = [2^{[x]}]$, where [.] denotes the greatest integer function $\forall\, x \in (-2, 2)$.

Sol. Graph of $y = 2^{[x]}$:

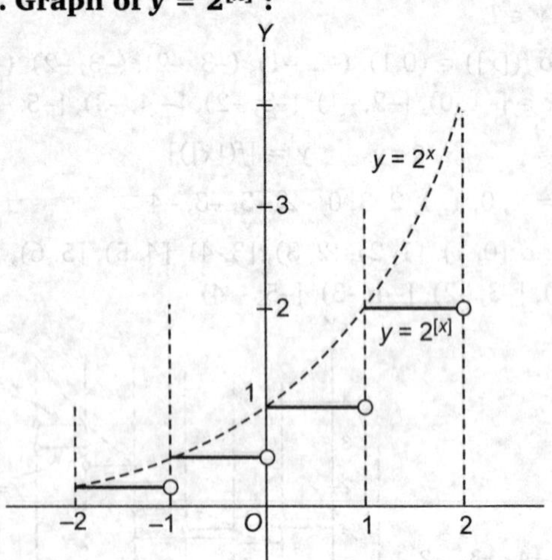

Graph of $y = [2^{[x]}]$:

Now when $x \in (-2, -1)$ then $y \in (0, 1)$ *i.e.*, $2^{[x]} \in (0, 1)$

$\therefore \quad y = [2^{[x]}] = 0$

and when $x \in [-1, 0)$ then $y \in (0, 1)$ *i.e.*, $2^{[x]} \in (0, 1)$

$\therefore \quad y = [2^{[x]}] = 0$

When $x \in [0, 1)$ then $y = 1$ *i.e.*, $2^{[x]} = 1$

$\therefore \quad y = [2^{[x]}] = [1] = 1$

and when $x \in [1, 2)$ then $y = 2$ *i.e.*, $2^{[x]} = 2$

$\therefore \quad y = [2^{[x]}] = 2$

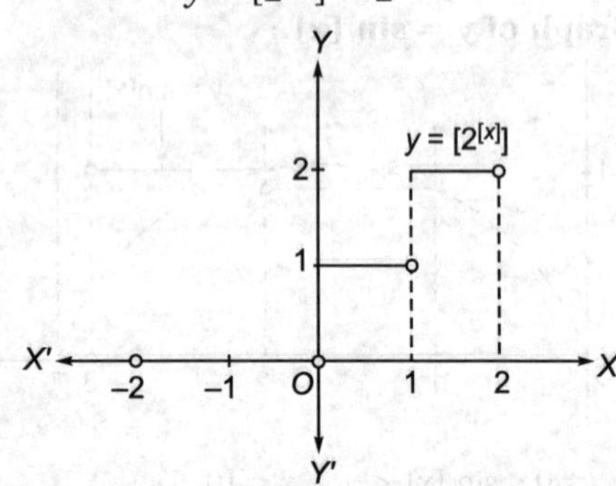

(xvii) Drawing the graph of [y] = f(x), from the known graph of y = f(x) ([.] denotes the greatest integer function) (candles graph) :

The graph of function $[y] = f(x)$ is obtained from the graph of the function $y = f(x)$. First find only those points for which $f(x) \in$ Integers.

Suppose say $f(x) = n$ integer

then, $\quad [y] = n$

or $\quad n \le y < n + 1$

then the graph of $[y] = f(x)$ as like a candle $\forall\, x \in [n, n+1)$.

Illustration 58. Draw the graph of $[y] = \cos x$, $x \in [-2\pi, 2\pi]$ where [.] denotes the greatest integer function.

Sol. Since $-1 \le \cos x \le 1$

$\Rightarrow \quad -1 \le [y] \le 1$

$\Rightarrow \quad [y] = -1, 0, 1$

$\therefore \quad -1 \le y < 0, 0 \le y < 1$

and $\quad 1 \le y < 2$

$\therefore \quad x = \pm\pi, \pm 3\pi, \ldots ; \quad x = \pm\frac{\pi}{2}, \pm\frac{3\pi}{2}, \ldots$

and $\quad x = 0, \pm 2\pi, \pm 4\pi, \ldots.$

Candle graph :

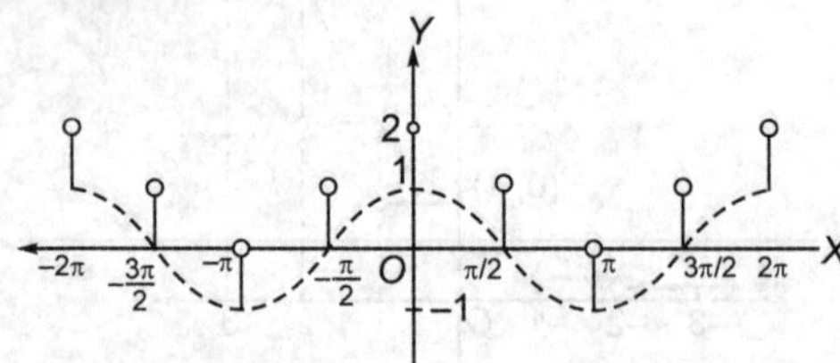

Illustration 59. Draw the graph of $[y] = \sin^{-1}\left(\frac{x}{2}\right)$, where [.] denotes the greatest integer function.

Sol. First draw the graph of $y = \sin^{-1}\left(\frac{x}{2}\right)$

$\Rightarrow \quad -1 \le \frac{x}{2} \le 1$

$\Rightarrow \quad -2 \le x \le 2$ and $-\frac{\pi}{2} \le y \le \frac{\pi}{2}$

$[y] = -2, -1, 0, 1$

$\therefore \quad \sin^{-1}\left(\frac{x}{2}\right) = -2, -1, 0, 1$

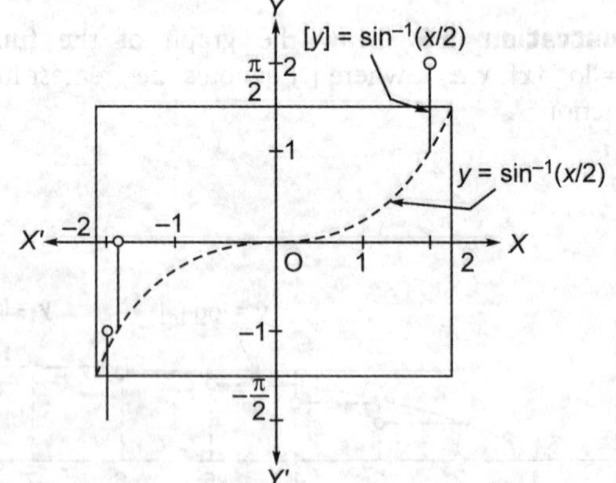

or $\quad \frac{x}{2} = -\sin 2, -\sin 1, 0, \sin 1$

$\Rightarrow \quad x = -2 \sin 2, -2 \sin 1, 0, 2 \sin 1$

Now when $x = -2 \sin 2, -2 \le y < -1$

$x = -2 \sin 1, -1 \le y < 0$

$x = 0, 0 \le y < 1$

$x = 2 \sin 1, 1 \le y < 2$

(xviii) Drawing the graph of [y] = f([x]), from the known graph of y = f(x) ([.] denotes the greatest integer function :

The graph of the function $[y] = f([x])$ is obtained from the graph of the function $y = f(x)$. First find the graph of $y = f([x])$ from the method (xv). Now find only those points for which $f([x]) \in I$.

Suppose say $f([x]) = n$ (Integer) then $[y] = n$

$\Rightarrow \quad n \le y < n + 1$

then the graph of $[y] = f([x])$ as like multi-candles in a Rectangular box $\forall\ y \in [n, n + 1)$.

Illustration 60. Draw the graph of $[y] = \sin^{-1}[x]$, where [.] denotes the greatest integer function.

Sol. $\because\ [x] = 0, -1 \quad \therefore\ [y] = 0, -\dfrac{\pi}{2}$

$[y] \ne -\dfrac{\pi}{2}, \quad \therefore\ [y] = 0 \Rightarrow 0 \le y < 1$

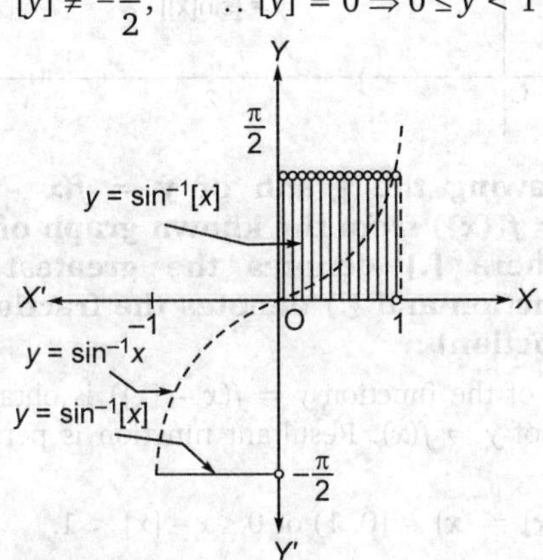

Illustration 61. Draw the graph of the function $[|x|] + [|y|] = 5$, where [.] denotes the greatest integer function.

Sol. First draw the graph of $[x] + [y] = 5$ in first quadrant

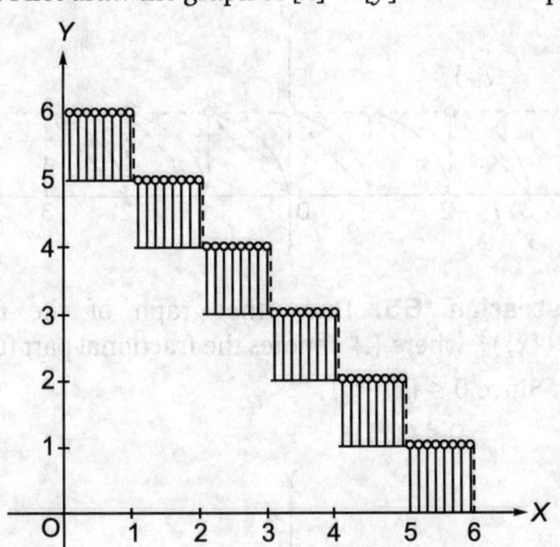

If $[x] = 0, 1, 2, 3, 4, 5$ then $[y] = 5, 4, 3, 2, 1, 0$

$\therefore\ x \in [0, 1), [1, 2), [2, 3), [3, 4), [4, 5), [5, 6)$ then

$y \in [5, 6), [4, 5), [3, 4), [2, 3), [1, 2), [0, 1)$

Graph of $[|x|] + [|y|] = 5$:

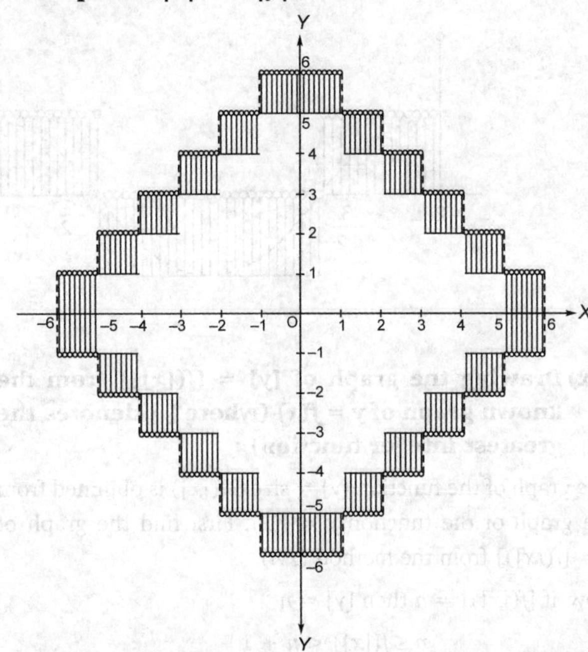

(xix) Drawing the graph of [y] = [f(x)], from the known graph of y = f(x) ([.] denotes the greatest integer function) :

The graph of the function $[y] = [f(x)]$ is obtained from the graph of the function $y = f(x)$. First find the graph of $y = [f(x)]$ from the method (xiv).

Now, if $n \le f(x) < n + 1$ then $n \le y < n + 1$

and $[y]$ implies only those values of x for which $f(x) \in$ Integer then the graph of $[y] = [f(x)]$ as like multi candles in a rectangular box $\forall\ y \in [n, n+1)$.

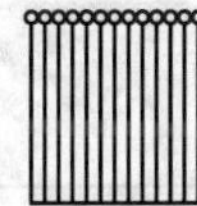

Illustration 62. Draw the graph of $[y] = [\cos x]$, $\forall\ x \in [-2\pi, 2\pi]$, where [.] denotes the greatest integer function.

Sol. Now $[y] = \cos x \Rightarrow -1 \le [y] \le 1 \Rightarrow [y] = -1, 0, 1$ then $\cos x = -1, 0, 1$

$\therefore$ when $y \in [-1, 0)$ then $x = \pm\pi$

and when $y \in [0, 1)$ then $x = \pm\dfrac{\pi}{2}, \pm\dfrac{3\pi}{2}$

and when $y \in [1, 2)$ then $x = 0, \pm 2\pi$

Now, if $[\cos x] = -1, 0, 1$

then $-1 \le \cos x < 0,\ 0 \le \cos x < 1,\ 1 \le \cos x < 2$

or $\quad x \in \left(\dfrac{\pi}{2}, \dfrac{3\pi}{2}\right) \cup \left(-\dfrac{3\pi}{2}, \dfrac{\pi}{2}\right);$

$x \in \left[-\dfrac{\pi}{2}, \dfrac{\pi}{2}\right] \cup \left[\dfrac{3\pi}{2}, 2\pi\right) \cup \left(-2\pi, -\dfrac{3\pi}{2}\right]$ and $x = 0, \pm 2\pi$

according as $-1 < y < 0,\ 0 \le y < 1,\ 1 \le y < 2$

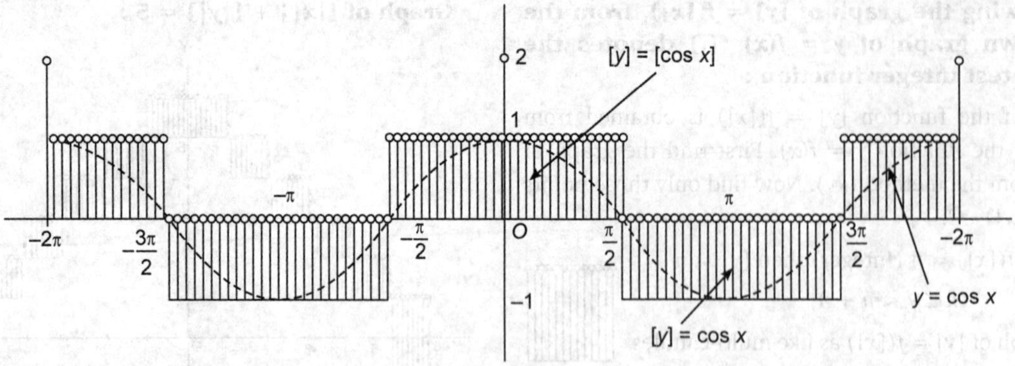

(xx) Drawing the graph of [y] = [f([x])] from the known graph of y = f(x) (where [.] denotes the greatest integer function) :

The graph of the function [y] = size 6 f([x]) is obtained from the graph of the function $y = f(x)$. First find the graph of $y = [f([x])]$ from the method (xvi).

Now, if $[f([x])] = n$ then $[y] = n$

$\therefore \quad n \le f([x]) < n + 1$

and $\quad n \le y < n + 1$

then the graph of $[y] = [f([x])]$ as like multi candles in a rectangular box $\forall\, y \in [n, n + 1)$.

Illustration 63. Draw the graph of $[y] = [\sin[x]]\ \forall\, x \in [0, \pi]$, where [.] denotes the greatest integer funcion.

Sol.

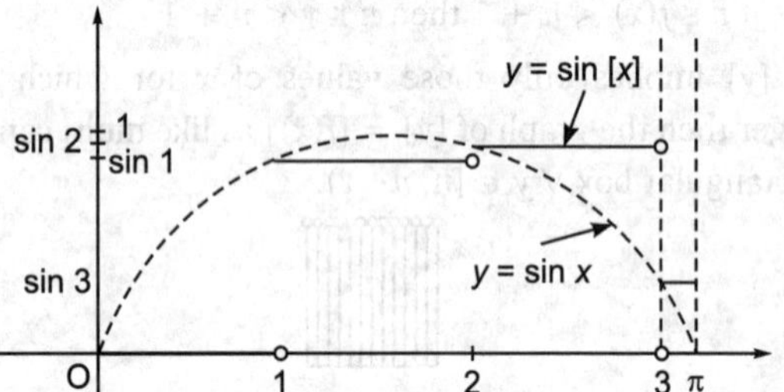

Now for $0 \le x \le \pi \Rightarrow y = \sin[x] \in [0, 1)$

$\therefore \quad [\sin [x]] = 0\ \forall\, x \in [0, \pi]$

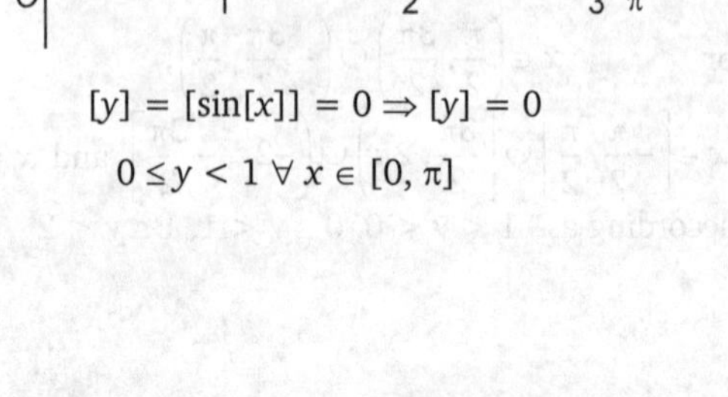

$\therefore \quad [y] = [\sin[x]] = 0 \Rightarrow [y] = 0$

$\Rightarrow \quad 0 \le y < 1\ \forall\, x \in [0, \pi]$

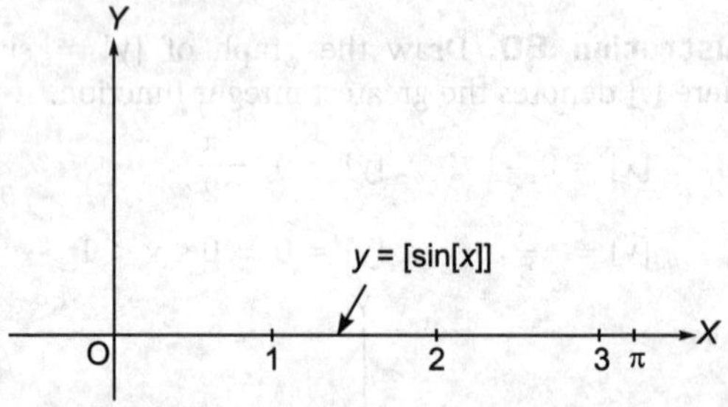

(xxi) Drawing the graph of y = f(x − [x]) or y = f({x}) from the known graph of y = f(x) (where [.] denotes the greatest integer function and {.} denotes the fractional part function) :

The graph of the function $y = f(x - [x])$ is obtained from the graph of $y = f(x)$. Resultant function is periodic with period 1.

Since $x - [x] = \{x\} \in [0, 1)$ or $0 \le x - [x] < 1$.

Illustration 64. Draw the graph of the function $y = \sqrt{\{x\}}$, where {.} denotes the fractional part function.

Sol. Since $0 \le \{x\} < 1$,

$\therefore \quad 0 \le \sqrt{\{x\}} < 1$

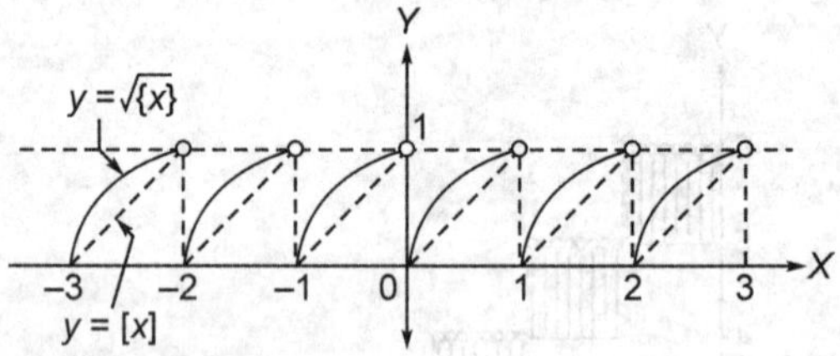

Illustration 65. Draw the graph of the function $y = (\{x\})^2$, where {.} denotes the fractional part function.

Sol. Since $0 \le \{x\} < 1$,

$\therefore \quad 0 \le (\{x\})^2 < 1$

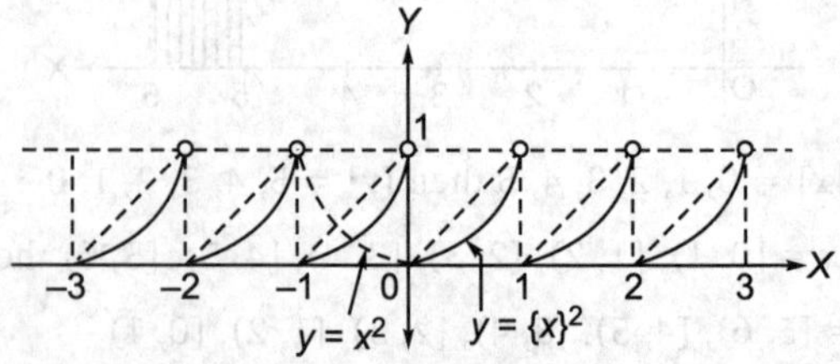

Illustration 66. Draw the graph of the function $y = \frac{e^x}{e^{[x]}} = e^{x-[x]}$ where [.] denotes the greatest integer function.

Sol. Since $0 \le x - [x] < 1$,

$\therefore \quad e^0 \le e^{x-[x]} < e^1$ or $1 \le y < e$

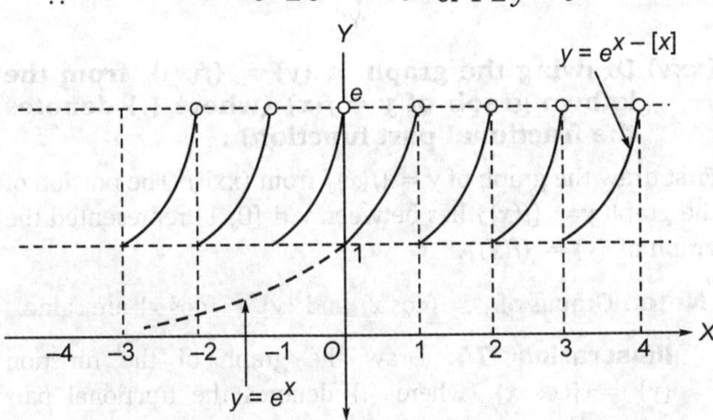

Illustration 67. Draw the graph of the function $y = \ln(x - [x])$ where [.] denotes the greatest integer function.

Sol. Since $0 < x - [x] < 1$,

$\therefore \quad -\infty < \ln(x - [x]) < 0$

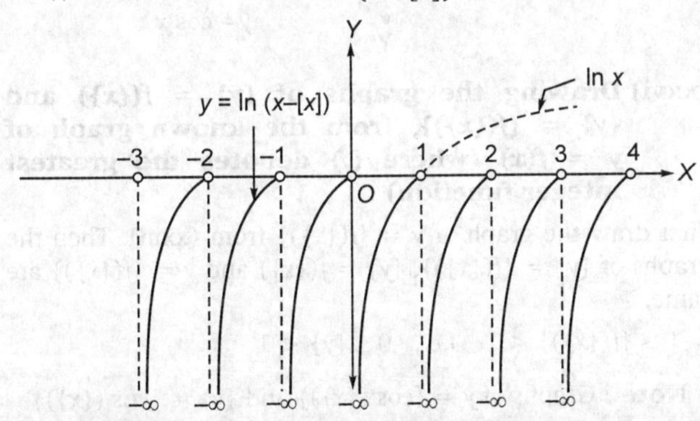

Illustration 68. Draw the graph of the function $y = \sin(\{x\})$, where {.} denotes the fractional part function.

Sol. Since $0 \le \{x\} < 1$, $0 \le \sin(\{x\}) < \sin 1$.

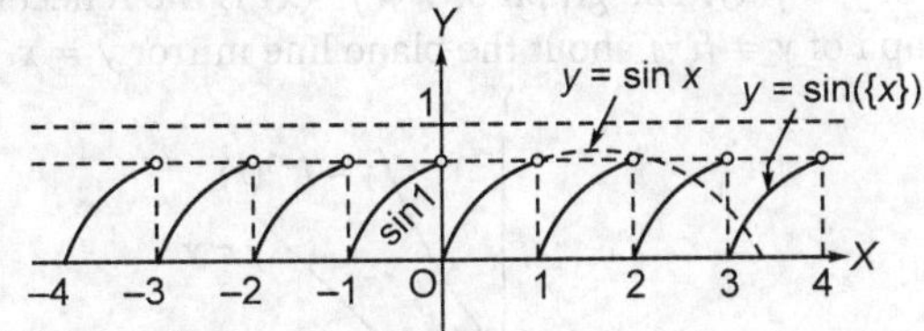

(xxii) Drawing the graph of $y = \{f(x)\}$, from the known graph of $y = f(x)$ (where {.} denotes the fractional part function) :

$\because \quad 0 \le \{f(x)\} < 1$

$\Rightarrow \quad 0 \le y < 1$

and if $f(x) < 0$

$\Rightarrow \quad -f(x) > 0$

$\because \quad \{n + f(x)\} = \{f(x)\}, n \in I$

Now first draw the horizontal lines for all integral value of y and draw vertical lines at points of intersection of $y = f(x)$ and horizontal lines and then translate the graph $y = f(x)$ in $0 \le y < 1$.

Illustration 69. Draw the graph of $y = \{\cos x\} \ \forall \ x \in \left[-2\pi, \frac{5\pi}{2}\right]$, (where {.} denotes the fractional part function).

Sol. The part of the graph $y = \cos x$ is same in $y \in [0, 1)$ and shift other portion of graph $y = \cos x$ in $y \in [0, 1)$ then we get the required graph of $y = \{\cos x\}$.

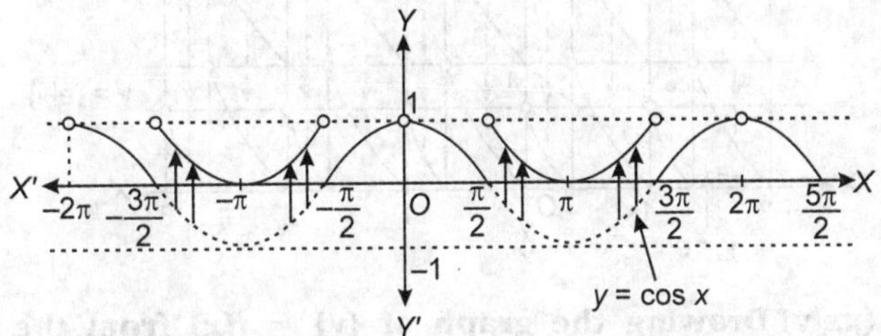

Illustration 70. Draw the graph of $y = \{x^2\} \ \forall \ x \in [-2, 2]$ (where {.} denotes the fractional part function).

Sol. The part of the graph $y = x^2$ is same in $y \in [0, 1)$ and shift other portion of the graph $y = x^2$ in $y \in [0, 1)$ then we get required graph of $y = \{x^2\}$.

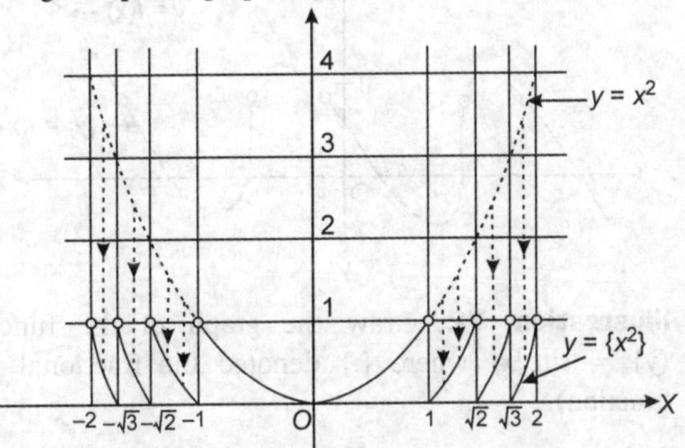

(xxiii) Drawing the graph of $y = \{f(\{x\})\}$, from the known graph of $y = f(x)$ (where {.} denotes the fractional part function) :

The graph of the function $y = \{f(\{x\})\}$ is obtained from the graph of the function $y = f(x)$. First find the graph of $y = f(\{x\})$ from the method (xxi).

Now, $\quad 0 \le \{f(\{x\})\} < 1$

i.e., $\quad 0 \le y < 1$

and $f(\{x\})$ is always periodic with period 1.

The part of the graph $y = f(\{x\})$ is same in $y \in [0, 1)$ and shift other portion of the graph $y = f(\{x\})$ in $y \in [0, 1)$ then we get the required graph $y = \{f(\{x\})\}$.

Illustration 71. Draw the graph of $y = \{e^{\{x\}}\}$, (where {.} denotes the fractional part function).

Sol. Graph of $y = e^{\{x\}}$ is periodic with period 1.

$\because$ Graph of $y = e^{\{x\}}$ lies in $1 \le y < e$ but the graph of $y = \{e^{\{x\}}\}$ lies in $0 \le y < 1$

Now shift the portion of the graph $y = e^{\{x\}}$ in $y \in [0, 1)$ then after shifting the graph of $y = e^{\{x\}}$ convert in $y = \{e^{\{x\}}\}$.

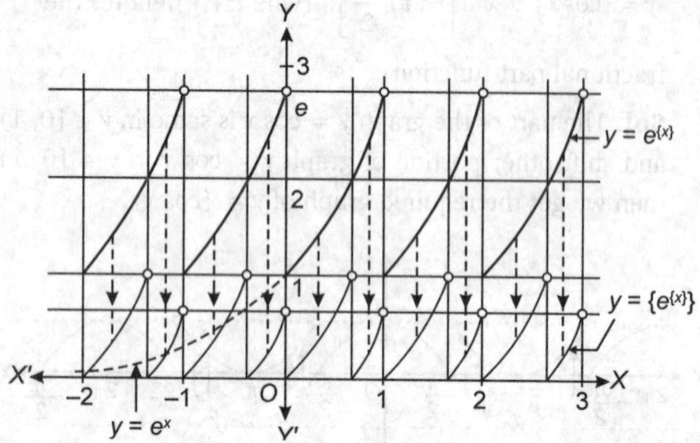

(xxiv) Drawing the graph of {y} = f(x) from the known graph of y = f(x) (where {.} denotes the fractional part function) :

First draw the graph of $y = f(x)$ the portion of the graph $y = f(x)$ lies between $y \in [0, 1)$ represented the graph $\{y\} = f(x)$.

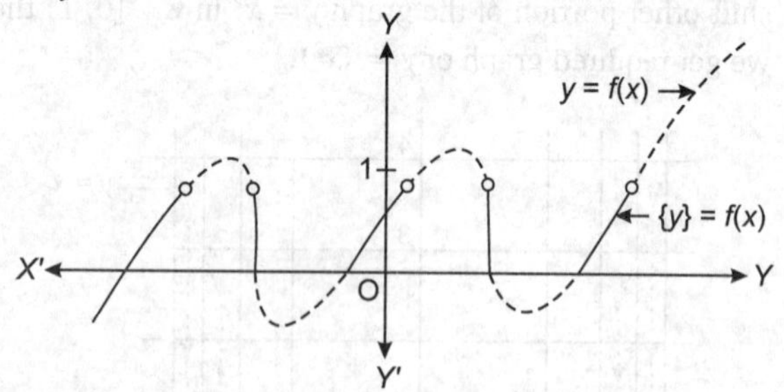

Illustration 72. Draw the graph of the function $\{y\} = \sin^{-1}x$ (where {.} denoted the fractional part function).

Sol.

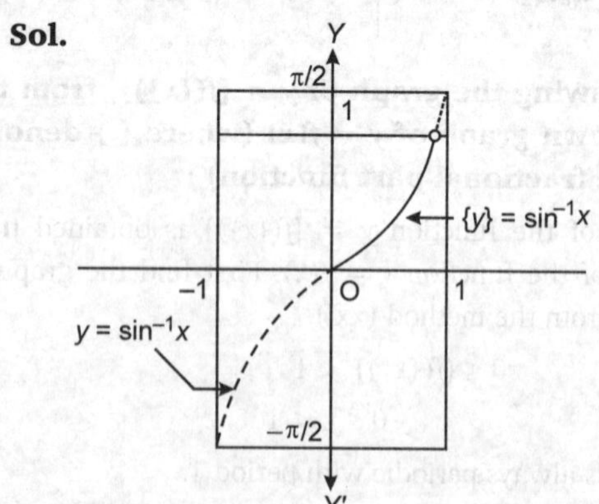

Illustration 73. Draw the graph of the function $\{y\} = \sin x$ (where {.} denotes the fractional part function).

Sol.

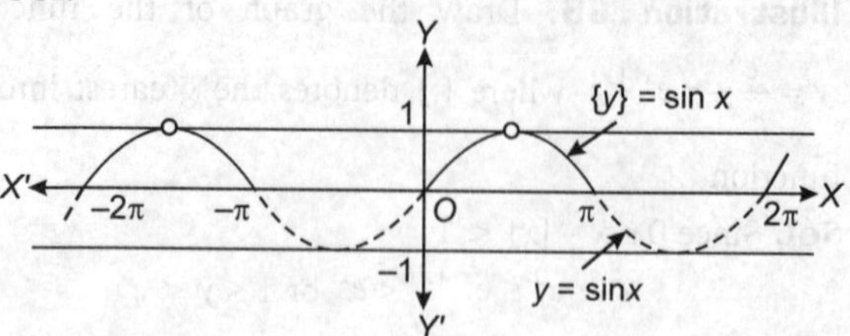

(xxv) Drawing the graph of {y} = {f(x)}, from the known graph of y = f(x) (where {.} denotes the fractional part function) :

First draw the graph of $y = \{f(x)\}$ from (xxii). The portion of the graph $y = \{f(x)\}$ lies between $y \in [0, 1)$ represented the graph of $\{y\} = \{f(x)\}$.

Note : Graphs of $y = \{\cos x\}$ and $\{y\} = \{\cos x\}$ are same.

Illustration 74. Draw the graph of the function $\{y\} = \{\cos x\}$ (where {.} denotes the fractional part function).

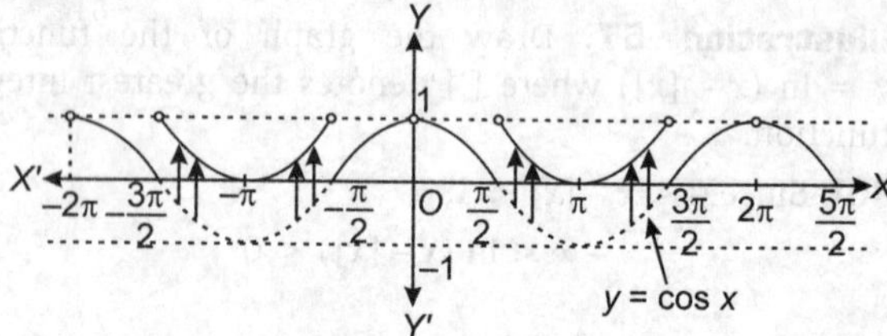

(xxvi) Drawing the graphs of {y} = f({x}) and {y} = {f({x})}, from the known graph of y = f(x) (where {.} denotes the greatest integer function) :

First draw the graph of $y = \{f(\{x\})\}$ from (xxiii). Then the graphs of $\{y\} = \{f(\{x\})\}$, $\{y\} = f(\{x\})$ and $y = \{f(\{x\})\}$ are same.

$\because\ 0 \le \{f(\{x\})\} < 1$ *i.e.,* $0 \le \{y\} < 1$

Note : Graphs of $y = \{\cos(\{x\})\}$ and $\{y\} = \{\cos(\{x\})\}$ are same.

(xxvii) Drawing the graph of $y = f^{-1}(x)$, from the known graph of y = f(x) :

The graph of the function $y = f^{-1}(x)$ is obtained from the graph of $y = f(x)$. The graph of $y = f^{-1}(x)$ is the reflection of the graph of $y = f(x)$ about the plane line mirror $y = x$.

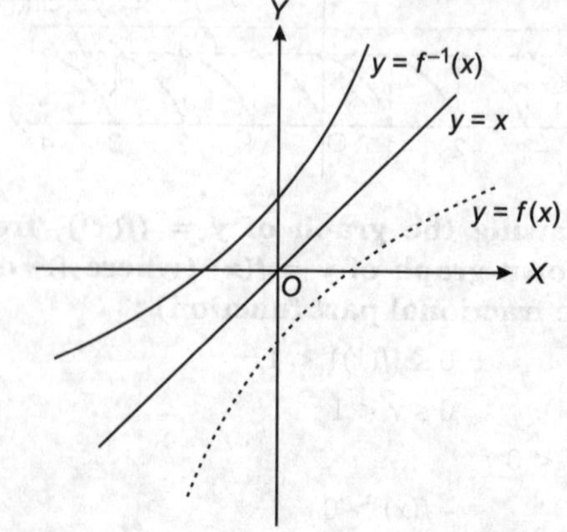

OR

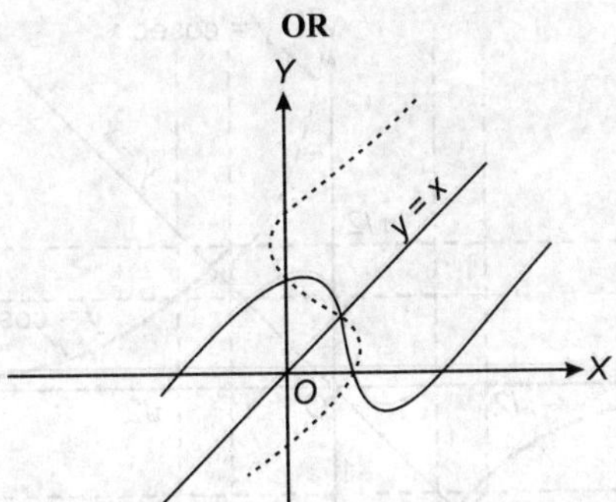

Illustration 75. Draw the graph of the function $y = f^{-1}(x)$, from the known graph of $y = f(x)$, if $f(x) = x^2 + 1, x \geq 0$.

Sol.

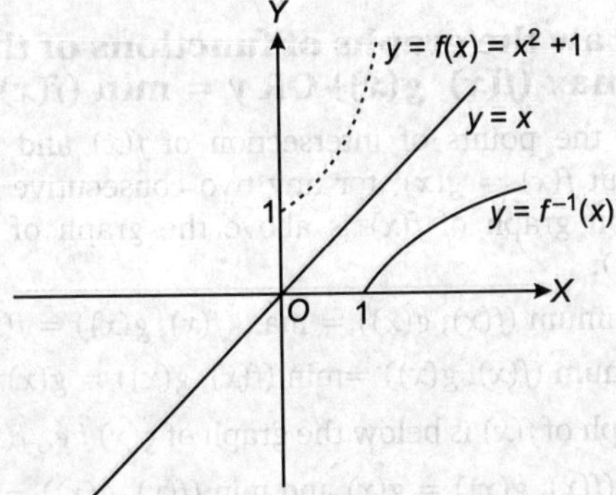

Illustration 76. Draw the graph of the function $y = f^{-1}(x)$, from the known graph of $y = f(x)$, if

$$f(x) = x^2 + 2ax + \left(\frac{1}{16}\right), 0 < a < \left(\frac{1}{4}\right).$$

Sol.

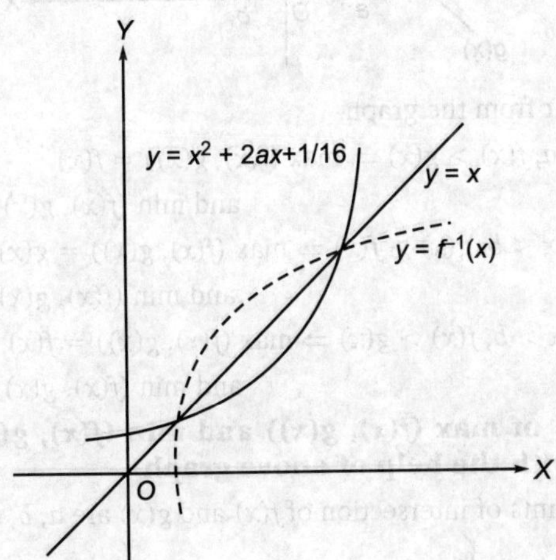

Illustration 77. Draw the graph of the function $y = \sin^{-1} x$, from the known graph of $y = \sin x$.

Sol. $y = \sin^{-1} x, -1 \leq x \leq 1, \quad -\frac{\pi}{2} \leq y \leq \frac{\pi}{2} \sin^{-1} x$ is the reflection of the graph of $\sin x, -\frac{\pi}{2} \leq x \leq \frac{\pi}{2}$ in $y = x$ line as a plane mirror.

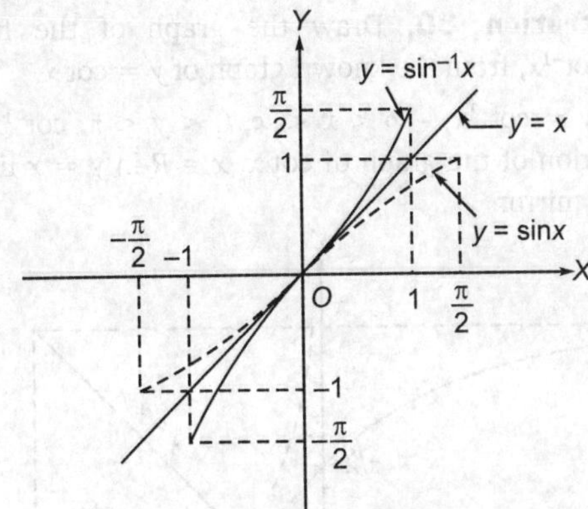

Illustration 78. Draw the graph of the function $y = \cos^{-1}x$, from the known graph of $y = \cos x$.

Sol.

$y = \cos^{-1} x, -1 \leq x \leq 1, -\frac{\pi}{2} \leq y \leq \frac{\pi}{2} \cdot \cos^{-1} x$ is the reflection of the graph of $\cos x, 0 \leq x \leq \pi$ in $y = x$ line as a plane mirror.

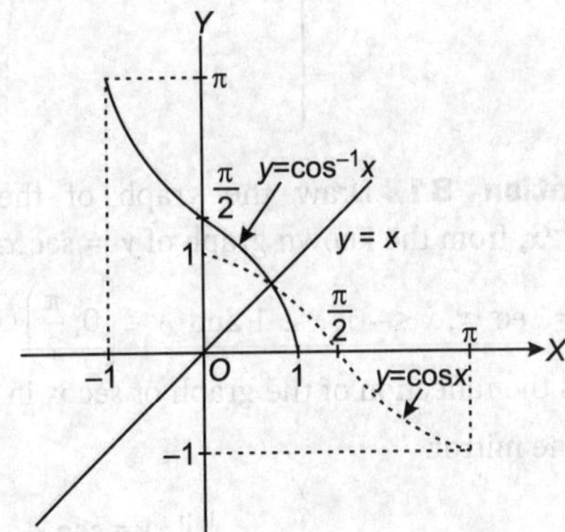

Illustration 79. Draw the graph of the function $y = \tan^{-1}x$, from the known graph of $y = \tan x$.

Sol. $y = \tan^{-1} x, -\infty < x < \infty, -\frac{\pi}{2} < y < \frac{\pi}{2} \cdot \tan^{-1} x$ is the reflection of the graph of $\tan x, x \in R$ in $y = x$ line as plane mirror.

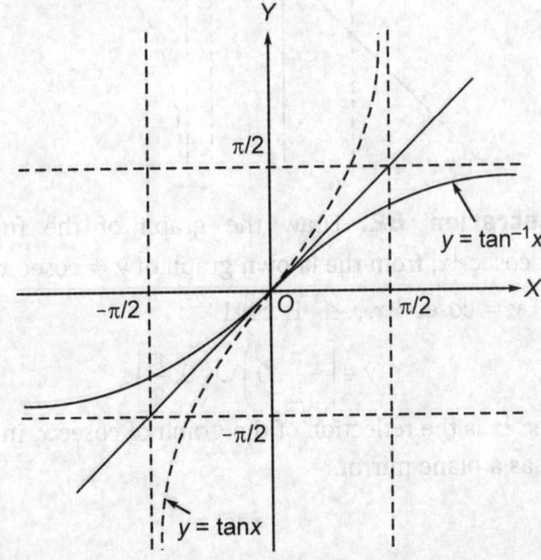

Illustration 80. Draw the graph of the function $y = \cot^{-1}x$, from the known graph of $y = \cot x$.

Sol. $y = \cot^{-1}x$, $-\infty < x < \infty$, $0 < y < \pi$. $\cot^{-1}x$ is the reflection of the graph of $\cot x$, $x \in R$ in $y = x$ line as a plane mirror.

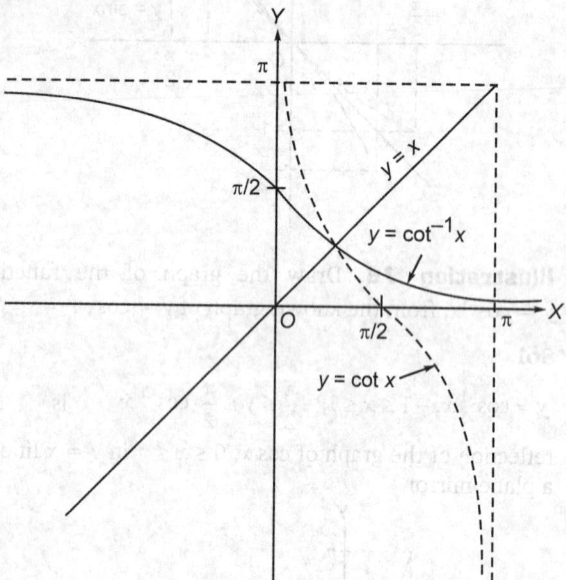

Illustration 81. Draw the graph of the function $y = \sec^{-1}x$, from the known graph of $y = \sec x$.

Sol. $y = \sec^{-1}x, x \le -1, x \ge 1$ and $y \in \left[0, \frac{\pi}{2}\right) \cup \left(\frac{\pi}{2}, \pi\right]$.

$\sec^{-1}x$ is the reflection of the graph of $\sec x$, in $y = x$ line as a plane mirror.

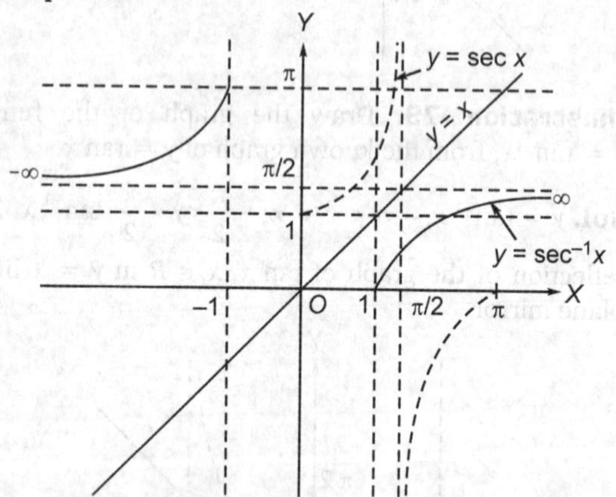

Illustration 82. Draw the graph of the function $y = \text{cosec}^{-1}x$, from the known graph of $y = \text{cosec}\, x$.

Sol. $y = \text{cosec}^{-1} x, x \le -1, x \ge 1$

and $$y \in \left[-\frac{\pi}{2}, 0\right) \cup \left(0, \frac{\pi}{2}\right]$$

$\text{cosec}^{-1}x$ is the reflection of the graph of $\text{cosec}\, x$, in $y = x$ line as a plane mirror.

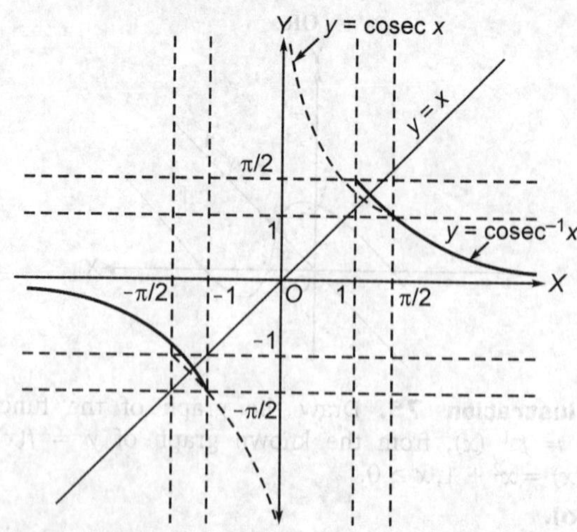

II. To draw the graphs of functions of the form $y = \max(f(x), g(x))$ OR $y = \min(f(x), g(x))$

First find the points of intersection of $f(x)$ and $g(x)$. For this we put $f(x) = g(x)$, for any two consecutive points of intersection graph of $f(x)$ is above the graph of $g(x)$ *i.e.*, $f(x) > g(x)$.

Then maximum $(f(x), g(x)) = \max(f(x), g(x)) = f(x)$ and minimum $(f(x), g(x)) = \min(f(x), g(x)) = g(x)$

And if graph of $f(x)$ is below the graph of $g(x)$ *i.e.*, $f(x) < g(x)$ then $\max(f(x), g(x)) = g(x)$ and $\min(f(x), g(x)) = f(x)$

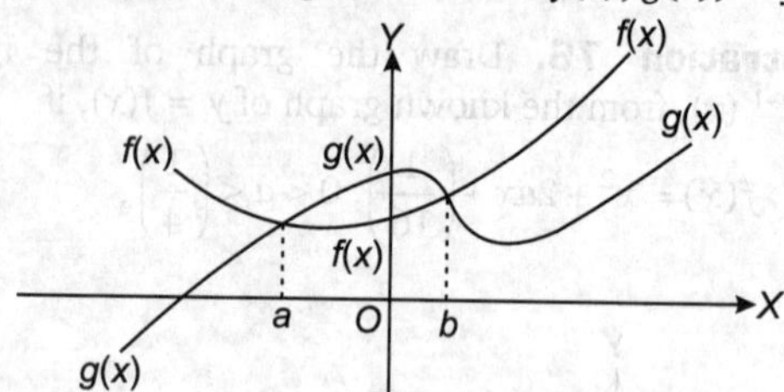

It is clear from the graph.

For $x < a$, $f(x) > g(x) \Rightarrow \max(f(x), g(x)) = f(x)$

and $\min(f(x), g(x)) = g(x)$

For $a < x < b$, $g(x) > f(x) \Rightarrow \max(f(x), g(x)) = g(x)$

and $\min(f(x), g(x)) = f(x)$

and for $x > b$, $f(x) > g(x) \Rightarrow \max(f(x), g(x)) = f(x)$

and $\min(f(x), g(x)) = g(x)$

Graphs of max $(f(x), g(x))$ and min $(f(x), g(x))$ for $x \in R$ with the help of above graph.

Since points of intersection of $f(x)$ and $g(x)$ are a, b, c, d, e.

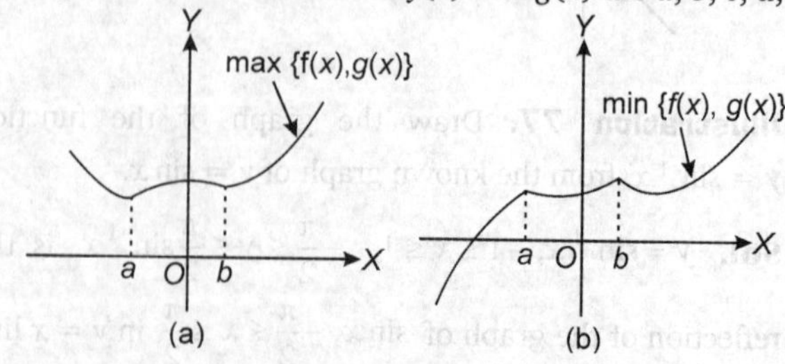

OR

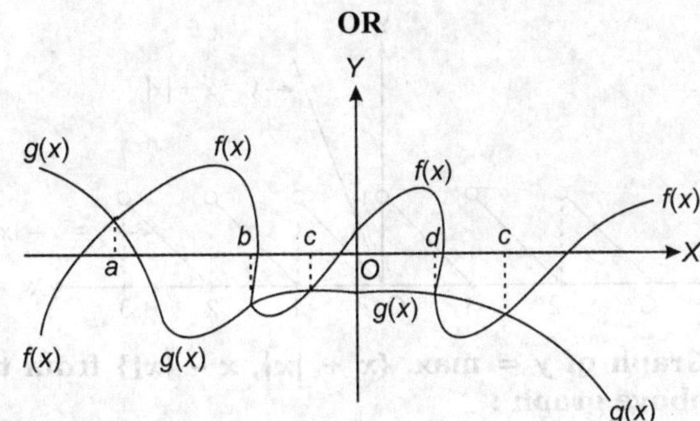

It is clear from the graph

For $x < a$, $g(x) > f(x) \Rightarrow \max(f(x), g(x)) = g(x)$

and $\min(f(x), g(x)) = f(x)$

For $a < x < b$, $f(x) > g(x) \Rightarrow \max(f(x), g(x)) = f(x)$

and $\min(f(x), g(x)) = g(x)$

For $b < x < c$, $g(x) > f(x) \Rightarrow \max(f(x), g(x)) = g(x)$

and $\min(f(x), g(x)) = f(x)$

For $c < x < d$, $f(x) > g(x) \Rightarrow \max(f(x), g(x)) = f(x)$

and $\min(f(x), g(x)) = g(x)$

For $d < x < e$, $g(x) > f(x) \Rightarrow \max(f(x), g(x)) = g(x)$

and $\min(f(x), g(x)) = f(x)$

For $x > e$, $f(x) > g(x) \Rightarrow \max(f(x), g(x)) = f(x)$

and $\min(f(x), g(x)) = g(x)$

Graphs of max $(f(x), g(x))$ and min $(f(x), g(x))$ for $x \in R$ with the help of following graphs :

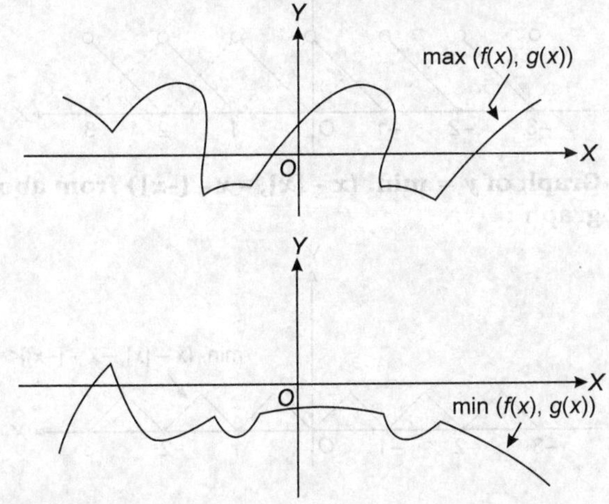

Theorems

(i) $\max(f(x), g(x)) = \dfrac{f(x)+g(x)}{2} + \left|\dfrac{f(x)-g(x)}{2}\right|$

(ii) $\min(f(x), g(x)) = \dfrac{f(x)+g(x)}{2} - \left|\dfrac{f(x)-g(x)}{2}\right|$

Proof (Analytically)

(i) If $f(x) > g(x)$, then $|f(x) - g(x)| = f(x) - g(x)$ ...(1)

$\max(f(x), g(x)) = f(x)$

$$= \frac{f(x)+g(x)}{2} + \frac{f(x)-g(x)}{2}$$

$$= \frac{f(x)+g(x)}{2} + \left|\frac{f(x)-g(x)}{2}\right| \quad \{\text{from (1)}\}$$

and if $f(x) < g(x)$, then $|f(x) - g(x)|$

$= -(f(x) - g(x))$ ...(2)

$\therefore \max(f(x), g(x)) = g(x)$

$$= \frac{f(x)+g(x)}{2} + \frac{(g(x)-f(x))}{2}$$

$$= \frac{f(x)+g(x)}{2} + \left|\frac{f(x)-g(x)}{2}\right| \quad \{\text{from (2)}\}$$

(ii) If $f(x) > g(x)$, then $|f(x) - g(x)| = f(x) - g(x)$ ...(3)

$\therefore \min(f(x), g(x)) = g(x)$

$$= \frac{f(x)+g(x)}{2} - \left(\frac{f(x)-g(x)}{2}\right)$$

$$= \frac{f(x)+g(x)}{2} - \left|\frac{f(x)-g(x)}{2}\right| \quad \{\text{from (3)}\}$$

and if $f(x) < g(x)$, then

$|f(x) - g(x)| = -(f(x) - g(x)) = g(x) - f(x)$ ...(4)

$\therefore \min(f(x), g(x)) = f(x)$

$$= \frac{f(x)+g(x)}{2} - \left(\frac{g(x)-f(x)}{2}\right)$$

$$= \frac{f(x)+g(x)}{2} - \left|\frac{f(x)-g(x)}{2}\right| \quad \{\text{from (4)}\}$$

Proof (Geometrically)

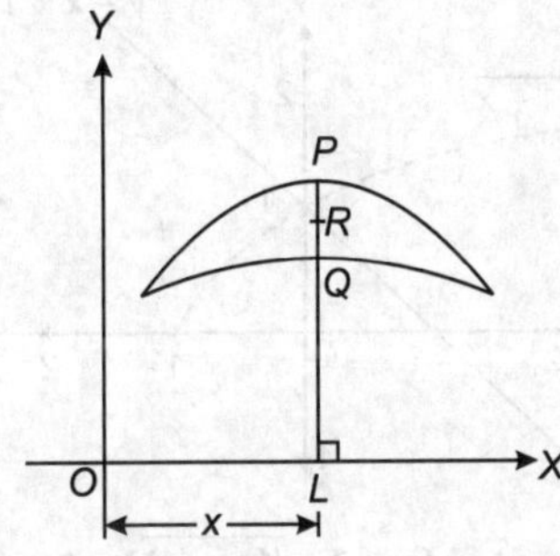

In the diagram, $PL \perp x$-axis.

Point $P \equiv (x, f(x))$ or $(x, g(x))$

and Point $Q \equiv (x, g(x))$ or $(x, f(x))$

$\therefore$ $PQ = |f(x) - g(x)|$

If R is the mid-point of P and Q, then

$$PR = QR = \frac{PQ}{2} = \left|\frac{f(x)-g(x)}{2}\right|$$

$\therefore \max(f(x), g(x)) = PL = RL + PR$

$$= \frac{f(x)+g(x)}{2} + \left|\frac{f(x)-g(x)}{2}\right|$$

and $\min(f(x), g(x)) = QL = RL - QR$

$$= \frac{f(x)+g(x)}{2} - \left|\frac{f(x)-g(x)}{2}\right|$$

Illustration 83. Draw the graph of the function

$$y = \max.\ \{2 - x, 2, 1 + x\}.$$

Sol. First draw the graphs of $y = 2 - x$, $y = 2$ and $y = 1 + x$.

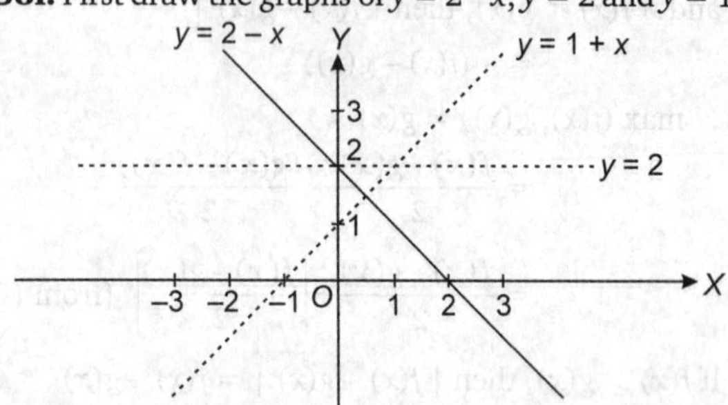

Graph of max. {2 – x, 2, 1 + x} from the above graph :

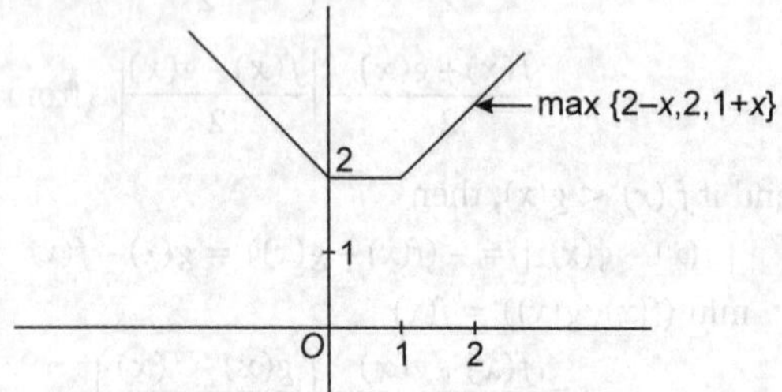

Illustration 84. Draw the graph of
$y = \min.\{x+1, \sqrt{(1-x)}\}$.

Sol. First draw the graph of $y = x + 1$ and $y = \sqrt{1-x}$.

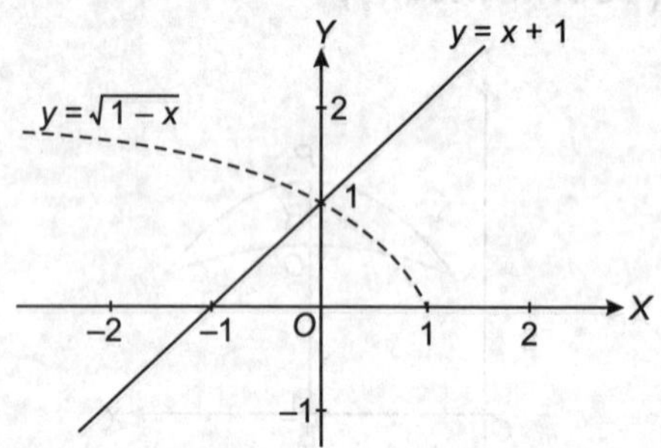

Graph of min. $\{x+1, \sqrt{(1-x)}\}$ **from the above graph :**

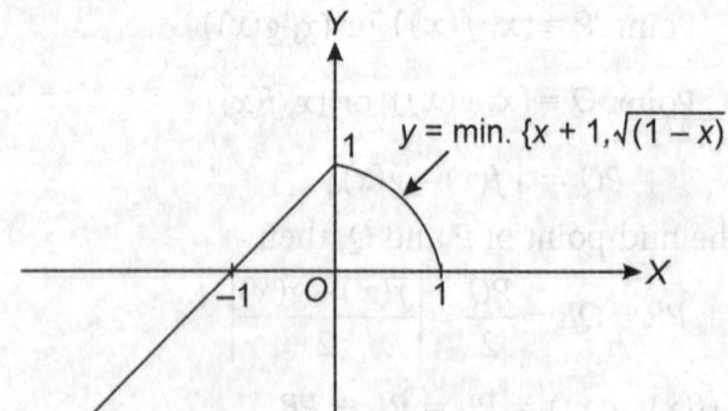

Illustration 85. Draw the graph of $y = \max.\ \{x + |x|, x - [x]\}$, where $[x]$ denotes the greatest integer $\leq x$.

Sol. First draw the graph of $y = x + |x|$ and $y = x - [x]$

$$\because \qquad y = x + |x| = \begin{cases} 2x, & x \geq 0 \\ 0, & x < 0 \end{cases}$$

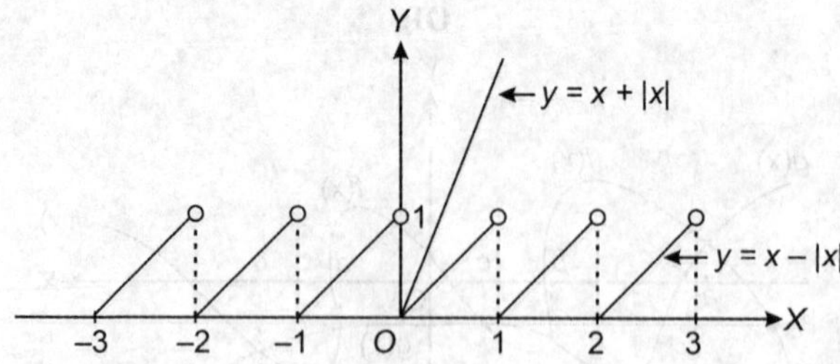

Graph of y = max. {x + |x|, x – [x]} from the above graph :

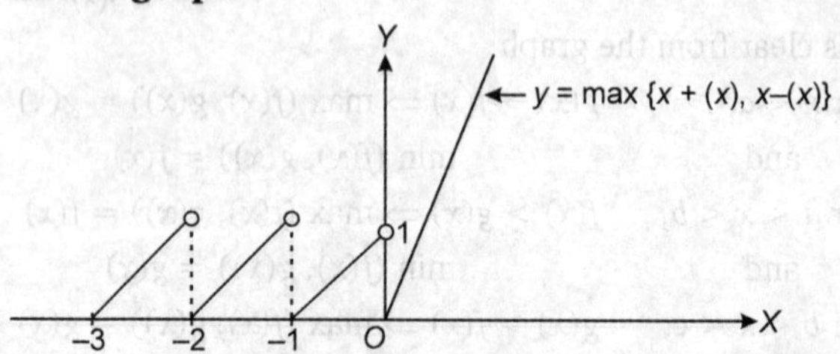

Illustration 86. Draw the graph of $y = \min.\ \{x - [x], -x - [-x]\}$, where $[x]$ denotes the greatest integer function $\leq x$.

Sol. First draw the graph of $y = x - [x]$ and $y = -x - [-x]$ (Graph of $y = -x - [-x]$ is the image of $y = x - [x]$ in the y-axis as a plane mirror)

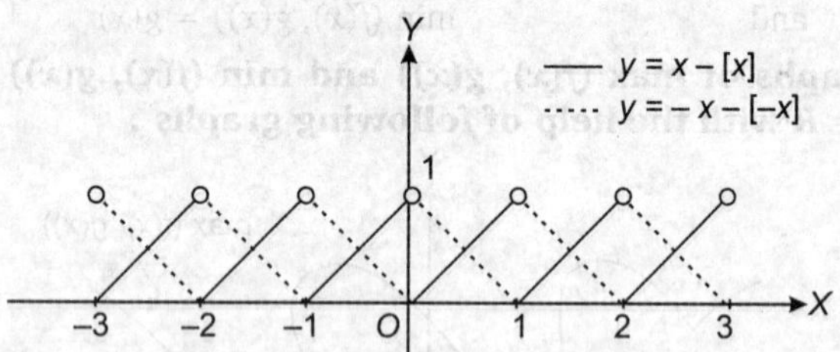

Graph of y = min. {x – [x], – x – [–x]} from above graph :

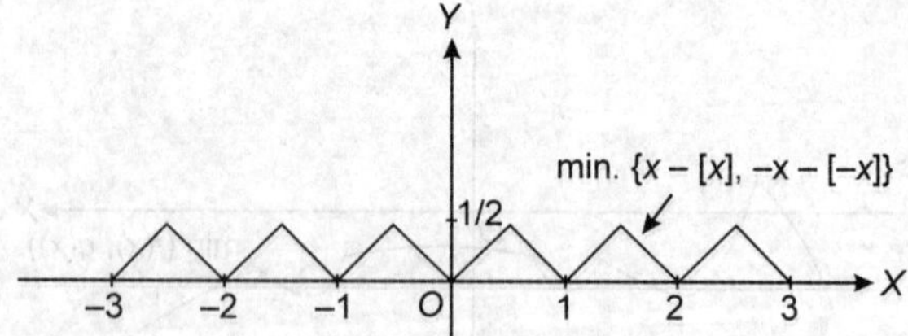

Illustration 87. Draw the graph of
$y = \min.\ \{|x|, |x-2|, 2 - |x-1|\}$.

Sol. First draw the graphs of $y = |x|$, $y = |x-2|$ and $y = 2 - |x-1|$.

How to draw the graph of y = 2 – |x – 1| ?

Graph of y = x – 1 :

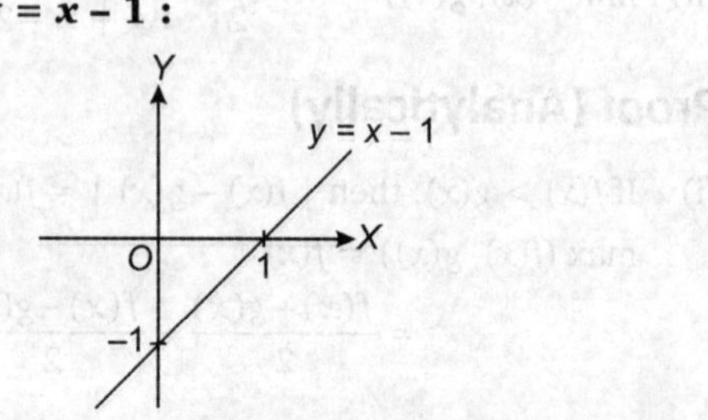

Graph of $y = |x-1|$:

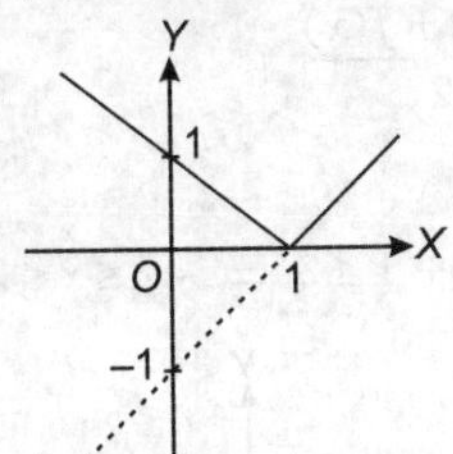

Graph of $y = -|x-1|$:

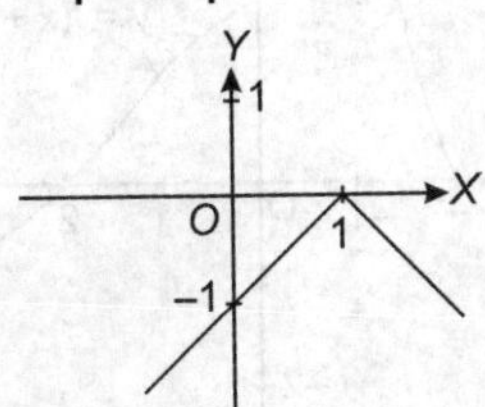

Graph of $y = 2 - |x-1|$:

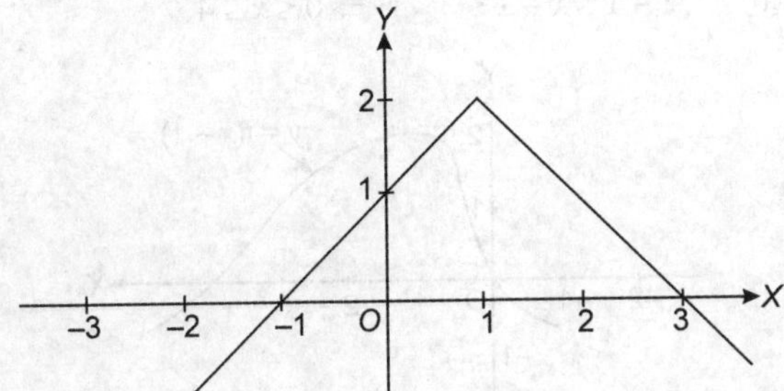

Graphs of $y = |x|, y = |x-2|$ and $y = 2 - |x-1|$

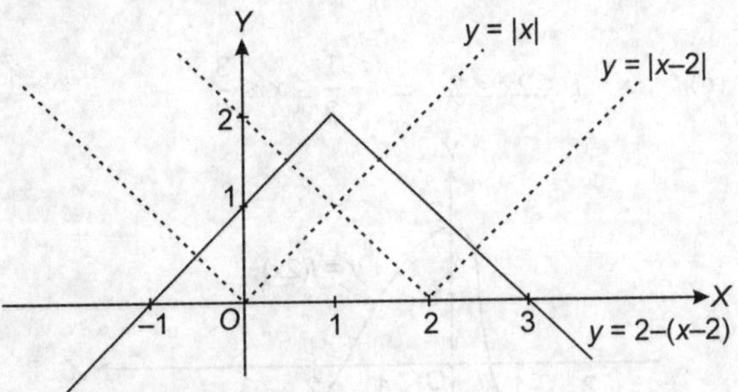

Hence graph of min. $\{|x|, |x-2|, 2-|x-1|\}$ from the above graph :

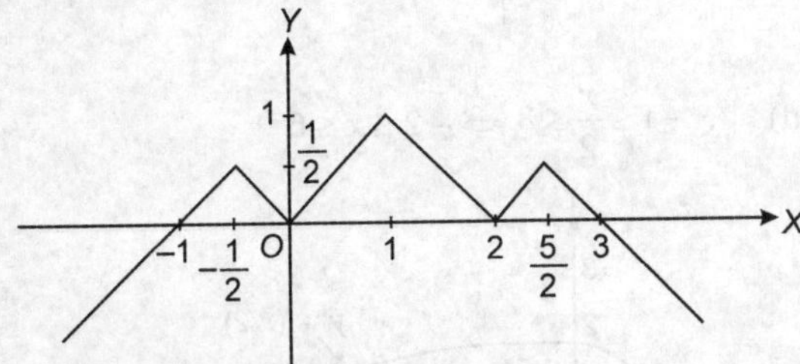

Illustration 88. Draw the graph of

$$y = \max.\left\{\sin x, \cos x, \frac{1}{2}\right\}.$$

Sol. First draw the graphs of $y = \sin x$, $y = \cos x$ and $y = \frac{1}{2}$.

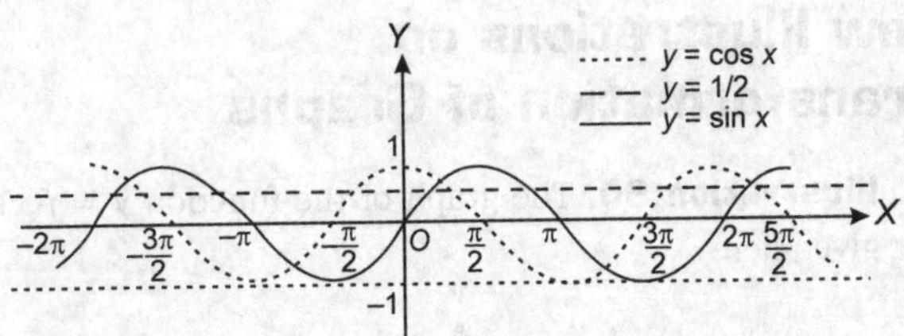

Graphs of $y = \max.\left\{\sin x, \cos x, \frac{1}{2}\right\}$ from the above graph :

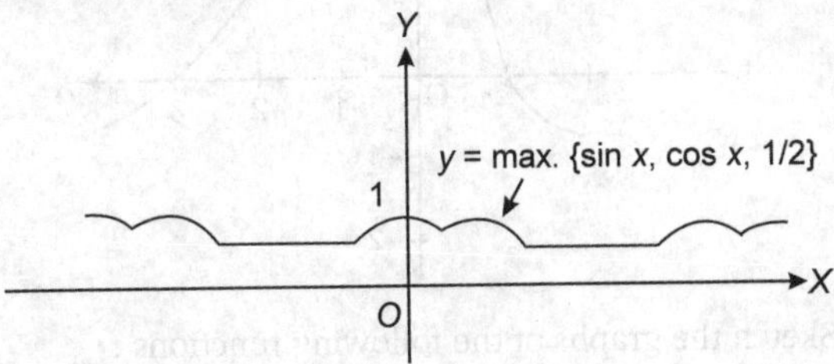

Illustration 89. Draw the graph of

$$y = \max.\ \{x^2, (1-x)^2, 2x(1-x)\}.$$

Sol. First draw the graph of $y = x^2$, $y = (1-x)^2$ and $y = 2x(1-x)$.

Since $\quad y = 2x(1-x)$

$$\Rightarrow \quad \left(y - \frac{1}{2}\right) = -2\left(x - \frac{1}{2}\right)^2$$

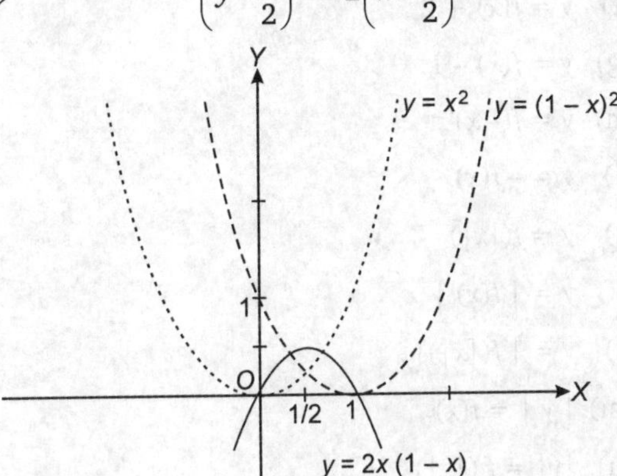

Graph of $y = \max.\ \{x^2, (1-x)^2, 2x(1-x)\}$ from the above graph :

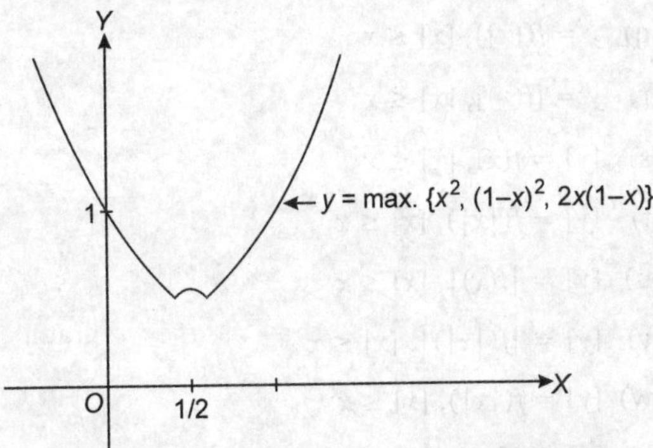

Few Illustrations on Transformation of Graphs

Illustration 90. The graph of the function $y = f(x)$ is given here

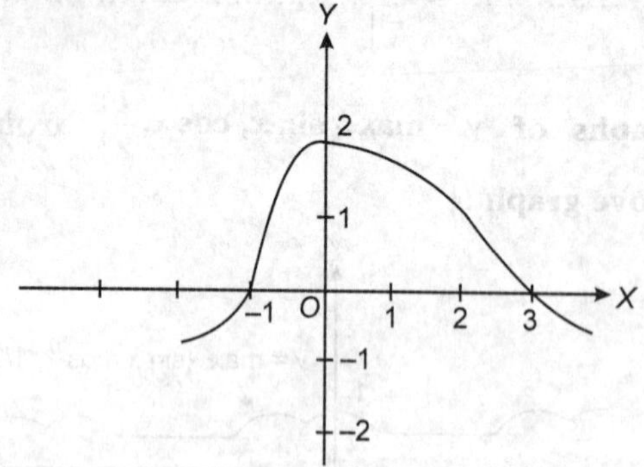

Sketch the graphs of the following functions :

(a) $y = f(x+1)$

(b) $y = f(x-1)$

(c) $y = f(2x)$

(d) $y = f\left(\frac{x}{2}\right)$

(e) $y = 2f(x)$

(f) $y = f(x) + 2$

(g) $y = f(x) - 1$

(h) $y = f(-x)$

(i) $y = -f(x)$

(j) $y = f(|x|)$

(k) $y = |f(x)|$

(l) $y = |f(|x|)|$

(m) $|y| = f(x)$

(n) $|y| = f(|x|)$

(o) $|y| = |f(x)|$

(p) $|y| = |f(|x|)|$

(q) $y = f([x]), [x] \leq x$

(r) $y = [f(x)], [x] \leq x$

(s) $[y] = f(x), [x] \leq x$

(t) $[y] = f(|x|), [x] \leq x$

(u) $[y] = |f(x)|, [x] \leq x$

(v) $[y] = |f(|x|)|, [x] \leq x$

(w) $[y] = f(|x|), [x] \leq x$

(x) $[y] = [f(x)], [x] \leq x$

(y) $y = |f(x)| / f(x)$

(z) $y = \dfrac{|f(x)| \pm f(x)}{2}$

Sol.

(a) $\because -1 \leq x + 1 \leq 3 \Rightarrow -2 \leq x \leq 2$

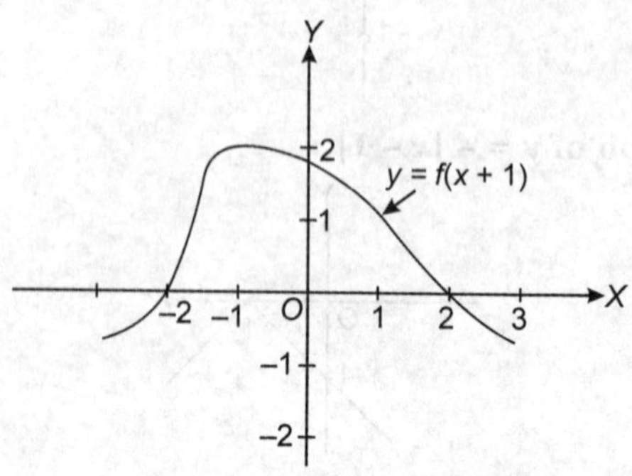

(b) $\because -1 \leq x - 1 \leq 3 \Rightarrow 0 \leq x \leq 4$

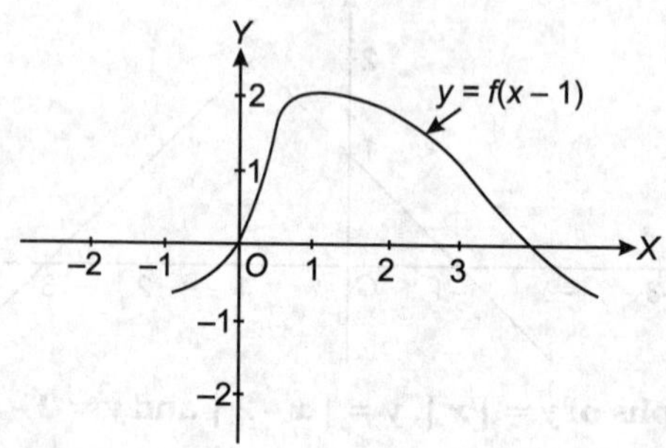

(c) $\because -1 \leq 2x \leq 3 \Rightarrow -\frac{1}{2} \leq x \leq \frac{3}{2}$

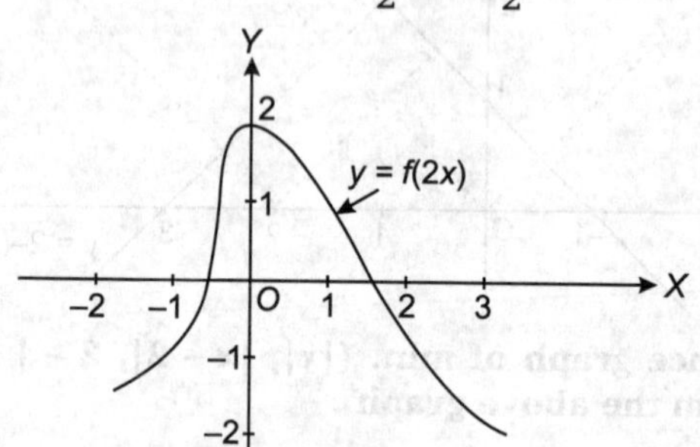

(d) $\because -1 \leq \frac{x}{2} \leq 3 \Rightarrow -2 \leq x \leq 6$

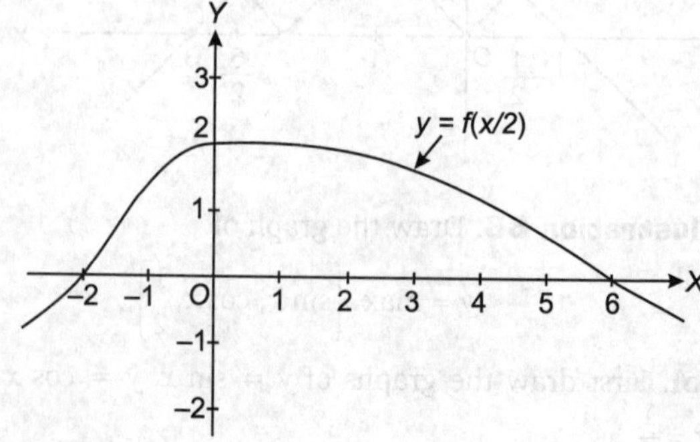

(e) $\because -1 \le f(x) \le 2 \Rightarrow -2 \le 2f(x) \le 4$

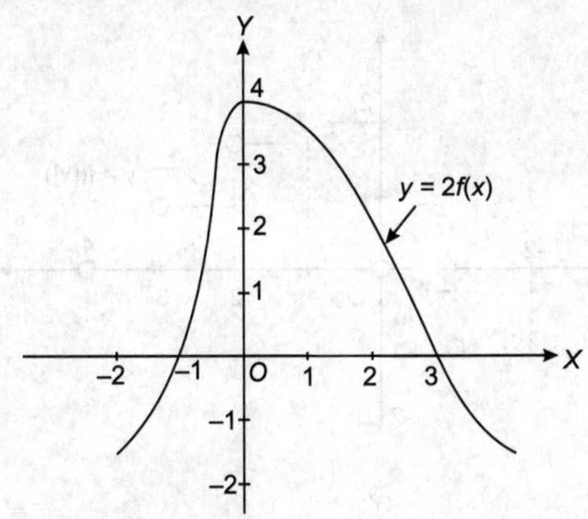

(f) $\because -1 \le f(x) \le 2$

$\Rightarrow \quad 1 \le f(x) + 2 \le 4$

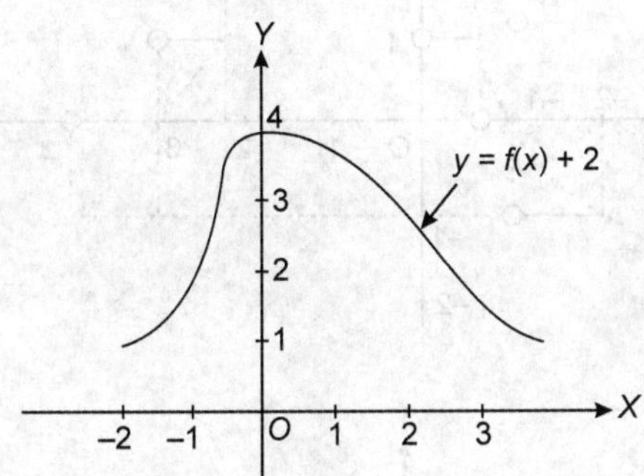

(g) $\because -1 \le f(x) \le 2 \quad \Rightarrow -2 \le f(x) - 1 \le 1$

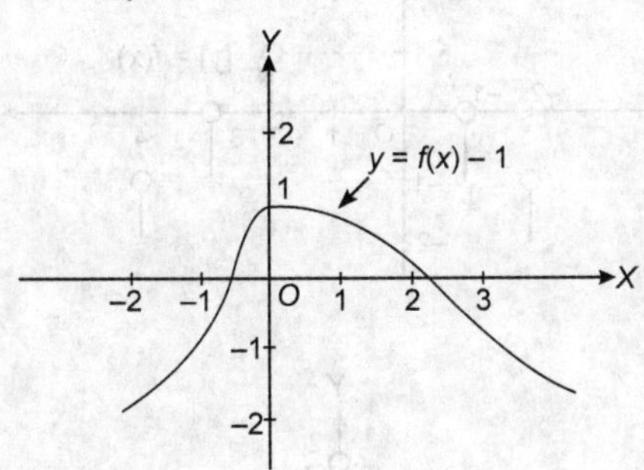

(h) $\because -1 \le x \le 3$

$\Rightarrow -3 \le x \le 1$

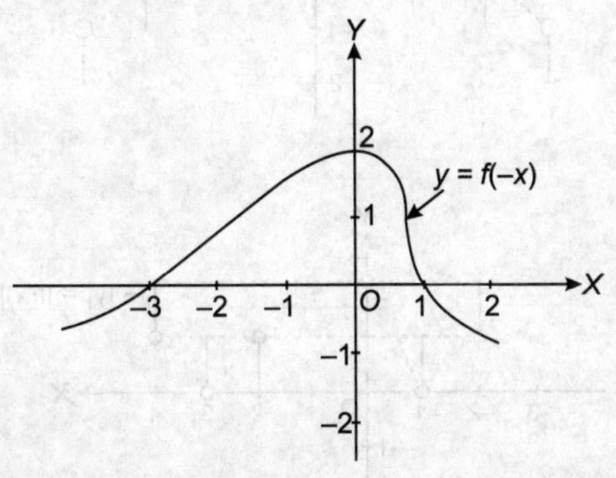

(i) $\because -1 \le f(x) \le 2 \Rightarrow -2 \le f(x) \le 1$

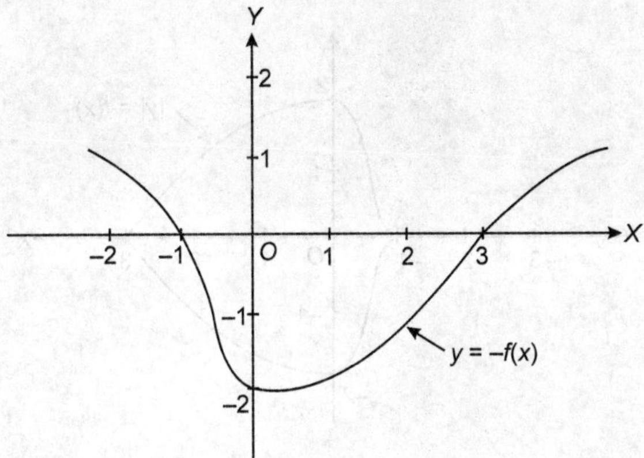

(j)

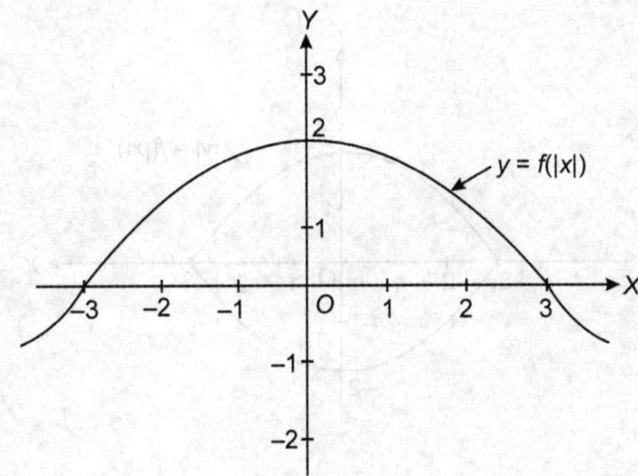

(k)

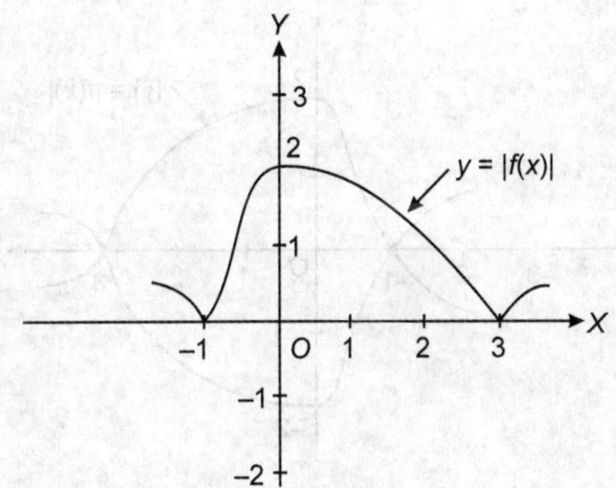

(l)

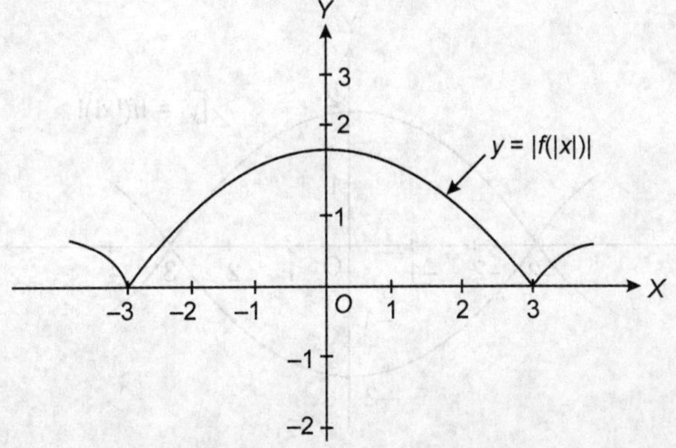

(m)

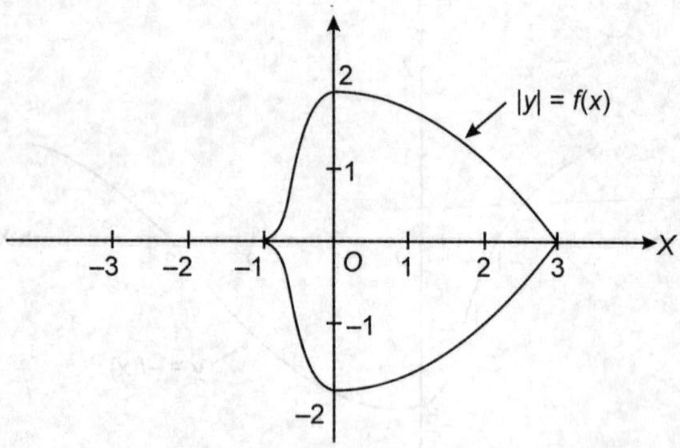

(n)

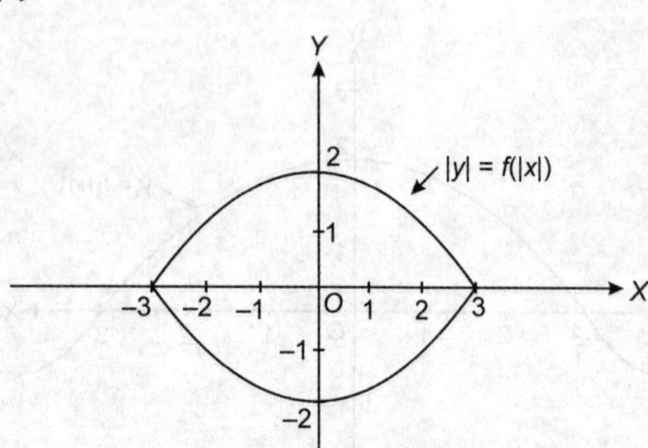

(o)

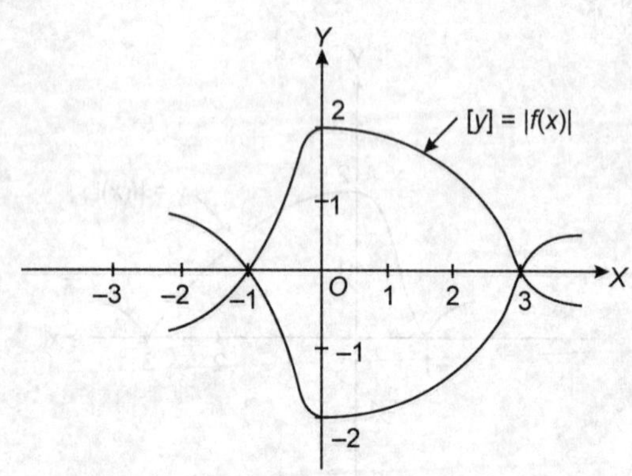

(p)

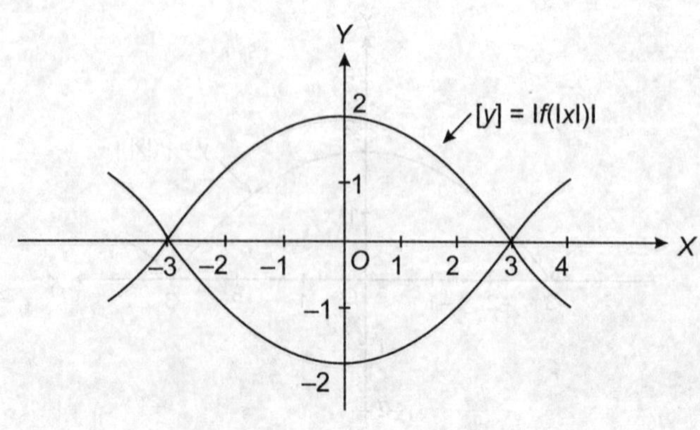

(q)

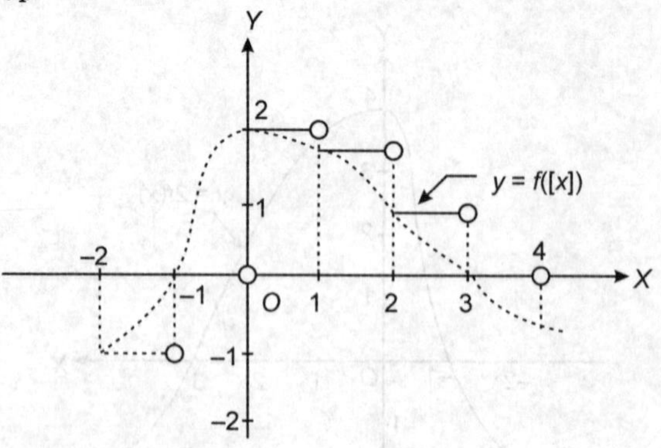

(r)

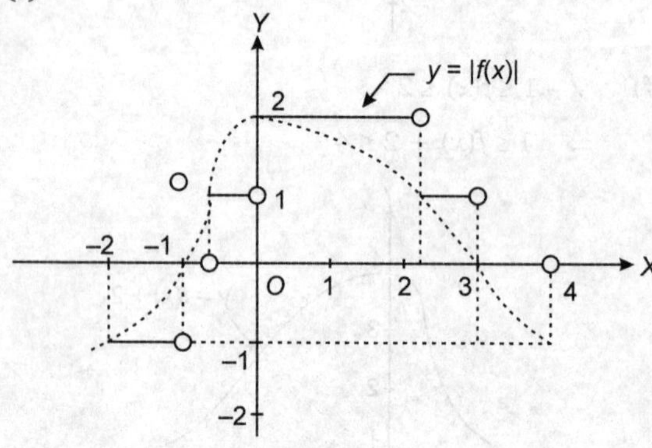

(s)

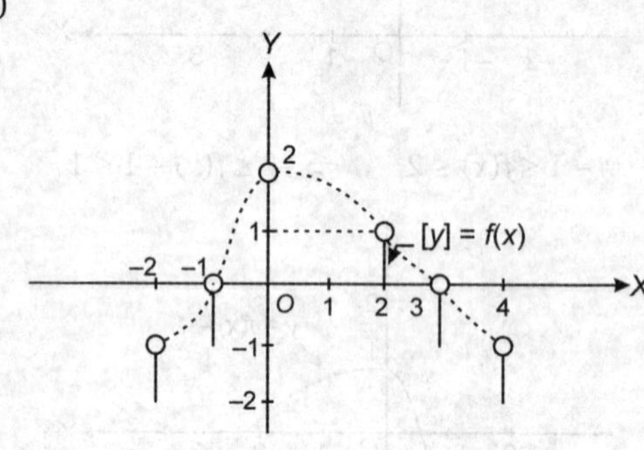

(t)

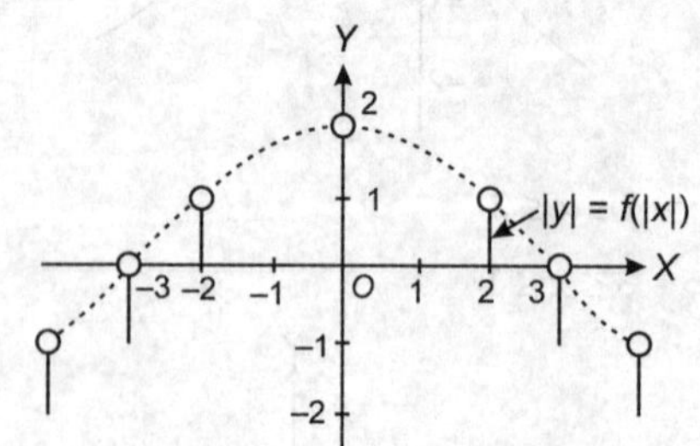

(u)

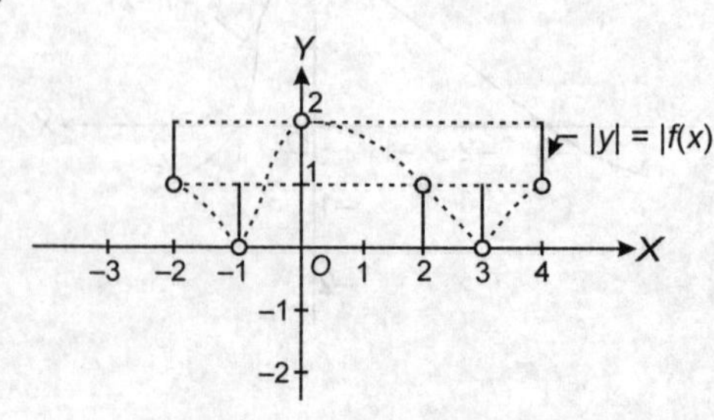

(v)

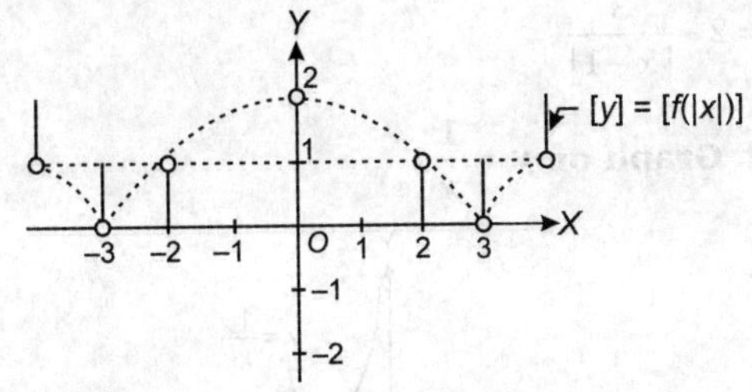

(w)

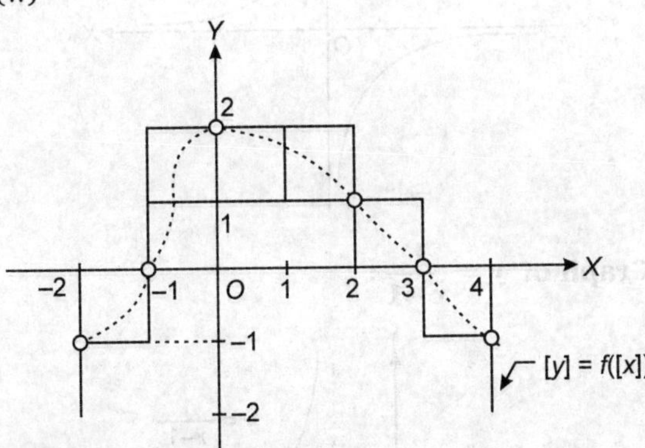

(x)

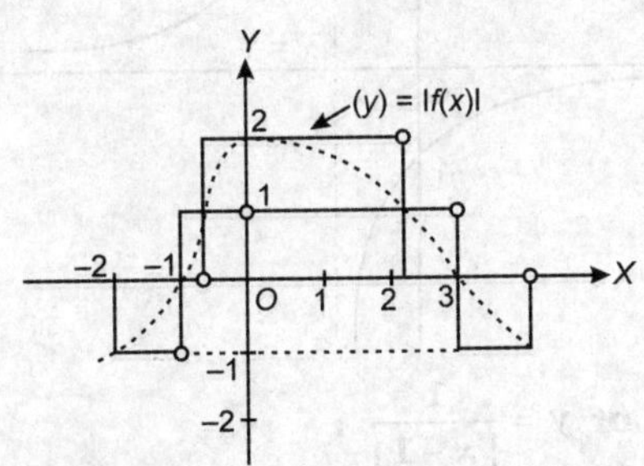

(y) If $f(x) > 0, y = 1$ and $f(x) < 0, y = -1$

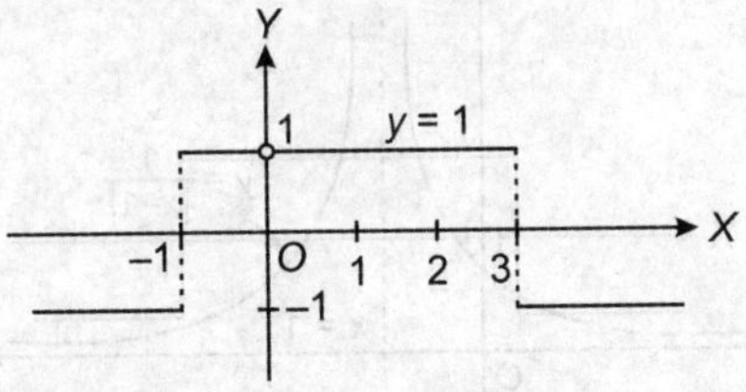

(z) If $f(x) \geq 0, y = f(x), 0$; if $f(x) \leq 0, y = 0, -f(x)$

Hence $y = |f(x)|$

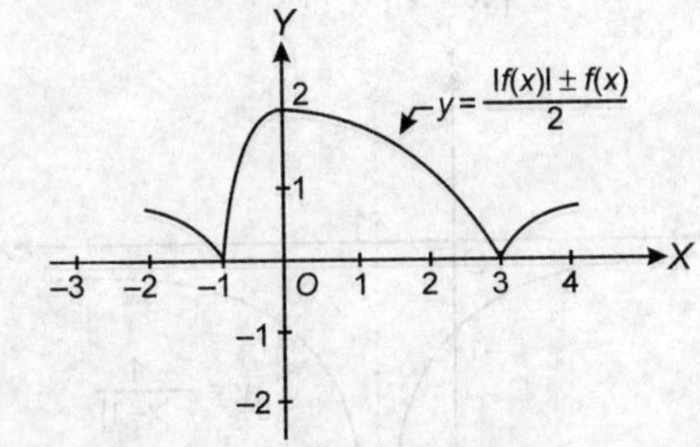

Illustration 91. Draw the graph of the function $y = |\, 2 - |x - 2|\,|$.

Sol. Graph of $y = x - 2$:

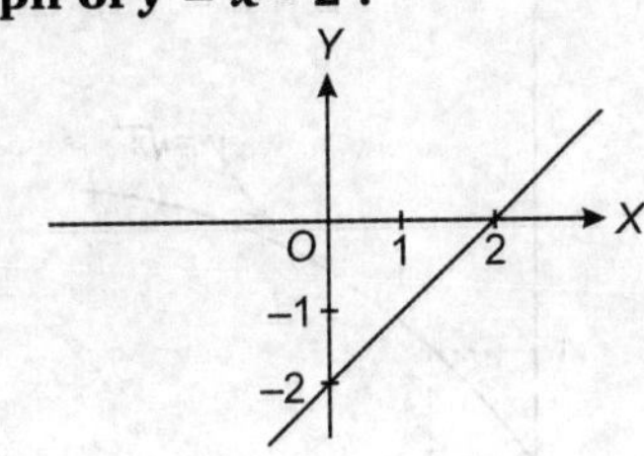

Graph of $y = |x - 2|$:

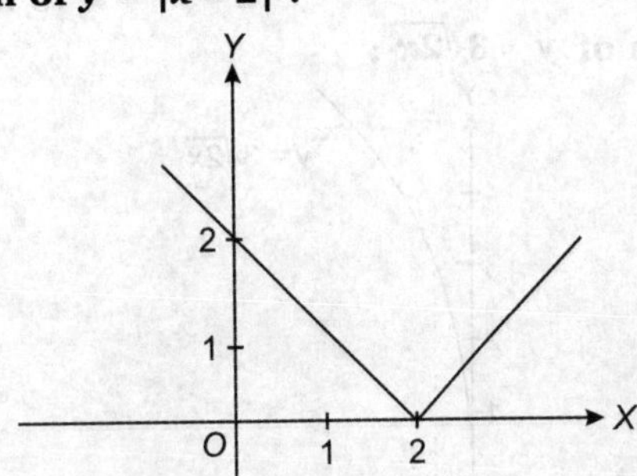

Graph of $y = -\,|\, x - 2\,|$:

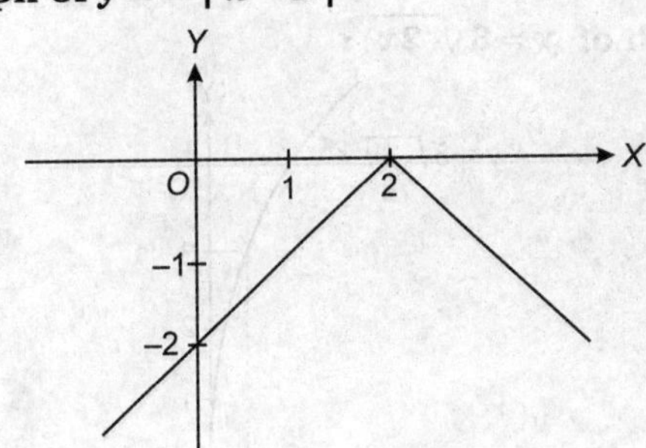

Graph of $y = 2 - |\, x - 2\,|$:

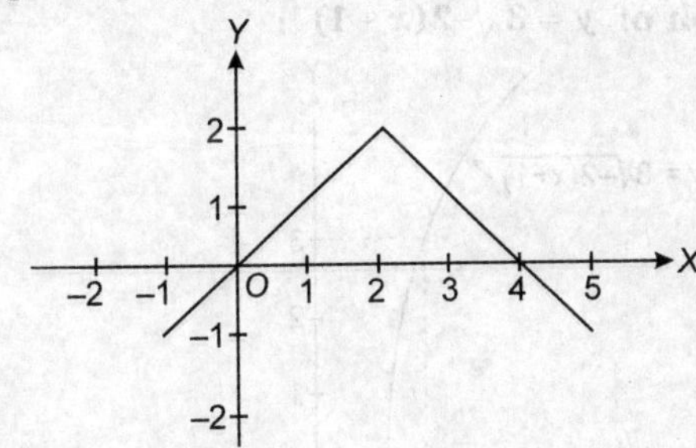

Graph of $y = |\, 2 - |\, x - 2\,||$:

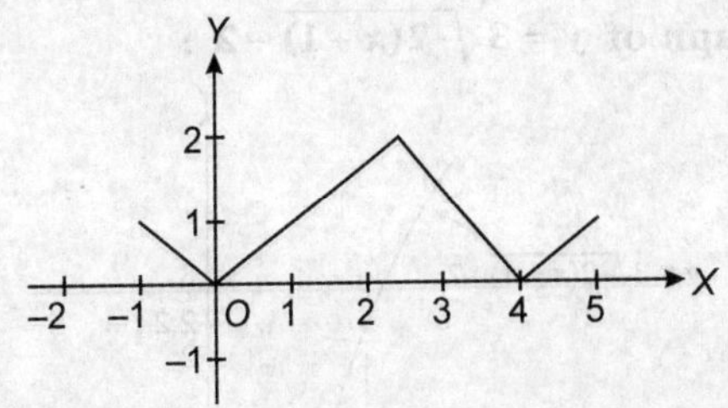

Illustration 92. Draw the graph of the function

$$y = 3\sqrt{\{-2(x+1)\}} - 2$$

Sol. Graph of $y = \sqrt{x}$:

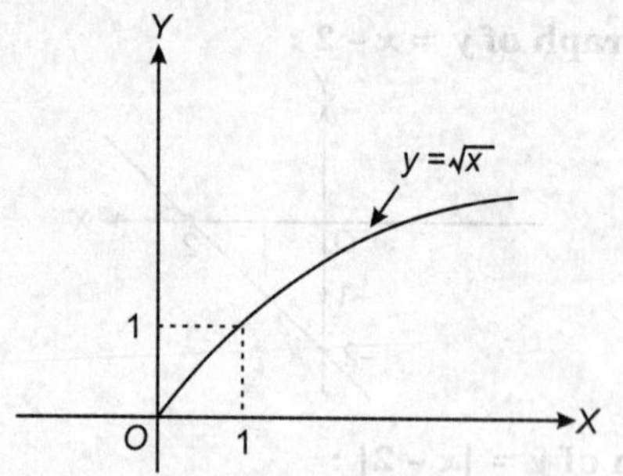

Graph of $y = 3\sqrt{2x}$:

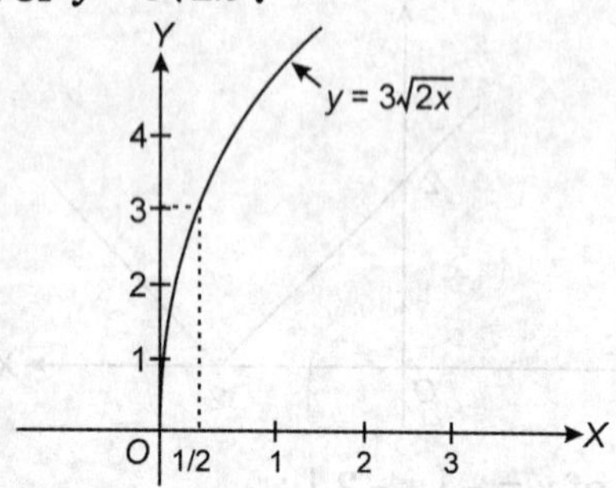

Graph of $y = 3\sqrt{-2x}$:

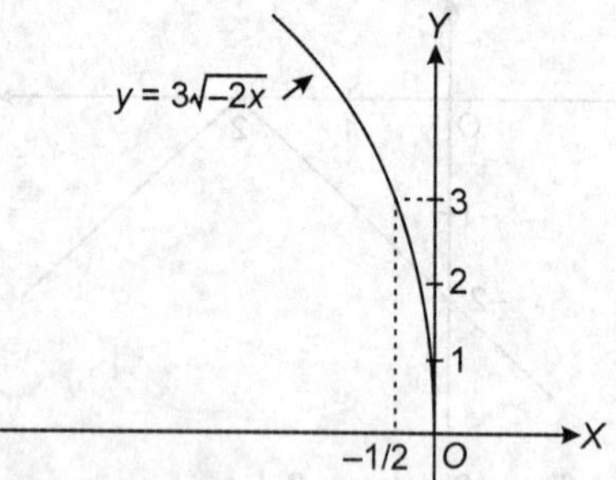

Graph of $y = 3\sqrt{-2(x+1)}$:

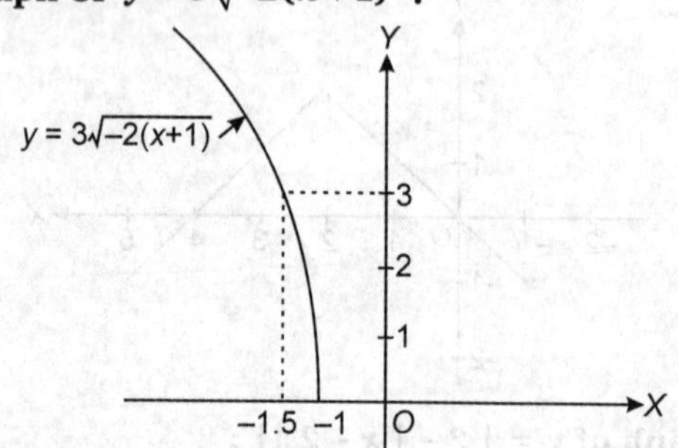

Graph of $y = 3\sqrt{-2(x+1)} - 2$:

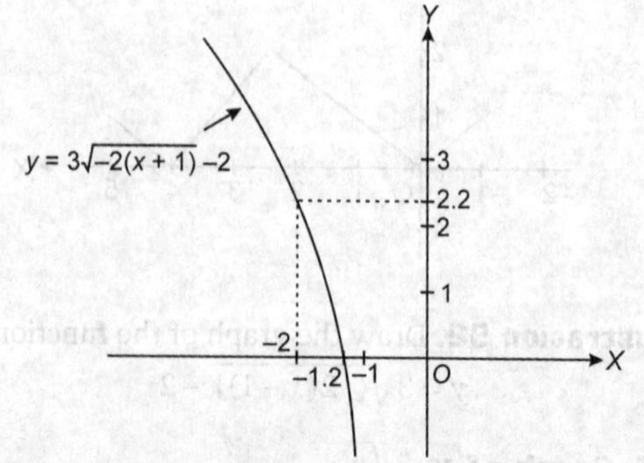

Illustration 93. Draw the graph of the function $y = 2 - \frac{3}{|x-1|}$.

Sol. Graph of $y = \frac{1}{x}$:

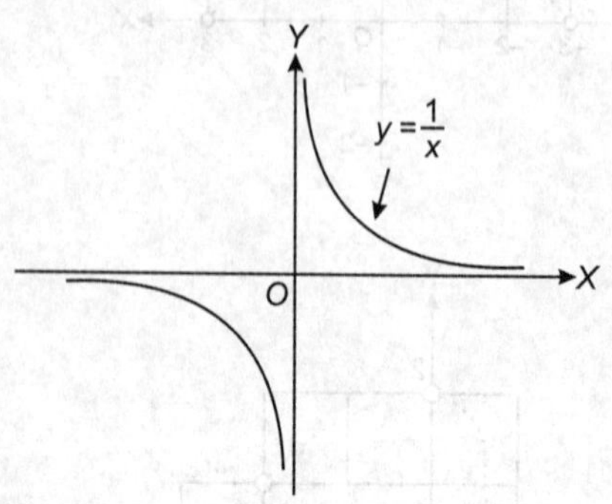

Graph of $y = \frac{1}{x-1}$:

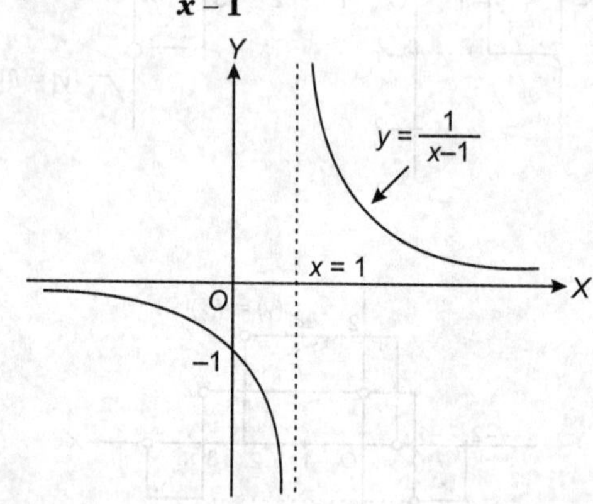

Graph of $y = \frac{1}{|x-1|}$:

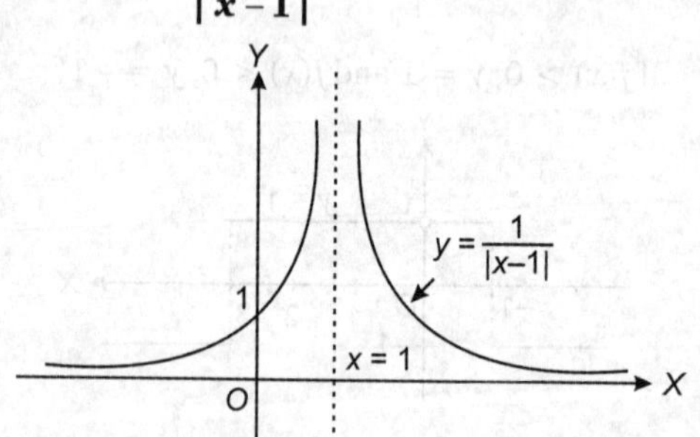

Graph of $y = -\frac{1}{|x-1|}$:

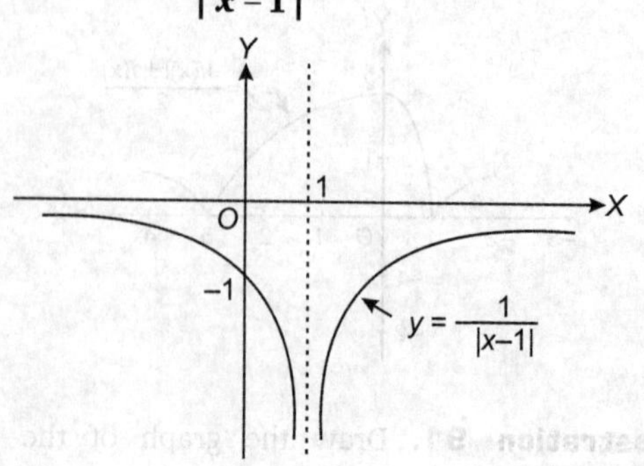

Graph of $y = -\dfrac{3}{|x-1|}$:

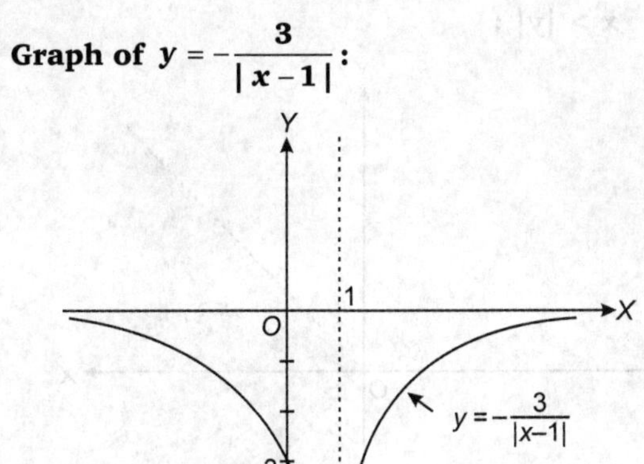

Graph of $y = 2 - \dfrac{3}{|x-1|}$:

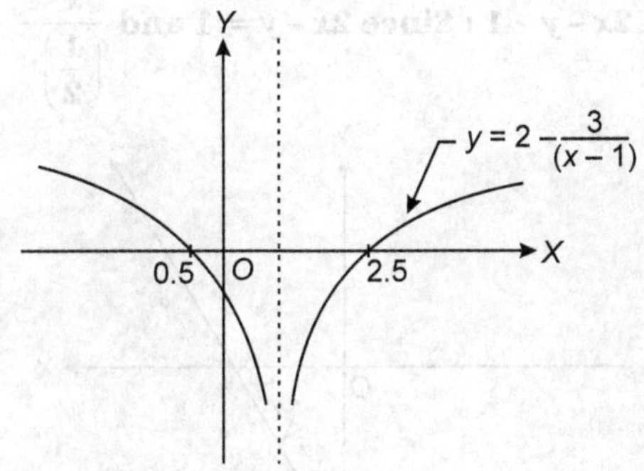

Illustration 94. Draw the graph of the function $y = |e^{|x|} - e^{-x}|$.

Sol. $y = |e^{|x|} - e^{-x}| = \begin{cases} |e^x - e^{-x}|, & x \geq 0 \\ 0 & , \quad x < 0 \end{cases}$

How to draw the graph of $y = e^x - e^{-x}$?

Draw graphs for the functions $y_1 = e^x$ and $y_2 = -e^{-x}$ and add graphically their ordinates.

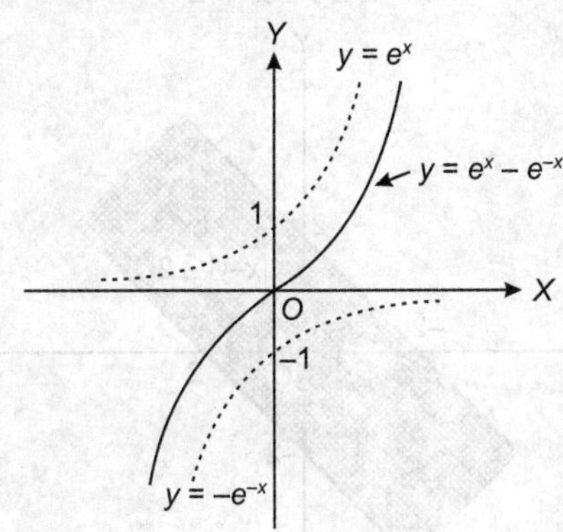

Also $\dfrac{d^2y}{dx^2} = e^x - e^{-x} > 0$ for $x > 0$ (Concave up)

< 0 for $x < 0$ (Concave down)

Graph of $y = |e^{|x|} - e^{-x}|$:

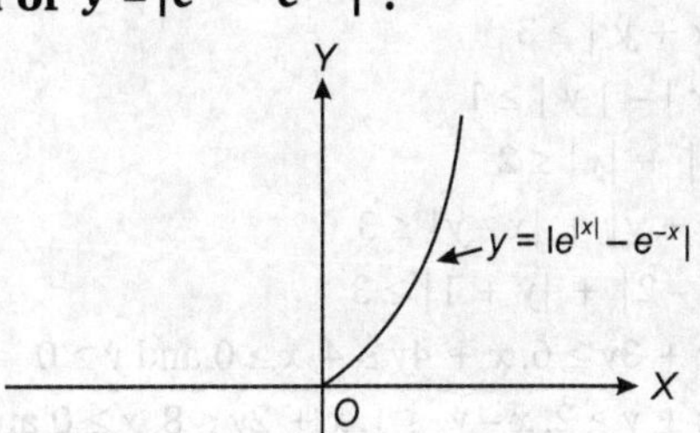

Illustration 95. Draw the graph of the function $y = \sqrt{3}\cos x + 3\sin x$.

Sol. The given function $y = \sqrt{3}\cos x + 3\sin x$ can be written as

$$y = 2\sqrt{3}\left(\frac{\sqrt{3}}{2\sqrt{3}}\cos x + \frac{3}{2\sqrt{3}}\sin x\right)$$

$$= 2\sqrt{3}\left(\frac{1}{2}\cos x + \frac{\sqrt{3}}{2}\sin x\right) = 2\sqrt{3}\cos\left(x - \frac{\pi}{3}\right)$$

$\because \quad -1 \leq \cos\left(x - \dfrac{\pi}{3}\right) \leq 1$

$\Rightarrow \quad -2\sqrt{3} \leq 2\sqrt{3}\cos\left(x - \dfrac{\pi}{3}\right) \leq 2\sqrt{3}$

$\Rightarrow \quad -2\sqrt{3} \leq y \leq 2\sqrt{3}$ and period 2π.

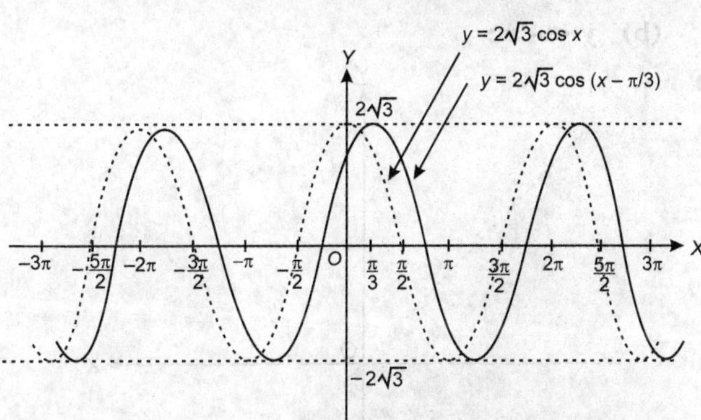

Graphs of a Linear Inequations

The inequalities of the form $ax + by + c > 0$, $ax + by + c < 0$, $ax + by + c \geq 0$ and $ax + by + c \leq 0$ are called linear inequations in two variables x and y.

Consider the equation $ax + by + c = 0$ draw the graph of the equation, which is a line. For strict inequation *i.e.*, $ax + by + c > 0$ or $ax + by + c < 0$ draw the line dotted (.....) otherwise make it thick (—). The shaded portion represents the solution set of the given inequation. The dotted line is not a part of the solution, but thick line is a part of the solution.

Illustration 96. Draw the graphs of the solution set of the following inequations :

(a) $y > x$ (b) $y < -x$

(c) $y \geq |x|$ (d) $x > |y|$

(e) $2x - y \geq 1$ (f) $|x - y| \leq 2$

(g) $| x + y | \geq 3$

(h) $| x | - | y | \geq 1$

(i) $|x| + |y| \leq 2$

(j) $|x + y| + |x - y| \leq 3$

(k) $|x - 2| + |y + 1| \geq 3$

(l) $2x + 3y \geq 6, x + 4y \leq 4, x \geq 0$ and $y \geq 0$

(m) $2x + y \geq 2, x - y \leq 1, x + 2y \leq 8, x \geq 0$ and $y \geq 0$

(n) $3y - x < 5, y + 2x < 10, 4y + x > 8$

Sol.

(a) $y > x$:

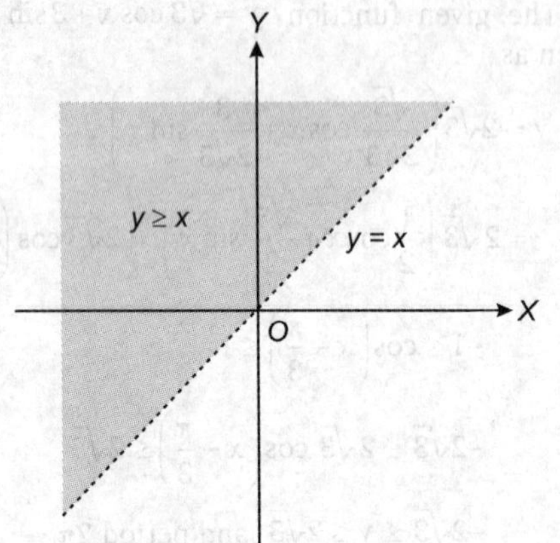

(b) $y < -x$:

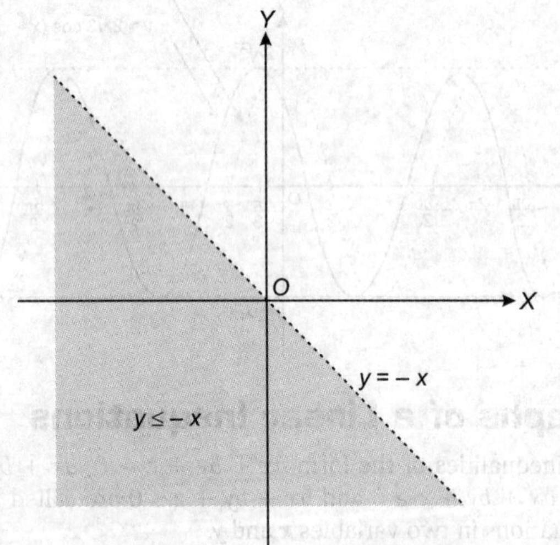

(c) $y \geq | x |$:

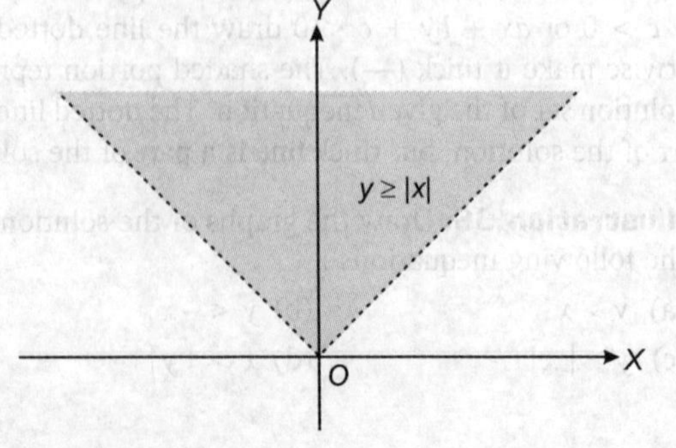

(d) $x > |y|$:

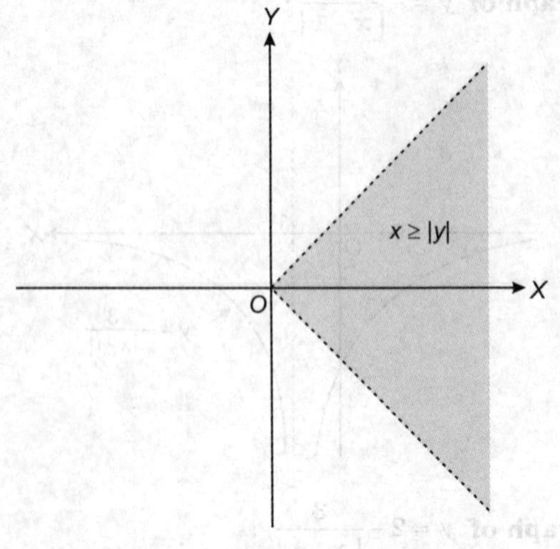

(e) $2x - y \geq 1$: Since $2x - y = 1$ and $\dfrac{x}{\left(\dfrac{1}{2}\right)} - \dfrac{y}{(1)} = 1$

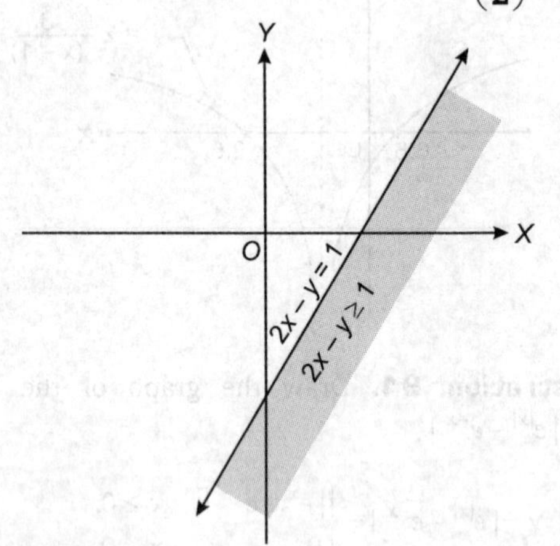

(f) $|x - y| \leq 2$:

i.e., $-2 \leq x - y \leq 2$

$\Rightarrow$ $-2 - x \leq -y \leq 2 - x$

or $x - 2 \leq y \leq x + 2$

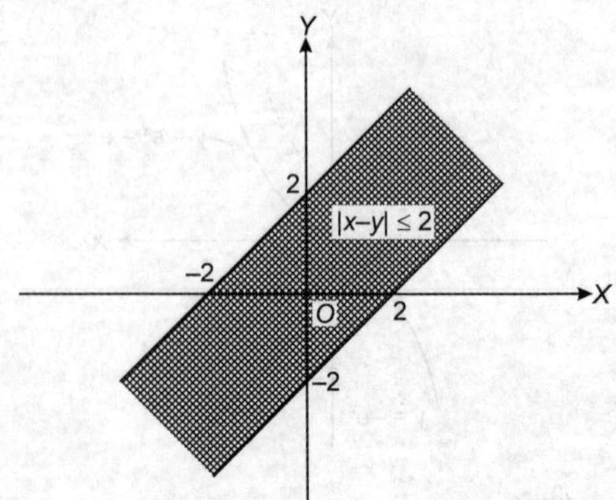

(g) $|x+y| \geq 3$:

$\Rightarrow$ $x + y \geq 3$ and $x + y \leq -30$

$\Rightarrow$ $y \geq 3 - x$ and $y \leq -3 - x$

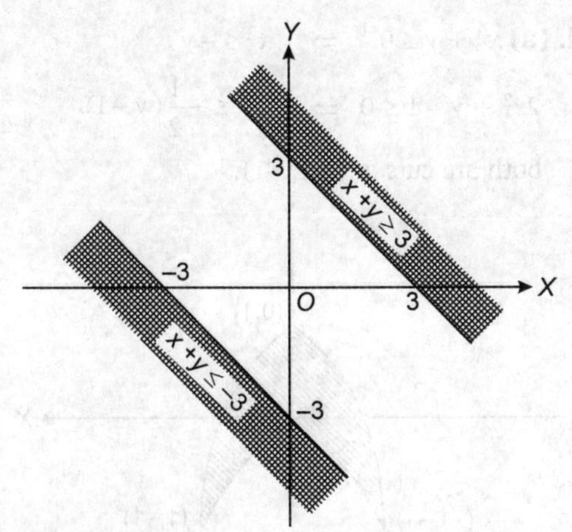

(h) $|x| - |y| \geq 1$ **:**

if $x \geq 0, y \geq 0,\quad x - y \geq 1 \Rightarrow y \leq x - 1$

$x \leq 0, y \geq 0,\quad -x - y \geq 1 \Rightarrow y \leq -1 - x$

$x \leq 0, y \leq 0,\quad -x + y \geq 1 \Rightarrow y \geq 1 + x$

$x \geq 0, y \leq 0,\quad x + y \geq 1 \Rightarrow y \geq 1 - x$

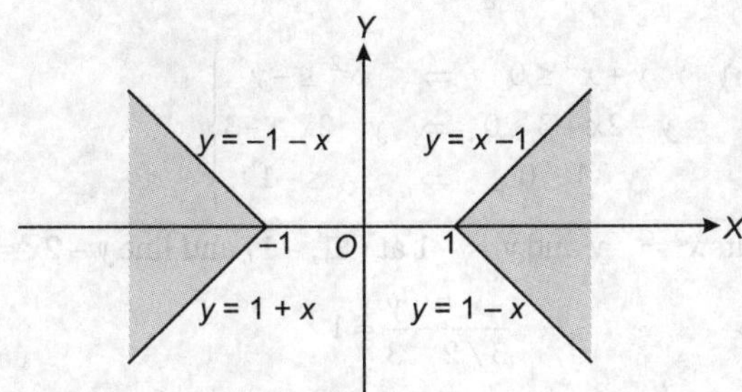

(i) $|x| + |y| \leq 2$ **:**

If $x \geq 0, y \geq 0,\quad x + y \leq 2 \Rightarrow y \leq 2 - x$

$x \leq 0, y \geq 0,\quad -x + y \leq 2 \Rightarrow y \leq 2 + x$

$x \leq 0, y \leq 0,\quad -x - y \leq 2 \Rightarrow y \geq -x - 2$

$x \geq 0, y \leq 0,\quad x - y \leq 2 \Rightarrow y \geq x - 2$

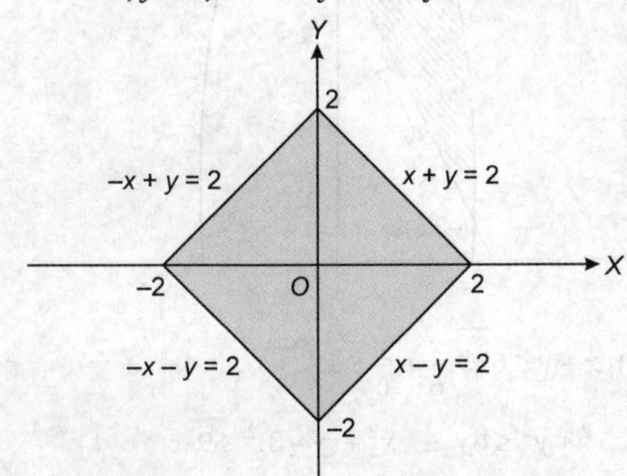

(j) $|x+y| + |x - y| \leq 3$ **:**

$$\left.\begin{array}{l}\text{If } x \geq y,\ x + y + x - y \leq 3 \ \Rightarrow\ x \leq \dfrac{3}{2} \\ \text{If } x \leq y,\ x + y - x + y \leq 3 \ \Rightarrow\ y \leq \dfrac{3}{2}\end{array}\right\}\text{I quadrant,}$$

$$\left.\begin{array}{l}x \geq -\dfrac{3}{2} \\ y \leq \dfrac{3}{2}\end{array}\right\}\text{II quad.,}\quad \left.\begin{array}{l}x \geq -\dfrac{3}{2} \\ y \geq -\dfrac{3}{2}\end{array}\right\}\text{III quad.,}\quad \left.\begin{array}{l}x \leq \dfrac{3}{2} \\ y \geq -\dfrac{3}{2}\end{array}\right\}\text{IV quad.}$$

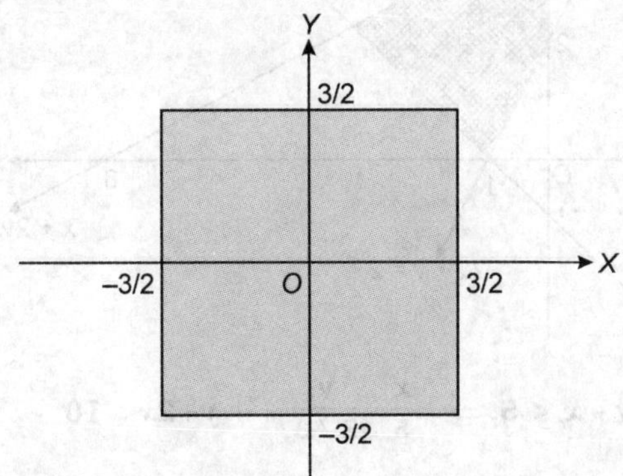

(k) $|x - 2| + |y + 1| \geq 3$ **:**

If $x \geq 2, y \geq -1 \Rightarrow x - 2 + y + 1 \geq 3 \Rightarrow x + y \geq 4$

If $x \leq 2, y \geq -1 \Rightarrow -x+2+y+1 \geq 3 \Rightarrow -x + y \geq 0$

If $x \geq 2, y \leq -1 \Rightarrow x - 2 - y - 1 \geq 3 \Rightarrow x - y \geq 6$

If $x \leq 2, y \leq -1 \Rightarrow -x + 2 - y - 1 \geq 3 \Rightarrow -x - y \geq 2$

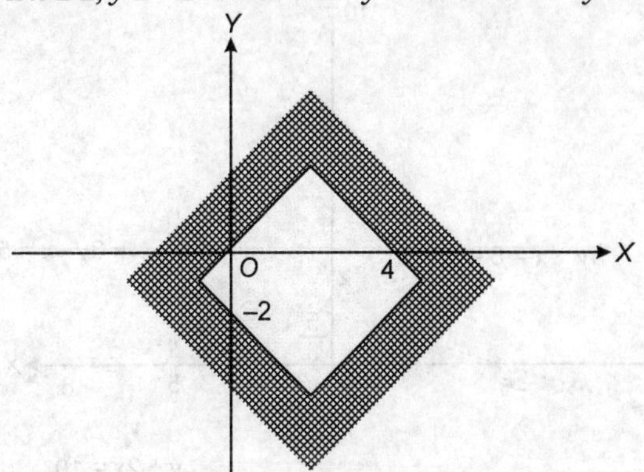

(l) $2x + 3x \geq 6 \Rightarrow \dfrac{x}{3} + \dfrac{y}{2} \geq 1$

$x + 4y \leq 4 \Rightarrow \dfrac{x}{4} + \dfrac{y}{1} \leq 1$ and $x \geq 0, y \geq 0$

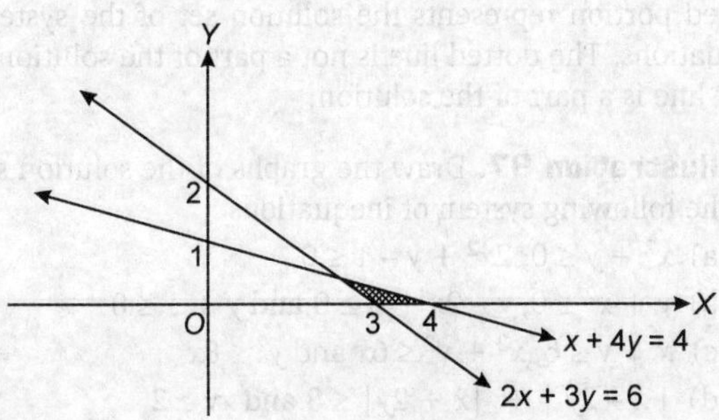

(m) $2x + y \geq 2 \Rightarrow \dfrac{x}{1} + \dfrac{y}{1} \geq 1\,;\ x - y \leq 1 \Rightarrow \dfrac{x}{1} + \dfrac{y}{-1} \leq 1$

and $\quad x + 2y \leq 8 \Rightarrow \dfrac{x}{8} + \dfrac{y}{4} \leq 1$

and $\quad x \geq 0, y \geq 0$

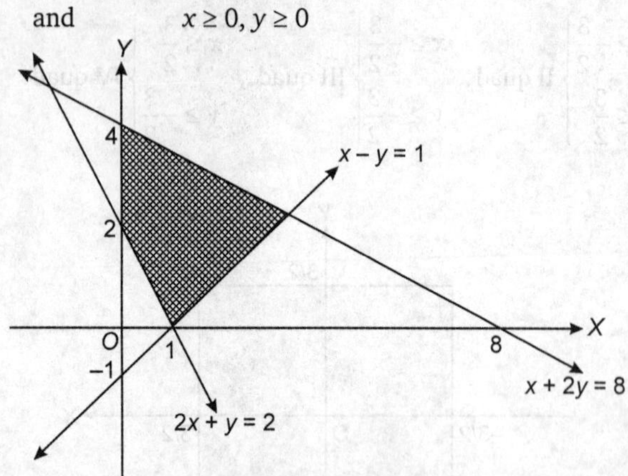

(n) $\mathbf{3y - x < 5, \Rightarrow -\frac{x}{5} + \frac{y}{5/3} < 1; \; y + 2x < 10}$

$\mathbf{\Rightarrow \frac{x}{5} + \frac{y}{10} < 1}$

$4x + x > 8 \qquad \Rightarrow \qquad \frac{x}{8} + \frac{y}{2} > 1$

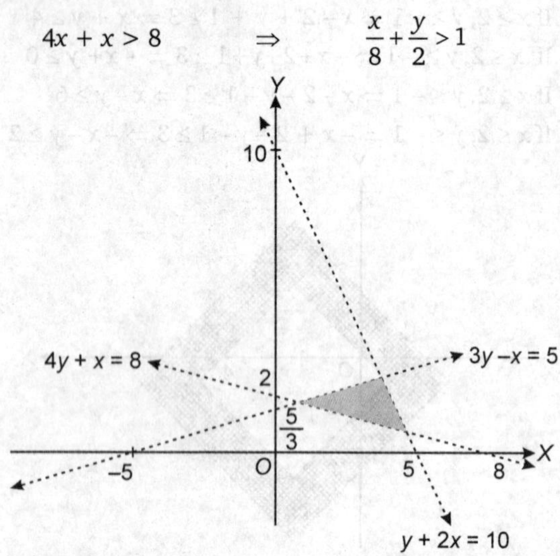

Graphs of the System of Inequations

First draw the graphs of all inequations in *x-y* plane. The shaded portion represents the solution set of the system of inequations. The dotted line is not a part of the solution, but thick line is a part of the solution.

Illustration 97. Draw the graphs of the solution set of the following system of inequations :

(a) $x^2 + y \leq 0, 2x^2 + y - 1 \leq 0$

(b) $y + x^2 \leq 0, y - 2x + 3 \geq 0$ and $y + 1 \leq 0$

(c) $x + y \leq 6, x^2 + y^2 \leq 6x$ and $y^2 \leq 8x$

(d) $| x - 2y | + |x + 2y| \leq 8$ and $xy \geq 2$

(e) $|x + y| + |x - y| \leq 4, |x| \leq 1$ and $y \geq \sqrt{(x^2 - 2x + 1)}$

(f) $4 \leq x^2 + y^2 \leq 2\,(|x| + |y|)$

(g) $|y + x| \leq 1, | y - x | \leq 1$ and $2x^2 + 2y^2 = 1$

(h) $2^{|x|}\,|y| + 2^{|x|-1} \leq 1, \; |x| \leq \frac{1}{2}$ and $|y| \leq \frac{1}{2}$

Sol. (a) $x^2 + y \leq 0 \quad \Rightarrow \quad x^2 \leq -y$

$2x^2 + y - 1 \leq 0 \;\Rightarrow\; x^2 \leq -\frac{1}{2}(y-1)$

both are cuts at $(\pm 1, -1)$.

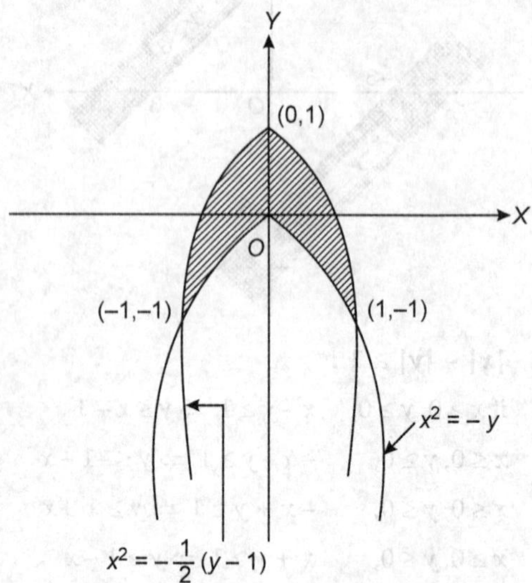

(b) $\left.\begin{aligned} y + x^2 \leq 0 &\quad\Rightarrow\quad x^2 \leq -y \\ y - 2x + 3 \geq 0 &\quad\Rightarrow\quad y - 2x \geq -3 \\ y + 1 \leq 0 &\quad\Rightarrow\quad y \leq -1 \end{aligned}\right\}$

cuts $x^2 = -y$ and $y = -1$ at $(\pm 1, -1)$ and line $y - 2x = -3$

$\Rightarrow \qquad \frac{x}{3/2} - \frac{y}{3} = 1$

(c) $x + y \leq 6 \Rightarrow \frac{x}{6} + \frac{y}{6} \leq 1$

$x^2 + y^2 \leq 6y \Rightarrow x^2 + (y-3)^2 \leq 9$

and $\quad y^2 \leq 8x$

$\Rightarrow$ circle $x^2 + (y-3)^2 = 9$ and line $\frac{x}{6} + \frac{y}{6} = 1$ cut at $(3, 3)$ and parabola $y^2 = 8x$ and line $\frac{x}{6} + \frac{y}{6} = 1$ cut at $(2, 4)$.

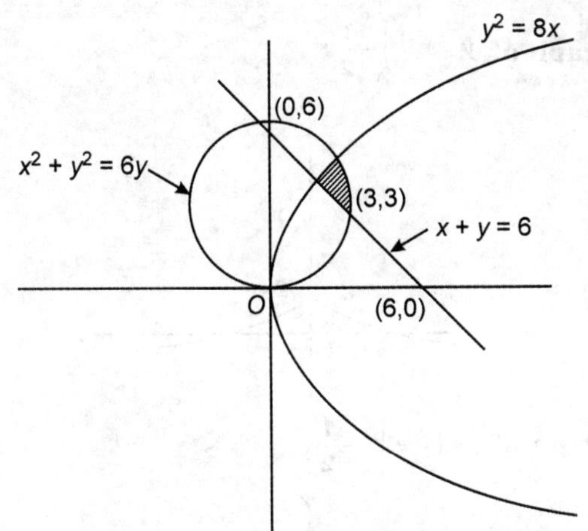

(d) $|x-2y|+|x+2y| \leq 8$ and $xy \geq 2$

$$\Rightarrow \quad |x-2y|+|x+2y| = 8$$

$$\pm(x-2y) \pm (x+2y) = 8$$

we get $x = \pm 4$ and $y = \pm 2$

Hence $-4 \leq x \leq 4$ and $-2 \leq y \leq 2$ for
$|x-2y|+|x+2y| \leq 8$

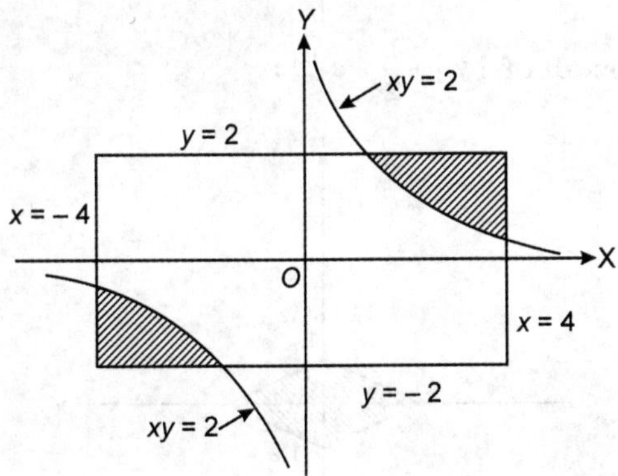

(e) $|x+y|+|x-y| \leq 4$

$$\Rightarrow |x+y|+|x-y| = 4 \pm (x+y) \pm (x-y) = 4$$

we get $x = \pm 2$ and $y = \pm 2$

Hence $\quad -2 \leq x \leq 2$

and $\quad -2 \leq y \leq 2$ for $|x+y|+|x-y| \leq 4$

$\Rightarrow \quad |x| \leq 1$

$\Rightarrow \quad -1 \leq x \leq 1$

and $\quad y \geq \sqrt{x^2 - 2x + 1} = \sqrt{(x-1)^2} = |x-1|$

$\therefore \quad y \geq |x-1|$

$\Rightarrow \quad -y \leq x - 1 \leq y$

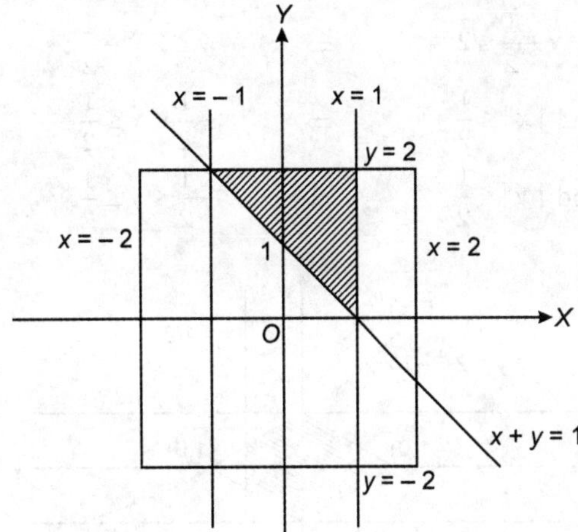

(f) $4 \leq x^2 + y^2 \leq 2(|x|+|y|)$

Now $\quad x^2 + y^2 \geq 4, |x|+|y| \geq 2$

and $\quad x^2 + y^2 - 2|x| - 2|y| \leq 0$

$\Rightarrow \quad (|x|-1)^2 + (|y|-1)^2 \leq 2$

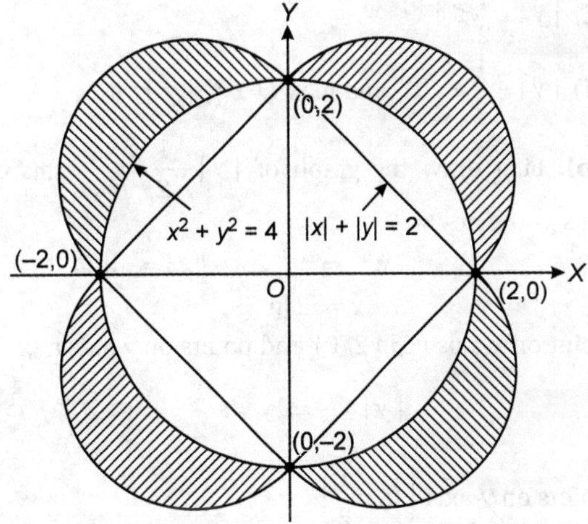

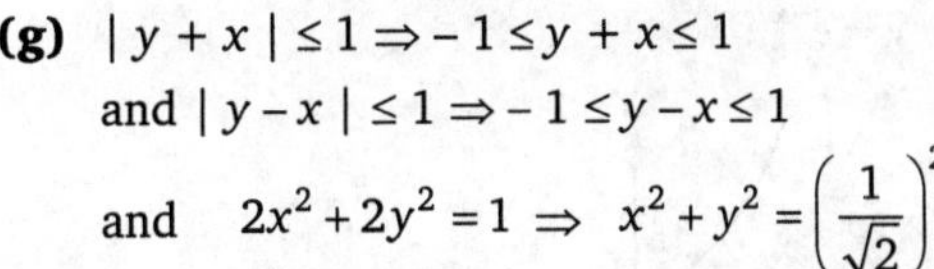

(g) $|y+x| \leq 1 \Rightarrow -1 \leq y + x \leq 1$

and $|y-x| \leq 1 \Rightarrow -1 \leq y - x \leq 1$

and $\quad 2x^2 + 2y^2 = 1 \Rightarrow x^2 + y^2 = \left(\frac{1}{\sqrt{2}}\right)^2$

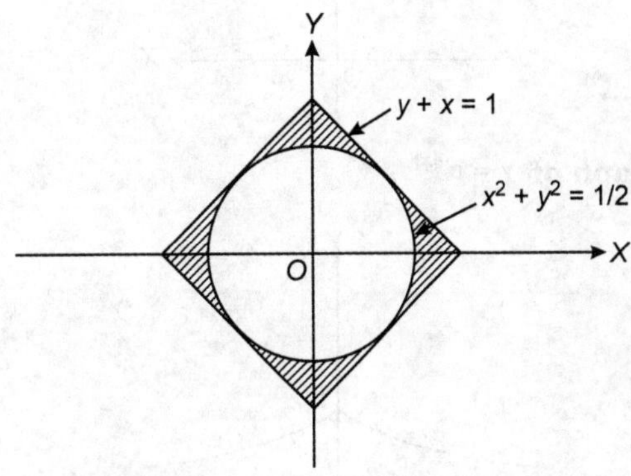

(h) $2^{|x|}|y| + 2^{|x|-1} \leq 1$

$\Rightarrow$ if $x \geq 0, y \geq 0$, then $2^x \cdot y + 2^{x-1} \leq 1$

$\Rightarrow y+\frac{1}{2}\le 2^{-x} \quad \Rightarrow \quad y\le 2^{-x}-\frac{1}{2}$

and $|x|\le\frac{1}{2} \quad \Rightarrow \quad -\frac{1}{2}\le x\le\frac{1}{2}$

and $|y|\le\frac{1}{2} \quad \Rightarrow \quad -\frac{1}{2}\le y\le\frac{1}{2}$

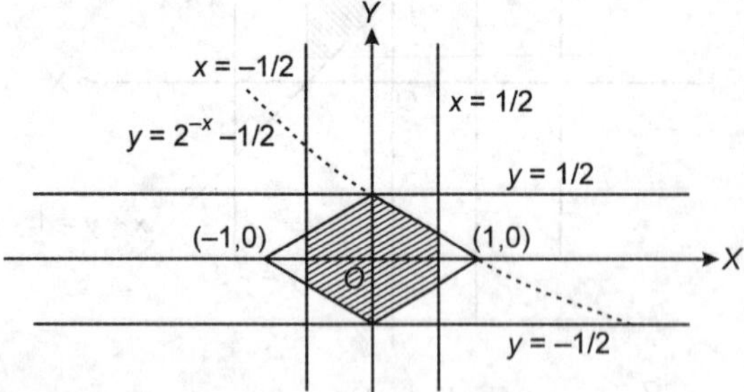

This graph is symmetrical about x-axis and y-axis.

Some Important Graphs

Illustration 98. Draw the graphs of

(i) $|y|+\frac{1}{2}\le e^{-|x|}$

(ii) $|y|+\frac{1}{2}\ge e^{-|x|}$ and max $(|x|, |y|)\le 2$

Sol. First draw the graph of $|y|+\frac{1}{2}=e^{-|x|}$ this curve cuts x-axis at $y = 0$

$\therefore \quad e^{-|x|}=2^{-1} \Rightarrow -|x|=-\ln 2$

$\therefore \quad x=\pm\ln 2$

Point on x-axis $(\pm \ln 2, 0)$ and points on y-axis at $x = 0$

$$|y|=\frac{1}{2}\Rightarrow y=\pm\frac{1}{2}$$

Points on y-axis $\left(0, \pm\frac{1}{2}\right)$

Graph of $y=e^{-x}$:

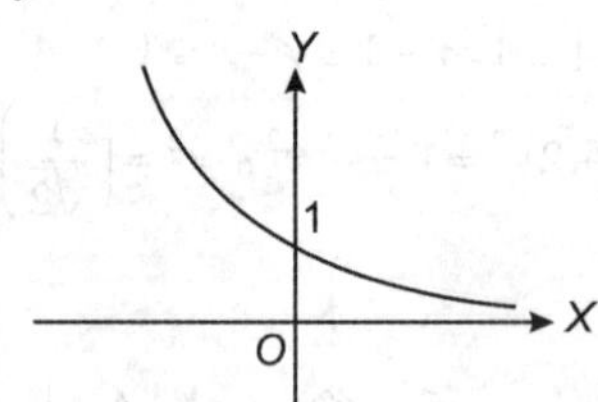

Graph of $y=e^{-|x|}$:

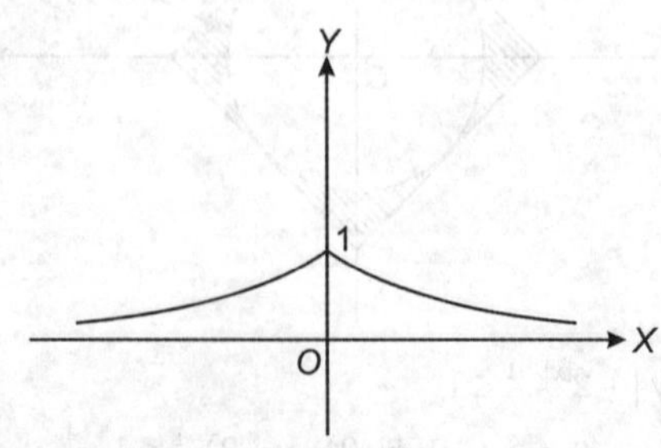

Graph of $y=e^{-|x|}-\frac{1}{2}$:

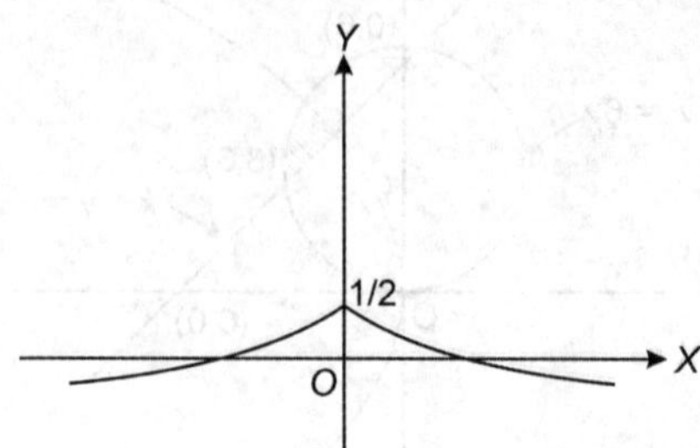

Graph of $|y|=e^{-|x|}-\frac{1}{2}$:

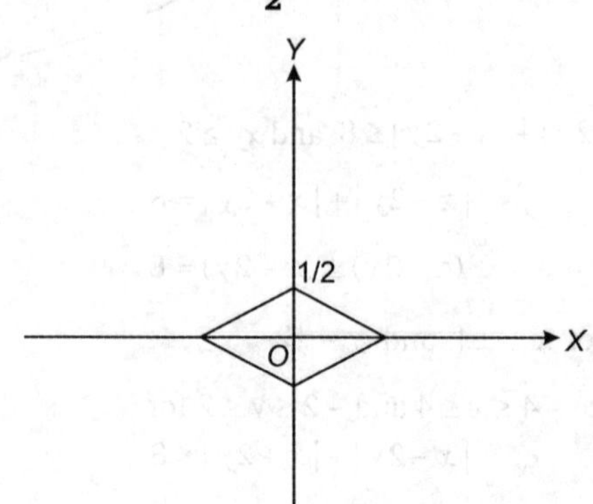

Graph of $|y|+\frac{1}{2}\le e^{-|x|}$:

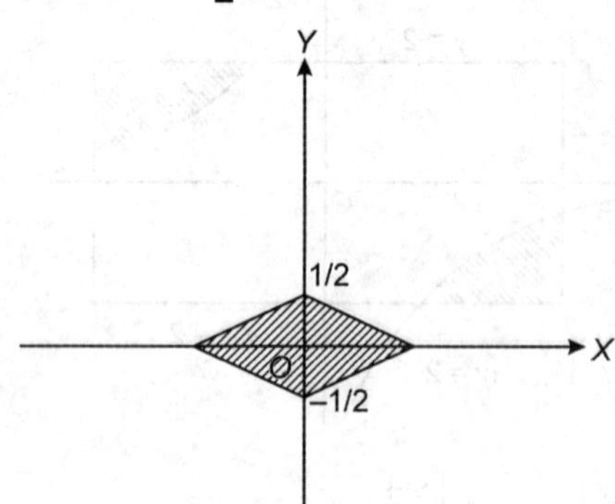

Graph of $|y|+\frac{1}{2}\ge e^{-|x|}$ and max $(|x|, |y|)\le 2$

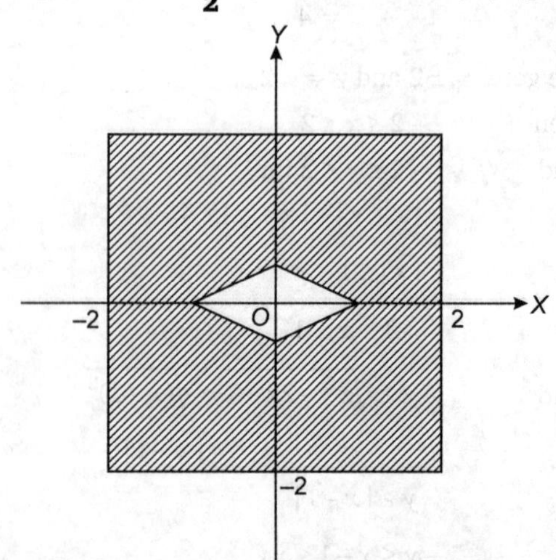

Illustration 99. Find all the points in the plane, whose co-ordinates x and y satisfy the inequality $\cos x - \cos y > 0$.

Sol. If $\cos x - \cos y = 0$ or $\cos y = \cos x$

$\therefore \quad y = 2n\pi \pm x, n \in I$

Since $\cos x - \cos y = 0$

i.e., $$2\sin\left(\frac{x+y}{2}\right)\sin\left(\frac{y-x}{2}\right) = 0$$

$\because \cos x - \cos y > 0$

$$\Rightarrow \quad 2\sin\left(\frac{x+y}{2}\right)\sin\left(\frac{y-x}{2}\right) > 0$$

$$\therefore \quad \sin\left(\frac{y+x}{2}\right) > 0, \ \sin\left(\frac{y-x}{2}\right) > 0$$

and $$\sin\left(\frac{y+x}{2}\right) < 0, \ \sin\left(\frac{y-x}{2}\right) < 0$$

$\Rightarrow \quad y + x > 2n\pi, \ y - x > 2n\pi$

and $\quad y + x < 2n\pi, \ y - x < 2n\pi$

$\Rightarrow \quad y > 2n\pi - x, \ y > 2n\pi + x$

and $\quad y < -x + 2n\pi, \ y < x + 2n\pi$ where $n \in I$

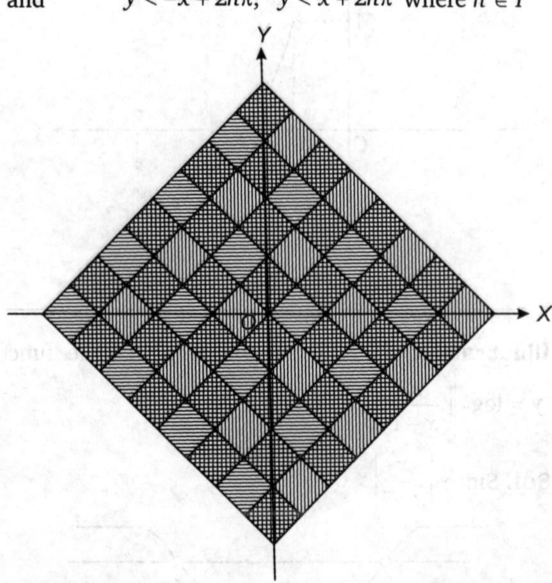

Illustration 100. Draw the graph of the function $\sin y = \sin x$.

Sol. Since $\sin y = \sin x$

$\therefore \quad y = n\pi + (-1)^n x, x \in I$

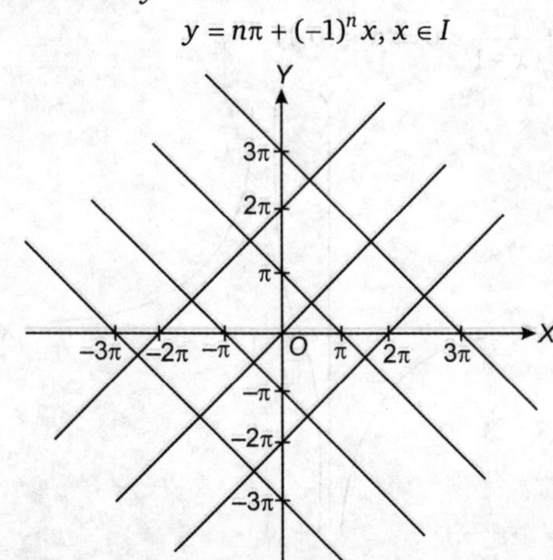

Illustration 101. Draw the graph of the function $y = 2^{\sin x}$.

Sol. Since $-1 \le \sin x \le 1$

$\Rightarrow \quad 2^{-1} \le 2^{\sin x} \le 2^1$

$\Rightarrow \quad \frac{1}{2} \le y \le 2, x \in R$ and period 2π.

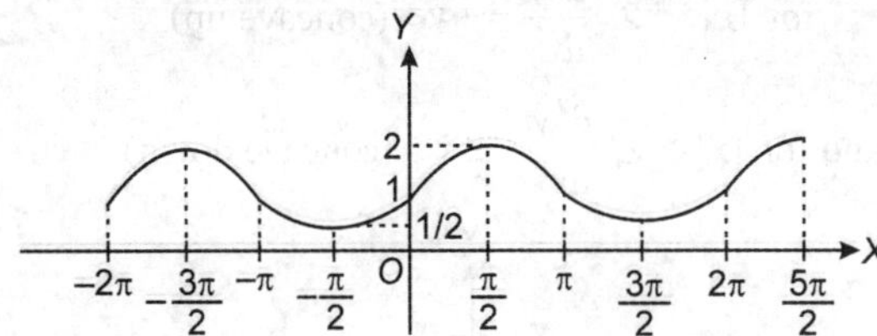

Illustration 102. Draw the graph of the function $y = \tan x \cot x$.

Sol. $y = \tan x \cot x = 1, \ x \ne n\pi, n\pi + \frac{\pi}{2}, n \in I.$

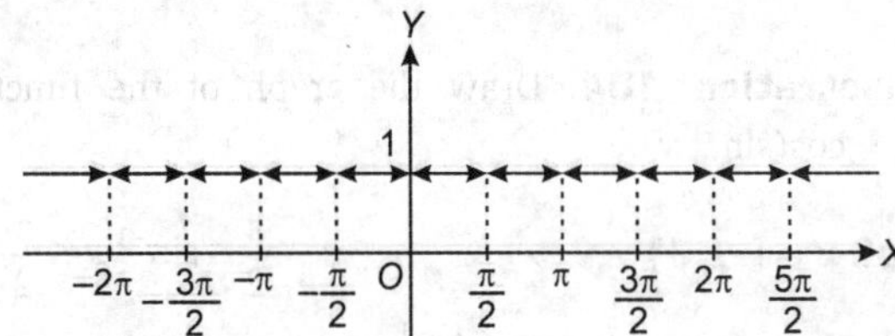

Illustration 103. Draw the graphs of the following functions :

(i) $y = \tan^{-1}(x^2 - 1)$ (ii) $y = \cot^{-1}(4 - x^2)$

Sol. (i) $y = \tan^{-1}(x^2 - 1)$

cut off x-axis $y = 0, x = \pm 1$

cut off y-axis $x = 0, \ y = -\frac{\pi}{2}$

and Range $y \in \left(-\frac{\pi}{2}, \frac{\pi}{2}\right)$ and $x \in R$, not periodic and

$$\frac{d^2y}{dx^2} = \frac{2(1-x^2)(1+3x^2)}{[1+(x^2-1)^2]^2}$$

for $0 < |x| < 1, \ \frac{d^2y}{dx^2} = +\text{ve}$ (concave up)

$|x| \ge 1, \ \frac{d^2y}{dx^2} = -\text{ve}$ (concave down)

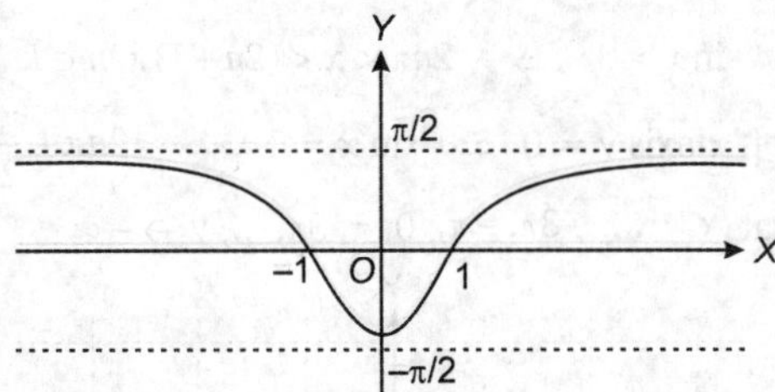

(ii) $y = \cot^{-1}(4 - x^2)$ domain $x \in R$ and Range $y \in (0, \pi)$, not periodic cut of y-axis $x = 0$,

$$y = \cot^{-1}(4) = \frac{\pi}{2} - \tan^{-1}(4) = \frac{\pi}{12}$$

(Approximate)

cut of x-axis $y = 0$, $\cot^{-1}(4-x^2) = 0$ (not defined)

the curve does not cut x-axis

and $$\frac{d^2y}{dx^2} = \frac{2(-3x^4+8x^2+17)}{[1+(4-x^2)^2]^2}$$

for $|x| < 2$, $\frac{d^2y}{dx^2} = +\text{ve}$ (concave up)

and for $|x| \geq 2$, $\frac{d^2y}{dx^2} = -\text{ve}$ (concave down)

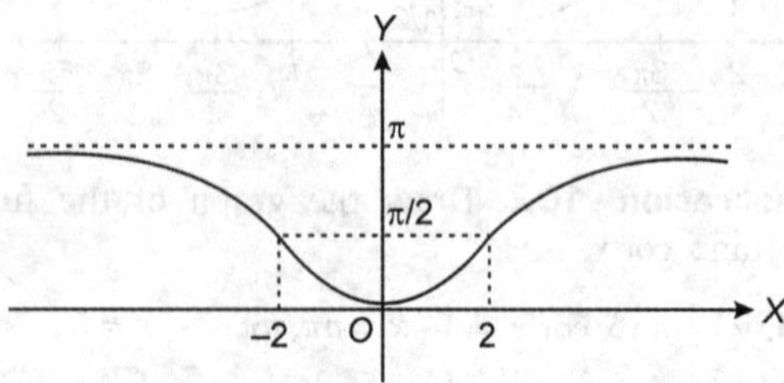

Illustration 104. Draw the graph of the function $y = \cos(\sin^{-1} x)$.

Sol. $x \in [-1, 1]$ $\quad \because -\frac{\pi}{2} \leq \sin^{-1} x \leq \frac{\pi}{2}$

$\therefore$ $$0 \leq \cos(\sin^{-1} x) \leq 1$$

$(\because \cos(\sin^{-1} x) = \cos(|\sin^{-1} x|))$

Hence $0 \leq y \leq 1 \Rightarrow y \in [0, 1]$

Now if $x = \sin\theta$ then $y = \cos\theta$

$\therefore$ $$x^2 + y^2 = 1$$

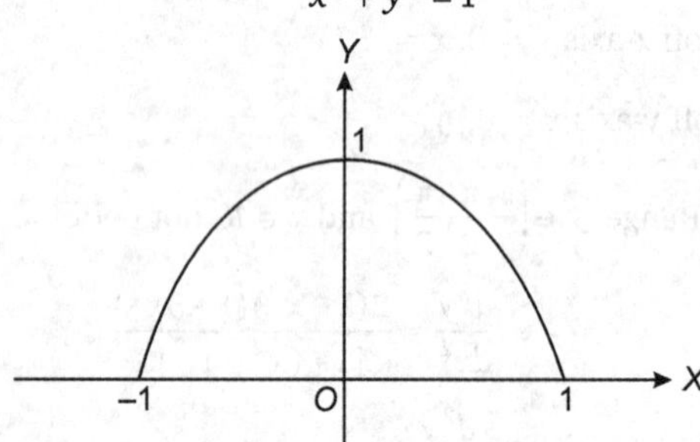

Illustration 105. Draw the graph of the function $y = \log_2 \sin x$.

Sol. $y = \log_2(\sin x)$

$\because \quad \sin x > 0 \Rightarrow 2n\pi < x < (2n+1)\pi, n \in I$

cut off x-axis $y = 0 \Rightarrow \sin x = 1 \Rightarrow x = 2n\pi + \frac{\pi}{2}, n \in I$

and at $x = ..., -3\pi, -\pi, 0, \pi, 3\pi, ..., y \to -\infty$

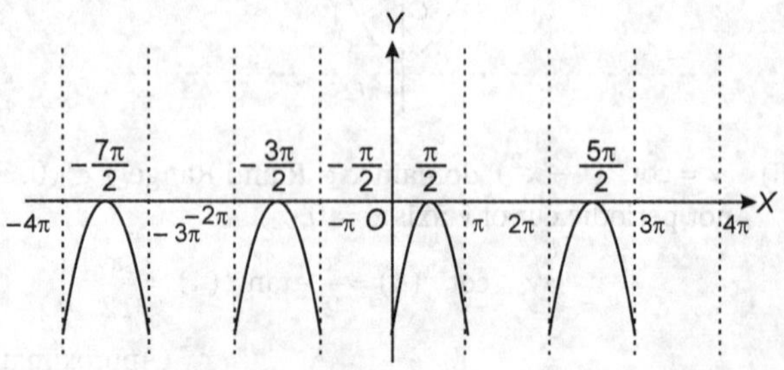

Illustration 106. Draw the graph of the function
$$y = \sin[\sin^{-1}(\log_{1/2} x)]$$

Sol. $x > 0$ and $-1 \leq \log_{1/2} x \leq 1$

$$\Rightarrow \left(\frac{1}{2}\right)^{-1} \geq x \geq \left(\frac{1}{2}\right)^{1}$$

$$\Rightarrow \frac{1}{2} \leq x < 2$$

and $y \in [-1, 1]$ and y is not periodic

cut of x-axis $y = 0$ then we get $x = 1$

and $y = \sin[\sin^{-1}(\log_{1/2} x)] = \log_{1/2} x = -\log_2 x$

$$\therefore \frac{dy}{dx} = -\frac{1}{x}\log_2 e \text{ and } \frac{d^2y}{dx^2} = \frac{1}{x^2}\log_2 e > 0$$

(concave up)

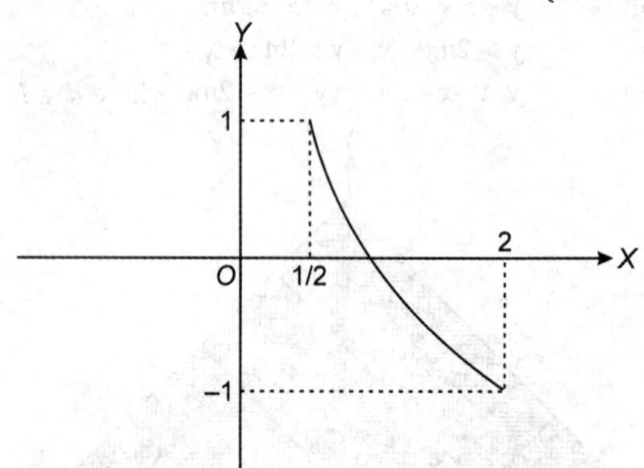

Illustration 107. Draw the graph of the function $y = \log_2\left|\left(\frac{x}{x-1}\right)\right|$.

Sol. Since $\left|\frac{x}{x-1}\right| > 0$

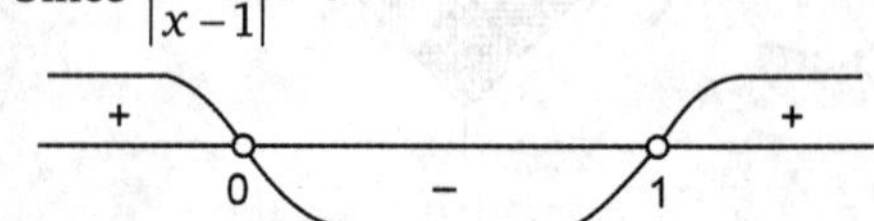

$$\therefore \frac{x}{x-1} > 0 \Rightarrow x \in (-\infty, 0) \cup (1, \infty)$$

and if $\frac{x}{x-1} < 0 \Rightarrow x \in (0,1)$

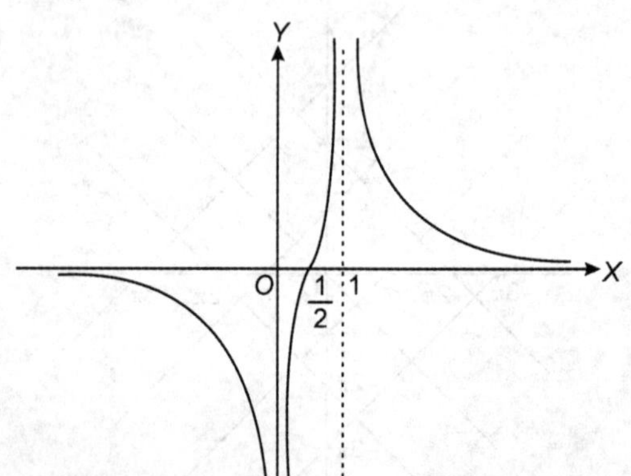

and $\dfrac{dy}{dx} = \log_2 e\left\{\dfrac{1}{x} - \dfrac{1}{x-1}\right\} = -\dfrac{1}{x(x-1)}\log_2 e$

$$\frac{d^2y}{dx^2} = \frac{(2x-1)}{(x^2-x)^2}\log_2 e$$

for $x > \frac{1}{x}$, $\dfrac{d^2y}{dx^2} > 0, 0 < x < \dfrac{1}{2}$, $\dfrac{d^2y}{dx^2} < 0$

and for $x < 0$, $\dfrac{d^2y}{dx^2} < 0$

at $x = 0, 1, y$ is not defined *i.e.*, $\to \infty$

Illustration 108. Draw the graph of the function $y = x^{1/(\log_{10} x)}$.

Sol. $y = x^{1/(\log_{10} x)} = x^{\log_x 10} = 10$ which is a straight line.

The domain of the given function $x \in (0, 1) \cup (1, \infty)$

Since at $x = 1, y = 1^\infty$ (which is not defined)

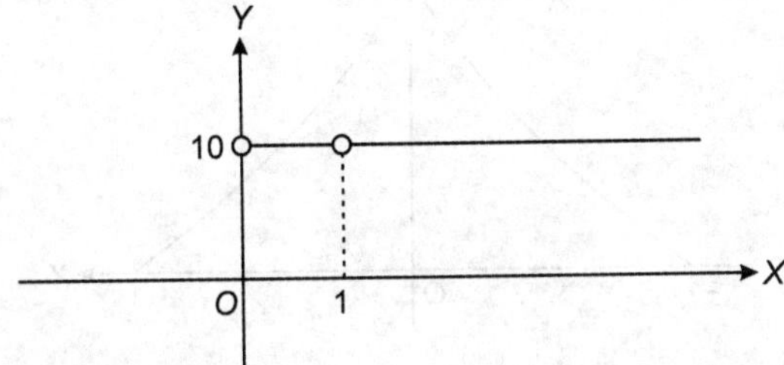

Illustration 109. Draw the graph of the function

$$y = \log_{1/3}\left(x - \frac{1}{3}\right) + \log_3 \sqrt{(9x^2 - 6x + 1)}.$$

Sol. $\because$ $9x^2 - 6x + 1 = (3x-1)^2 = 9\left(x - \dfrac{1}{3}\right)^2$

$\therefore$ $\sqrt{(9x^2 - 6x + 1)} = 3\left|x - \dfrac{1}{3}\right|$

$\Rightarrow$ $\log_3\sqrt{(9x^2 - 6x + 1)}$

$$= \log_3 3 + \log_3\left|x - \frac{1}{3}\right| = 1 + \log_3\left|x - \frac{1}{3}\right|$$

$\therefore$ $y = \log_{1/3}\left(x - \dfrac{1}{3}\right) + 1 + \log_3\left|x - \dfrac{1}{3}\right|$

$$= -\log_3\left(x - \frac{1}{3}\right) + 1 + \log_3\left|x - \frac{1}{3}\right|$$

Domain of y, $x > \dfrac{1}{3} \Rightarrow y = 1$

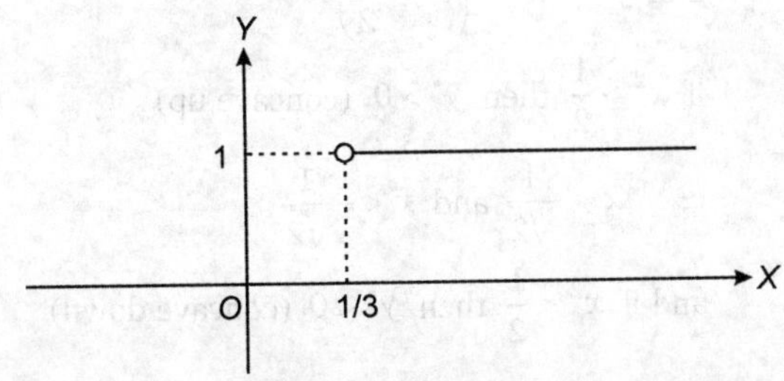

Illustration 110. Draw the graph of the function $y = x^3 + ax$ where $a \in I^-$.

Sol. $\because$ $y = x^3 + ax$

$\therefore$ $\dfrac{dy}{dx} = 3x^2 + a$ and $\dfrac{d^2y}{dx^2} = 6x$

for $x > 0$, $\dfrac{d^2y}{dx^2} = 0$ (Concave up)

and for $x < 0$, $\dfrac{d^2y}{dx^2} < 0$ (Concave down)

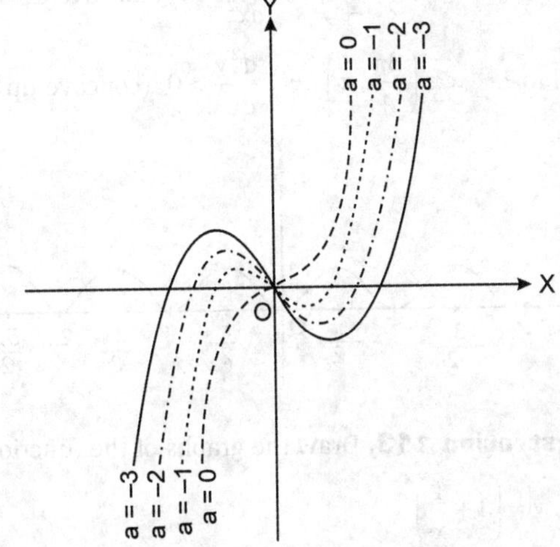

cut off x-axis $y = 0 \Rightarrow x^3 + ax = 0$ or $x = 0$, $x = \pm\sqrt{-a}$

and cut off y-axis $x = 0$

$\Rightarrow$ $y = 0$

Illustration 111. Draw the graph of the function

$$2|y| = ||x-1| - |x+1|| \forall x \le 0$$

Sol. for $-1 \le x < 0$

$$2|y| = -2x \Rightarrow |y| = -x$$

and for $x < -1$

$$2|y| = 2 \Rightarrow |y| = 1$$

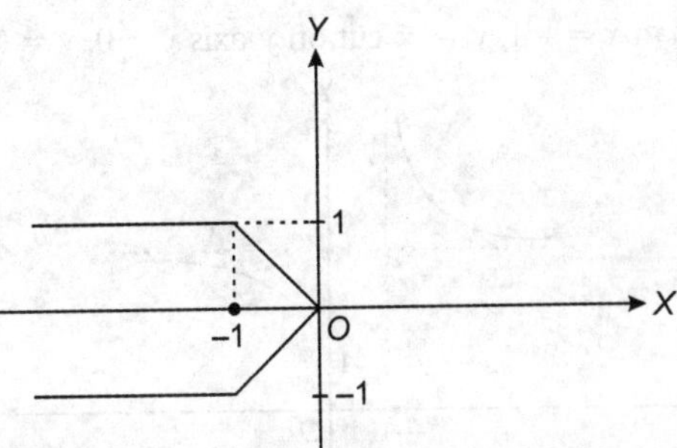

Illustration 112. Draw the graph of the function $y = \sin^2 x$.

Sol. Period of $\sin^2 x$ is π.

cut off x-axis $y = 0 \Rightarrow x = n\pi, n \in I$

cut off y-axis $x = 0 \Rightarrow y = 0$

Domain of y is $x \in R$

and Range of y is [0, 1] $\quad (\because 0 < \alpha < 1 \text{ then } \alpha^2 < \alpha)$

and $\quad \dfrac{dy}{dx} = \sin 2x \Rightarrow \dfrac{d^2y}{dx^2} = 2\cos 2x$

when $\quad x \in \left(0, \dfrac{\pi}{4}\right) \Rightarrow \dfrac{d^2y}{dx^2} > 0$ (concave up)

for $\quad x \in \left(\dfrac{\pi}{4}, \dfrac{\pi}{2}\right) \Rightarrow \dfrac{d^2y}{dx^2} < 0$ (concave down)

for $\quad x \in \left(\dfrac{\pi}{2}, \dfrac{3\pi}{4}\right) \Rightarrow \dfrac{d^2y}{dx^2} < 0$ (concave down)

and for $\quad x \in \left(\dfrac{3\pi}{4}, \pi\right) \Rightarrow \dfrac{d^2y}{dx^2} > 0$ (concave up)

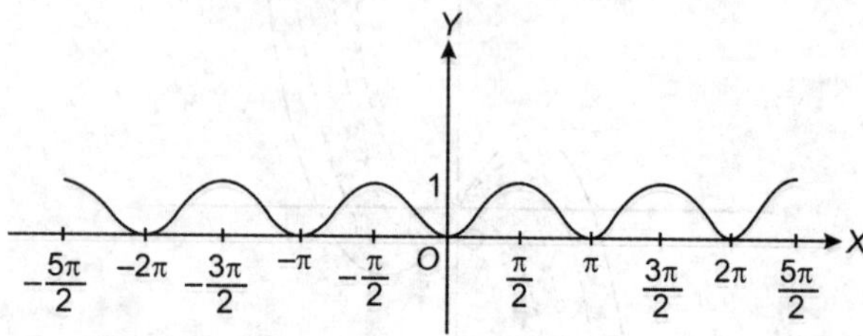

Illustration 113. Draw the graphs of the functions

(i) $y = \left(1 + \dfrac{1}{x}\right)^x$

(ii) $y = 2^{1/x}$

(iii) $y = 1 - \sqrt[3]{x^2}$

(iv) $y = (1 - x^{2/3})^{3/2}$

(v) $y = e^{-x^2}$ (Gaussian Curve)

(vi) $\sqrt{|x|} + \sqrt{|y|} = \sqrt{a}\ (a > 0)$

Sol. (i) $y = \left(1 + \dfrac{1}{x}\right)^x$

$$\lim_{x\to\infty}\left(1 + \frac{1}{x}\right)^x = e \text{ and } \lim_{x\to\infty}\left(1 + \frac{1}{x}\right)^x = e$$

and at $x = -1$, $y \to \infty$ cut off y-axis $x = 0$, $y = 1$

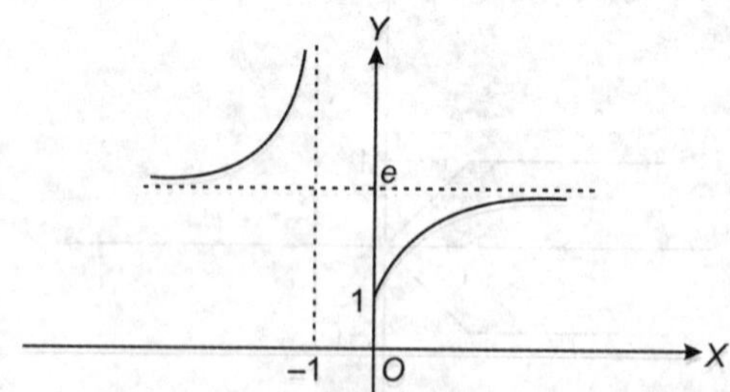

(ii) $y = 2^{1/x}$

$y = \lim\limits_{x\to 0+} 2^{1/2} = \lim\limits_{h\to 0} 2^{1/h} = 2^{\infty} = \infty$

and $\lim\limits_{x\to 0-} 2^{1/x} = \lim\limits_{h\to 0} 2^{-1/h} = 2^{-\infty} = 0$

and $y > 0$

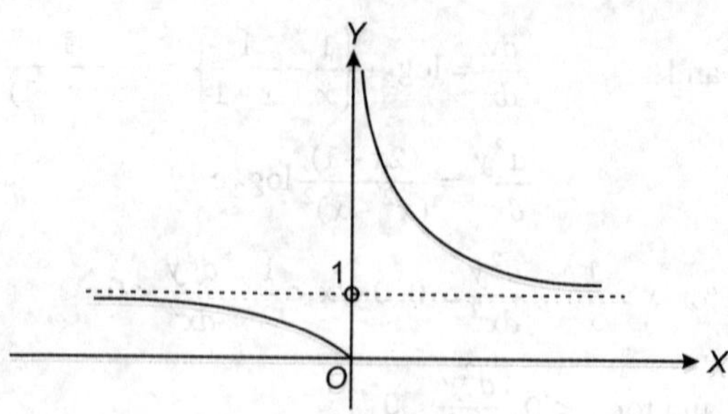

(iii) $y = 1 - \sqrt[3]{x^2}$

$= 1 - x^{2/3}$ which is symmetrical about y-axis cut off x-axis $y = 0$, $x = \pm 1$ and $\dfrac{d^2y}{dx^2} = \dfrac{2}{9}x^{-4/3} > 0$

and y-axis $x = 0$, $y = 1$

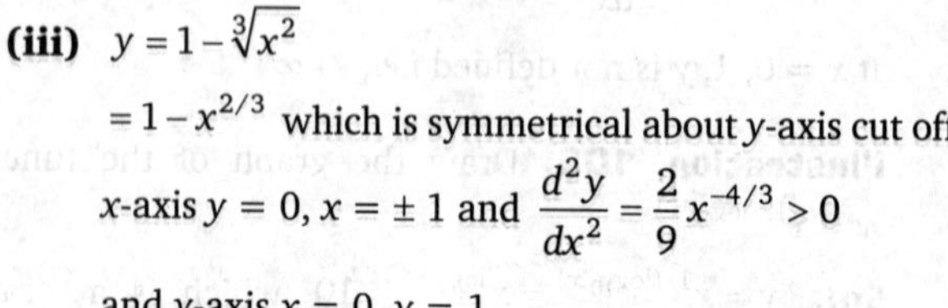

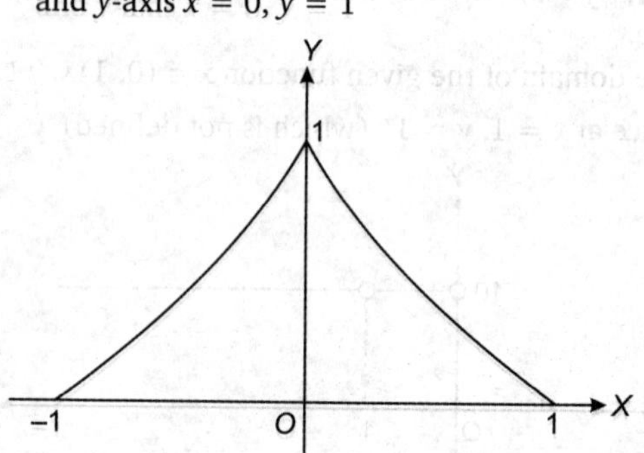

(iv) $y = (1 - x^{2/3})^{3/2}$ which is symmetrical about y-axis cut off x-axis $y = 0$, $x = \pm 1$ and y-axis $x = 0$, $y = 1$ y has no derivative at $x = 0$, since $y' =$ becomes infinite at $x = 0$, but the function has a maximum at this point.

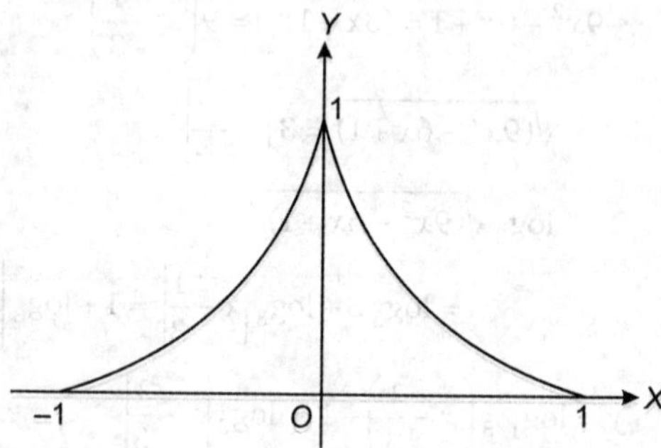

(v) $y = e^{-x^2}$

$y > 0$,

cut off y-axis $x = 0$, $y = 1$ symmetrical about y-axis

and $y'' = 4e^{-x^2}\left(x^2 - \dfrac{1}{2}\right)$

if $x^2 > \dfrac{1}{2}$ then $y'' > 0$ (concave up)

$\Rightarrow \quad x \not> \dfrac{1}{\sqrt{2}}$ and $x < -\dfrac{1}{\sqrt{2}}$

and if $x^2 < \dfrac{1}{2}$ then $y'' < 0$ (concave down)

$\Rightarrow \quad -\frac{1}{\sqrt{2}} < x < \frac{1}{\sqrt{2}}$

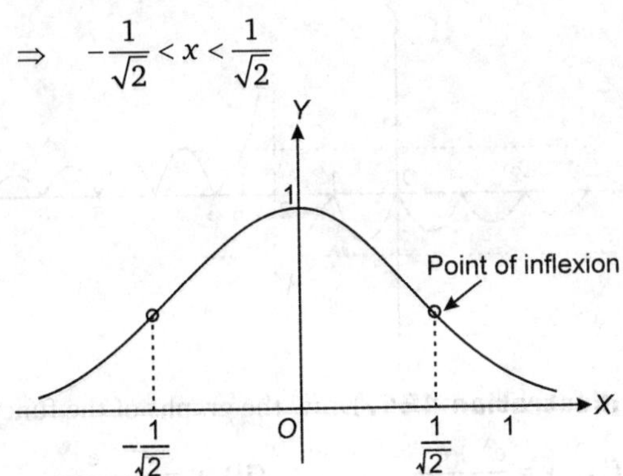

(vi) $\sqrt{|x|} + \sqrt{|y|} = \sqrt{a}\ (a > 0)$

Curve is symmetrical about x-axis and y-axis

cut off x-axis put $y = 0$, $\therefore x = \pm a$

and cut off y-axis put $x = 0$, $\therefore y = \pm a$

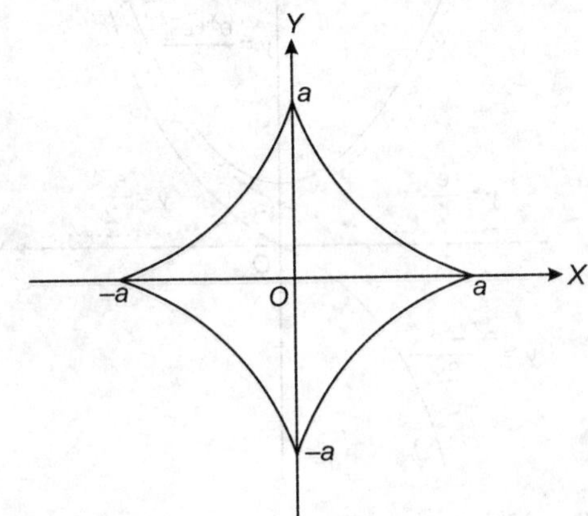

Illustration 114. Draw the graph of the function $y = \sqrt{(\sin x)}$.

Sol. y is defined if $\sin x \geq 0$

$\therefore \quad x \in [2n\pi, (2n+1)\pi], n \in I$

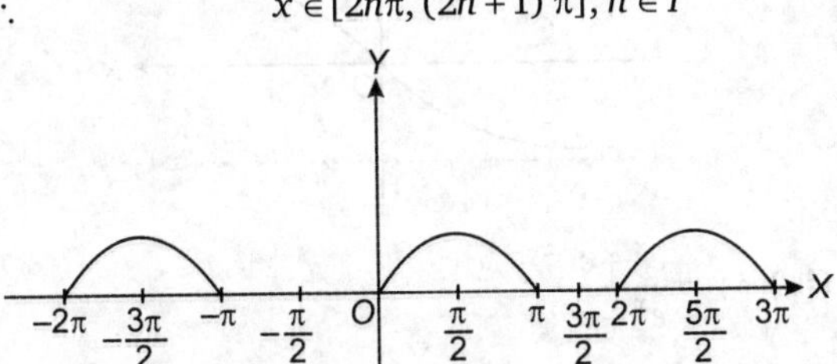

Illustration 115. Draw the graph of solution set of the equation

$$|x+y| = |y| - x.$$

Sol. Case I : If $x + y \geq 0$ and $y \geq 0$

then $x + y = y - x$ impossibe

Case II : If $x + y < 0$ and $y \geq 0$

then $-x - y = y - x$ impossible

Case III : If $x + y < 0$ and $y < 0$

then $-x - y = -y - x$ which is true

Case IV : If $x + y \geq 0$ and $y < 0$

then $x + y = -y - x$ impossible

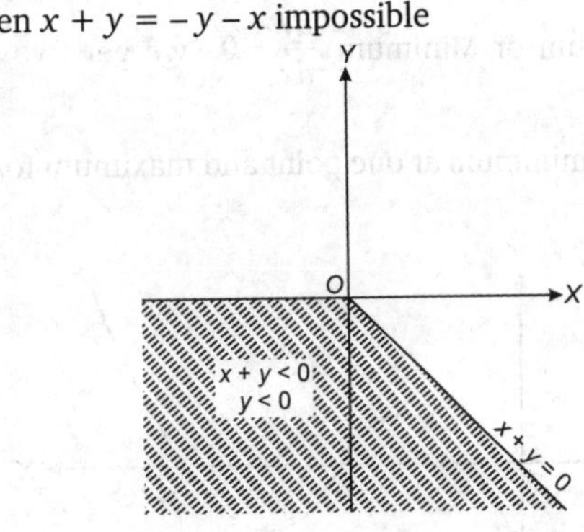

Illustration 116. Draw the graph of the function $y = |x| + \{-x\} + [x]\ \forall\ x \in [0, 3]$, where [.] and {.} denotes the greatest integer function and fractional part functions respectively.

Sol. for $0 \leq x < 1$, $|x| = x$, $[x] = 0$ and $\{-x\} = 1 - x$

for $1 \leq x < 2$, $|x| = x$, $[x] = 1$ and $\{-x\} = 2 - x$

and for $2 \leq x < 3$, $|x| = x$, $[x] = 2$ and $\{-x\} = 3 - x$

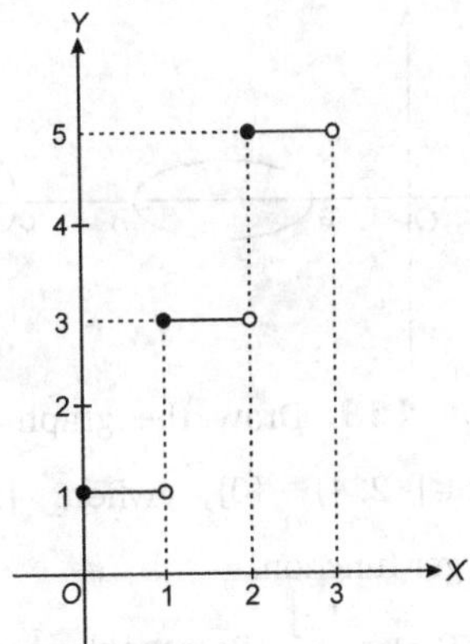

Illustration 117. Draw the graph of the function $y = \sin(\pi\{|x| + [x]\})$, where [.] denotes the greatest integer function.

Sol. for $x \geq 0$, $\pi\{|x| + [x]\} = \pi(x + [x])$ and for $x < 0$, $\pi\{|x| + [x]\} = \pi(-x + [x])$

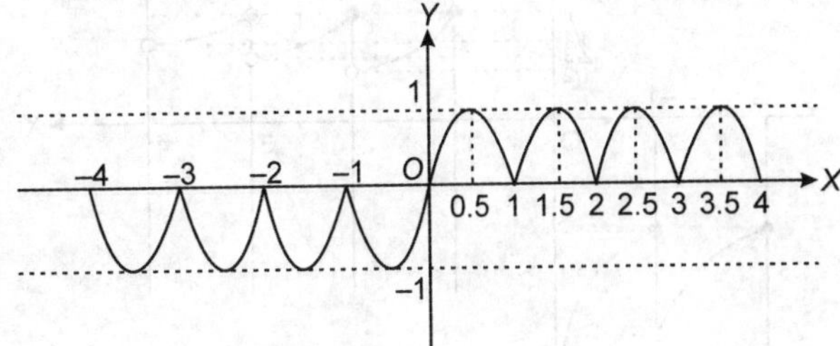

Illustration 118. Draw the graphs of the functions

(i) $y = (x-a)(x-b)(x-c);\ a, b, c > 0$

(ii) $y^2 = (x-a)(x-b)(x-c);\ 0 < a < b < c.$

Sol. (i) $y = (x-a)(x-b)(x-c)$

cut off x-axis, $y = 0$ then $x = a, b, c$

and cut off y-axis, $x = 0$ then $y = -abc$

for Maximum or Minimum $\frac{dy}{dx} = 0$, we get two values of x.

Hence y is minimum at one point and maximum for other point.

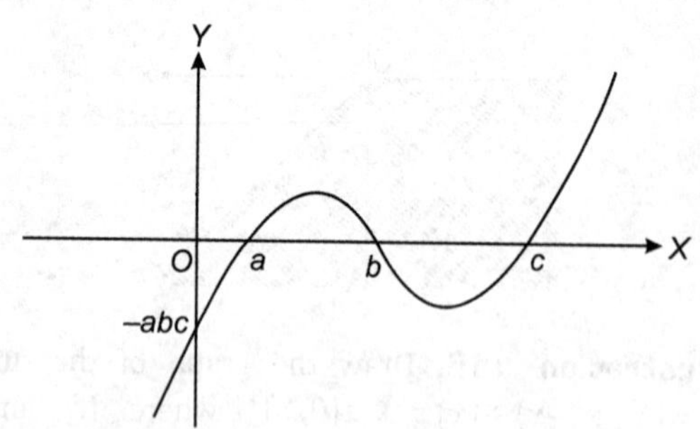

(ii) $y^2 = (x-a)(x-b)(x-c)$ where $0 < a < b < c$

Curve is symmetrical about x-axis,

cut off x-axis, $y = 0$ then $x = a, b, c$

cut off y-axis, $x = 0$ then $y^2 = -abc$ (Impossible)

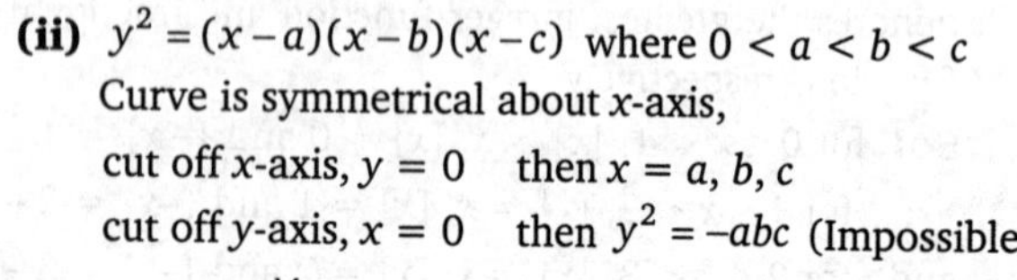

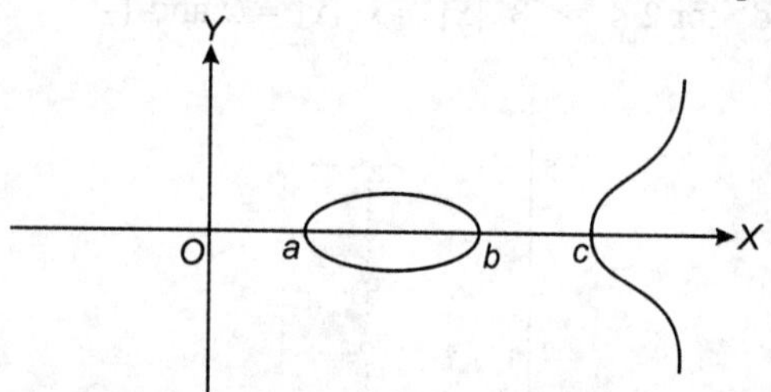

Illustration 119. Draw the graph of the function $y = \frac{[x]}{|x|}\ \forall\ x \in [-2, 4) \sim \{0\}$, where [.] denotes the greatest integer function.

Sol. for $x > 0$, $y = \frac{[x]}{x}$ (Rectangular hyperbola)

and for $x < 0$, $y = -\frac{[x]}{x}$ (Rectangular hyperbola)

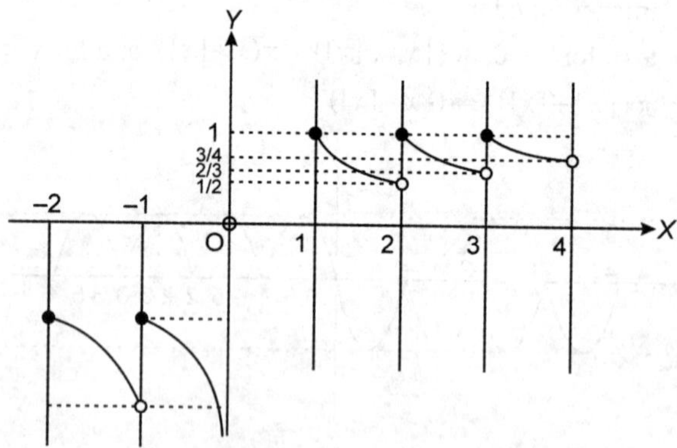

Illustration 120. Draw the graph of the function $y = \frac{\{x\}}{x-2}$, where {.} denotes the fractional part function.

Sol. for $x = 2$, y is not defined

for $x > 2$, $y > 0$ $\{\because 0 \le (x) < 1\}$

and for $x < 2$, $y < 0$

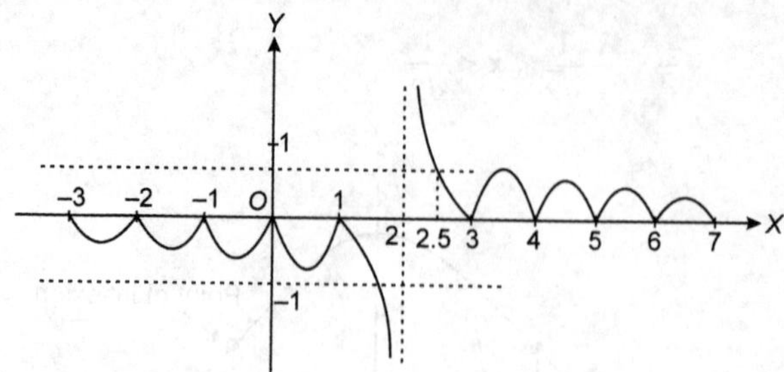

Illustration 121. Draw the graphs of the functions

(i) $y = \frac{e^x + e^{-x}}{2}$ **(ii)** $y = \frac{e^x - e^{-x}}{2}$

(iii) $y = \frac{e^x - e^{-x}}{e^x + e^{-x}}$ **(iv)** $y = \frac{e^x + e^{-x}}{e^x - e^{-x}}$

Sol. (i) and **(ii)**

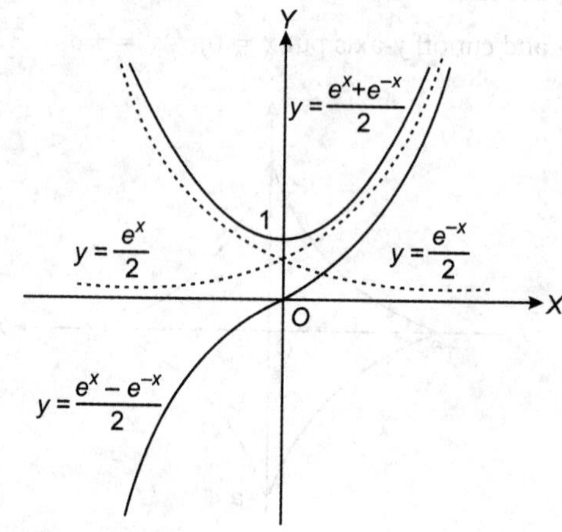

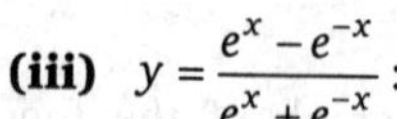

(iii) $y = \frac{e^x - e^{-x}}{e^x + e^{-x}}$:

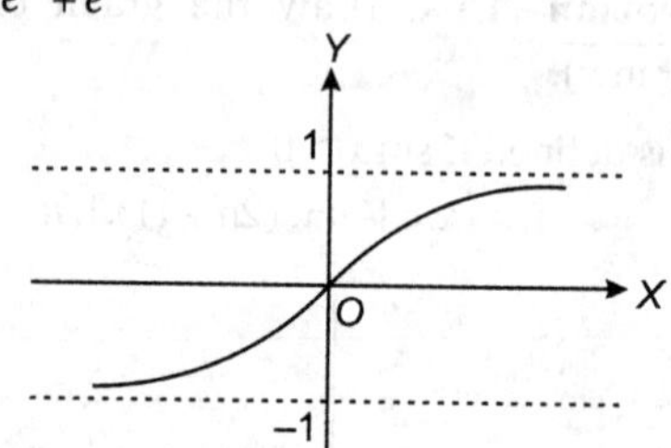

(iv) $y = \frac{e^x + e^{-x}}{e^x - e^{-x}}$:

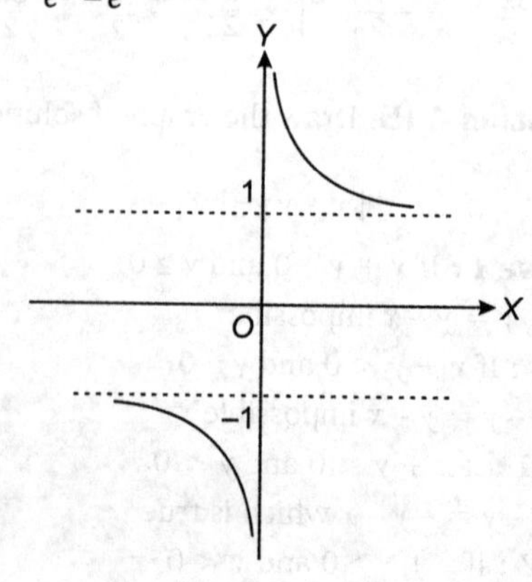

Some Important Graphs to Remember

(a) Cubical parabola

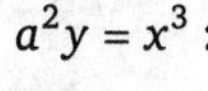

$a^2y = x^3$:

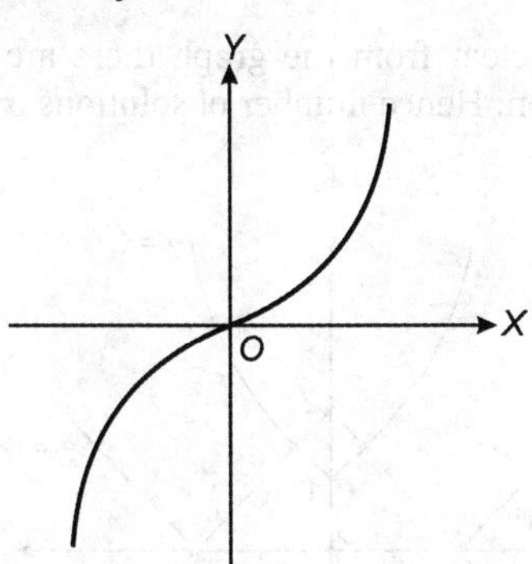

(b) Semi-cubical parabola

$ay^2 = x^3$:

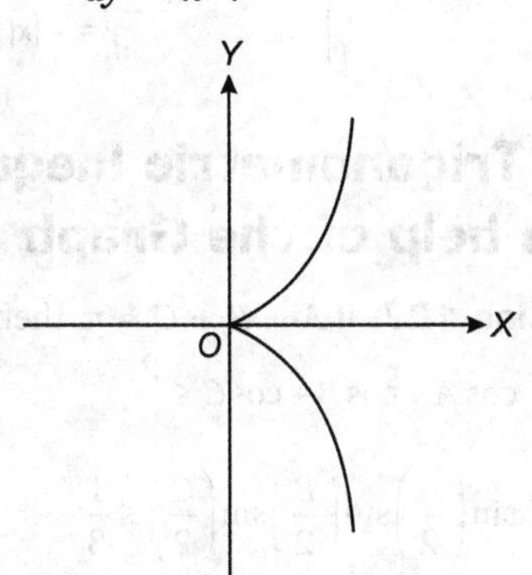

(c) Hypocycloid $x^{2/3} + y^{2/3} = a^{2/3}$

or $x = a\cos^3\theta,\ y = a\sin^3\theta$:

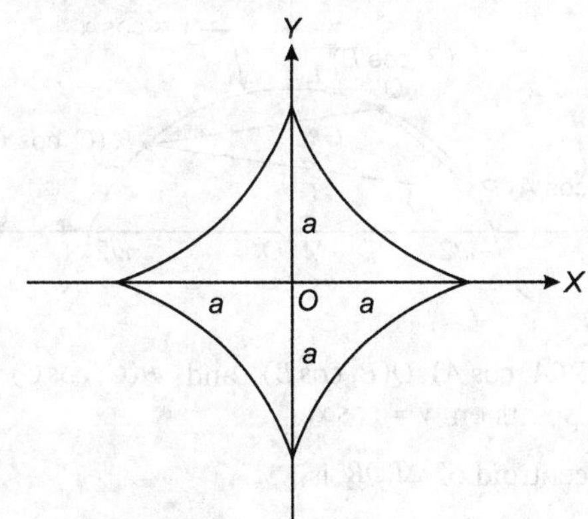

(d) $y = ax^3 + bx^2 + cx + d,\ a > 0$:

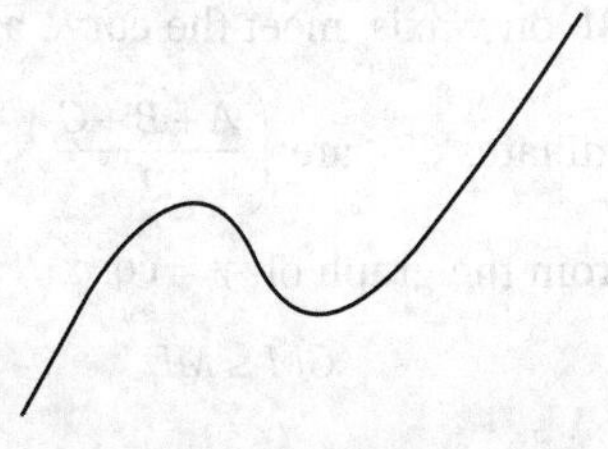

(e) $y = ax^3 + bx^2 + cx + d,\ a < 0$:

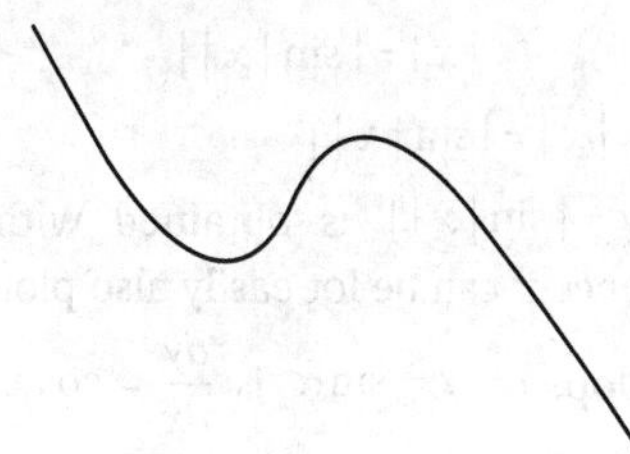

(f) $y = ax^4 + bx^3 + cx^2 + dx + e,\ a > 0$:

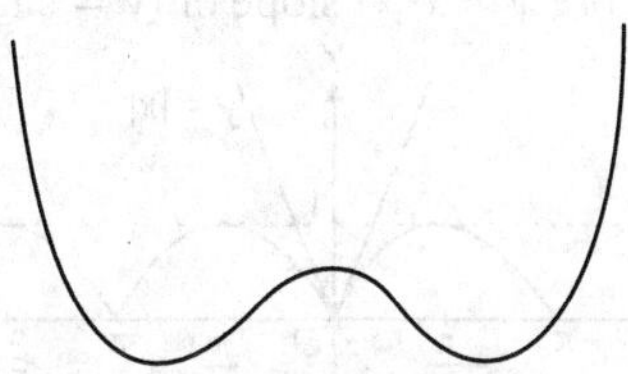

(g) $y = ax^4 + bx^3 + cx^2 + dx + e,\ a < 0$:

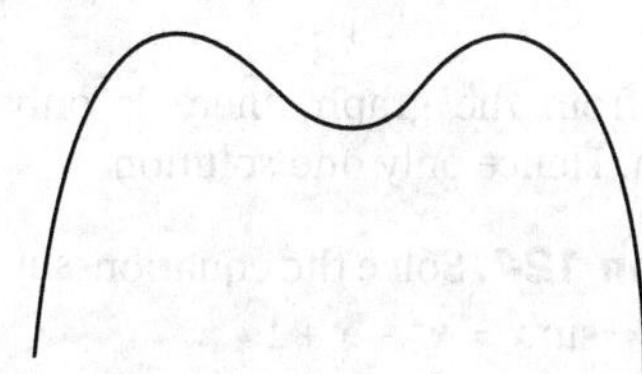

Solution of Equation by Graphs

Suppose equation is $f(x) - g(x) = 0$ or $f(x) = g(x) = y$ (say) then draw the graphs of $y = f(x)$ and $y = g(x)$. If graphs of $y = f(x)$ and $y = g(x)$ cuts at one, two, three, ... , no points then number of solutions are one, two, three, ..., zero respectively.

Also find $f'(x)$ and $g'(x)$

If $f'(x) > g'(x) \Rightarrow y = f(x)$ is above $y = g(x)$

and If $f'(x) < g'(x) \Rightarrow y = f(x)$ is below $y = g(x)$

Illustration 122. Find the number of solutions of the equations

$$y = |x| - [|x|] \text{ and } y = \sqrt{(1 - x^2)}$$

where [.] denotes the greatest integer function.

Sol. Graph of $y = |x| - [|x|]$ is obtained with the help of $y = x - [x]$. Hence it can be plot easily. Also plotting upper part of circle $x^2 + y^2 = 1$

i.e., $y = \sqrt{1 - x^2}$

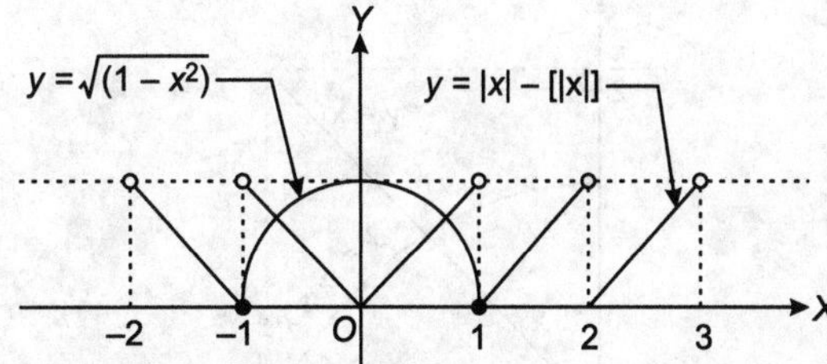

It is clear from the graph there are four points of intersection. Hence number of solutions is four.

Illustration 123. Find the number of solutions of the equation

$$|x| = |\sin|x||$$

Sol. Let $y = |x| = |\sin|x||$

Graph of $y = |\sin|x||$ is obtained with the help of $y = \sin x$. Hence it can be lot easily also plotting $y = |x|$.

$\because$ Slope of $y = \sin x$ is $\dfrac{dy}{dx} = \cos x < 1$

and slope of $y = x$ is $\dfrac{dy}{dx} = 1$

$\therefore$ slope of ($y = x$) > slope of ($y = \sin x$)

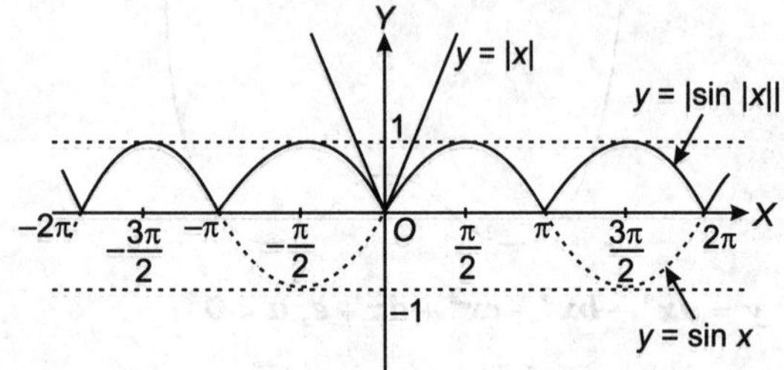

It is clear from the graph, there is only one point of intersection. Hence only one solution.

Illustration 124. Solve the equation $\sin x = x^2 + x + 1$.

Sol. Let $y = \sin x = x^2 + x + 1$

first draw the graphs of $y = \sin x$ and $y = x^2 + x + 1$.

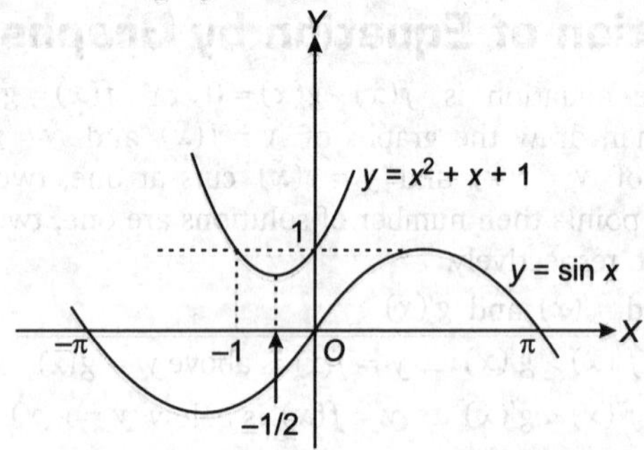

It is clear from the graphs, the curves do not intersect and so the equation has no solutions.

Illustration 125. Show that there is a number x between 0 and $\dfrac{\pi}{2}$ such that $\cos x = x$.

Sol. Let $y = \cos x = x$ first draw the graphs of $y = \cos x$ and $y = x$ when x lies between 0 and $\dfrac{\pi}{2}$.

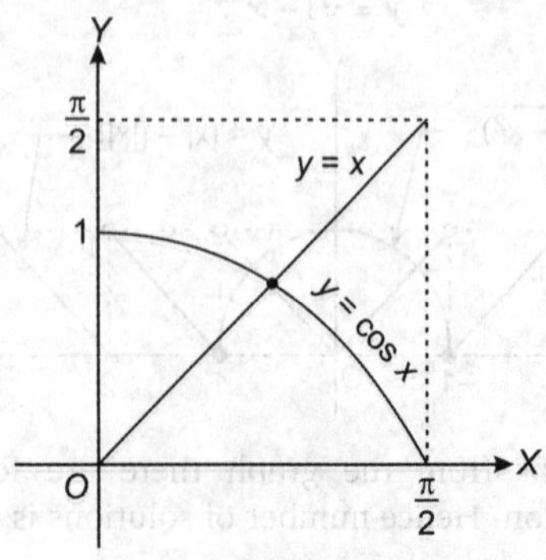

It is clear from the graph there is only one point of intersection. Hence only one solution.

Illustration 126. Find the number of solutions of the equation $3^{|x|} = |2 - |x||$

Sol. It is clear from the graph there are two points of intersection. Hence number of solutions is two.

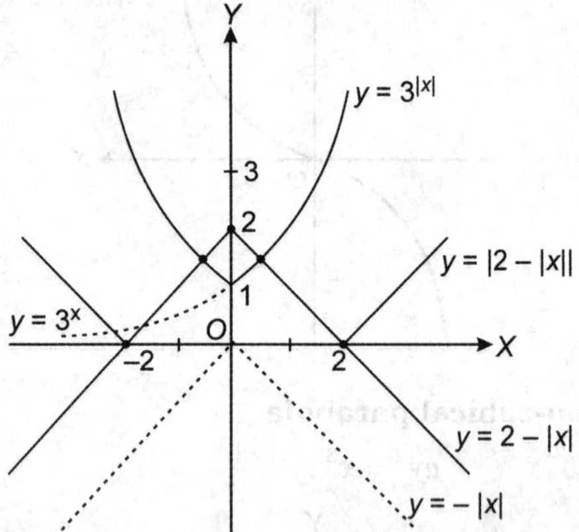

Proving Trigonometric Inequalities with the help of the Graph

Illustration 127. If $A + B + C = \pi$, then prove that

$$\cos A + \cos B + \cos C \le \frac{3}{2}$$

or $$\sin\left(\frac{A}{2}\right)\sin\left(\frac{B}{2}\right)\sin\left(\frac{C}{2}\right) \le \frac{1}{8}$$

Sol. First draw the graph of $y = \cos x$

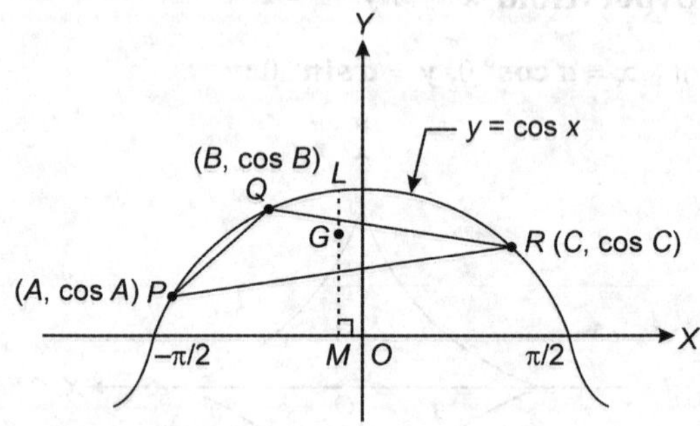

Let $P(A, \cos A)$, $Q(B, \cos B)$ and $R(C, \cos C)$ be any three points on $y = \cos x$.

then centroid of ΔPQR is

$$G\left(\frac{A+B+C}{3}, \frac{\cos A + \cos B + \cos C}{3}\right)$$

Draw $\perp GM$ on x-axis, meet the curve at L

$\therefore$ co-ordinates of L are $\left(\dfrac{A+B+C}{3}, \cos\left(\dfrac{A+B+C}{3}\right)\right)$

It is clear from the graph of $y = \cos x$

$$GM \le ML$$

$$\Rightarrow \quad \frac{\cos A + \cos B + \cos C}{3} \le \cos\left(\frac{A+B+C}{3}\right)$$

$$= \cos\left(\frac{\pi}{3}\right) = \frac{1}{2}$$

$$\therefore \quad \cos A + \cos B + \cos C \le \frac{3}{2}$$

from trigonometry,

$$1 + 4\sin\left(\frac{A}{2}\right)\sin\left(\frac{B}{2}\right)\sin\left(\frac{C}{2}\right) \le \frac{3}{2}$$

Hence, $\sin\left(\frac{A}{2}\right)\sin\left(\frac{B}{2}\right)\sin\left(\frac{C}{2}\right) \le \frac{1}{8}$

Illustration 128. If $A + B + C = \pi$, then prove that

$$\tan^2\left(\frac{A}{2}\right) + \tan^2\left(\frac{B}{2}\right) + \tan^2\left(\frac{C}{2}\right) \ge 1$$

Sol. First draw the graph of $y = \tan^2\left(\frac{x}{2}\right)$ its period is 2π

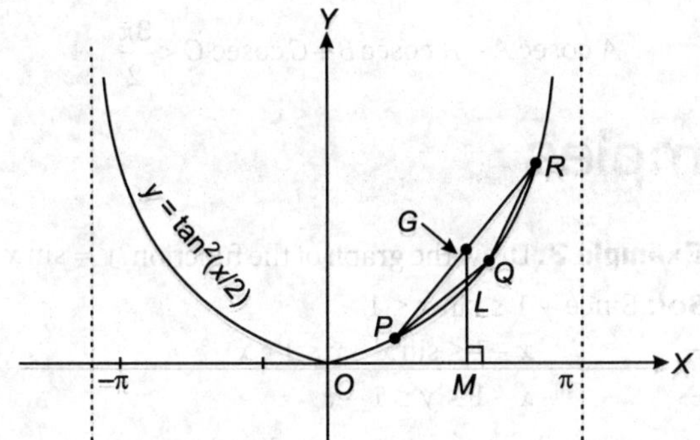

Let $P\left[A, \tan^2\left(\frac{A}{2}\right)\right]$, $Q\left[B, \tan^2\left(\frac{B}{2}\right)\right]$ and $R\left(C, \tan^2\frac{C}{2}\right)$

be any three points on $y = \tan^2\left(\frac{x}{2}\right)$

then centroid of ΔPQR is

$$G\left(\frac{A+B+C}{3}, \frac{\tan^2\left(\frac{A}{2}\right) + \tan^2\left(\frac{B}{2}\right) + \tan^2\left(\frac{C}{2}\right)}{3}\right)$$

Draw $\perp GM$ on x-axis, meet the curve at L

$\therefore$ co-ordinates of L are $\left(\frac{A+B+C}{3}, \tan^2\left(\frac{A+B+C}{6}\right)\right)$

It is clear from the graph of $y = \tan^2\left(\frac{x}{2}\right)$

$$GM \ge ML$$

$$\Rightarrow \quad \frac{\tan^2\left(\frac{A}{2}\right) + \tan^2\left(\frac{B}{2}\right) + \tan^2\left(\frac{C}{2}\right)}{3}$$

$$\ge \tan^2\left(\frac{A+B+C}{6}\right) = \tan^2\left(\frac{\pi}{6}\right) = \left(\frac{1}{\sqrt{3}}\right)^2 = \frac{1}{3}$$

Hence $\tan^2\left(\frac{A}{2}\right) + \tan^2\left(\frac{B}{2}\right) + \tan^2\left(\frac{C}{2}\right) \ge 1$.

Illustration 129. If $f(x)$ is a monotonically increasing function $\forall\ x \in R$, $f''(x) > 0$ and $f^{-1}(x)$ exists, then prove that

$$\frac{f^{-1}(x_1) + f^{-1}(x_2) + f^{-1}(x_3)}{3} < f^{-1}\left(\frac{x_1 + x_2 + x_3}{3}\right)$$

Sol. $\because$ $f''(x) > 0 \Rightarrow \therefore f(x)$ is concave up also given $f(x)$ is a monotonically increasing function then $f^{-1}(x)$ is symmetrical about the line $y = x$.

Let $P(x_1, f^{-1}(x_1))$, $Q(x_2, f^{-1}(x_2))$ and $R(x_3, f^{-1}(x_3))$ be any three points on the curve $y = f^{-1}(x)$ then co-ordinate of centroid of ΔPQR are

$$G\left(\frac{x_1 + x_2 + x_3}{3}, \frac{f^{-1}(x_1) + f^{-1}(x_2) + f^{-1}(x_3)}{3}\right)$$

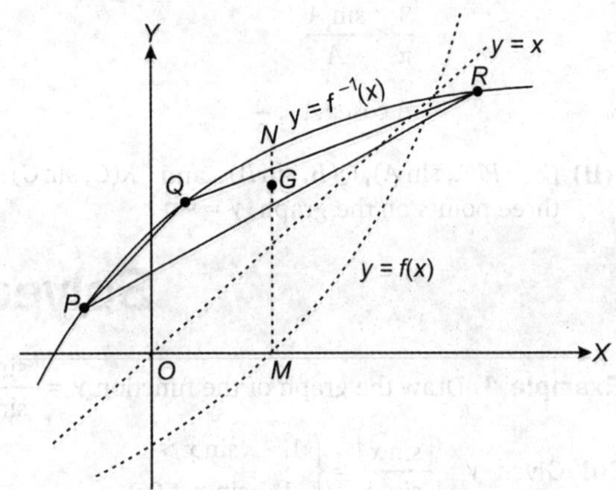

Now draw a line NM which is perpendicular to to x-axis and passes through G.

The co-ordinate of N are

$$\left(\frac{x_1 + x_2 + x_3}{3}, f^{-1}\left(\frac{x_1 + x_2 + x_3}{3}\right)\right)$$

($\because$ N lies on $y = f^{-1}(x)$)

from the graph $MG < MN$

i.e., ordinate of G < ordinate of N

$$\Rightarrow \quad \frac{f^{-1}(x_1) + f^{-1}(x_2) + f^{-1}(x_3)}{3}$$

$$< f^{-1}\left(\frac{x_1 + x_2 + x_3}{3}\right)$$

Illustration 130. (i) If $0 < A < \frac{\pi}{6}$, then show that $A \operatorname{cosec} A < \frac{\pi}{3}$

(ii) If $0 < A, B, C < \frac{\pi}{2}$, then show that

$$A \operatorname{cosec} A + B \operatorname{cosec} B + C \operatorname{cosec} C < \frac{3\pi}{2}$$

Sol. (i) Let $P(A, \sin A)$ and $Q\left(\frac{\pi}{6}, \sin\frac{\pi}{6}\right)$ be any two points on the graph $y = \sin x$.

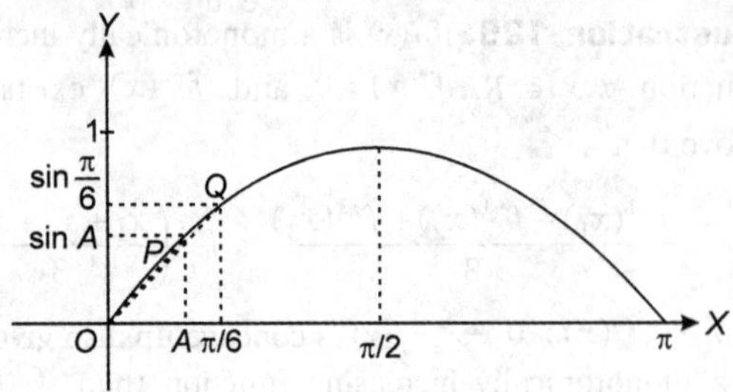

It is clear from the figure

slope of OQ < slope of OP

$$\Rightarrow \quad \frac{\sin\frac{\pi}{6}}{\frac{\pi}{6}} < \frac{\sin A}{A} \qquad \left(\because \text{slope} = \frac{\text{per}}{\text{base}}\right)$$

$$\Rightarrow \quad \frac{3}{\pi} < \frac{\sin A}{A}$$

or $$A \operatorname{cosec} A < \frac{\pi}{3}$$

(ii) Let $P(A, \sin A), Q(B, \sin B)$ and $R(C, \sin C)$ are three points on the graph $y = \sin x$.

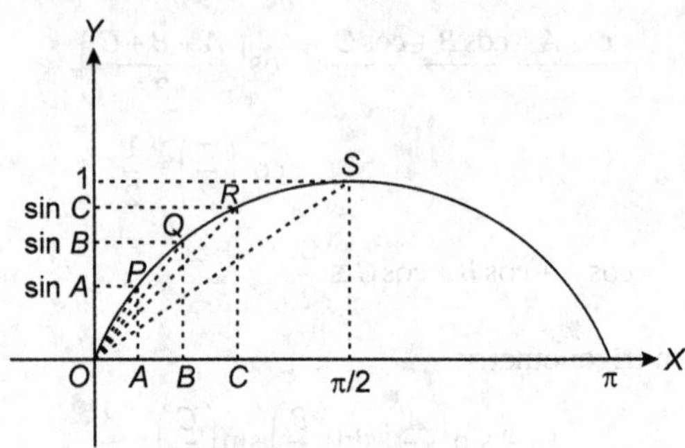

It is clear from the graph

slope of OP > slope of OS

$\Rightarrow$ slope of OQ > slope of OS and slope of OR >slope of OS

i.e., $\frac{\sin A}{A} > \frac{1}{\pi/2}$, $\frac{\sin B}{B} > \frac{1}{\pi/2}$ and $\frac{\sin C}{c} > \frac{1}{\pi/2}$

or $A \operatorname{cosec} A < \frac{\pi}{2}$, $B \operatorname{cosec} B < \frac{\pi}{2}$ and $C \operatorname{cosec} C < \frac{\pi}{2}$.

Adding all we get

$$A \operatorname{cosec} A + B \operatorname{cosec} B + C \operatorname{cosec} C < \frac{3\pi}{2}$$

Solved Examples

Example 1. Draw the graph of the function $y = \frac{|\sin x|}{\sin x}$

Sol. Given $y = \frac{|\sin x|}{\sin x} = \begin{cases} 1, & \sin x > \\ -1, & \sin x < 0 \end{cases}$

or $$= \begin{cases} 1, & x \in (2n\pi, (2n+1)\pi) \\ -1, & x \in ((2n+1)\pi, (2n+2)\pi, n \in I) \end{cases}$$

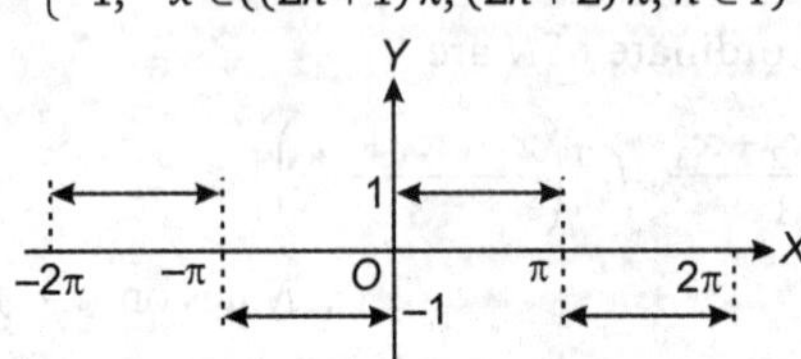

Example 2. Draw the graph of the function $y = x + \sin x$.

Sol. Since $-1 \le \sin x \le 1$

$\therefore \quad x - 1 \le x + \sin x \le x + 1 \Rightarrow x - 1 \le y \le x + 1$

Hence graph of $y = x + \sin x$ lies between $y = x - 1$ and $y = x + 1$.

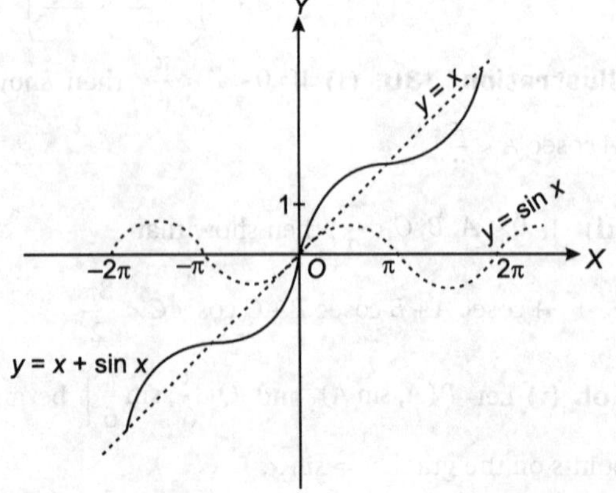

Example 3. Draw the graph of the function $y = \sin x - x$.

Sol. Since $-1 \le \sin x \le 1$

$\Rightarrow \quad -x - 1 \le \sin x - x \le 1 - x$

$\Rightarrow \quad -x - 1 \le y \le 1 - x$

Hence graph of $y = \sin x - x$ lies between $y = -x - 1$ and $y = 1 - x$.

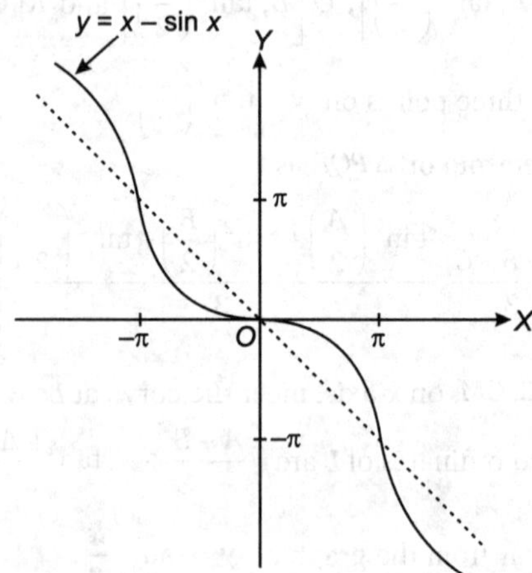

Example 4. Draw the graph of $y = [\sin x] + \sin x$, where [.] denotes the greatest integer function.

Sol. $\because -1 \le \sin x \le 1$

$\therefore \quad [\sin x] = -1, 0, 1$

and $[\sin x] + \sin x = 2$ only when $x = 2n\pi + \frac{\pi}{2}, n \in I$

and $$-2 \le [\sin x] + \sin x \le 2$$

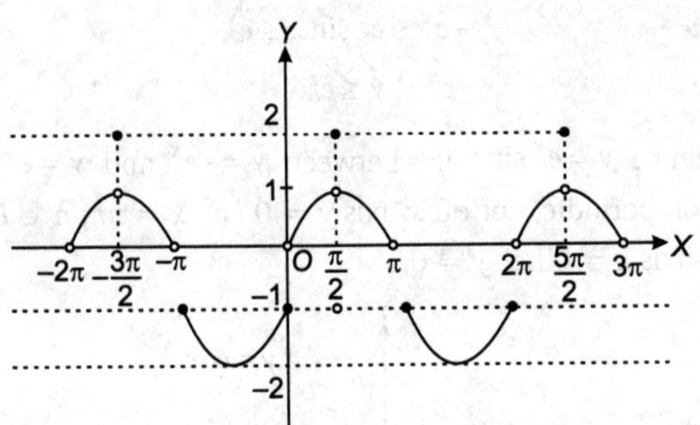

Example 5. Draw the graph of $y=[\sin x]+|\sin x|$, where [.] denotes the greatest integer function.

Sol. $\because -1 \le \sin x \le 1$

$\therefore \quad 0 \le |\sin x| \le 1$

and $[\sin x] = -1, 0, 1$

then $-1 \le [\sin x]+|\sin x| \le 2$

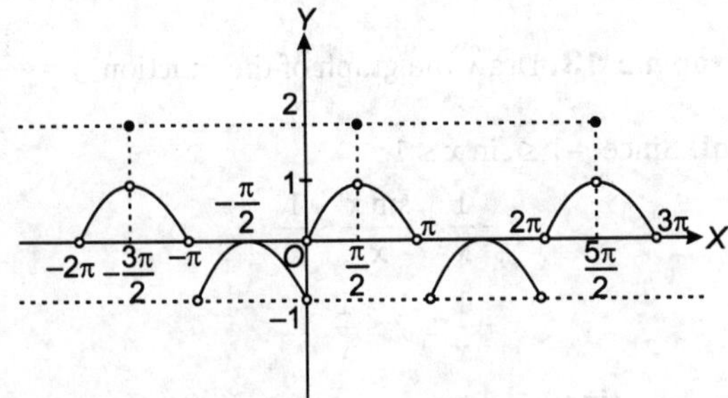

Example 6. Draw the graph of the function $y=4\sin^4 x+4\cos^4 x$.

Sol. Since $y=4\sin^4 x+4\cos^4 x$

$$= 4[\sin^4 x+\cos^4 x]$$
$$= 4\{(\sin^2 x+\cos^2 x)^2-2\sin^2 x\cos^2 x)\}$$
$$= 4\{1-2\sin^2 x\cos^2 x\}$$
$$= 4-2\sin^2 2x$$
$$= 4-(1-\cos 4x)$$
$$= 3+\cos 4x$$

$\Rightarrow \quad -1 \le \cos 4x \le 1$

$\Rightarrow \quad 2 \le 3+\cos 4x \le 4$

or $\quad 2 \le y \le 4$ and period of y is $\pi/2$.

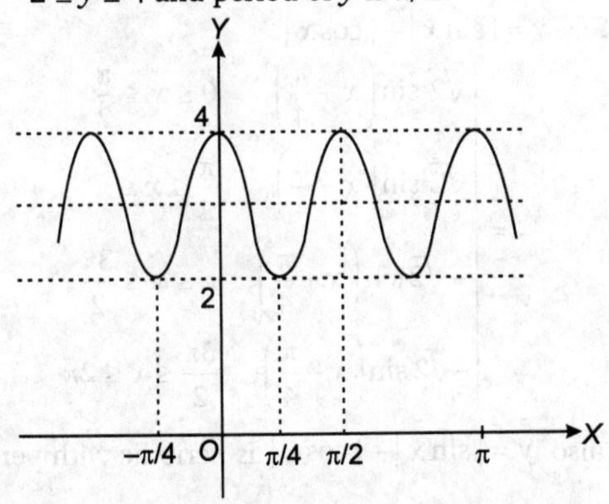

Example 7. Draw the graph of the function $y=\sin(2^x)$.

Sol. Since $-1 \le \sin(2^x) \le 1$

$\Rightarrow \quad -1 \le y \le 1$ and non periodic.

and cut off x-axis put $y = 0$

$\therefore \quad \sin(2^x)=0$

$\Rightarrow \quad 2^x = n\pi, n \in N$

$\therefore \quad x = \log_2(n\pi), n \in N$

i.e., $\quad x = \log_2 \pi, \log_2 2\pi, \log_2 3\pi, \ldots$

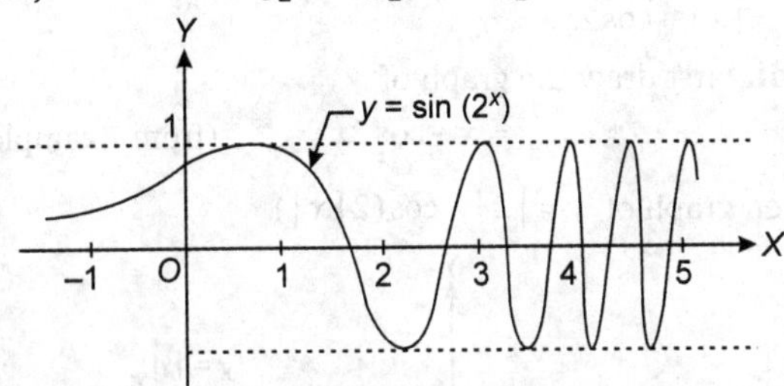

Example 8. Draw the graph of the function $y=\sin^3 x$.

Sol. Since $-1 \le \sin x \le 1$

$\Rightarrow \quad -1 \le y \le 1$ and periodic with period 2π

and cut off x-axis, $\sin^3 x = 0$ *i.e.*, $x = n\pi, n \in I$

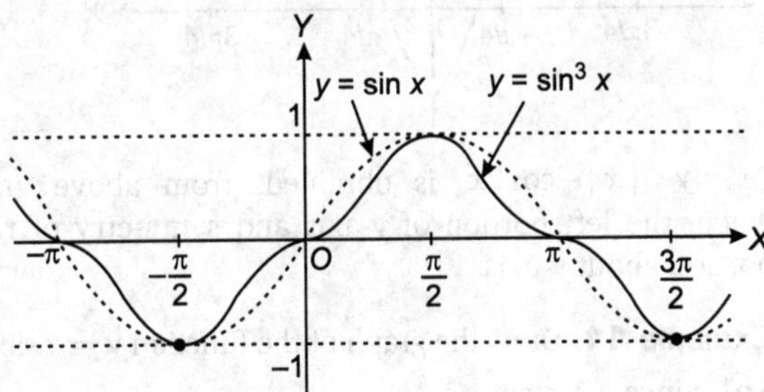

Example 9. Draw the graph of the function $y = x-\cos 2x$.

Sol. Since $-1 \le \cos 2x \le 1 \Rightarrow -1 \le -\cos 2x \le 1$

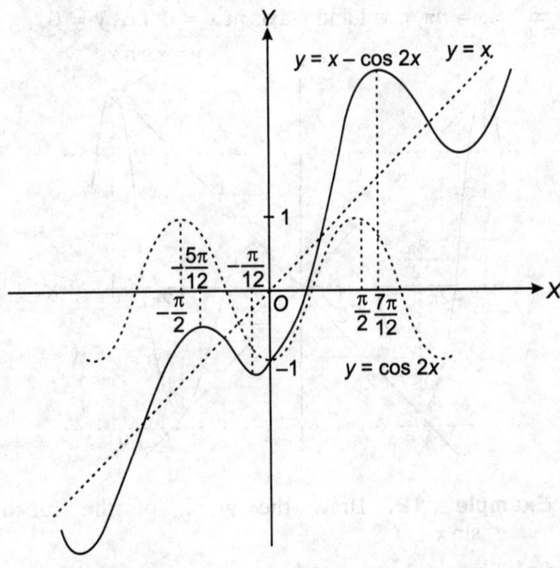

or $\quad x-1 \le x - \cos 2x \le x+1$

or $\quad x-1 \le y \le x+1$

Hence $y = x - \cos 2x$ between $y = x - 1$ and $y = x + 1$ and non-periodic lies and cut off x-axis, $0 = x - \cos 2x$

$\Rightarrow \quad x = \cos 2x$

also $y = x - \cos 2x$ cut off y-axis at $x = 0$

then $y = 0 - \cos 0 = 0 - 1 = -1$

Example 10. Draw the graph of the function $y = |x| - \cos 2x$.

Sol. First draw the graph of

$$y = x - \cos 2x \qquad \text{(from Example 9)}$$

then graph of $y = |x| - \cos(2|x|)$

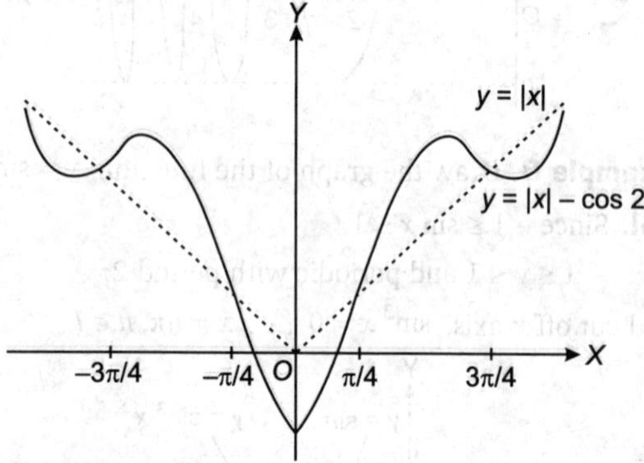

i.e., $y = |x| - \cos 2x$ is obtained, from above graph delete the left portion of y-axis and symmetry of right portion about y-axis.

Example 11. Draw the graph of the function $y = x \sin x$.

Sol. Since $-1 \le \sin x \le 1$

$\Rightarrow -x \le x \sin x \le x$ or $-x \le y \le x$

Hence graph of $y = x \sin x$ lies between $y = -x$ and $y = x$ and non periodic, and cut off x-axis at $y = 0$, *i.e.*, $x \sin x = 0$.

$\Rightarrow \quad x = n\pi, n \in$ I and y-axis at $x = 0$ *i.e.*, $y = 0$

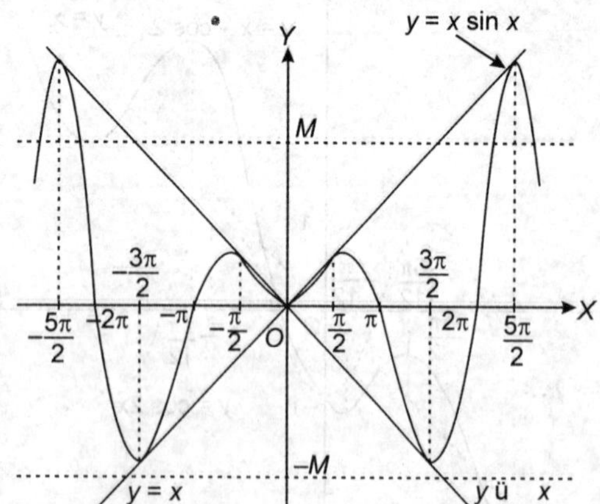

Example 12. Draw the graph of the function $y = e^x \sin x$.

Sol. Since $-1 \le \sin x \le 1$

$\Rightarrow \quad -e^x \le e^x \sin x \le e^x$

$\Rightarrow \quad -e^x \le y \le e^x$

Hence $y = e^x \sin x$ lies between $y = -e^x$ and $y = e^x$ and non-periodic, cut off x-axis $y = 0$ *i.e.*, $x = n\pi$, $n \in I$ and y-axis $x = 0$ *i.e.*, $y = 0$.

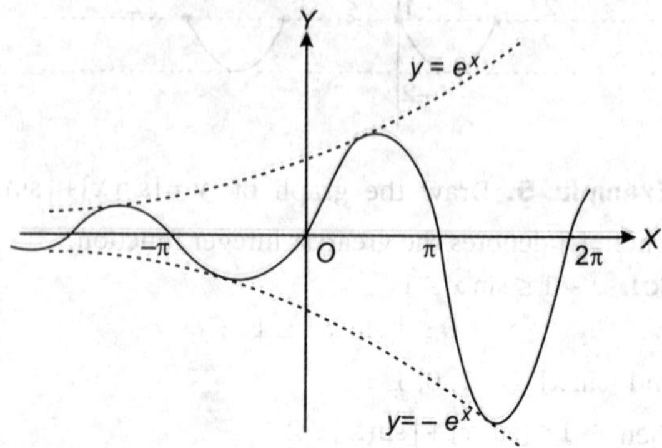

Example 13. Draw the graph of the function $y = \dfrac{\sin x}{x}$.

Sol. Since $-1 \le \sin x \le 1$

$$\Rightarrow \quad -\frac{1}{x} \le \frac{\sin x}{x} \le \frac{1}{x}$$

or

$$-\frac{1}{x} \le y \le \frac{1}{x}$$

and $\lim\limits_{x\to 0} \dfrac{\sin x}{x} = 1$, and y at $(x = 0) = \dfrac{0}{0}$ not defined.

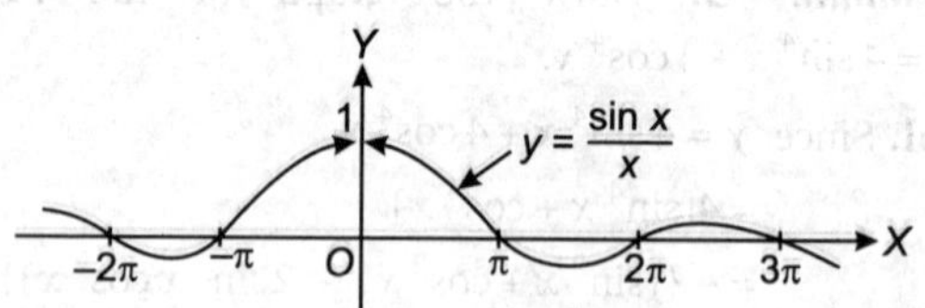

Hence $y = \dfrac{\sin x}{x}$ lies between $y = -\dfrac{1}{x}$ and $y = \dfrac{1}{x}$, y is non periodic cut off x-axis $\dfrac{\sin x}{x} = 0$,

$x = n\pi, n \ne 0, n \in I$ and curve not cut the y-axis.

Example 14. Draw the graph of the function

$$y = |\sin x| + |\cos x|$$

Sol. $\because \; y = |\sin x| + |\cos x|$

$$= \begin{cases} \sqrt{2}\sin\left(x + \dfrac{\pi}{4}\right), & 0 \le x \le \dfrac{\pi}{2} \\ \sqrt{2}\sin\left(x - \dfrac{\pi}{4}\right), & \dfrac{\pi}{2} \le x \le \pi \\ -\sqrt{2}\sin\left(x + \dfrac{\pi}{4}\right), & \pi \le x \le \dfrac{3\pi}{2} \\ -\sqrt{2}\sin\left(x - \dfrac{\pi}{4}\right), & \dfrac{3\pi}{2} \le x \le 2\pi \end{cases}$$

and also $y = |\sin x| + |\cos x|$ is periodic with period $\dfrac{\pi}{2}$.

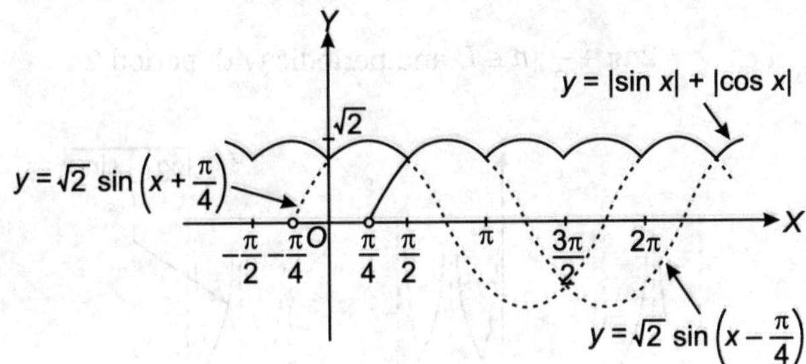

Example 15. Draw the graph of the function

$$y = \frac{\sin x}{\sqrt{(1+\tan^2 x)}} + \frac{\cos x}{\sqrt{(1+\cot^2 x)}}$$

Sol. Since $y = \frac{\sin x}{\sqrt{(1+\tan^2 x)}} + \frac{\cos x}{\sqrt{(1+\cot^2 x)}}$

$\Rightarrow \quad y = \sin x \,|\cos x| + \cos x \,|\sin x|$

$$= \begin{cases} \sin 2x, & 0 \le x \le \frac{\pi}{2} \\ 0, & \frac{\pi}{2} \le x \le \pi \\ -\sin 2x, & \pi \le x \le \frac{3\pi}{2} \\ 0, & \frac{3\pi}{2} \le x \le 2\pi \end{cases}$$

and also, y is periodic with period 2π.

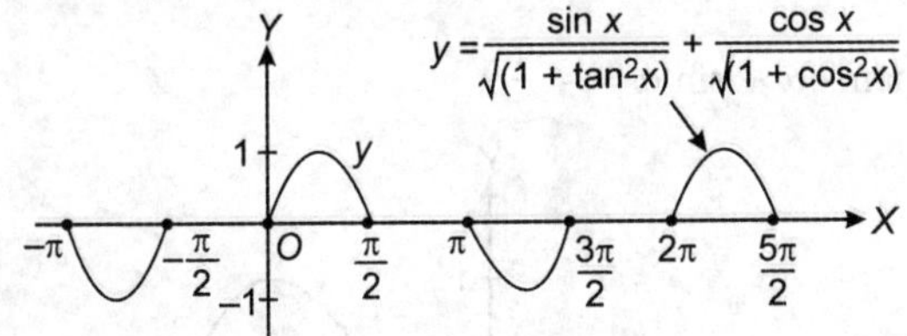

Example 16. Draw the graph of the function $y = \ln \tan x + \ln \cot x$.

Sol. y is not defined at $x = \frac{n\pi}{2}, n \in I$ and periodic with period π, also $y = \ln(\tan x \cot x) = \ln 1 = 0$.

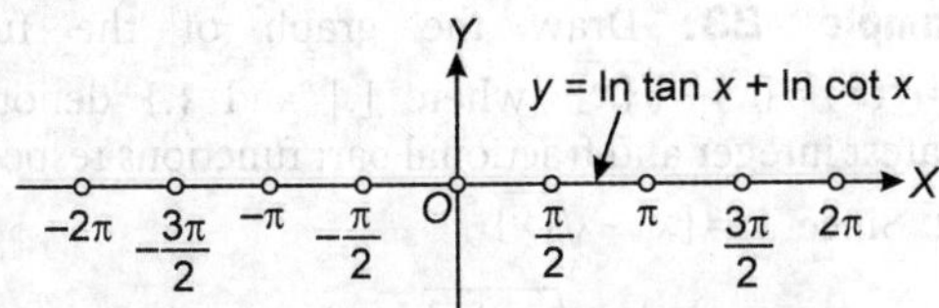

Example 17. Draw the graph of the function

$$y = \frac{(x-1)}{|x|-1} + \frac{|x+1|}{(x+1)} + \frac{1}{x}$$

Sol. It is clear y is not defined at $x = 0, \pm 1$

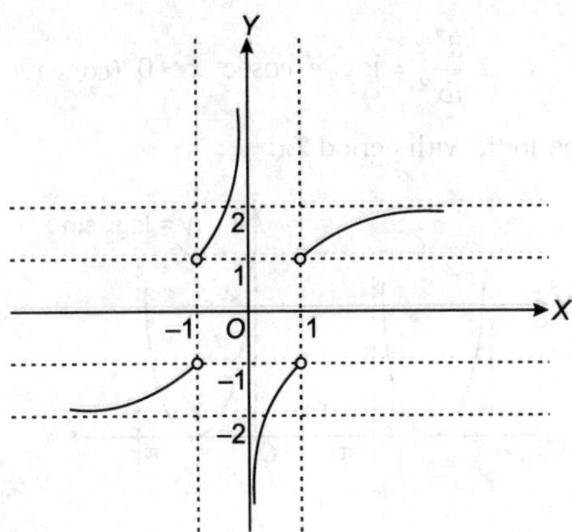

Example 18. Draw the graph of the function

$$y = \frac{\cos\left(|x| + \frac{\pi}{2}\right)}{\sin x}$$

Sol. Since $y = \frac{\cos\left(|x| + \frac{\pi}{2}\right)}{\sin x}$

$\because$ y is not defined at $\sin x = 0$

i.e., $x = n\pi, n \in I$ and non-periodic.

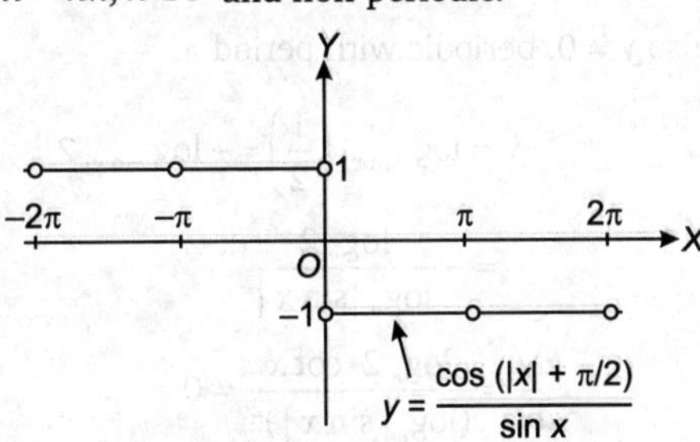

Example 19. Draw the graph of the function $y = \log_{1/2} \sin x$.

Sol. Since $y = \log_{1/2} \sin x$ is defined when $\sin x > 0$

$\Rightarrow \quad 2n\pi < x < (2n+1)\pi, n \in I$

cut off x-axis, $y = 0 \Rightarrow \sin x = \left(\frac{1}{2}\right)^{\circ} = 1$

$\therefore \quad x = 2p\pi + \frac{\pi}{2}, p \in I$

and also, $\quad y = -\log_2 \sin x$

$\therefore \quad \frac{dy}{dx} = -\log_2 e \cdot \cot x$

$\Rightarrow \quad \left.\frac{dy}{dx}\right|_{x = 2p\pi + \frac{\pi}{2}} = 0$, x-axis is a tangent.

and $\frac{d^2y}{dx^2} = \log_2 e \cdot \text{cosec}^2 x > 0$ (concave up).

and periodic with period 2π.

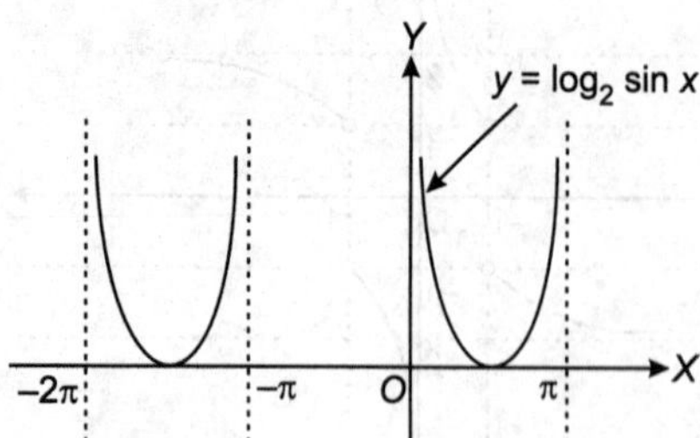

Example 20. Draw the graph of the function $y = \log_{|\sin x|}\left(\frac{1}{2}\right)$.

Sol. Since $y = \log_{|\sin x|}\left(\frac{1}{2}\right)$ is defined when $|\sin x| > 0$ and $|\sin x| \neq 1$

$\Rightarrow \quad \sin x < 0$ and $\sin x > 0$, $\sin x \neq \pm 1$

$\therefore \quad x \neq n\pi$

and $\quad x \neq 2n\pi + \frac{\pi}{2}; n \in I$

and also $y \neq 0$, periodic with period π.

Now $\quad y = \log_{|\sin x|}\left(\frac{1}{2}\right) = -\log_{|\sin x|} 2$

$$= -\frac{\log_e 2}{\log_e |\sin x|}$$

$\Rightarrow \quad \frac{dy}{dx} = \frac{\log_e 2 \cdot \cot x}{(\log_e |\sin x|)^2} \neq 0$

and $\quad \frac{d^2y}{dx^2} = +\text{ve}$ (concave up)

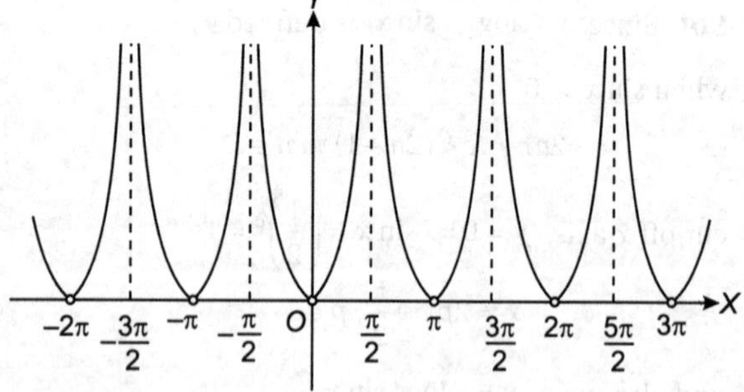

Example 21. Draw the graph of the function $y = \frac{1}{\log_{1/2} \sin x}$.

Sol. Since $y = \frac{1}{\log_{1/2} \sin x}$ is defined when

$\sin x > 0$ and $\log_{1/2} \sin x \neq 0$

$\Rightarrow \quad x \in (2n\pi, (2n+1)\pi), n \in I$ and $\sin x \neq 1$

i.e., $x \neq 2n\pi + \frac{\pi}{2}, n \in I$ and periodic with period 2π.

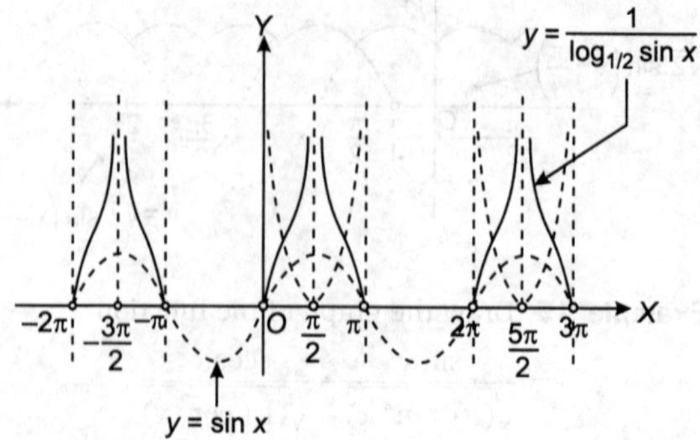

also $\quad \frac{d^2y}{dx^2} = +\text{ve} \quad$ (concave up)

Example 22. Draw the graph of the function $y = \sin(x - \sin x)$.

Sol. Since $y = \sin(x - \sin x)$ is odd function which is symmetrical in opposite quadrant and periodic with period 2π and cut off x-axis,

$$y = 0 \Rightarrow \sin(x - \sin x) = 0 \Rightarrow x - \sin x = n\pi$$

$\Rightarrow \quad x = n\pi, n \in I$ and cut off y-axis at $x = 0$, then $y = 0$.

If y has maximum value at $x = \alpha$ then

$\alpha - \sin\alpha = \frac{\pi}{2}$ and has minimum value at $x = -\alpha$

then $-\alpha + \sin\alpha = -\frac{\pi}{2}$

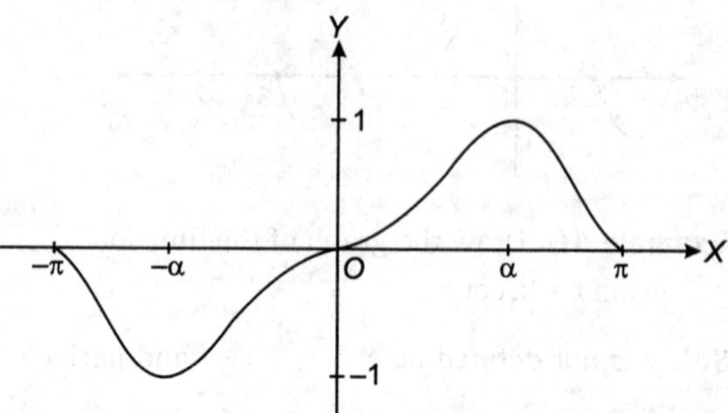

Example 23. Draw the graph of the function $y = f(x) = [x] + \sqrt{\{x\}}$ where [.] and {.} denotes the greatest integer and fractional part functions respectively.

Sol. Since $y = [x] + \sqrt{\{x\}}$

$$y = [x] + \sqrt{x - [x]}$$

If $x \in [n, n+1), n \in I$

then $\quad y = n + \sqrt{(x - n)}$

If let $y_1 = \sqrt{x - n}$ represent the upper part of the parabola $y_1^2 = x - n$ with vertex $(n, 0)$ and $f(n) = n + \sqrt{n - n} = n$

$\therefore \quad$ point (n, n) lie on $y = [x] + \sqrt{\{x\}}$ and not periodic

and $\frac{d^2y}{dx^2} < 0$ (concave down)

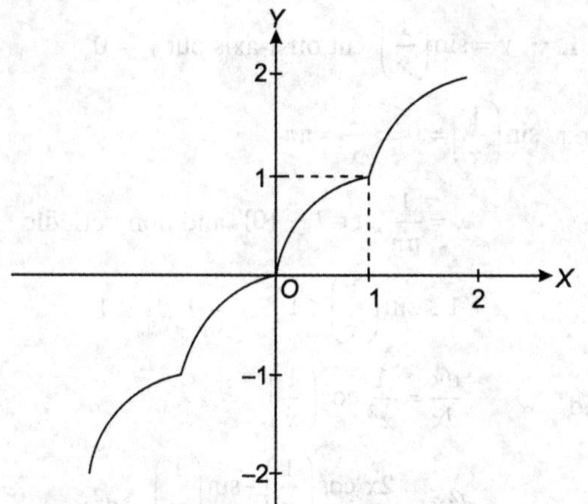

Example 24. Draw the graph of the function $y = \dfrac{2x-1}{x^2-2x-3}$.

Sol. Since $y = \dfrac{(2x-1)}{x^2-2x-3} = \dfrac{(2x-1)}{(x-3)(x+1)}$ is defined when $x \neq -1, 3$.

cut off x-axis $y = 0 \Rightarrow x = 1/2$

and cut off y-axis $x = 0 \Rightarrow y = 1/3$

also $\dfrac{d^2y}{dx^2} > 0$ for $x \in \left(-1, \dfrac{1}{2}\right)$

and $x \in (3, \infty)$ (concave up)

and $\dfrac{d^2y}{dx^2} < 0$ for $x \in (-\infty, -1)$

and $x \in \left(\dfrac{1}{2}, 3\right)$ (concave down)

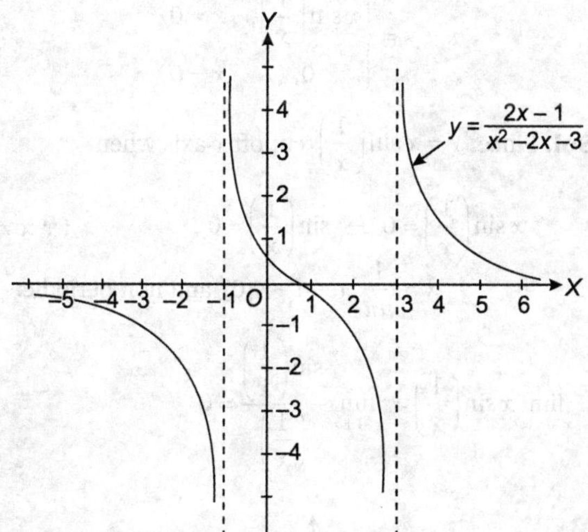

Example 25. Draw the graph of the function $y = \dfrac{x+1}{(x^2+3)}$.

Sol. Since $y = \dfrac{x+1}{(x^2+3)}$, cut off x-axis at $y = 0$, $x = -1$ and cut off y-axis at $x = 0$, $y = 1/3$.

and $\dfrac{dy}{dx} = -\dfrac{(x+3)(x-1)}{(x^2+3)^2}$,

for max. or min. $\dfrac{dy}{dx} = 0 \Rightarrow x = -3, 1$

$$\left.\frac{d^2y}{dx^2}\right|_{x=1} < 0 \text{ and } \left.\frac{d^2y}{dx^2}\right|_{x=-3} > 0$$

$\therefore$ minimum at $x = -3$ and maximum at $x = 1$

$\Rightarrow$ minimum and maximum values are $-1/6$ and $1/2$ respectively.

Here $\dfrac{d^2y}{dx^2} = \dfrac{2(x^3+3x^2-9x-3)}{(x^2+3)^3}$

$\Rightarrow \dfrac{d^2y}{dx^2} < 0$ for $x = 1 - h$ and $x = 1 + h$ (concave down)

and $\dfrac{d^2y}{dx^2} > 0$ for $x = -3 - h$ and $x = -3 + h$ (concave up)

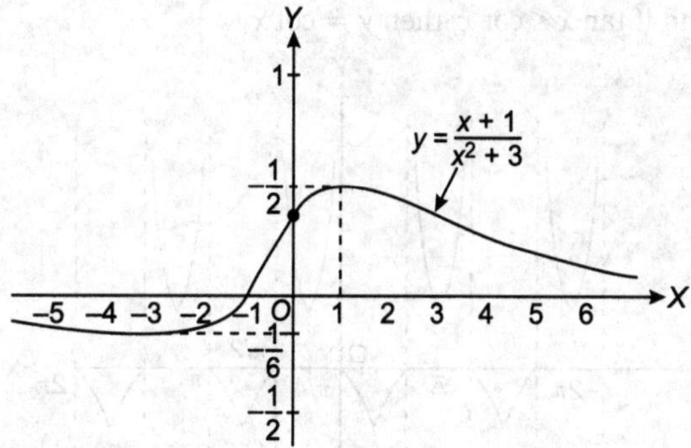

Example 26. Draw the graph of the function $y = \dfrac{x^3}{(1-x^2)}$.

Sol. Since $y = \dfrac{x^3}{(1-x^2)}$ is defined when $x \neq \pm 1$.

and cut off x-axis $y = 0$ then $x = 0$

and cut off y-axis $x = 0$ then $y = 0$

and $\dfrac{dy}{dx} = 0$ then $x = 0, \pm\sqrt{3}$

$\Rightarrow \dfrac{d^2y}{dx^2} = \dfrac{2x(x^2+3)}{(1-x^2)^3}$, Minimum at $x = \sqrt{3}$ and maximum at $x = -\sqrt{3}$

$\Rightarrow \dfrac{d^2y}{dx^2} > 0$ when $x \in [0, 1)$

and $x \in (-\infty, -1)$ (concave up)

and $\dfrac{d^2y}{dx^2} < 0$ when $x \in (-1, 0]$

and $x \in (1, \infty)$ (Concave down)

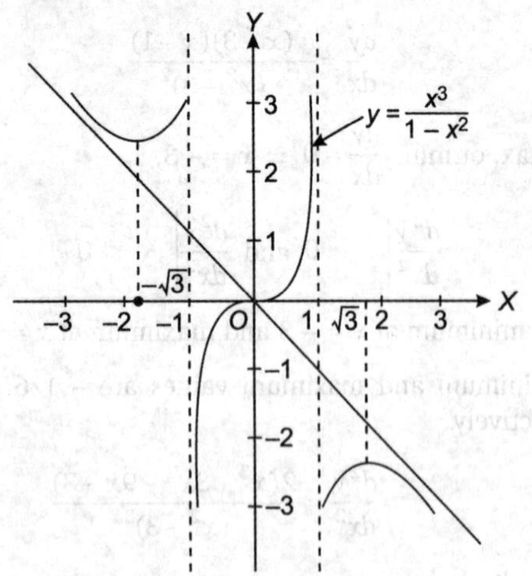

Example 27. Draw the graph of the function

$$y = \frac{\tan x + \cot x}{2} + \left|\frac{\tan x - \cot x}{2}\right|$$

Sol. If $\tan x \geq \cot x$, then $y = \tan x$
and if $\tan x \leq \cot x$, then $y = \cot x$

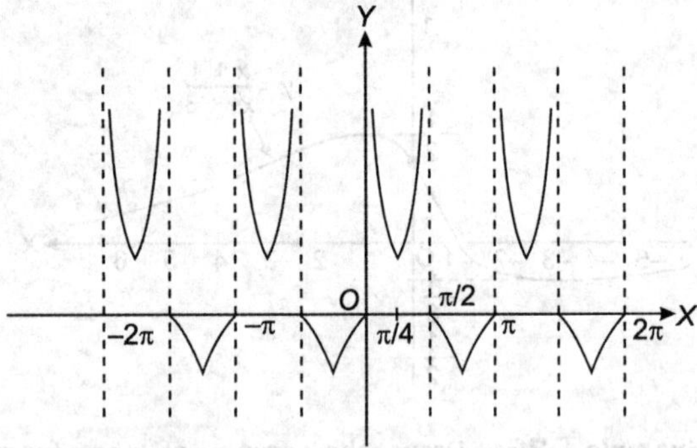

Example 28. Draw the graph of

$$y = \frac{\tan x + \cot x}{2} - \left|\frac{\tan x - \cot x}{2}\right|.$$

Sol. If $\tan x \geq \cot x$, then $y = \cot x$
and if $\tan x \leq \cot x$, then $y = \tan x$

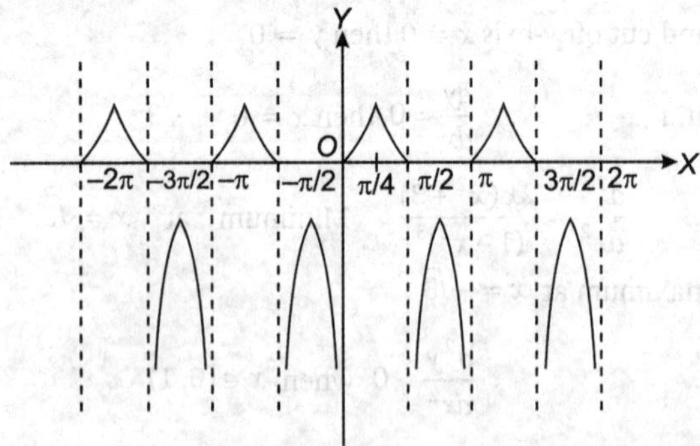

Example 29. Draw the graph of the function

$$y = \begin{cases} \sin\left(\frac{1}{x}\right), & x \neq 0 \\ 0, & x = 0 \end{cases}$$

Sol. $\because$ $y = \sin\left(\frac{1}{x}\right)$ cut off x-axis put $y = 0$

then $\sin\left(\frac{1}{x}\right) = 0 \Rightarrow \frac{1}{x} = n\pi$

$\therefore$ $x = \frac{1}{n\pi}, n \in I \sim \{0\}$ and non periodic.

$\because$ $-1 \leq \sin\left(\frac{1}{x}\right) \leq 1$ $\therefore -1 \leq y \leq 1$

and $\frac{dy}{dx} = \frac{1}{x^2}\cos\left(\frac{1}{x}\right)$

also $\frac{d^2y}{dx^2} = \frac{2x\cos\left(\frac{1}{x}\right) - \sin\left(\frac{1}{x}\right)}{x^4}$, $\frac{d^2y}{dx^2} > 0$

for $x > \frac{2}{\pi}$ (concave up)

< 0 for $x < -\frac{2}{\pi}$ (concave down)

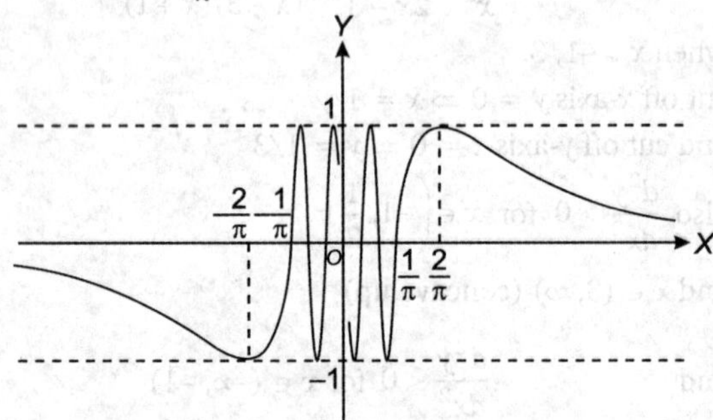

Example 30. Draw the graph of the function

$$y = \begin{cases} x\sin\left(\frac{1}{x}\right), & x \neq 0 \\ 0, & x = 0 \end{cases}$$

Sol. Since $y = x\sin\left(\frac{1}{x}\right)$ cut off x-axis when

$x\sin\left(\frac{1}{x}\right) = 0 \Rightarrow \sin\left(\frac{1}{x}\right) = 0$ $(\because x \neq 0)$

$x = \frac{1}{n\pi}, n \in I \sim \{0\}$ and non periodic.

$$\lim_{x \to \pm\infty} x\sin\left(\frac{1}{x}\right) = \lim_{x \to \pm\infty} \frac{\sin\left(\frac{1}{x}\right)}{\left(\frac{1}{x}\right)} = 0$$

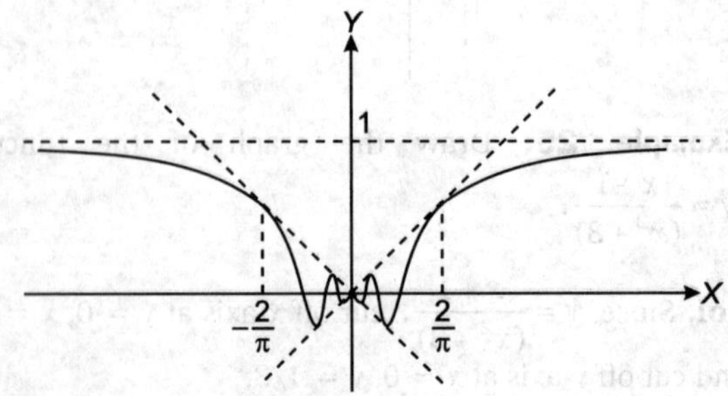

Example 31. Draw the graph of the function

$$y = \cos\left(\frac{\pi}{x}\right), \text{ as } x \to 0$$

Sol. Since $y = \cos\left(\frac{\pi}{x}\right)$ is defined when $x \neq 0$

cut off x-axis $y = 0 \Rightarrow \cos\left(\frac{\pi}{x}\right) = 0$

$$\Rightarrow \quad \frac{\pi}{x} = n\pi + \frac{\pi}{2}, n \in I$$

$$\Rightarrow \quad \frac{1}{x} = n + \frac{1}{2}$$

$\therefore \quad x = \frac{2}{2n+1}$, $n \in I$ and non periodic.

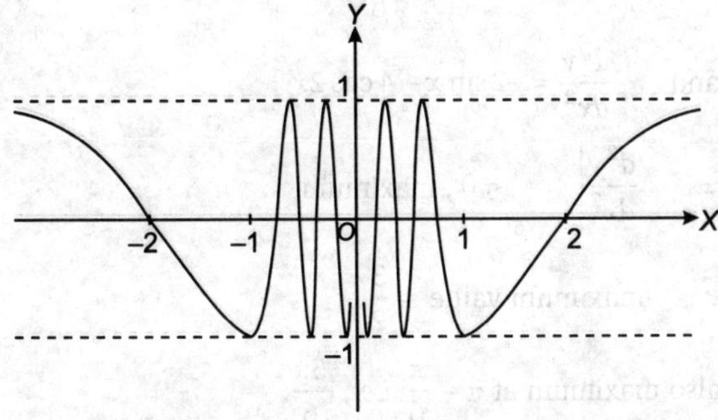

Example 32. Draw the graph of the function

$$y = \sqrt{(1+x^2)} \sin\left(\frac{1}{x}\right).$$

Sol. Since $-1 \leq \sin\left(\frac{1}{x}\right) \leq 1$

$$\Rightarrow \quad -\sqrt{(1+x^2)} \leq \sqrt{(1+x^2)} \sin\left(\frac{1}{x}\right) \leq \sqrt{(1+x^2)}$$

$$\Rightarrow \quad -\sqrt{(1+x^2)} \leq y \leq \sqrt{(1+x^2)}$$

$\therefore$ y lies between $y = \sqrt{(1+x^2)}$ and $y = -\sqrt{(1+x^2)}$ and y is defined when $x \neq 0$.

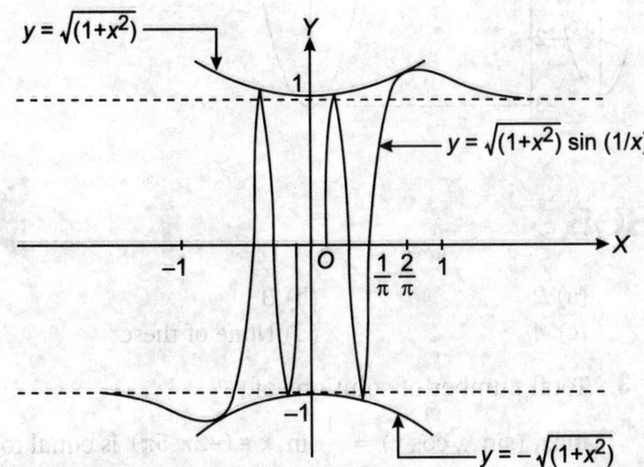

Example 33. Draw the graph of

$$y = \left|\frac{\sec x + \operatorname{cosec} x}{2}\right| + \left|\frac{\sec x - \operatorname{cosec} x}{2}\right|$$

Sol. If $\sec x \geq \operatorname{cosec} x$, then $y = \sec x$ and if $\sec x \leq \operatorname{cosec} x$, then $y = \operatorname{cosec} x$

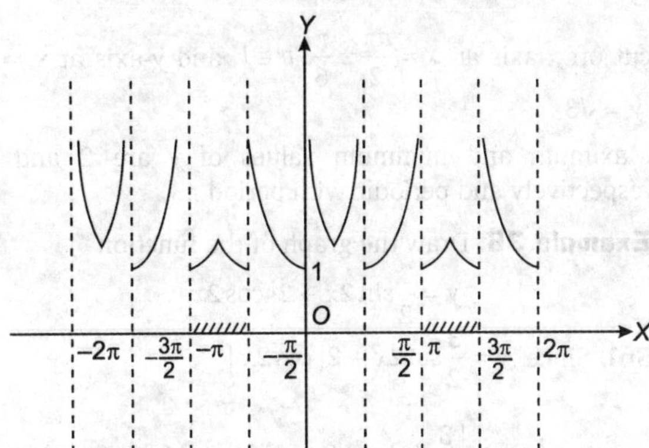

Example 34. Draw the graph of the function

$$|y| = \frac{\sec x + \operatorname{cosec} x}{2} - \left|\frac{\sec x - \operatorname{cosec} x}{2}\right|$$

Sol. If $\sec x \geq \operatorname{cosec} x$, then $|y| = \operatorname{cosec} x$ and if $\sec x \leq \operatorname{cosec} x$, then $|y| = \sec x$

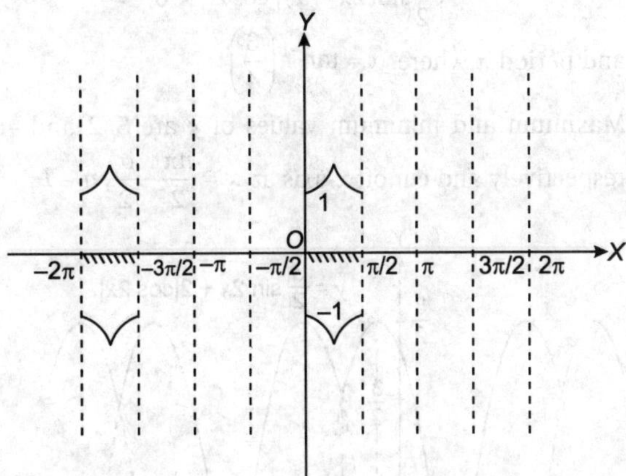

Example 35. Draw the graph of the function

$$y = |\sin x| + \sqrt{3} \cos 2x.$$

Sol. Since $y = |\sin 2x| + \sqrt{3} \cos 2x$

If $\sin 2x > 0$, $y = \sin 2x + \sqrt{3} \cos 2x$

$$= 2\sin\left(2x + \frac{\pi}{3}\right)$$

and if $\sin 2x < 0$, $y = -\sin 2x + \sqrt{3} \cos 2x$

$$= -2\sin\left(2x - \frac{\pi}{3}\right)$$

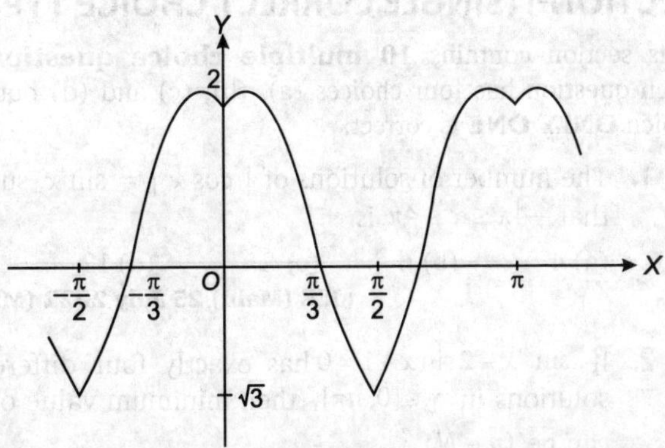

cut off x-axis at $x = \frac{n\pi}{2} \pm \frac{\pi}{6}, n \in I$ and y-axis at $x = 0$, $y = \sqrt{3}$

Maximum and minimum values of y are 2 and −2 respectively and periodic with period π.

Example 36. Draw the graph of the function

$$y = \frac{3}{2}\sin 2x + 2|\cos 2x|.$$

Sol. Since $y = \frac{3}{2}\sin 2x + 2|\cos 2x|$

$$= \begin{cases} \frac{3}{2}\sin 2x + 2\cos 2x, & \cos 2x > 0 \\ \frac{3}{2}\sin 2x - 2\cos 2x, & \cos 2x < 0 \end{cases}$$

$$= \begin{cases} \frac{5}{2}\sin(2x + \alpha), \cos 2x > 0 \\ \frac{5}{2}\sin(2x - \alpha), \cos 2x < 0 \end{cases}$$

and period π where $\alpha = \tan^{-1}\left(\frac{3}{4}\right)$

Maximum and minimum values of y are 5/2 and −5/2 respectively and cut off x-axis at $x = \frac{n\pi}{2} - \frac{\alpha}{2}, n \in I$.

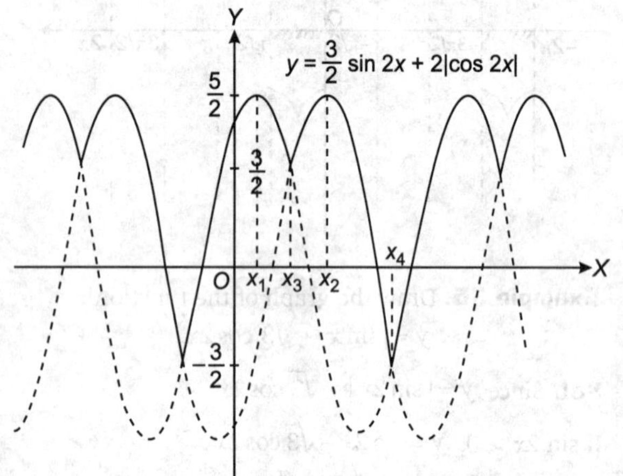

Example 37. Draw the graph of the function $y = 2\sin x + \cos 2x$.

Sol. Since $y = 2\sin x + \cos 2x$ period of y is 2π.

and $\frac{dy}{dx} = 2\cos x - 2\sin 2x$

$$= 2\cos x(1 - 2\sin x)$$

for max. or min. $\frac{dy}{dx} = 0$

$\therefore \quad \cos x = 0, \sin x = \frac{1}{2} \quad \Rightarrow \quad x = n\pi + \frac{\pi}{2}$

and $\quad x = p\pi + (-1)^n \frac{\pi}{6}, p, n \in I$

and $\quad \frac{d^2y}{dx^2} = -2\sin x - 4\cos 2x$

$\Rightarrow \quad \left.\frac{d^2y}{dx^2}\right|_{x=\pi/6} < 0$, maximum,

$\therefore \quad$ maximum value $= \frac{3}{2}$

also maximum at $\pi - \frac{\pi}{6}$ *i.e.*, $\frac{5\pi}{6}$

and $\left.\frac{d^2y}{dx^2}\right|_{x=-\pi/2} > 0$, minimum, minimum value = −3

also minimum at $3\pi/2$ and at $\pi/2$, $y = 1$.

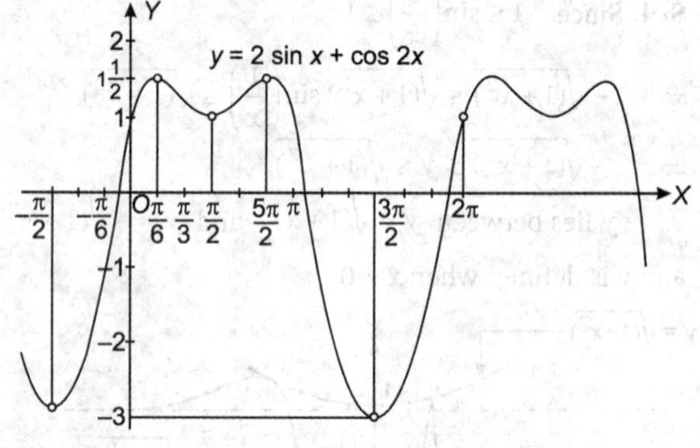

Exercises

SECTION-I (SINGLE CORRECT CHOICE TYPE)

This section contains **10 multiple choice questions**. Each question has four choices (a), (b), (c) and (d) out of which **ONLY ONE** is correct.

1. The number of solutions of $|\cos x| = \sin x$, such that $-4\pi \le x \le 4\pi$ is

(a) 4 (b) 6 (c) 8 (d) 12

[JEE (Main) 25 July 2022 (MS)]

2. If $\sin^2 x - 2\sin x - 1 = 0$ has exactly four different solutions in $x \in [0, n\pi]$, then minimum value of n can be $(n \in N)$:

(a) 2 (b) 3
(c) 4 (d) None of these

3. Total number of solutions of

max. $(\sin x, \cos x) = \frac{1}{2}$ in $x \in (-2\pi, 5\pi)$ is equal to:

(a) 3 (b) 5
(c) 7 (d) 9

4. Total number of solutions of $\sin \pi x = |\ln|x||$ is :

(a) 8 (b) 10 (c) 9 (d) 6

5. Total number of solutions of $2^{\cos x} = |\sin x|$ in $[-2\pi, 5\pi]$ is equal to:
(a) 12 (b) 14
(c) 15 (d) 16

6. Total number of solutions of $x^2 - 4 - [x] = 0$ is (where [.] denotes the greatest integer function) :
(a) 1 (b) 2
(c) 3 (d) 4

7. Number of solutions of the equation $|\sin^{-1}(\sin x)| = \cos x$, for $x \in \left[-\frac{\pi}{2}, \frac{\pi}{2}\right]$ is :
(a) 0 (b) 2
(c) 4 (d) 6

8. Total number of solutions of $\sin\{x\} = \cos\{x\}$ (where {.} denotes the fractional part function) in $x \in [0, 2\pi]$ is equal to :
(a) 5 (b) 7
(c) 8 (d) None of these

9. Total number of solutions of the system of equations $|y| = \cos x$ and $y = \cot^{-1}(\cot x)$ in $x \in \left(-\frac{3\pi}{2}, \frac{5\pi}{2}\right)$ is equal to :
(a) 2 (b) 4
(c) 6 (d) None of these

10. Number of solutions of the equation $2\sin^{-1}\left(\frac{2x}{1+x^2}\right) - \pi x^3 = 0$ is equal to :
(a) 0 (b) 1 (c) 2 (d) 3

SECTION-II (SUBJECTIVE TYPE) (LEVEL I)

In this section, there are **11** subjective questions.

11. Sketch the graphs of the following functions :
(i) $f(x) = x + 1$ (ii) $f(x) = x^2 - x + 5$
(iii) $f(x) = \sin x + 2$ (iv) $f(x) = \frac{2}{x-3}$
(v) $f(x) = \sqrt{(2x+3)}$ (vi) $f(x) = \sqrt{(9-x^2)}$
(vii) $f(x) = \cos\left(\frac{x\pi}{2}\right)$ (viii) $f(x) = -\sin\left(\frac{\pi x}{3}\right)$
(ix) $f(x) = 1 - \sqrt{x}$ (x) $f(x) = (x-1)^{2/3}$

12. Sketch the graphs of the following functions :
(i) $f(x) = ||x| + 1|$ (ii) $f(x) = |\ln x|$
(iii) $f(x) = \tan|x|$ (iv) $f(x) = e^{|x|} - 1$
(v) $|f(x)| = \ln x$ (vi) $f(x) = \left(\frac{x+1}{x+4}\right)$
(vii) $f(x) = x + |x|$
(viii) $f(x) = -\sin|x|$
(ix) $f(x) = \cos\left(x - \frac{\pi}{4}\right) - 1$
(x) $f(x) = \ln x^2$

13. Sketch the graphs of the following functions :
(i) $f(x) = \ln(1+x^2)$ (ii) $f(x) = \frac{1}{x^2} + 1$
(iii) $f(x) = \sqrt[3]{(x-1)} + 1$ (iv) $f(x) = \frac{2}{x + \sqrt{x^2}}$
(v) $f(x) = |(x+2)|x$ (vi) $f(x) = 2x^4 + 5$
(vii) $f(x) = |x^2 + 5x + 6|$ (viii) $f(x) = \sin x + \cos x$
(ix) $f(x) = \frac{1}{2-|x|}$ (x) $f(x) = \frac{|x-1|}{|x|-1}$

14. Sketch the graphs of the following functions :
(i) $f(x) = \begin{cases} x, & 0 \le x \le 1 \\ 2-x, & 1 < x \le 2 \end{cases}$
(ii) $f(x) = \begin{cases} 3-x, & x \le 1 \\ 2-x, & x > 1 \end{cases}$
(iii) $f(x) = \begin{cases} \frac{1}{x}, & x < 0 \\ x, & x \ge 0 \end{cases}$
(iv) $f(x) = \begin{cases} 1-x^2, & 0 < x \le 1 \\ 2-x^2, & 1 < x \le 2 \end{cases}$
(v) $f(x) = \begin{cases} 4^x - 1, & x < 0 \\ \sqrt{(4x-x^2)}, & x \ge 0 \end{cases}$
(vi) $f(x) = \begin{cases} 1-\sqrt{(1-x^2)}, & x \le 1 \\ 1+\log_{1/2} x, & x > 1 \end{cases}$

15. Draw the graph of the function $f(x) = \left|\left|\frac{1}{x}\right| - 2\right|$.

16. Draw the graph of the function $f(x) = \frac{|\ln x|}{\ln x}$.

17. Find the number of solutions of the equation $|\sin x| = |x|$.

18. Find the number of solutions of the equation $y = |\sin x|$ and $x^2 + y^2 = 1$.

19. Draw the graph of the function $f(x) = |\sin x| + \sin x$ on the interval $[0, 3\pi]$.

20. Draw the graph of the function $y = x^2 \text{sgn}(x)$, where $\text{Sgn}(x) = \begin{cases} 1, & x > 0 \\ 0, & x = 0 \\ -1, & x < 0 \end{cases}$

21. Draw the graph of the function $y = |x^2 - 3|x| + 2|$.

SECTION-III (SUBJECTIVE TYPE) (LEVEL II)

In this section, there are **14** subjective questions.

22. Sketch the graphs of the following functions :

(i) $f(x) = 2x + \sin x + 5$ (ii) $f(x) = \dfrac{1}{1+\sqrt{x}}$

(iii) $f(x) = 1 + |\sin x|$ (iv) $f(x) = \dfrac{|(x-1)(x-2)|}{(x-1)(x-2)}$

(v) $f(x) = \dfrac{8}{2+x^2}$ (vi) $f(x) = \cos(\sin x)$

(vii) $f(x) = xe^x$ (viii) $f(x) = \dfrac{x}{1+x^2}$

(ix) $f(x) = |2 - 2^x|$ (x) $f(x) = 2^{1/x}$

23. Sketch the graphs of the following functions :

(i) $f(x) = (|x+1| + 1)(x-3)$

(ii) $f(x) = \log_2(1-x^2)$

(iii) $f(x) = \sqrt{(x-1)} + \sqrt{(3-x)}$

(iv) $f(x) = \left(1+\dfrac{1}{x}\right)^x$ (v) $f(x) = \sin^2 x - 2\sin x$

(vi) $f(x) = x^{1/\log_{10} x}$ (vii) $f(x) = 2^{x(x+1)}$

(viii) $|x| + |y| = 2$ (ix) $f(x) = 4 - \dfrac{6}{|1+x|}$

(x) $f(x) = \dfrac{x^3}{3} - x^2 - 1$

24. Sketch the graphs of the following functions :

(i) $f(x) = \log_2 \cos x$ (ii) $f(x) = \log_2 |\cos x|$

(iii) $f(x) = \cos(\log_2 x)$ (iv) $f(x) = x^{\log_x 2}$

(v) $f(x) = -\cos^{-1}(-x^2)$ (vi) $f(x) = \cot^{-1}(3-x)$

(vii) $f(x) = [|x-1|]$, [.] denotes the greatest integer function.

(viii) $f(x) = x + [x]$, [.] denotes the greatest integer function.

(ix) $f(x) = \sin\dfrac{\pi}{2}(x - [x])$, [.] denotes the greatest integer function.

(x) $f(x) = [\log_2 x]$, [.] denotes the greatest integer function.

25. Sketch the graphs of the following functions :

(i) $f(x) = \text{Sgn}(\ln x)$ (ii) $f(x) = \text{Sgn}(1 - |x|)$

(iii) $f(x) = \text{Sgn}(\sin x)$ (iv) $f(x) = \text{Sgn}(x - |x|)$

(v) $f(x) = \text{Sgn}([x])$, where [.] denotes the greatest integer function.

26. Sketch the graphs of the following functions :

(i) $f(x) = \begin{cases} \sin x, & -\pi \le x \le 0 \\ 2, & 0 < x \le 1 \\ \dfrac{1}{x-1}, & 1 < x \le 4 \end{cases}$

(ii) $f(x) = \begin{cases} -2, & x > 0 \\ \dfrac{1}{2}, & x = 0 \\ -x^3, & x < 0 \end{cases}$

(iii) $f(x) = \begin{cases} x^2 + x + 1, & -1 \le x < 0 \\ \sin^2 x, & 0 \le x \le \pi \\ \dfrac{x-1}{x+1}, & \pi < x \le 5 \end{cases}$

27. Sketch the graphs of the following functions :

(i) $f(x) = [x] + [x]^2$

(ii) $f(x) = [|x|] + \sqrt{\{|x|\}}$

where [.] and {.} are the greatest integer function and fractional part functions respectively.

28. Find the number of solutions of the equation $[\sin^{-1} x] = x - [x]$ where [.] denotes the greatest integer function.

29. Find the number of solutions of the following equations.

(i) $\cos x = \tan 4x;\ 0 < x < \pi$

(ii) $\log_{0.5} |x| = 2|x|$

30. Sketch the graphs of the following functions.

(i) $[x] + [y] = 3$ in first quadrant, [.] denotes the greatest integer function.

(ii) $f(x) = 2 + (-x)^{[x]}$, [.] denotes the greatest integer function.

(iii) $f(x) = [2e^x],\ x \ge 0$, [.] denotes the greatest integer function.

(iv) $f(x) = [\sin^{-1} \sin x]$, [.] denotes the greatest integer function.

(v) $f(x) = \cos\{\pi(|x| + [x])\}$, where [.] denotes the greatest integer function.

31. Draw the graph of the function $f(x) = 1 - 2^{1+\sin(x+1)}$.

32. Draw the graph of the function

$$f(x) = |\sin 2x| + \sqrt{3}|\cos 2x|.$$

33. Draw the graph of the function $f(x) = (x)^{[x]}$, $x = [x] + (x)$ and $x > 0$ and $x \le 3$ where [.] denotes the greatest integer function.

34. Draw the graph of the function $y = [\sin x] - |\sin x|$, where [.] denotes the greatest integer function.

35. Draw the graph of the function $y = \dfrac{e^x}{[x]+1}, x \ge 0$, [.] denotes the greatest integer function.

Answers

1. (c)	**2.** (c)	**3.** (c)	**4.** (d)	**5.** (b)	**6.** (b)	**7.** (b)	**8.** (b)	**9.** (a)	**10.** (d)
17. (1)	**18.** (2)	**28.** (1)	**29.** (i) 5	(ii) 2					

SOLUTIONS

1. Let $y = |\cos x| = \sin x$

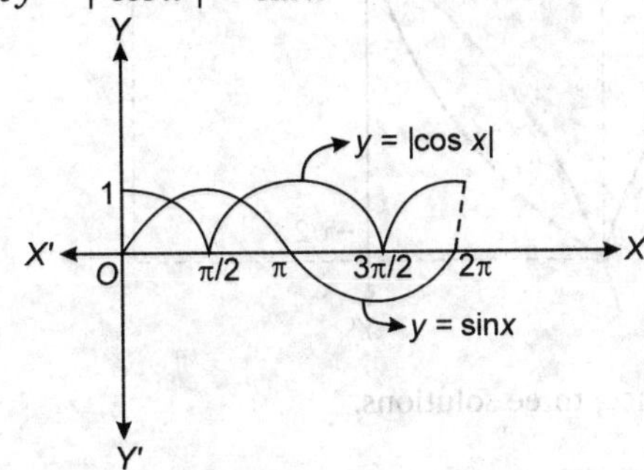

The number of solutions of the equation $|\cos x| = \sin x$, for all $x \in [-4\pi, 4\pi]$ will be equal to 4 times the number of solutions of the same equation $|\cos x| = \sin x$ for $x \in [0, 2\pi]$. Clearly, from the above figure, two solutions of the equation $|\cos x| = \sin x$ in $[0, 2\pi]$. Thus, total 8 solutions in $[-4\pi, 4\pi]$.

2. We have,

$$\sin^2 x - 2\sin x - 1 = 0$$

$$\Rightarrow \quad \sin x = \frac{2 \pm \sqrt{4+4}}{2} = 1 \pm \sqrt{2}$$

$$\Rightarrow \quad \sin x = 1 - \sqrt{2} \qquad (\because \sin x = 1 + \sqrt{2} > 1)$$

$$(\therefore \sin x \neq 1 + \sqrt{2})$$

Let $y = \sin x = 1 - \sqrt{2}$

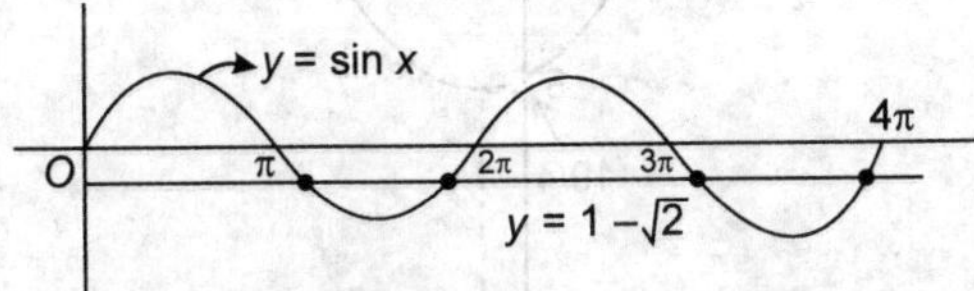

There are four solutions in $[0, 4\pi]$. Thus $n_{\min} = 4$.

3. Let $y = \max. (\sin x, \cos x)$ is periodic with period 2π. Graphs of $y = \max. (\sin x, \cos x)$ and $y = \frac{1}{2}$ meet 2 times in $[0, 2\pi]$.

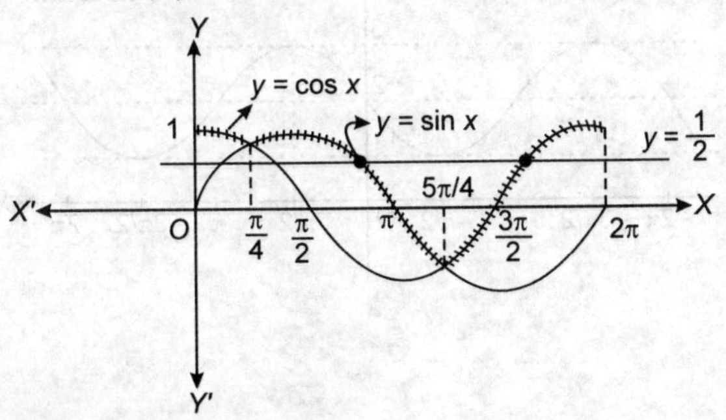

Thus, there are 7 solutions.

4. There are exactly six solutions.

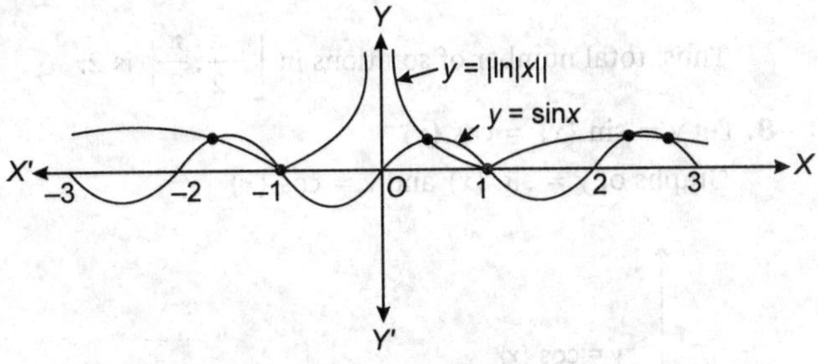

5. Let $y = 2^{\cos x} = |\sin x|$ $\because$ Graphs of $y = 2^{\cos x}$ and $y = |\sin x|$ meet four times in $[0, 2\pi]$

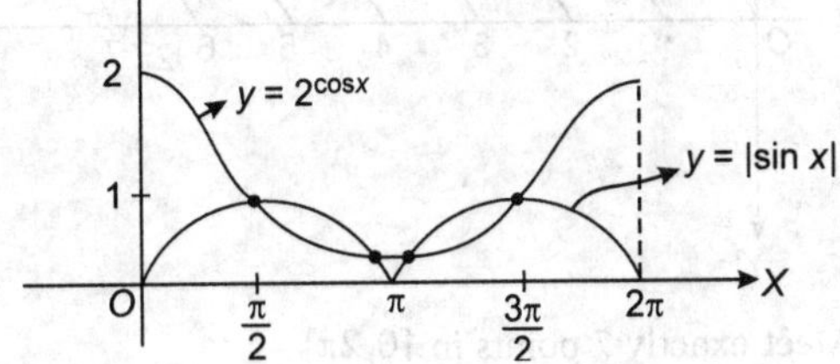

Thus, total number of solutions = 4 + 4 + 4 + 2 = 14.

6. We have, $x^2 - 4 - [x] = 0 \Rightarrow x^2 - 4 = [x]$

Let $\quad y = x^2 - 4 = [x]$

Graphs $y = x^2 - 4$ and $y = [x]$ meet at two points.

Thus, total number of solutions = 2.

7. Let $y = |\sin^{-1}(\sin x)| = \cos x$

$\because \quad |\sin^{-1}(\sin x)| = |x|$ for $x \in \left[-\frac{\pi}{2}, \frac{\pi}{2}\right]$

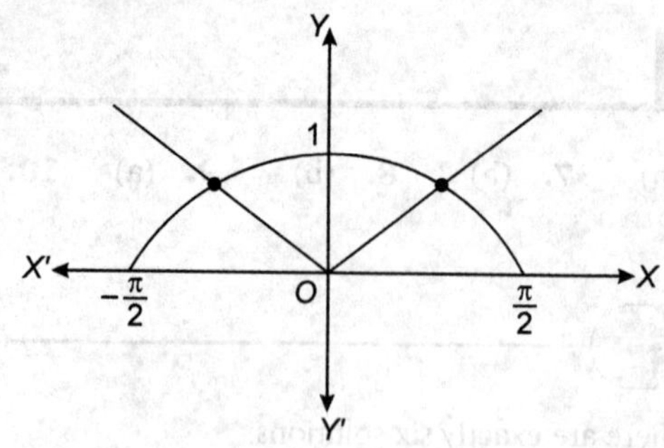

Thus, total number of solutions in $\left[-\frac{\pi}{2}, \frac{\pi}{2}\right]$ is 2.

8. Let $y = \sin\{x\} = \cos\{x\}$

Graphs of $y = \sin\{x\}$ and $y = \cos\{x\}$

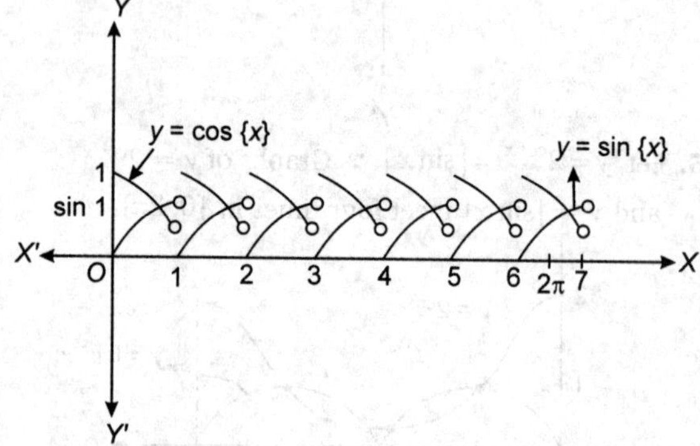

Meet exactly 7 points in $[0, 2\pi]$.

9.

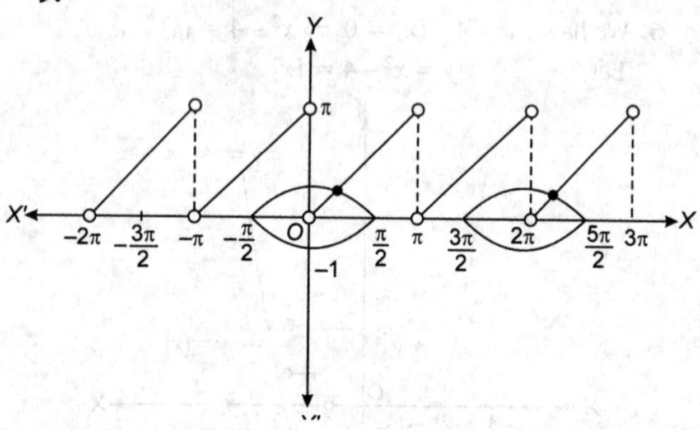

Hence, two solutions.

10. We have,

$$2\sin^{-1}\left(\frac{2x}{1+x^2}\right) - \pi x^3 = 0$$

$$\Rightarrow \quad 2\sin^{-1}\left(\frac{2x}{1+x^2}\right) = \pi x^3$$

Let $y = \sin^{-1}\left(\frac{2x}{1+x^2}\right) = \frac{\pi x^3}{2}$

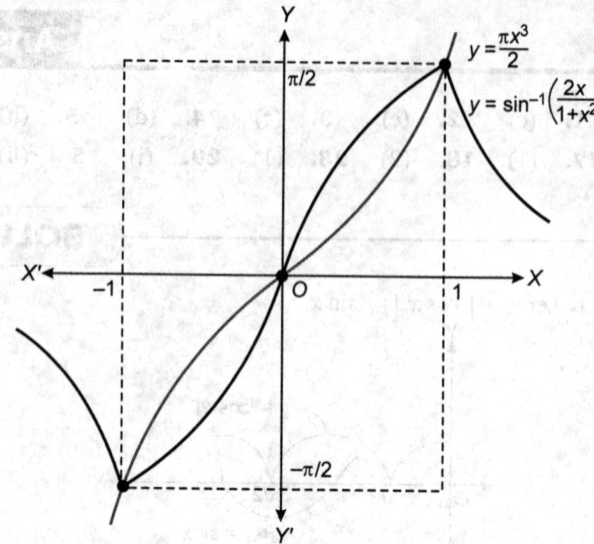

Hence, three solutions.

11. (i)

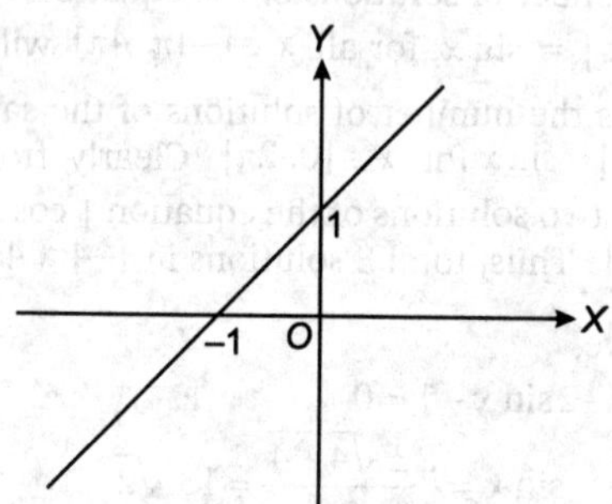

(ii)

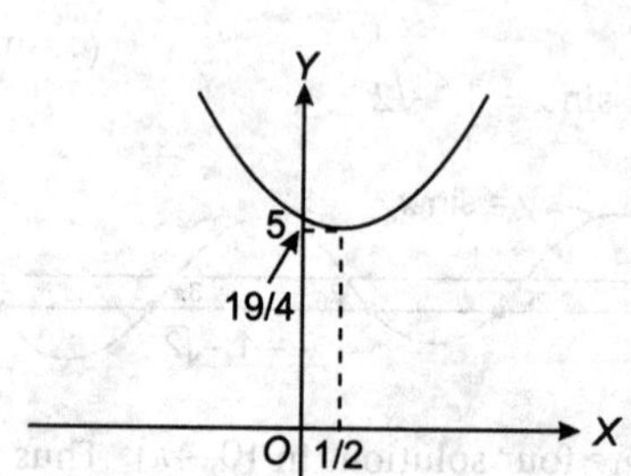

(iii)

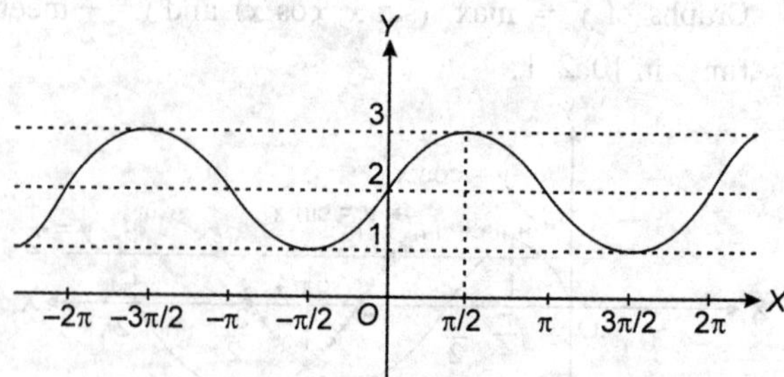

(iv)

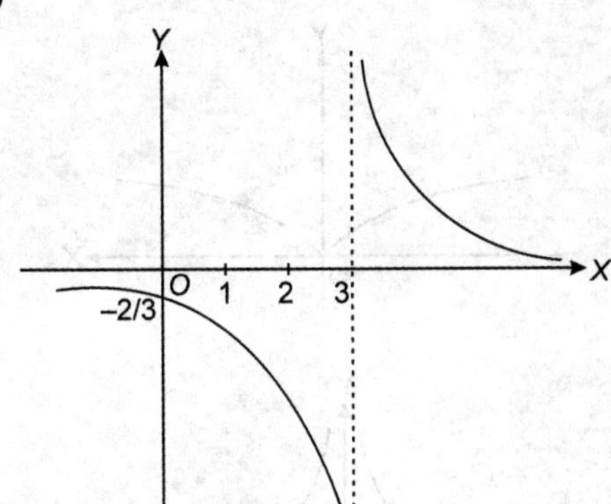

(v)

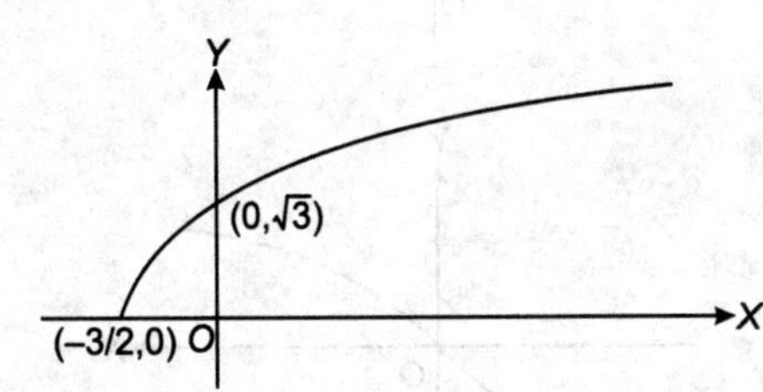

(vi)

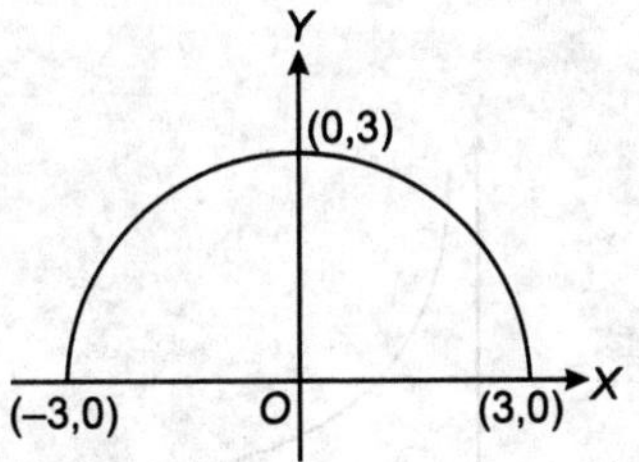

(vii)

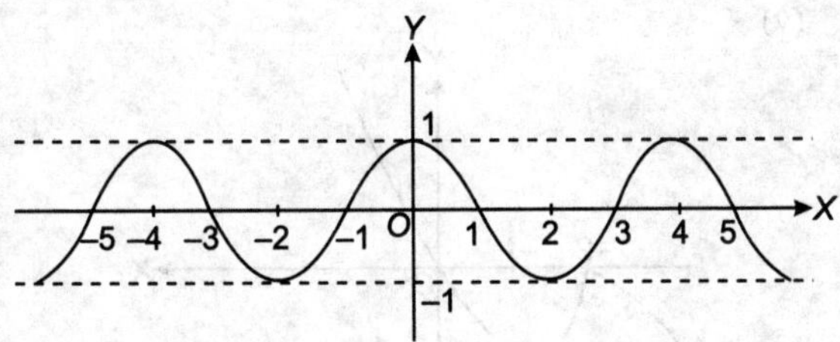

(viii)

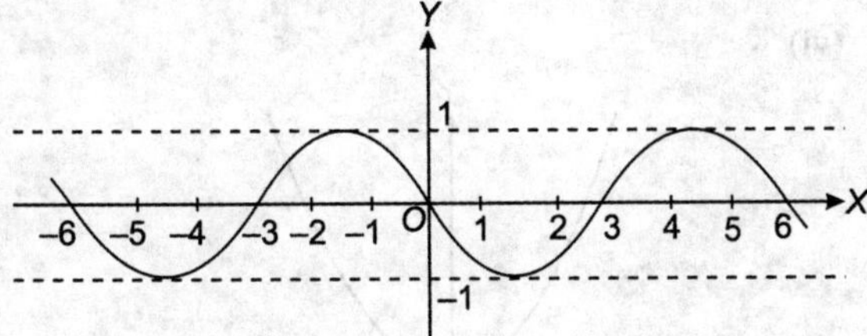

(ix)

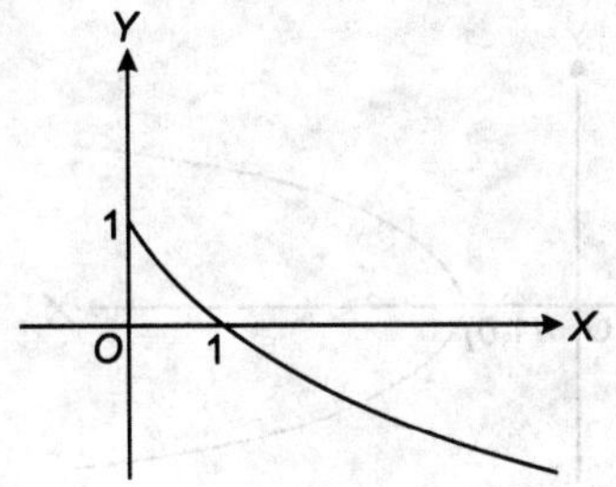

(x)

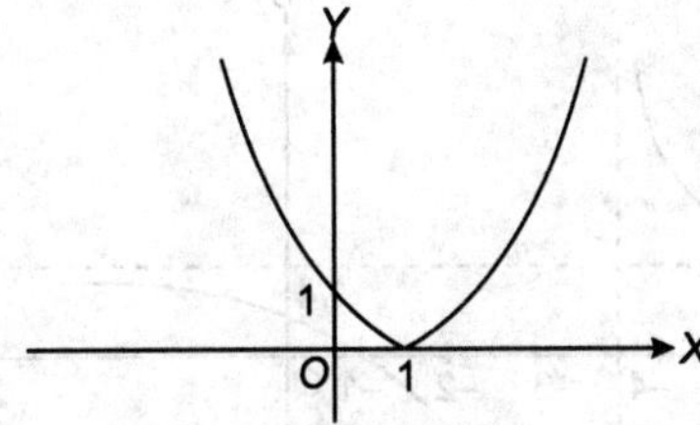

12. (i)

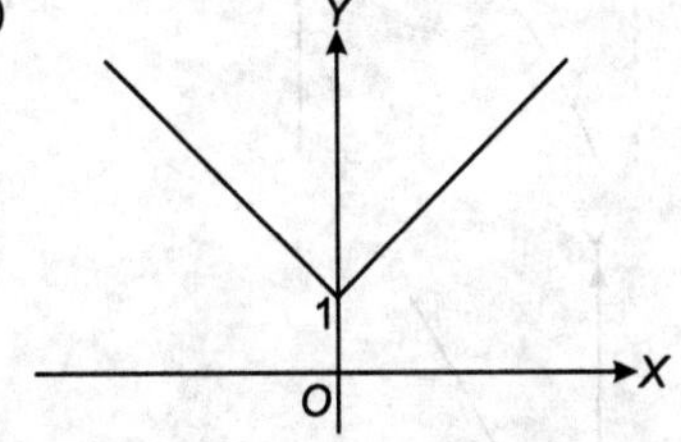

(ii)

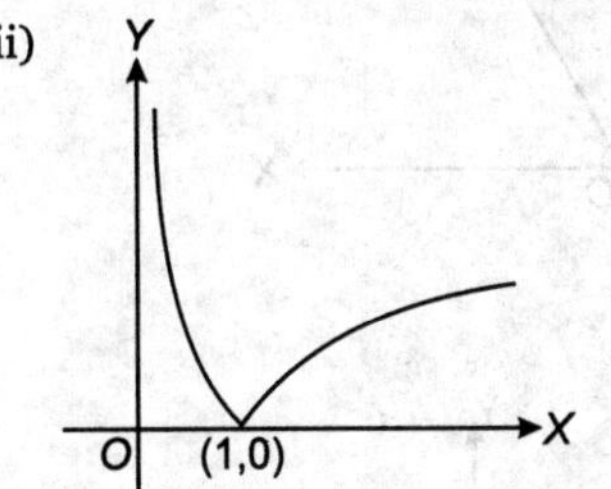

(iii)

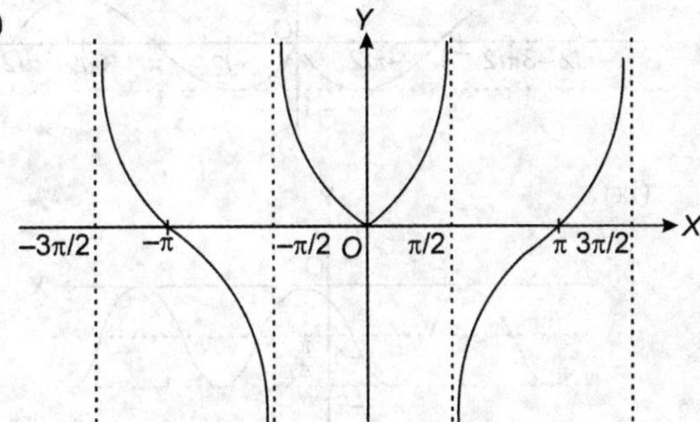

(iv)

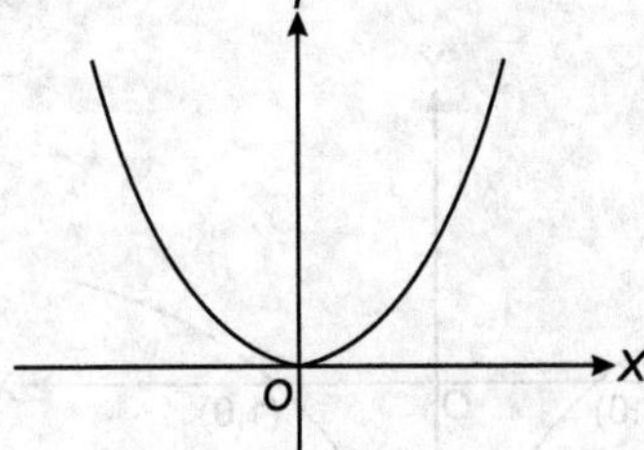

(v)

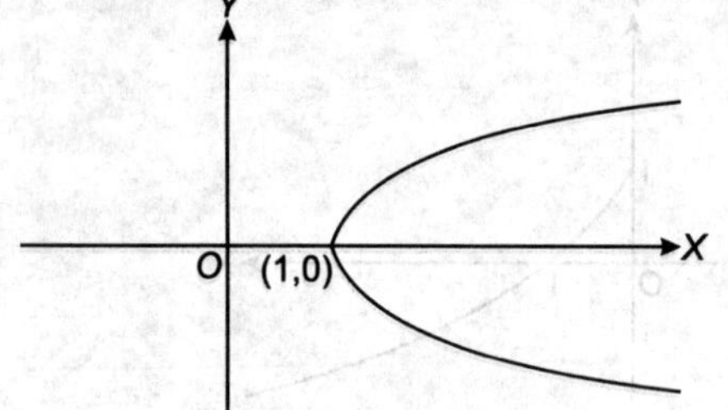

(vi)

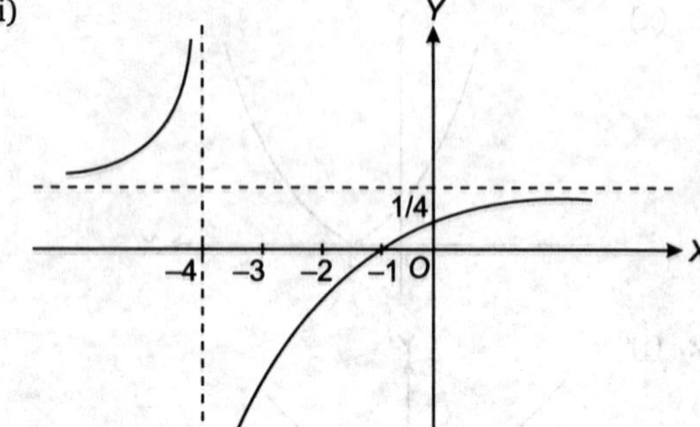

(vii)

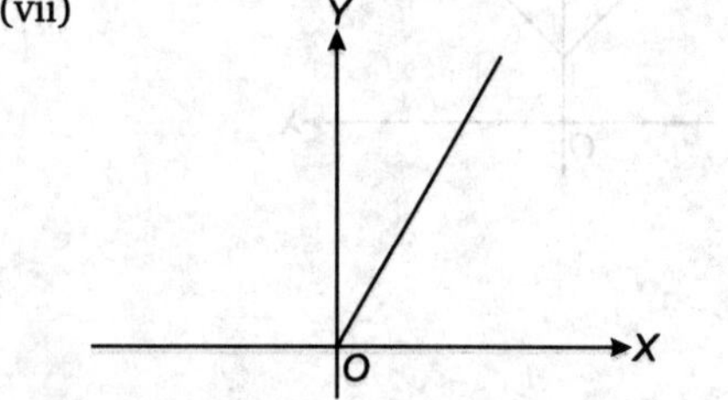

(viii)

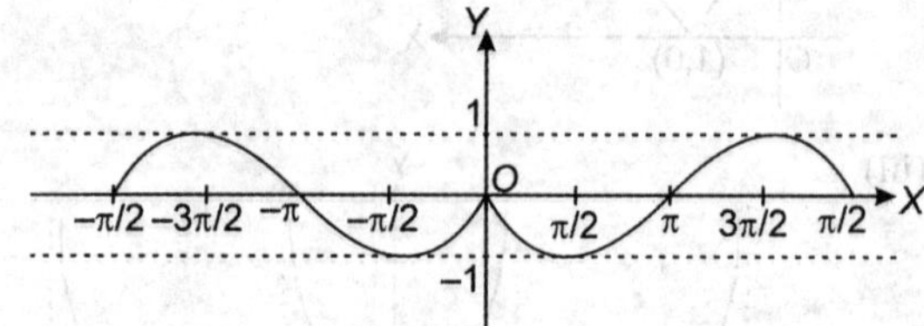

(ix)

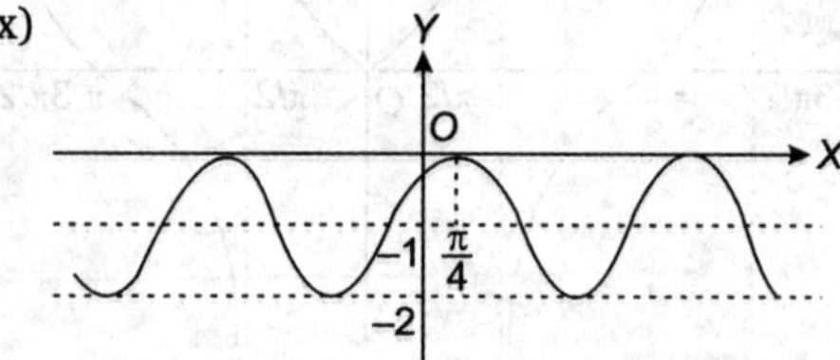

(x)

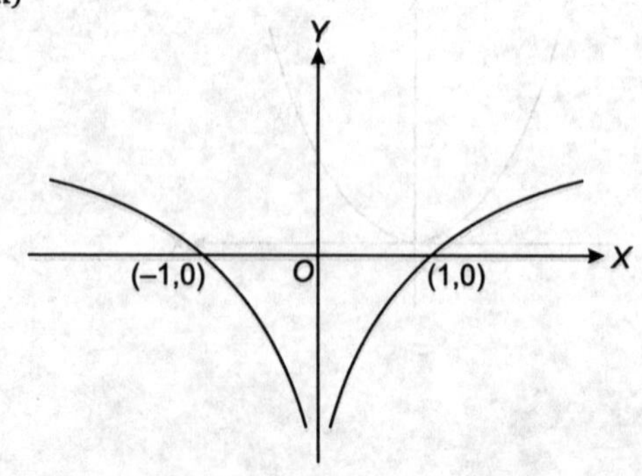

13. (i)

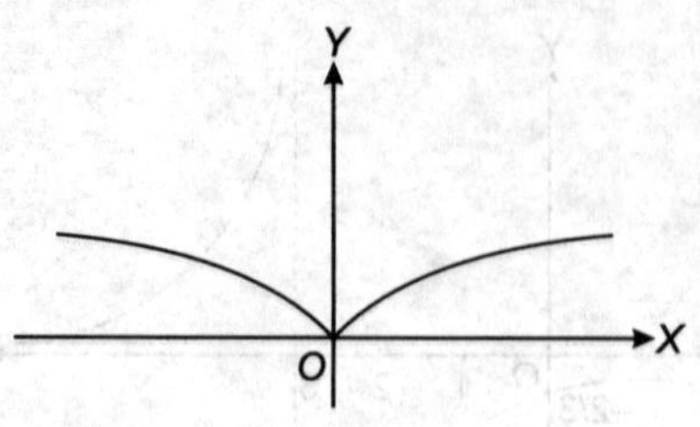

(ii)

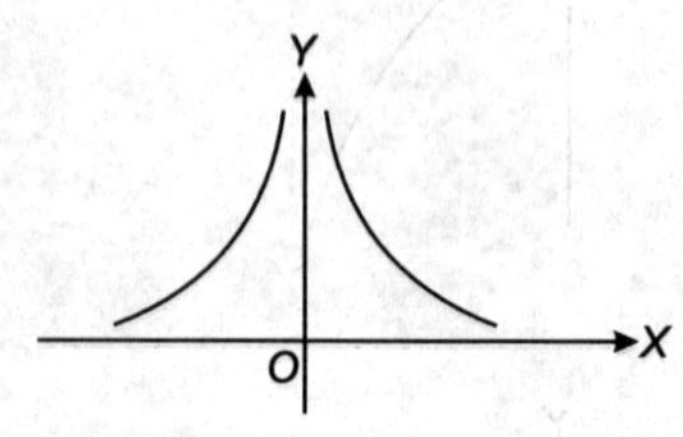

(iii)

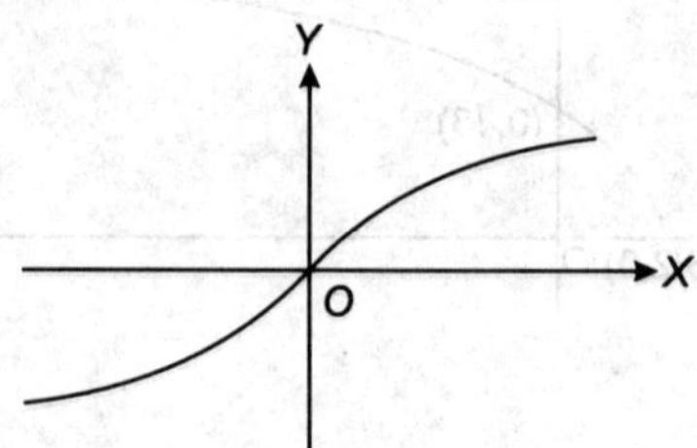

(iv)

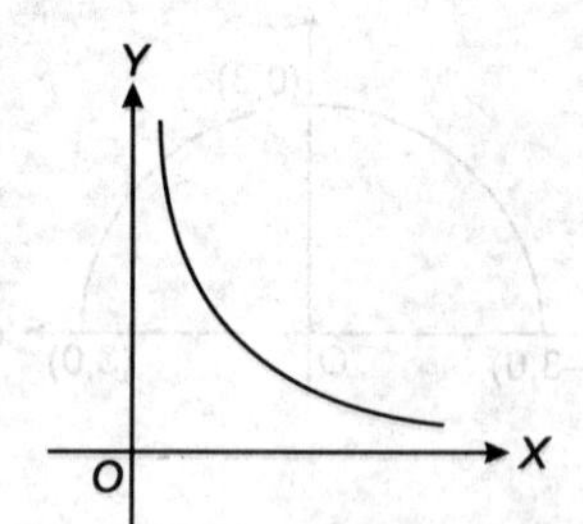

(v)

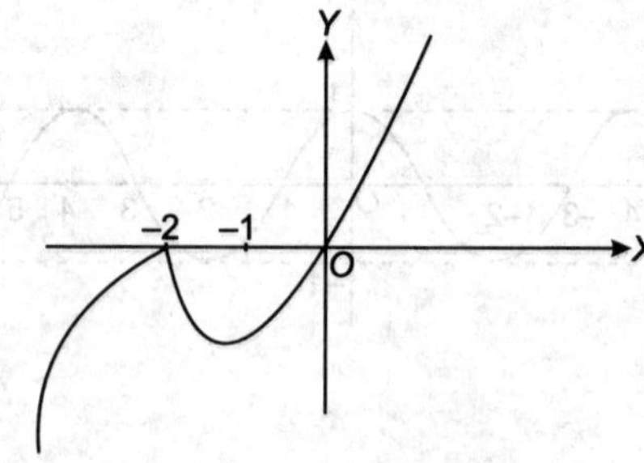

(vi)

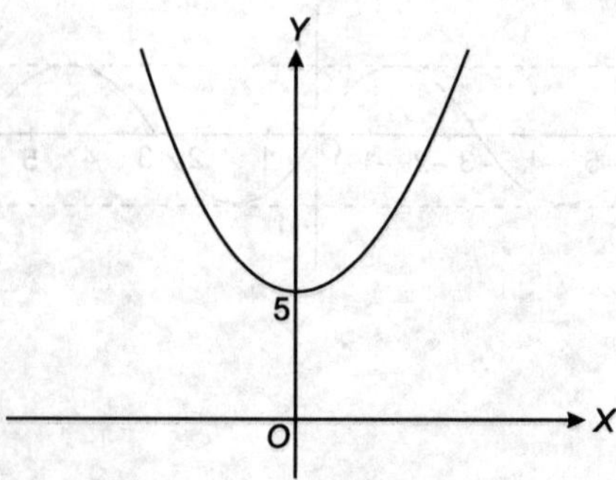

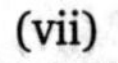

(vii)

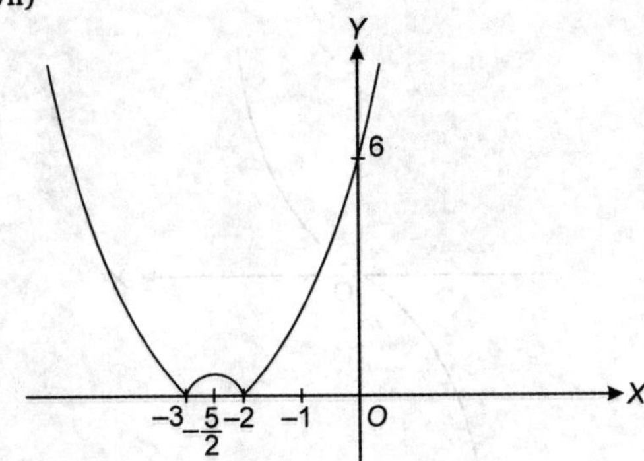

(viii)

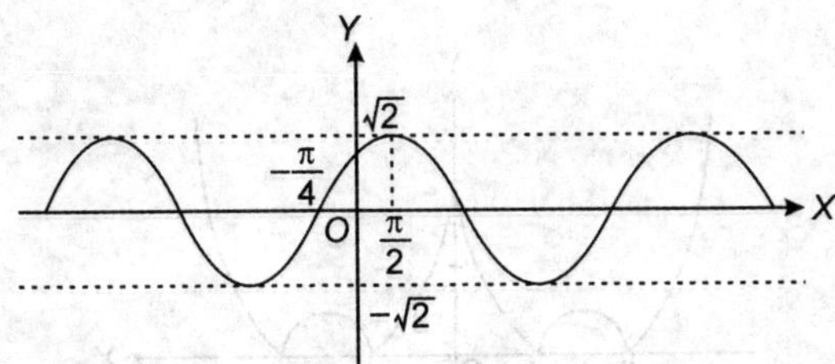

(ix)

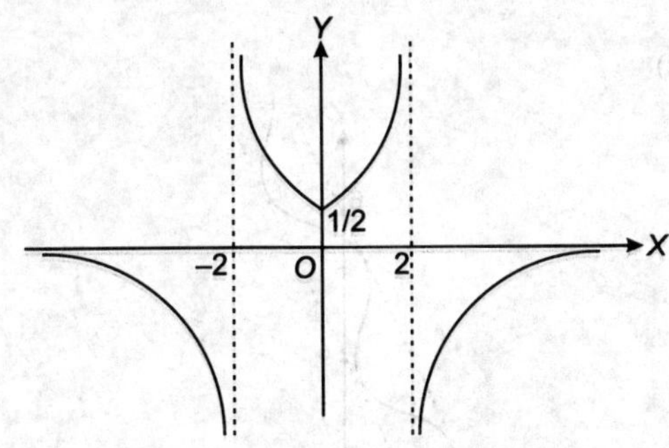

(x)

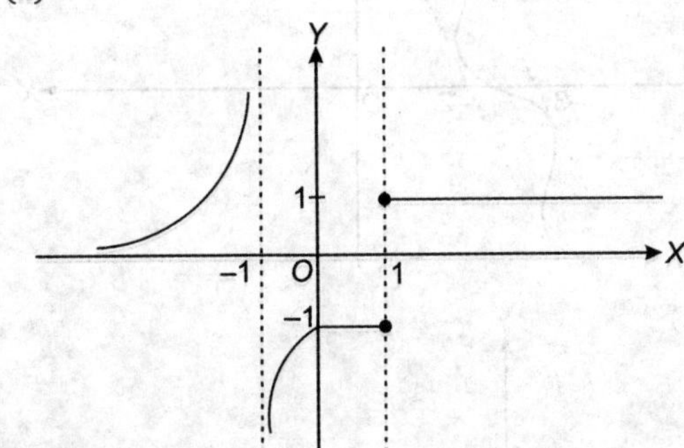

14. (i)

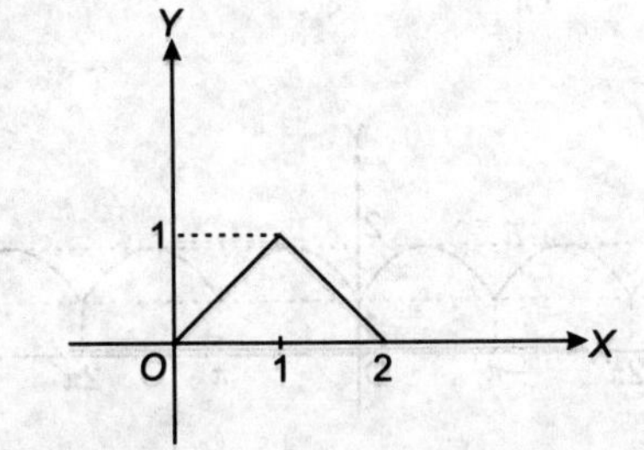

(ii)

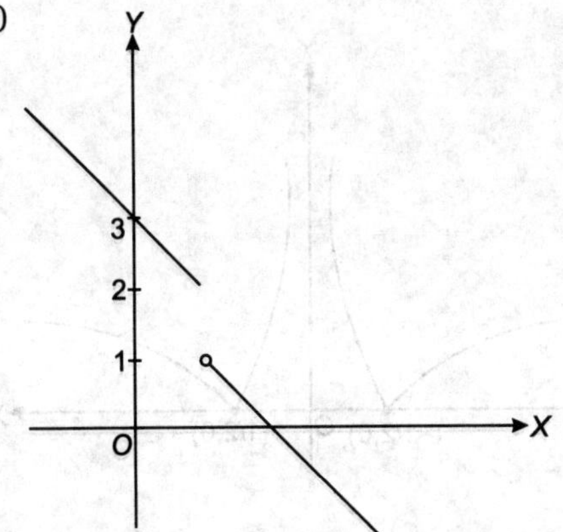

(iii)

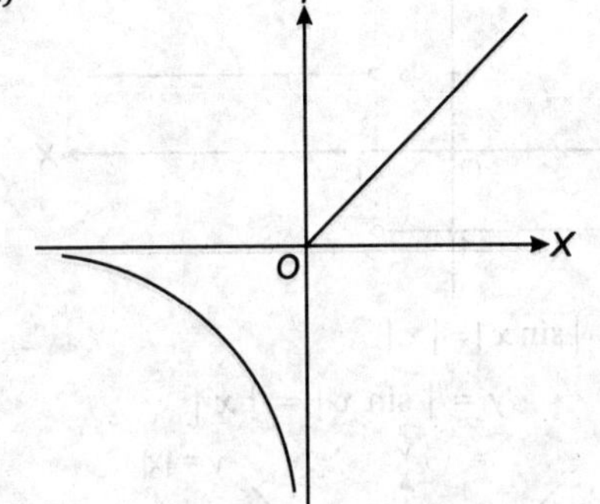

(iv)

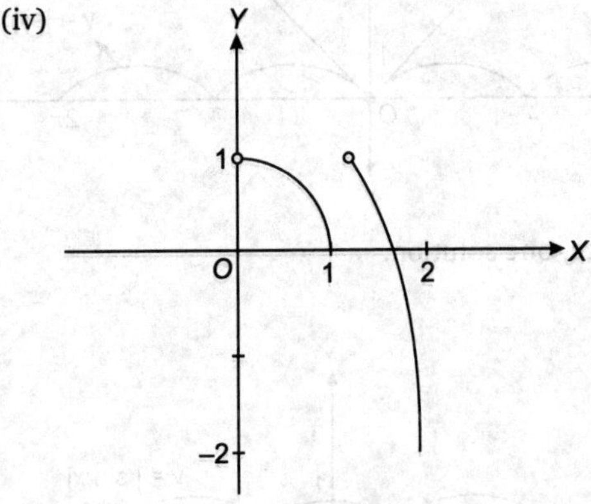

(v)

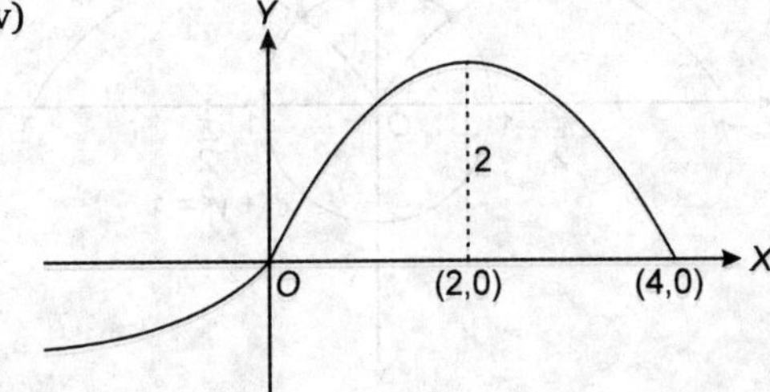

(vi)

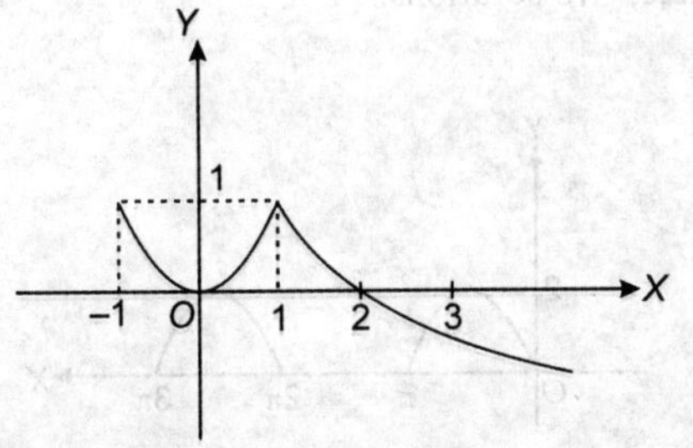

15.

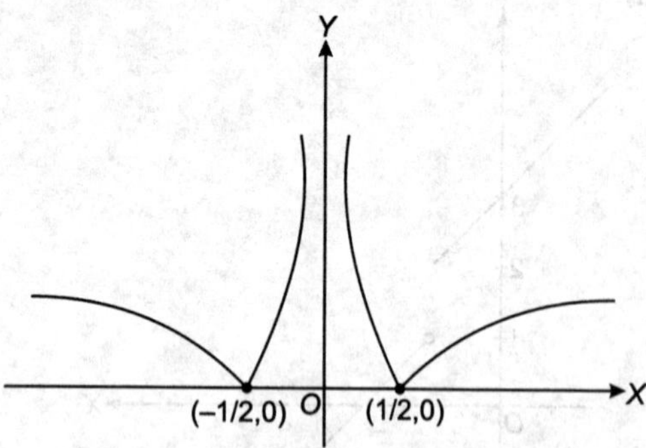

16.

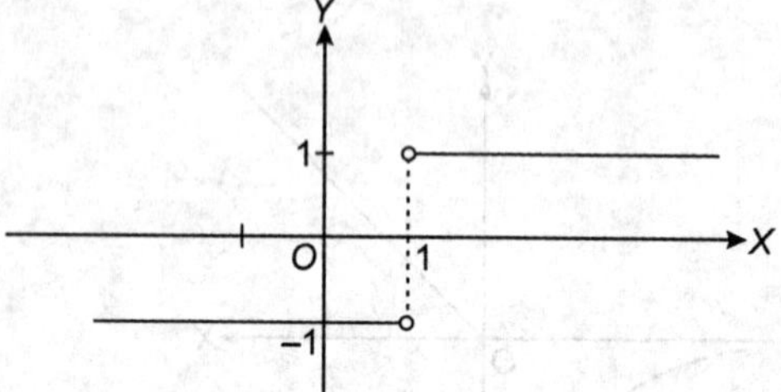

17. We have, $|\sin x| = |x|$

Let $\quad y = |\sin x| = |x|$

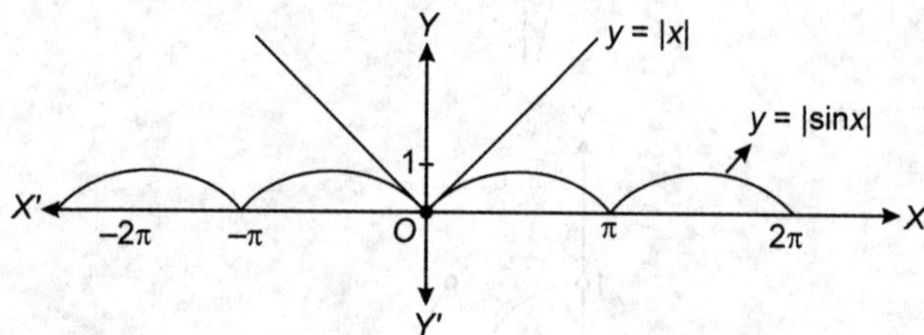

Hence, one solution.

18.

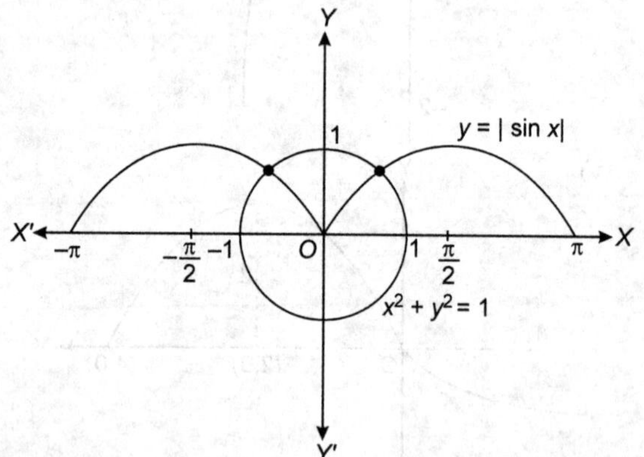

Hence, two solutions.

19.

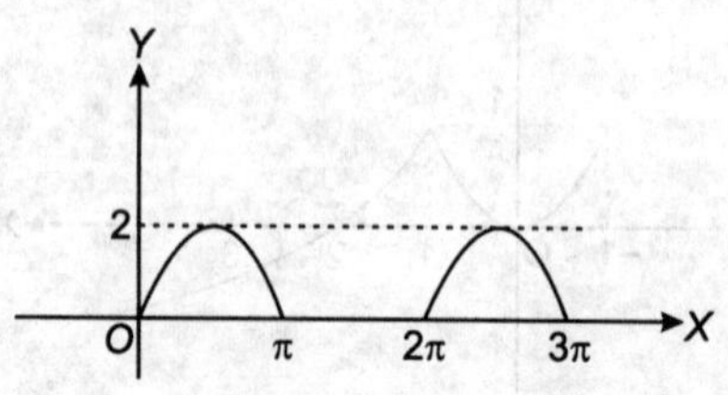

20.

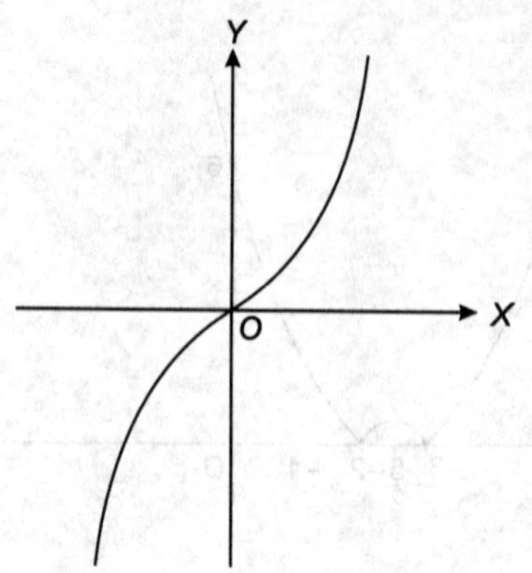

21.

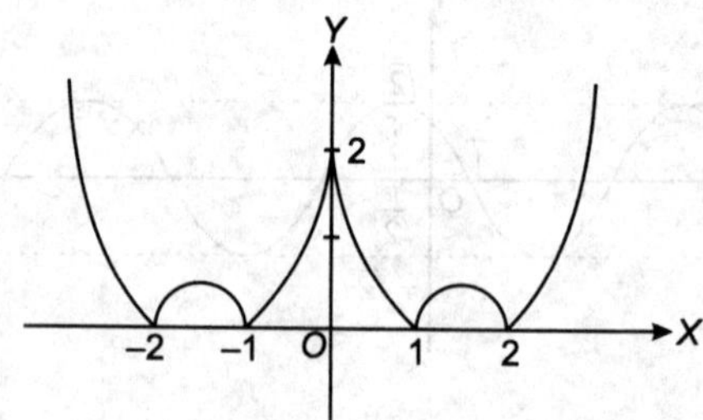

22. (i)

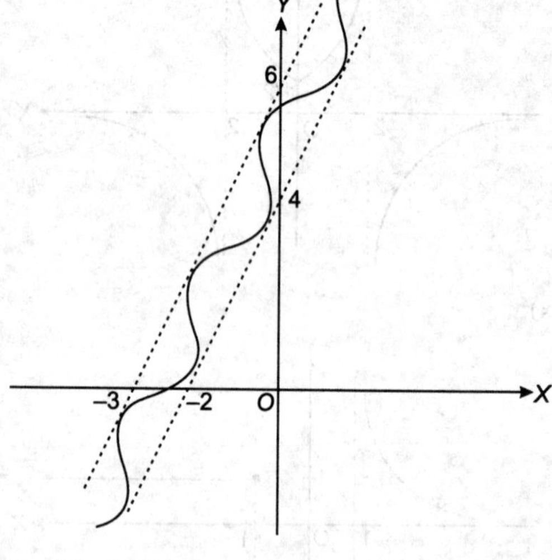

(ii)

(iii)

(iv)

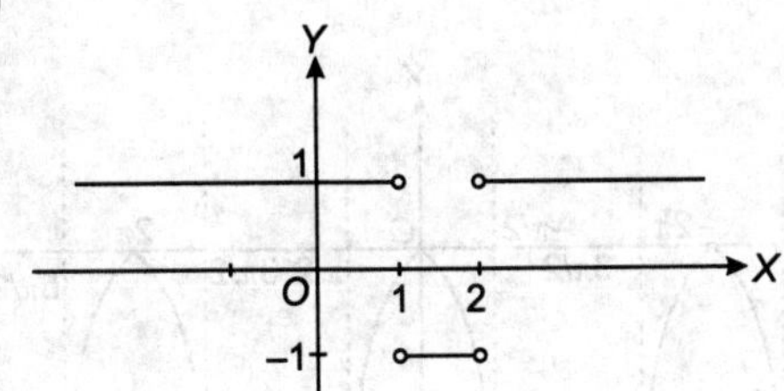

(v)

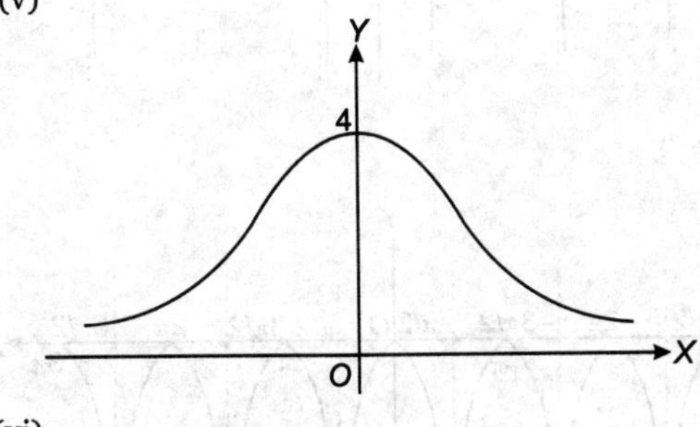

(vi)

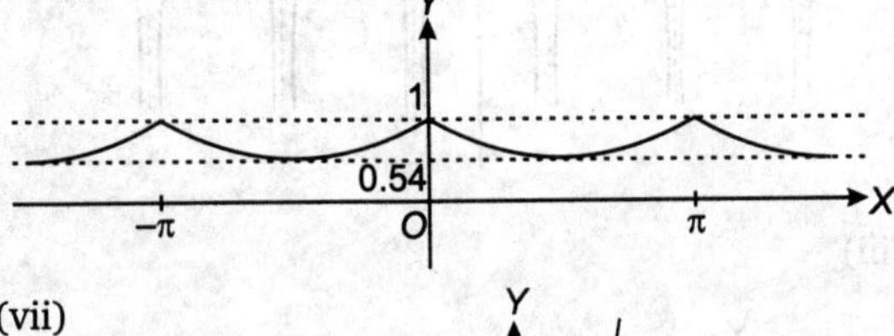

(vii)

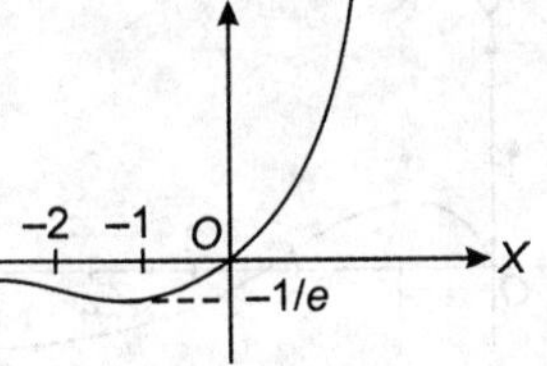

(viii)

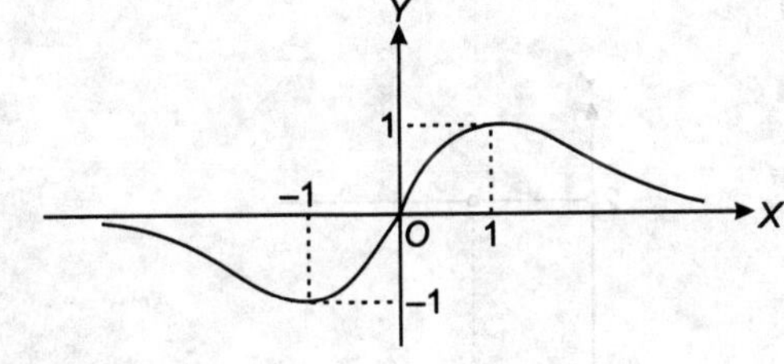

(ix)

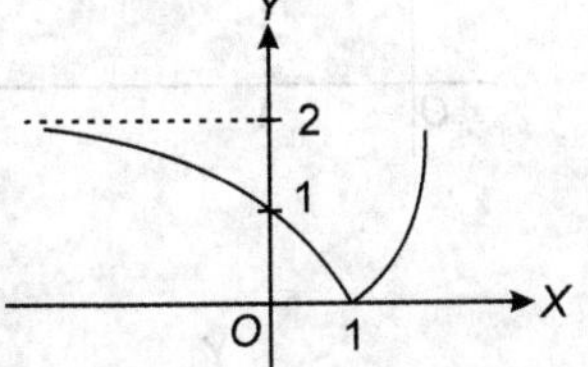

(x)

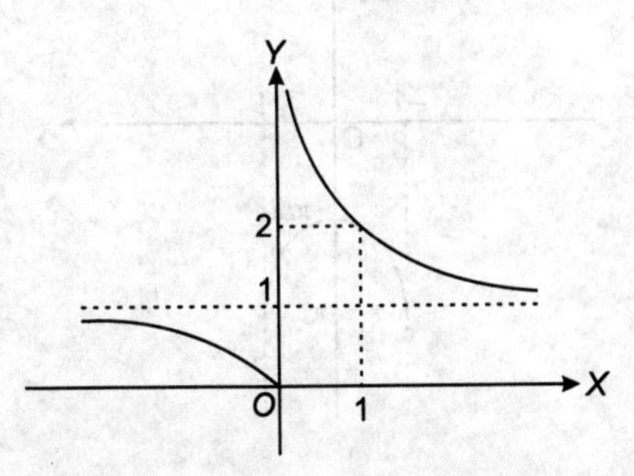

23. (i)

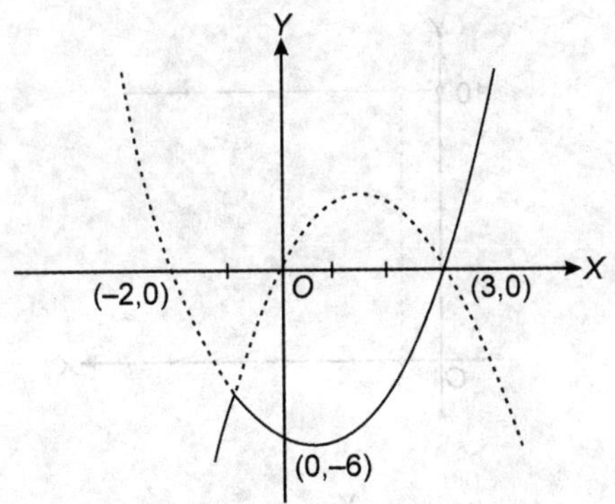

(ii)

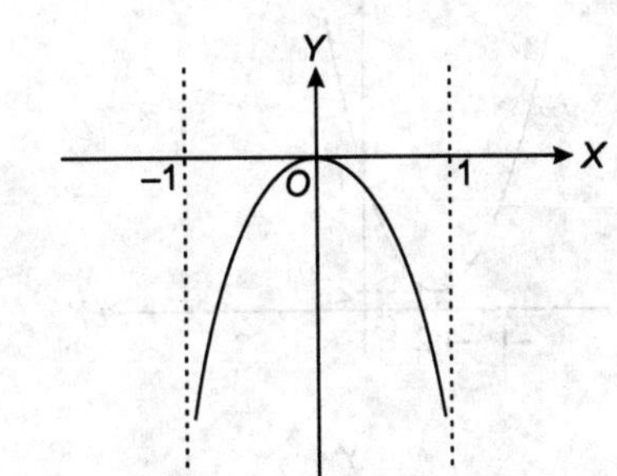

(iii)

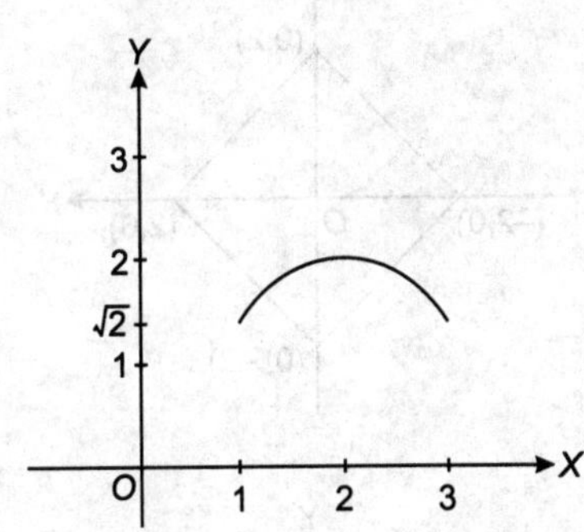

(iv)

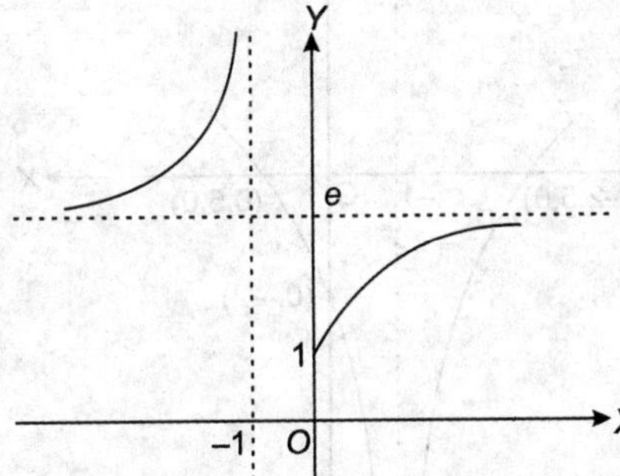

(v)

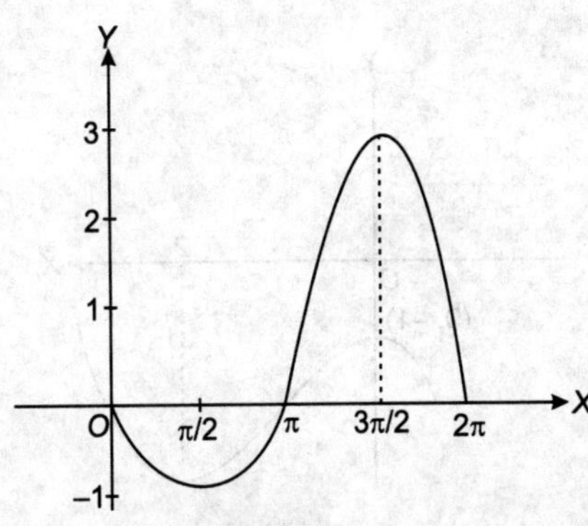

(vi)

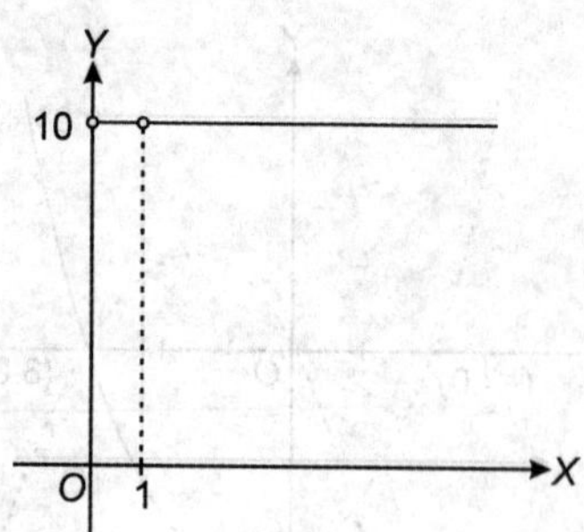

(vii)

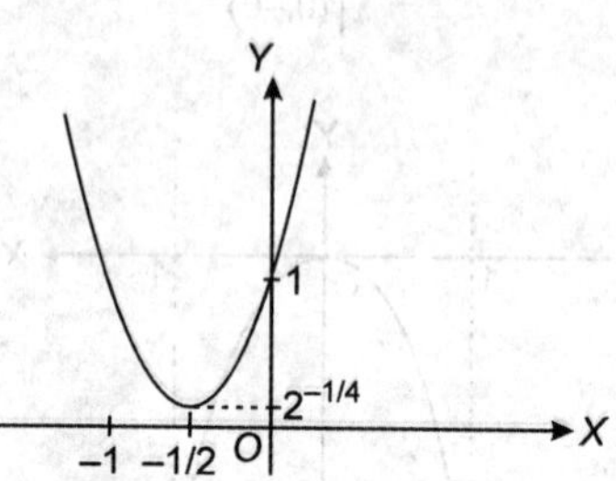

(viii)

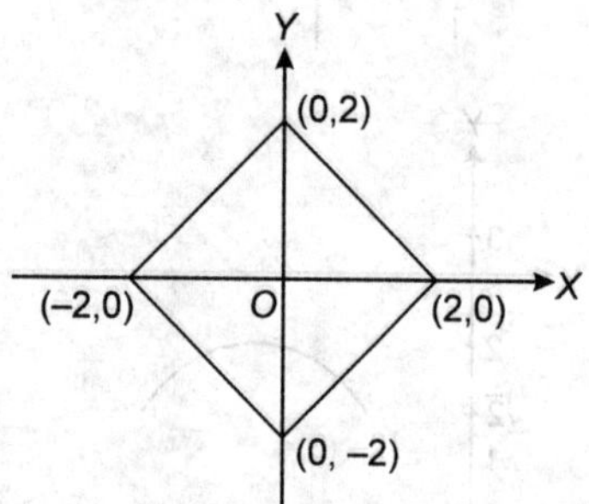

(ix)

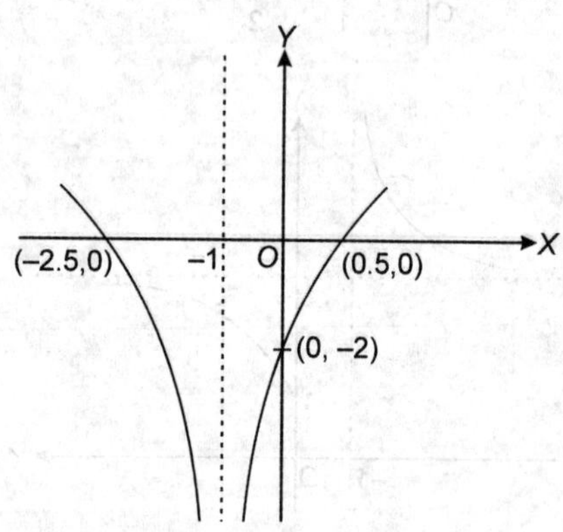

(x)

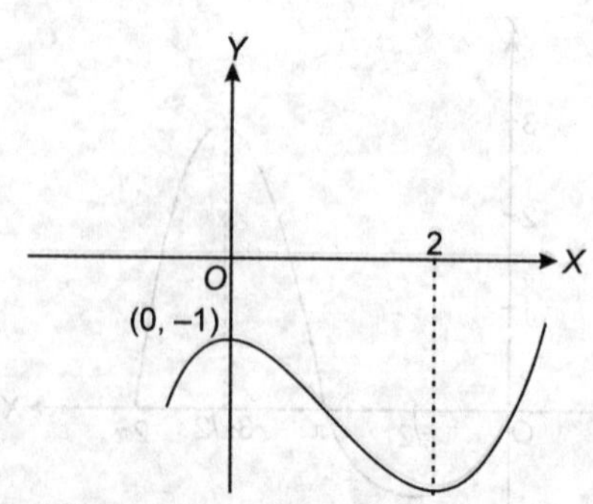

24. (i)

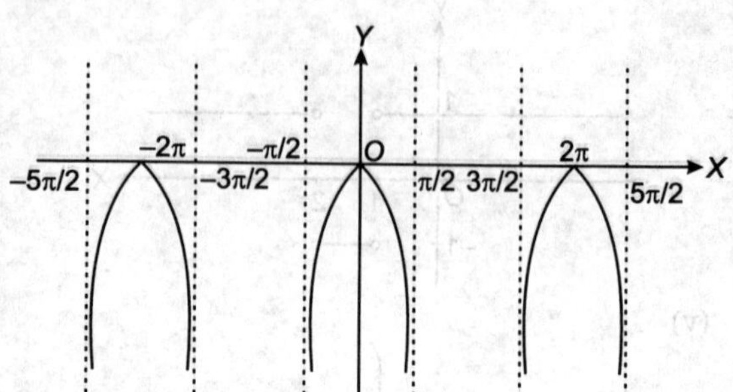

(ii)

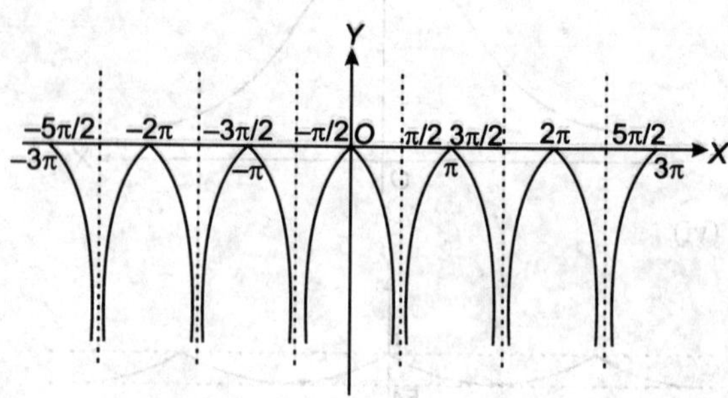

(iii)

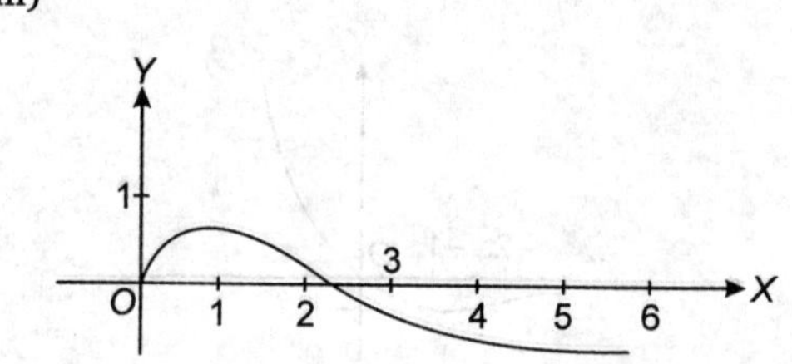

(iv)

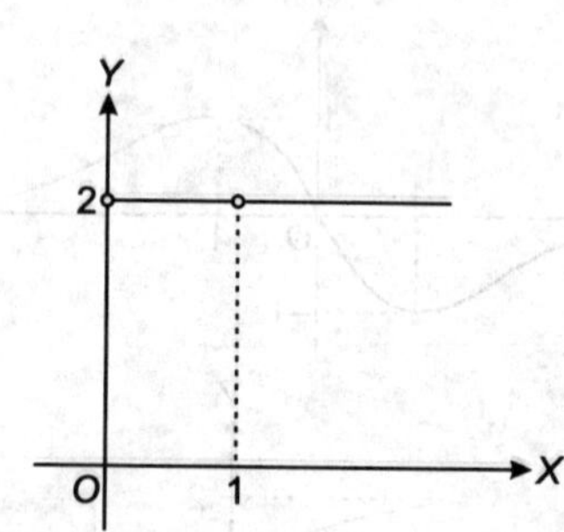

(v)

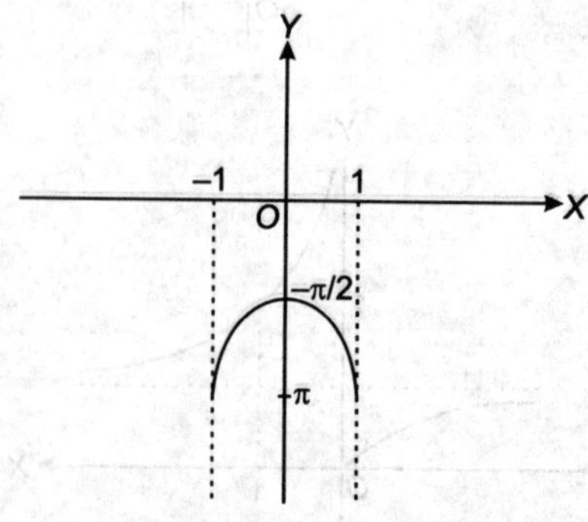

(vi)

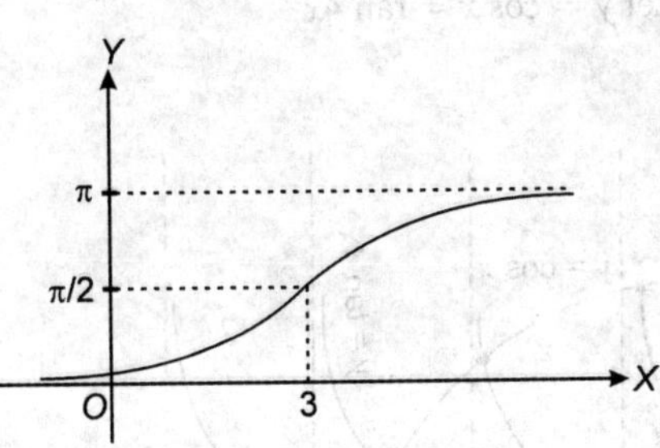

(vii)

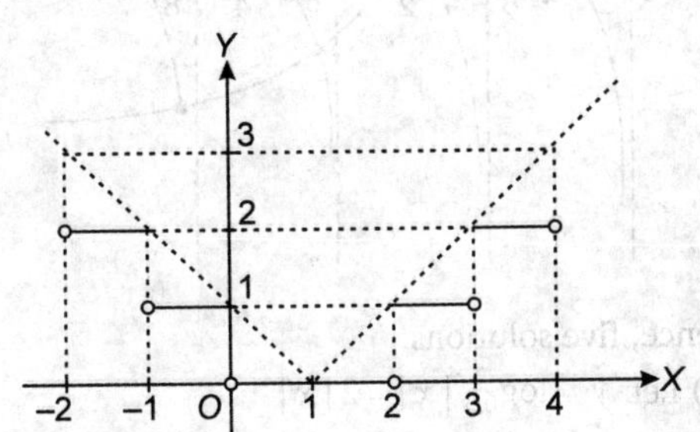

(viii)

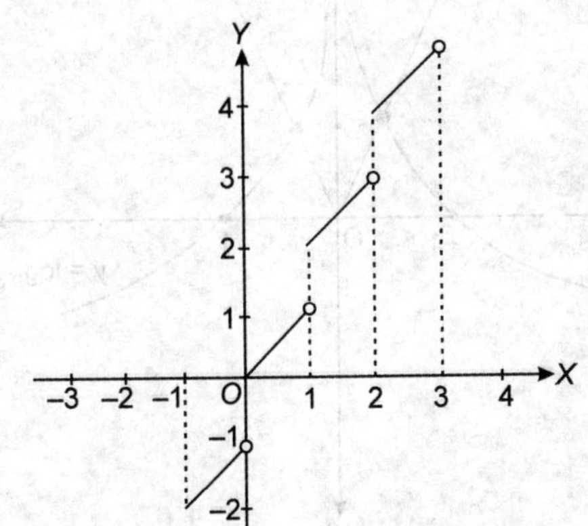

(ix)

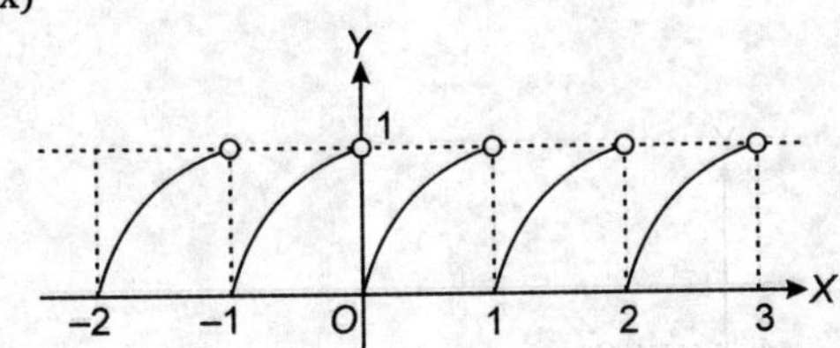

(x)

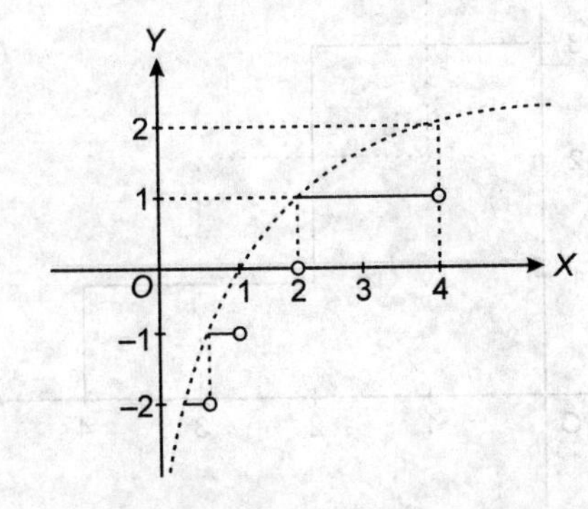

25. (i)

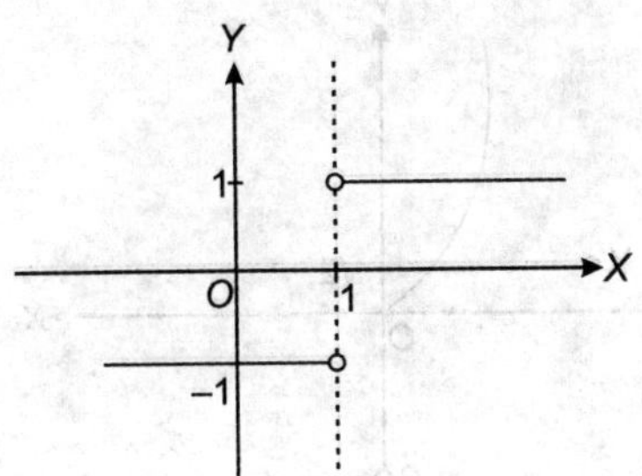

(ii)

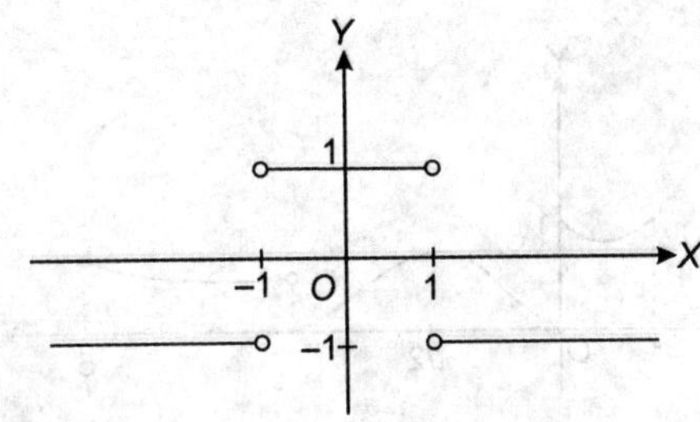

(iii)

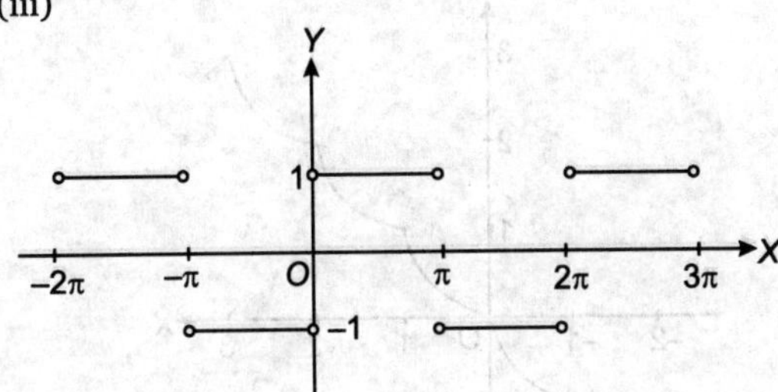

(iv)

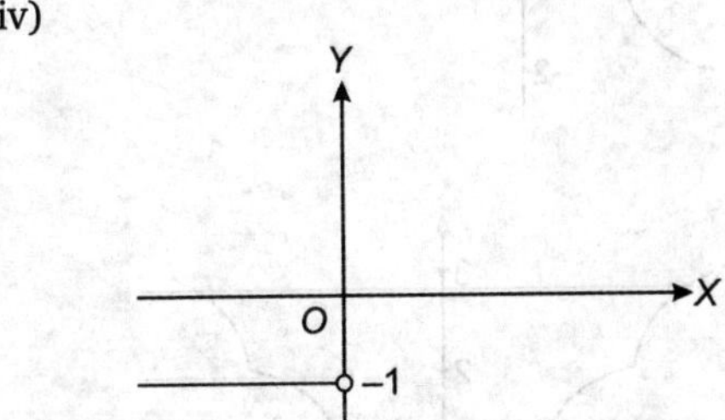

(v)

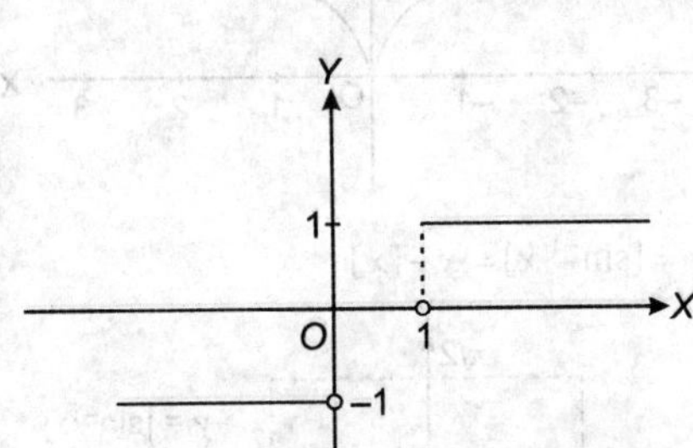

26. (i)

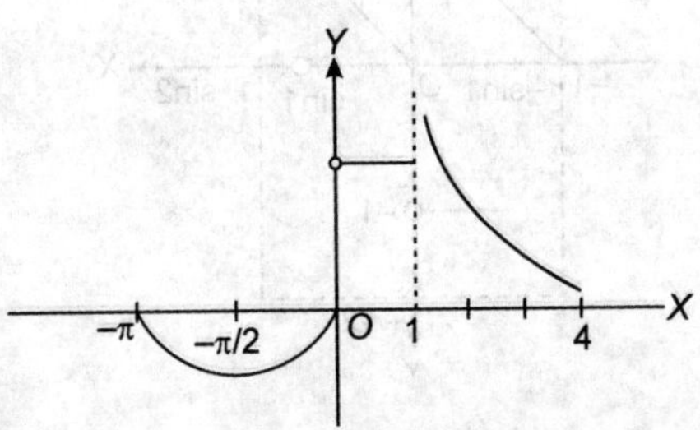

(ii)

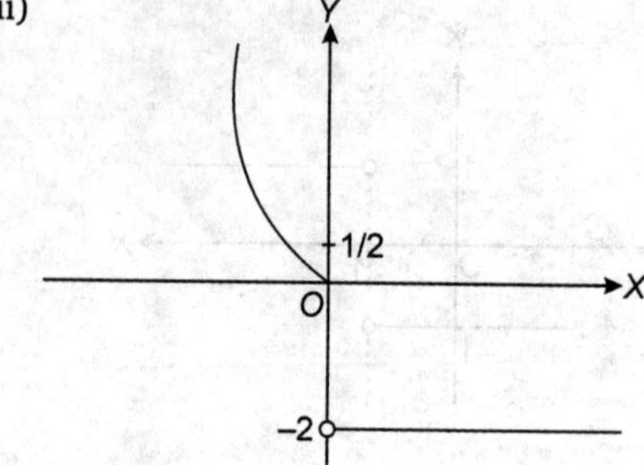

(iii)

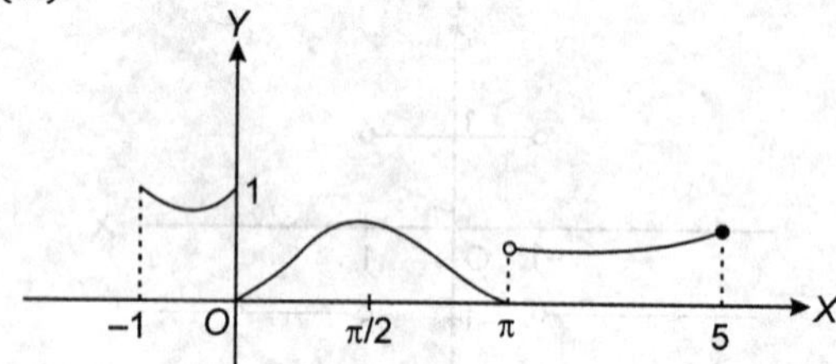

27. (i)

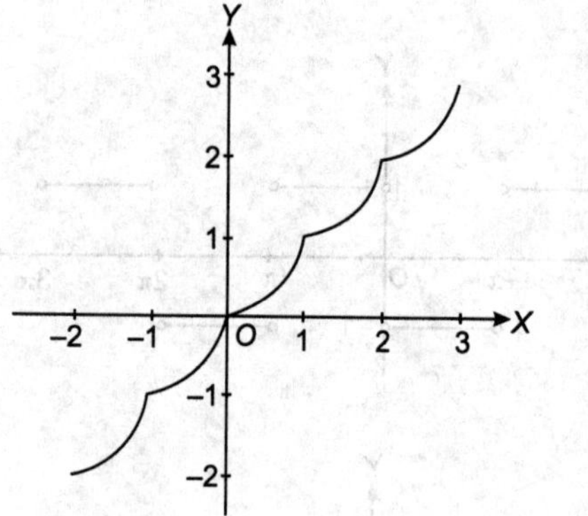

(ii)

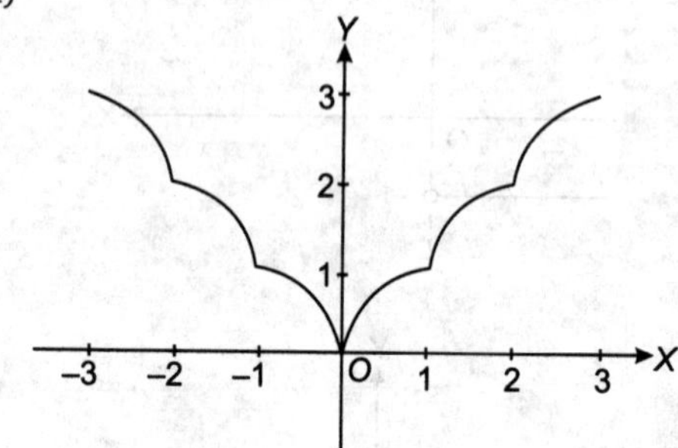

28. Let $y = [\sin^{-1} x] = x - [x]$

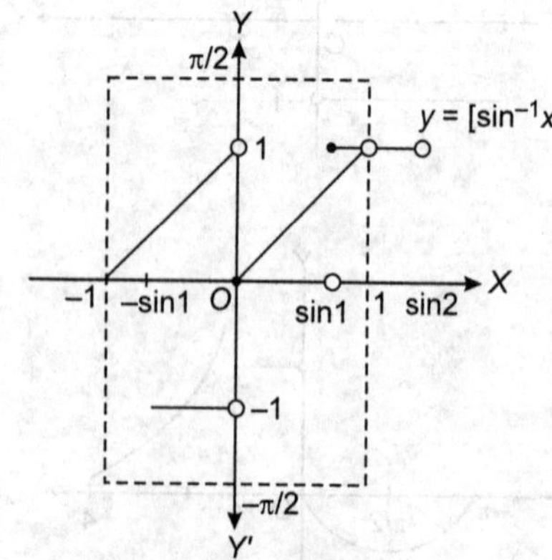

Hence, one solution.

29. (i) Let $y = \cos x = \tan 4x$

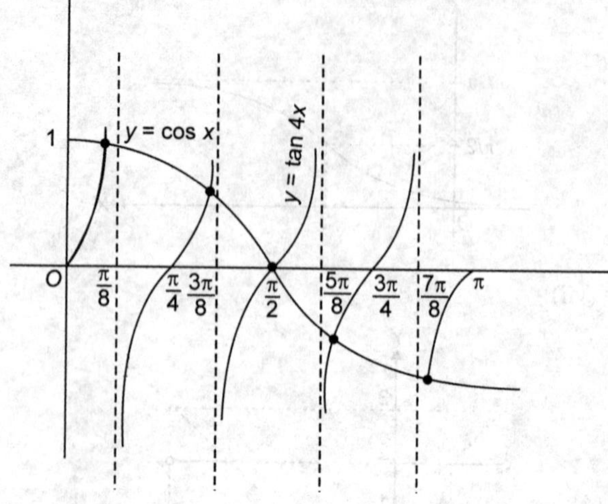

Hence, five solutions.

(ii) Let $y = \log_{0.5}|x| = 2|x|$

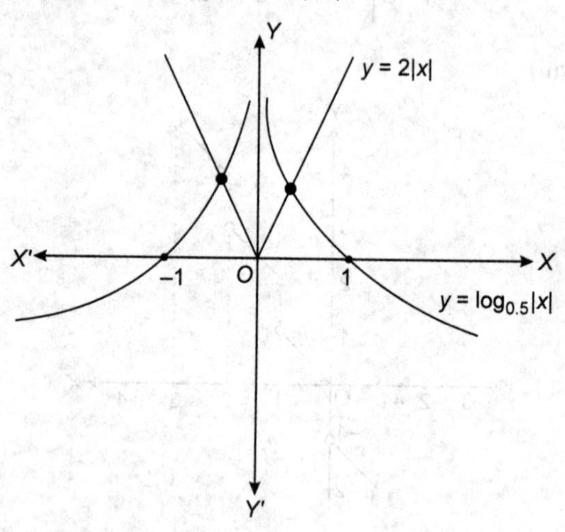

Hence, two solutions.

30. (i)

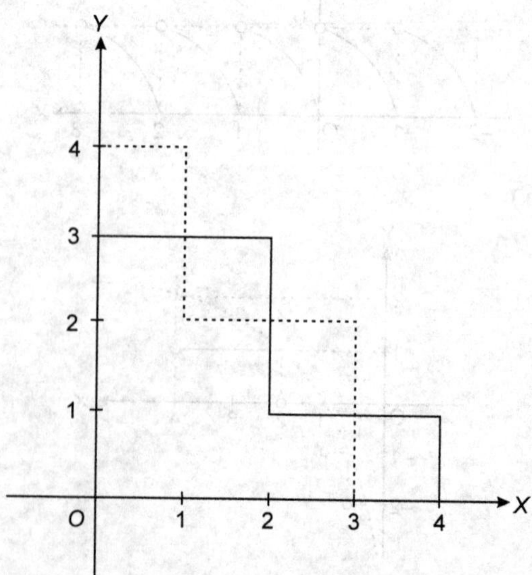

(ii)

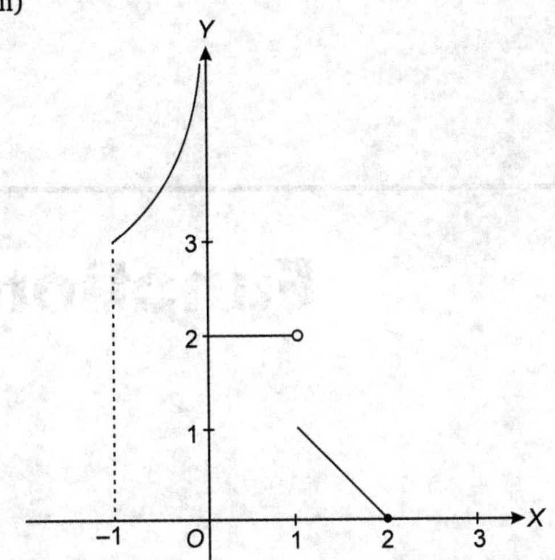

(iii)

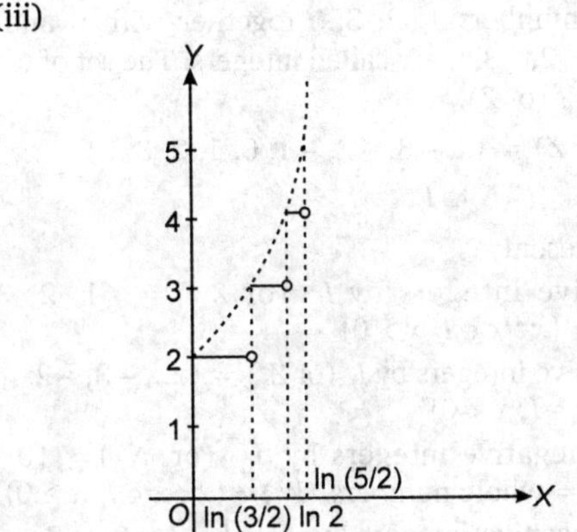

(iv)

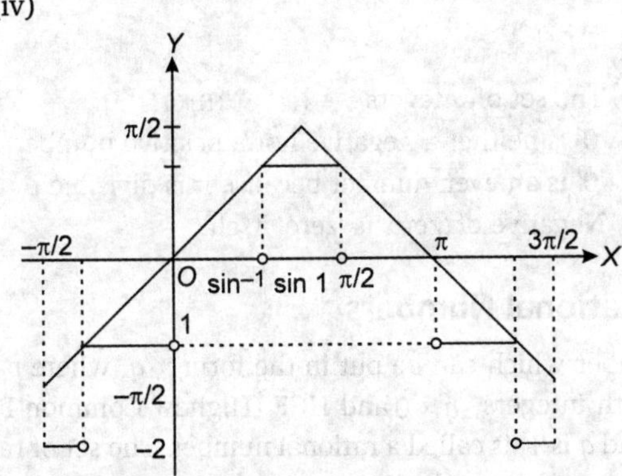

(v)

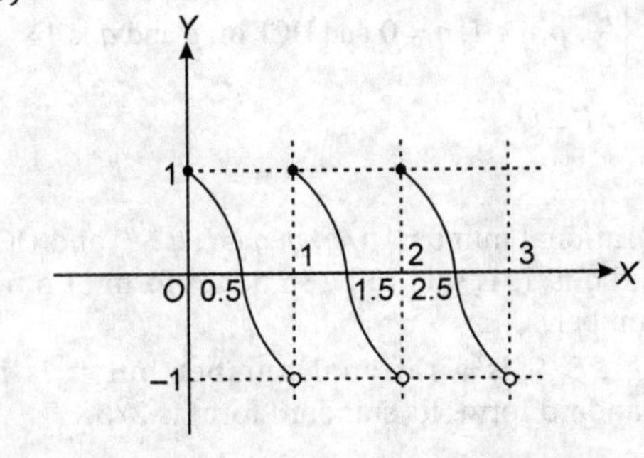

31.

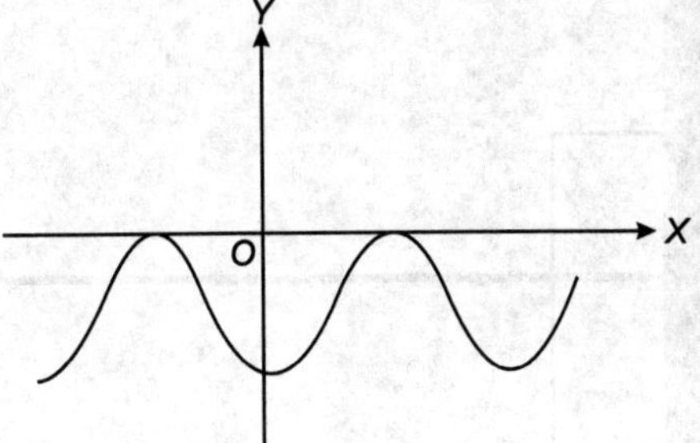

32.

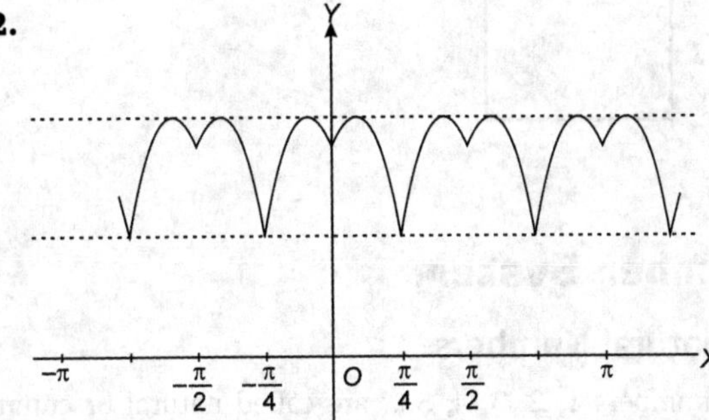

33.

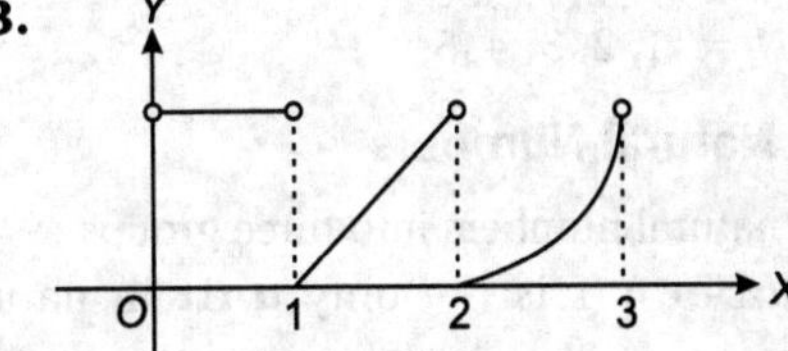

34.

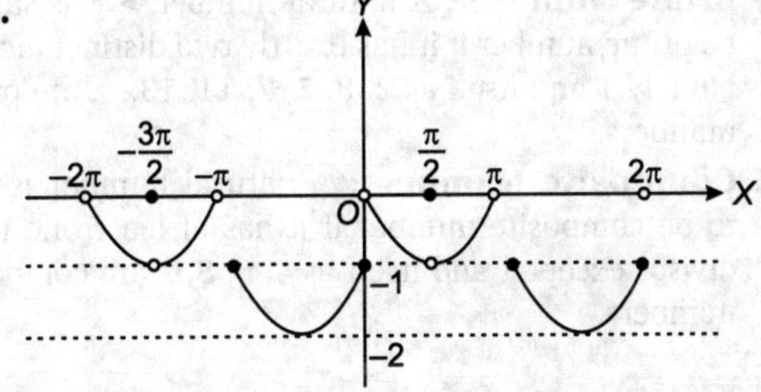

35.

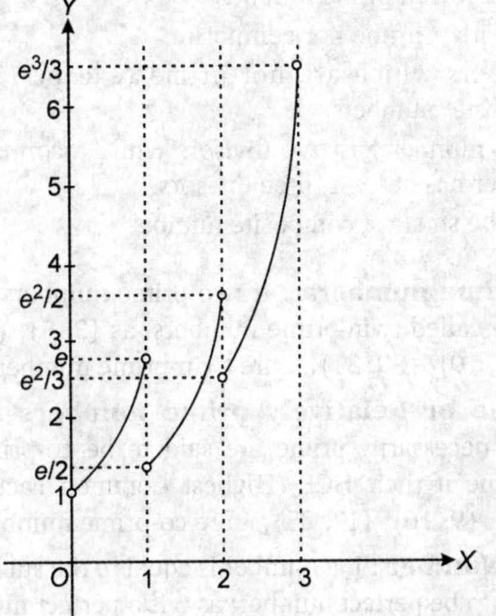

Chapter

Functions

Number System

(i) Natural Numbers

The numbers 1, 2, 3, 4, 5,... are called natural or counting numbers. The set of natural numbers is denoted by N.

i.e., $N = \{1, 2, 3, 4, 5, ...\}$

Classification of Natural Numbers

(a) We can divide natural numbers into three groups :

- **(i) Unit Number :** 1 is the only **unit** in natural numbers.
- **(ii) Prime Number :** A natural number > 1 is said to be prime number if it has exactly two distinct factors, namely 1 and itself as 2, 3, 5, 7, 11, 13,.... are prime numbers.
- **(iii) Composite Number :** A natural number is said to be composite number if it has at least one more divisor except 1 and itself as 4, 6, 8,... are composite numbers.

Note :

- (i) 2 is only even prime number.
- (ii) 1 is neither prime nor composite.
- (iii) Numbers which are not prime (except 1) are composite numbers.
- (iv) Prime number has two divisors while a composite number has at least three divisors.
- (v) '4' is the smallest composite number.

(b) Twin-prime numbers : If two prime numbers differ by 2, they are called twin-prime numbers, as {3, 5}, {5, 7}, {11, 13}, {17, 19}, {29, 31},... are twin-prime numbers.

(c) Co-prime or Relatively prime numbers : Two numbers not necessarily prime are said to be co-prime or relatively prime if their HCF (Highest Common Factor) is one, as {1, 2}, {9, 16}, {12, 25},... are co-prime numbers.

(d) Perfect Number : If a number is equal to the sum of its factors is said to be perfect number; as 6 is a perfect number. ($\because 1 + 2 + 3 = 1 \times 2 \times 3 = 6$)

(ii) Integers

The natural numbers 1, 2, 3,... together with 0 and their negative – 1, – 2, – 3,... are called integers. The set of integers is denoted by I (or Z).

i.e., I (or Z) = {... – 3, – 2, – 1, 0, 1, 2, 3,...)

Clearly, $N \subset I$

Here, we represent;

- (a) Positive integers by I^+ (or Z^+) = {1, 2, 3,....} $= N = \{x : x \in I, x > 0\}$
- (b) Negative integers by I^- (or Z^-) = (...., – 3, – 2, – 1} = $\{x : x \in I, x < 0\}$
- (c) Non-negative integers by I_0^+ (or N_0) = {0, 1, 2, 3,...} = whole numbers $(W) = \{x : x \in I, x \geq 0\}$
- (d) Non-positive integers by $I_0^- = \{ ...., -3, -2, -1, 0\}$ $= \{x : x \in I, x \leq 0\}$

Note :

- (i) The set of integers $I = \{(I^+) \cup (0) \cup (I^-)\}$.
- (ii) '0' is neither a negative nor a positive number.
- (iii) '0' is an even number because it is divisible by 2.
- (iv) Negative of 'zero' is 'zero' itself.

(iii) Rational Numbers

A number which can be put in the form p/q, where p and q are both integers, $q \neq 0$ and HCF (Highest Common Factor) of p and q is 1, is called a rational number. The set of rational numbers is denoted by Q.

i.e., $Q = \left\{\frac{p}{q}; p, q \in I, q \neq 0 \text{ and HCF of } p \text{ and } q \text{ is } 1\right\}$

Clearly, $N \subset I \subset Q$

Note :

(i) A rational number p/q; $p, q \in I, q > 0$ and HCF of p and q is 1, is said to be standard form of a rational number.

e.g., 6/9 is a rational number but it is not in standard form. Its standard form is 2/3.

(ii) If $q = 1$, then $Q = \{p;\, p \in I\}$
i.e., every integer is a rational number.

(iii) All recurring or terminating decimals are rational numbers.

e.g., (a) $Q = 0.\dot{3} = 0.3333...$ (Recurring decimal)

$\Rightarrow \quad 10Q - Q = 3$

$\Rightarrow Q = \frac{3}{9} = \frac{1}{3}$, which is a rational number.

(b) $Q = -\frac{3}{2} = -1.5$ (Terminating decimal)

(iv) "Two consecutive rational numbers" is meaningless.

(iv) Irrational Numbers

All those values which can not be expressed as recurring or terminating decimals are irrational numbers and is denoted by Q^c (i.e., complement of Q).

e.g., $\sqrt{2}, \pi + \sqrt{2}, 2 + \sqrt{5}, 3 - 2\sqrt{2}, \frac{1}{\sqrt{3}}, ...$ etc.

(v) Real Numbers

The complete set of rational and irrational numbers is the set of real numbers and is denoted by R.

Thus $\quad R = Q \cup Q^c$

Clearly, $\quad N \subset I \subset Q \subset R$

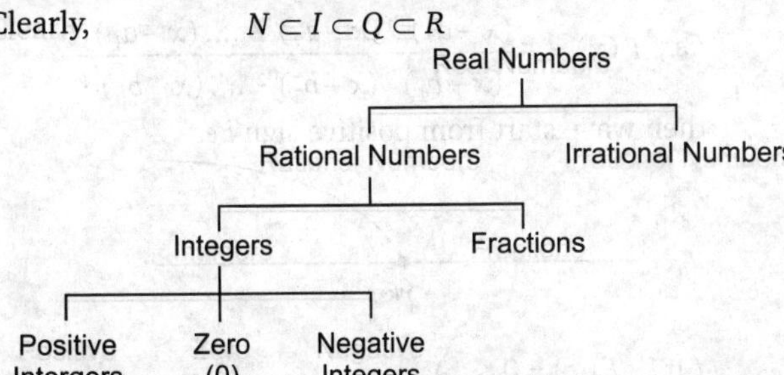

Note :

(i) We will take every real number as a point on a line and call this line number line or real line.

$-\pi \quad -\sqrt{6} \quad -\sqrt{3} \; -1 \quad -1 \quad -1/2 \quad 0 \quad 1 \quad \sqrt{2} \quad 2 \quad \sqrt{5} \quad \pi$

(ii) The set R represent the set of continuous values (not discrete values).

INTRODUCTORY EXERCISE 2.1

Only One Option is Correct

1. If $\lambda = \sqrt{2 + \sqrt{2 + \sqrt{2 + ... \infty}}} = \sqrt{2\sqrt{2\sqrt{2 ..... \infty}}}$, then 2λ is :

(a) first prime number (b) first composite number
(c) irrational number (d) none of these

2. If $a^b + b^a = 4$, then $\{a, b\}$ are :

(a) twin prime numbers
(b) relative prime numbers
(c) prime numbers
(d) irrational numbers

3. If $q = 0.\dot{6}$ (recurring decimal) and $q = 0.\dot{1}$ (recurring decimal) and if $N = p/q$, then N is :

(a) a prime number
(b) an odd integer
(c) a perfect number
(d) a unit natural number

4. If α, β, γ are the roots of the equation $x^3 - nx^2 + nx + 5x - n = 0$, where $n \in N$, then n is :

(a) a prime number
(b) an odd integer
(c) a perfect number
(d) none of these

5. The more accurate value of π is :

(a) $\frac{22}{7}$ (b) 3.141
(c) $\frac{355}{113}$ (d) 3.1412

Intervals

The set of numbers between any two real numbers is called interval. The following are four types of interval :

(i) Closed Interval : $x \in [a, b] = \{x \in R : a \le x \le b\}$

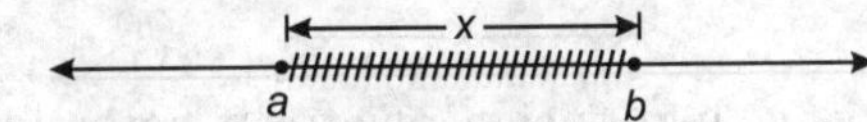

(ii) Open Interval : $x \in (a, b)$ or $]\,a, b\,[= \{x \in R : a < x < b\}$

(iii) Closed-open Interval : $x \in [a, b)$ or $[\,a, b\,[= \{x \in R : a \le x < b\}$

(iv) Open-Closed Interval : $x \in (a, b]$ or $]\,a, b\,] = \{x \in R : a < x \le b\}$

What is Infinity?

Infinity (∞) is a symbol and not a number and we can not draw ∞ on the paper. We will assume the existence of two symbols $-\infty$ (minus infinity) and ∞ (plus infinity) outside the set of real numbers R.

i.e., $\quad x \in R \Rightarrow -\infty < x < \infty$ or $x \in (-\infty, \infty)$

Note :

(i) If $x = \infty$ or $x \to \infty$ (same thing) then x is larger than any real number however large and x is not a fixed number.

(ii) If $x = -\infty$ or $x \to -\infty$ (same thing) then x is smaller than any real number however small and x is not a fixed number.

Infinite Intervals are Defined as Follows

(i) $x \in (a, \infty)$ or $x \in]a, \infty[= \{x \in R : x > a\}$

$-\infty$ ⟵ a ⟶ ∞

(ii) $x \in [a, \infty)$ or $x \in [a, \infty[= \{x \in R : x \geq a\}$

$-\infty$ ⟵ a ⟶ ∞

(iii) $x \in (-\infty, b)$ or $x \in]-\infty, b[= \{x \in R : x < b\}$

$-\infty$ ⟵ b ⟶ ∞

(iv) $x \in (-\infty, b]$ or $x \in]-\infty, b] = \{x \in R : x \leq b\}$

$-\infty$ ⟵ b ⟶ ∞

Basic Properties of Inequalities

(i) If $a \leq b$, then either $a < b$ or $a = b$
(ii) If $a < b$ and $b < c$, then $a < c$
(iii) If $a < b$, then $a + m < b + m$, where $m \in R$
(iv) If $a < b$, then $ma < mb$ for $m > 0$ and $ma > mb$ for $m < 0$
(v) If $a < b$ and $c < d$, then $a + c < b + d$ and $a - d < b - c$
(vi) If $0 < a < b$, then $a^x < b^x$ if $x > 0$ and $a^x < b^x$ if $x < 0$.
(vii) If $0 < x < y$, then $a^x < a^y$ if $a > 1$ and $a^x > a^y$ if $0 < a < 1$.

Note :

(i) The inequality sign reverse if both sides are multiplied by a negative number :
i.e., if $a < b \Rightarrow -a > -b$

(ii) If a and b both are positive or both negative, then if $a < b \Rightarrow \frac{1}{a} > \frac{1}{b}$

(iii) $a + \frac{1}{a} \geq 2$ for $a > 0$ and equality holds for $a = 1$

(iv) $a + \frac{1}{a} \leq -2$ for $a < 0$ and equality holds for $a = -1$

(v) If $\frac{a}{b} > 0$, then $\frac{b}{a} > 0$ and if $\frac{a}{b} < 0$, then $\frac{b}{a} < 0$.

(vi) If $a < b$, then $a^2 < b^2$ does not follow always.
$\because 3 < 4 \Rightarrow 9 < 16$ (true) but $-4 < 3 \Rightarrow 16 < 9$ (false)
so if $x > 3 \Rightarrow x^2 > 9$ but for $x < -3 \Rightarrow x^2 \geq 0$

(vii) (a) If $3 < x < 4 \Rightarrow 9 < x^2 < 16$
(b) If $-3 < x < 4 \Rightarrow 0 \leq x^2 < 16$
(c) If $-4 < x < 3 \Rightarrow 0 \leq x^2 < 16$
(d) If $-4 < x < -3 \Rightarrow 9 < x^2 < 16$

Wavy Curve Method

Generalised Method of Intervals

In order to solve inequalities of the form

$$\frac{P(x)}{Q(x)} > 0, \ \frac{P(x)}{Q(x)} < 0, \ \frac{P(x)}{Q(x)} \geq 0 \text{ and } \frac{P(x)}{Q(x)} \leq 0,$$

where $P(x)$ and $Q(x)$ are polynomial functions. Assume,

$$P(x) = (x - a_1)^{m_1}(x - a_2)^{m_2}(x - a_3)^{m_3} \ldots (x - a_k)^{m_k}$$

and $Q(x) = (x - b_1)^{n_1}(x - b_2)^{n_2}(x - b_3)^{n_3} \ldots (x - b_p)^{n_p}$,

where $m_1, m_2, m_3, \ldots, m_k, n_1, n_2, n_3, \ldots, n_p$ are natural numbers and $a_1, a_2, a_3, \ldots a_k, b_1, b_2, b_3, \ldots, b_p$ are distinct real numbers.

We use the following method :

Let $F(x) = \frac{P(x)}{Q(x)}$

(i) $F(x) = 0$ for $x = a_1, a_2, a_3, \ldots a_k$ and $F(x)$ is not defined for $x = b_1, b_2, b_3, \ldots, b_p$.
Apart from these $(k + p)$ real numbers. Now arrange $a_1, a_2, a_3, \ldots, a_k, b_1, b_2, b_3, \ldots, b_p$ in an increasing order say $c_1, c_2, c_3, \ldots, c_{k+p}$
i.e., $c_1 < c_2 < c_3 < \ldots < c_{k+p}$
Plot them on the real line and draw a curve starting from right of c_{k+p} along the real line.

(ii) For $F(x) \geq 0$ or $F(x) \leq 0$ arrange black circles (•) for the roots of $P(x) = 0$ and white circles (o) for the roots of $Q(x) = 0$ and for $F(x) > 0$ or $F(x) < 0$ arrange white circles (o) for the roots of $P(x) = 0$ and $Q(x) = 0$ on the real line.

(iii) If in all brackets before x positive sign and expression has also positive sign

i.e., $F(x) = + \frac{(x - a_1)^{m_1}(x - a_2)^{m_2} \ldots (x - a_k)^{m_k}}{(x - b_1)^{n_1}(x - b_2)^{n_2} \ldots (x - b_p)^{n_p}}$

then wave start from positive sign *i.e.,*

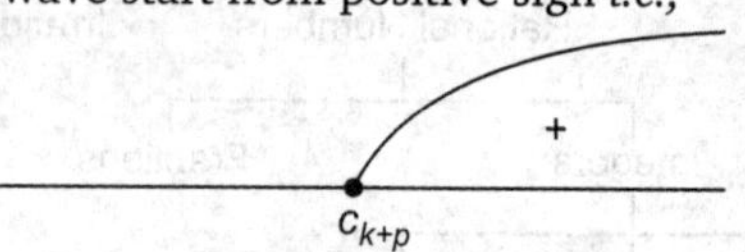

$\begin{pmatrix} \text{If} & F(x) = 0 \\ \text{at} & x = c_{k+p} \end{pmatrix}$

and if in all brackets before x positive sign and expression has negative sign

i.e., $F(x) = - \frac{(x - a_1)^{m_1}(x - a_2)^{m_2} \ldots (x - a_k)^{m_k}}{(x - b_1)^{n_1}(x - b_2)^{n_2} \ldots (x - b_p)^{n_p}}$

then wave start from negative sign *i.e.,*

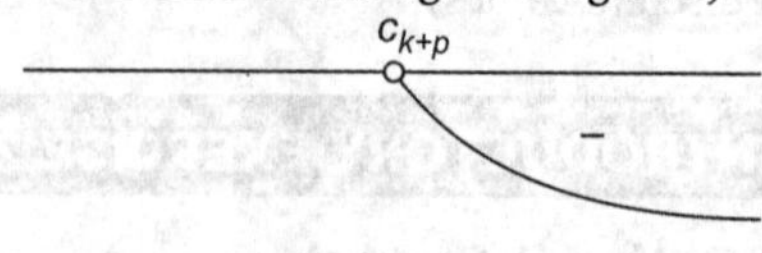

$\begin{pmatrix} \text{If} & F(x) \neq 0 \\ \text{at} & x = c_{k+p} \end{pmatrix}$

(iv) If roots occur even times then sign remains same from right to left side of the roots

i.e., + + or − −

and if roots occur odd times then, sign will change from right to left side of the roots.

i.e.,

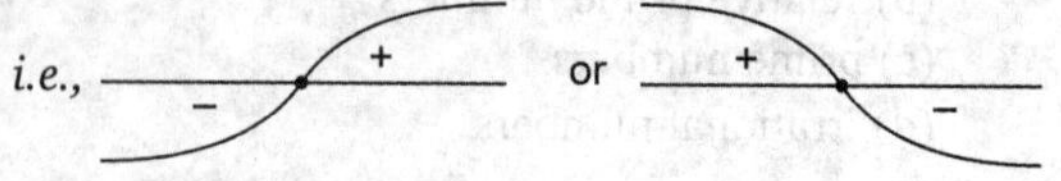

Frequently Used Inequalities

(i) $(x-a)(x-b) < 0$ or $\frac{(x-a)}{(x-b)} < 0$ or $\frac{(x-b)}{(x-a)} < 0 \Rightarrow x \in (a, b)$, where $a < b$

(ii) $(x-a)(x-b) > 0$ or $\frac{(x-a)}{(x-b)} > 0$ or $\frac{(x-b)}{(x-a)} > 0 \Rightarrow x \in (-\infty, a) \cup (b, \infty)$, where $a < b$

(iii) (a) $(x-a)(x-b) \le 0 \Rightarrow x \in [a, b]$, where $a < b$ (b) $\frac{(x-a)}{(x-b)} \le 0 \Rightarrow x \in [a, b)$, where $a < b$

(c) $\frac{(x-b)}{(x-a)} \le 0 \Rightarrow x \in (a, b]$, where $a < b$

(iv) (a) $(x-a)(x-b) \ge 0 \Rightarrow x \in (-\infty, a] \cup [b, \infty)$, where $a < b$

(b) $\frac{(x-a)}{(x-b)} \ge 0 \Rightarrow x \in (-\infty, a] \cup (b, \infty)$, where $a < b$

(c) $\frac{(x-b)}{(x-a)} \ge 0 \Rightarrow x \in (-\infty, a) \cup [b, \infty)$, where $a < b$

Illustration 1. Let $F(x) = \dfrac{(x-1)^{10000}\left(x-\frac{2}{3}\right)^{57}(x-7)^4\left(x+\frac{8}{7}\right)^{927}}{(x+3)^{17}(x-2)^6(x+6)^2}$.

Solve the following inequalities :

(i) $F(x) > 0$ (ii) $F(x) \ge 0$ (iii) $F(x) < 0$ (iv) $F(x) \le 0$

Sol : The critical points are $-6, -3, -\frac{8}{7}, \frac{3}{3}, 1, 2, 7$

(i) For $F(x) > 0$: Critical points are : -6 -3 $-8/7$ $\frac{2}{3}$ 1 2 7

Circles : o o o o o o o

Roots repeated : even odd odd odd even even even times

The relevant wavy curve of the function $F(x) \ge 0$ is :

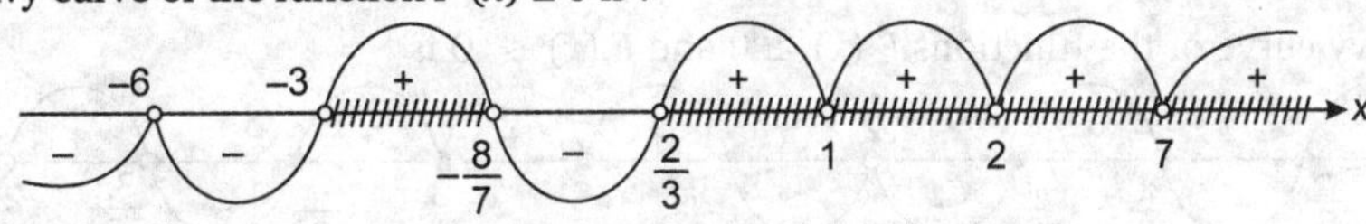

$\therefore\ x \in \left(-3, -\frac{8}{7}\right) \cup \left(\frac{2}{3}, 1\right) \cup (1, 2) \cup (2, 7) \cup (7, \infty)$

(ii) For $F(x) \ge 0$: Critical point are : -6 -3 $-8/7$ $2/3$ 1 2 7

Circles : o o • • • o •

Roots repeated : even odd odd odd even even even times

The relevant wavy curve of the function $F(x) \ge 0$ is :

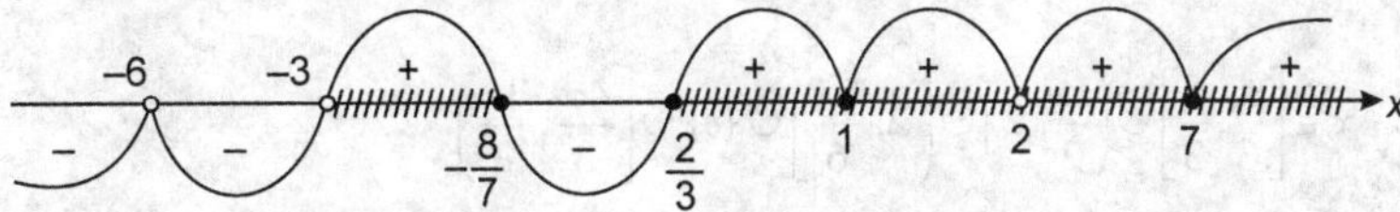

$\therefore\ x \in \left(-3, -\frac{8}{7}\right] \cup \left[\frac{2}{3}, 1\right] \cup [1, 2) \cup (2, 7] \cup [7, \infty)$ or $x \in \left(-3, -\frac{8}{7}\right] \cup \left[\frac{2}{3}, 2\right) \cup (2, \infty)$

(iii) For $F(x) < 0$: Critical points are : -6 -3 $-8/7$ $2/3$ 1 2 7

Circles 0 0 0 0 0 0 0

Roots repeated : even odd odd odd even even even times

The relevant wavy curve of the function $F(x) < 0$ is :

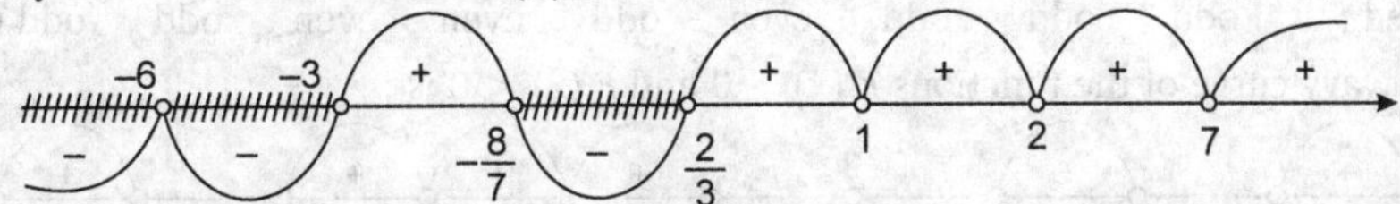

$\therefore\ x \in (-\infty, -6) \cup (-6, -3) \cup \left(-\frac{8}{7}, \frac{2}{3}\right)$

(iv) For $F(x) \le 0$: Critical points are : –6 – 3 – 8/7 2/3 1 2 7

Circles : o o • • • o •

Roots repeated : even odd odd odd even even even times

The relevant wavy curve of the function $F(x) \le 0$ is :

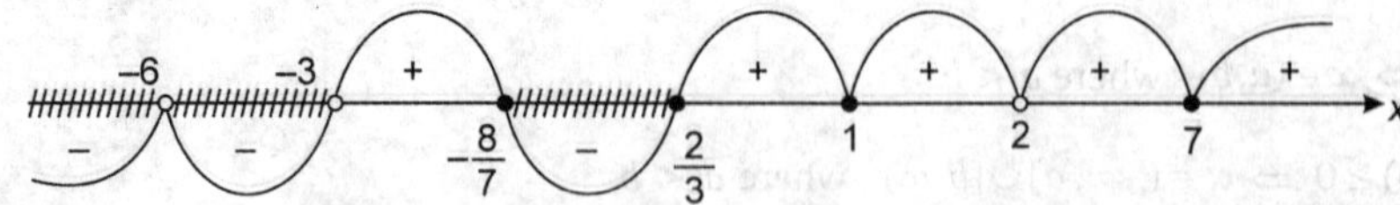

$$\therefore \quad x \in (-\infty, -6) \cup (-6, -3) \cup \left[-\frac{8}{7}, \frac{2}{3}\right] \cup \{1, 7\}$$

Illustration 2. Let $F(x) = \dfrac{\left(\sin x - \frac{1}{2}\right)(\ln x - 1)^4 (x-2)^3 (\tan x - \sqrt{3})^5}{(e^x - e^2)(x-3)^6(\sin^2 x - 1)}$

Solve the following inequalities for $x \in (0, 2\pi)$

(i) $F(x) \ge 0$ (ii) $F(x) \le 0$ (iii) $F(x) > 0$ (iv) $F(x) < 0$

Sol : Here critical points are (solving $N^r = 0$ and $D^r = 0$)

$$\frac{\pi}{6}, \frac{\pi}{3}, \frac{\pi}{2}, 2, \frac{5\pi}{6}, e, 3, \frac{4\pi}{3}, \frac{3\pi}{2} \text{ in } (0, 2\pi)$$

(i) $F(x) \ge 0$ and (ii) For $F(x) \le 0$:

Critical points : $\frac{\pi}{6}$ $\frac{\pi}{3}$ $\frac{\pi}{2}$ 2 $\frac{5\pi}{6}$ e 3 $\frac{4\pi}{3}$ $\frac{3\pi}{2}$

Circles : • • o o • • o • o

Roots repeated : odd odd odd even odd even even odd odd times

The relevant wavy curve of the functions $F(x) \ge 0$ and $F(x) \le 0$ is

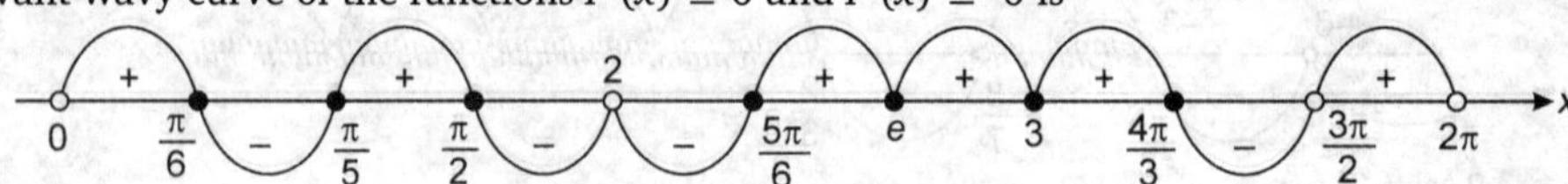

(i) For $F(x) \ge 0$: $x \in \left(0, \frac{\pi}{6}\right] \cup \left[\frac{\pi}{3}, \frac{\pi}{2}\right) \cup \left[\frac{5\pi}{6}, e\right] \cup [3, 3) \cup \left(3, \frac{4\pi}{3}\right] \cup \left(\frac{3\pi}{2}, 2\pi\right)$

or $x \in \left(0, \frac{\pi}{6}\right] \cup \left[\frac{\pi}{3}, \frac{\pi}{2}\right) \cup \left[\frac{5\pi}{6}, \frac{4\pi}{3}\right] \cup \left(\frac{3\pi}{2}, 2\pi\right) - \{3\}$

(ii) For $F(x) \le 0$:

$$x \in \left[\frac{\pi}{6}, \frac{\pi}{3}\right] \cup \left(\frac{\pi}{2}, 2\right) \cup \left(2, \frac{5\pi}{6}\right] \cup \{e\} \cup \left[\frac{4\pi}{3}, \frac{3\pi}{2}\right)$$

or $x \in \left[\frac{\pi}{6}, \frac{\pi}{3}\right] \cup \left(\frac{\pi}{2}, \frac{5\pi}{6}\right) \cup \left[\frac{4\pi}{3}, \frac{3\pi}{2}\right) \cup \{e\} - \{2\}$

(iii) $F(x) > 0$ and (iv) $F(x) < 0$:

Critical points : $\frac{\pi}{6}$ $\frac{\pi}{3}$ $\frac{\pi}{2}$ 2 $\frac{5\pi}{6}$ e 3 $\frac{4\pi}{3}$ $\frac{3\pi}{2}$

Circles : o o o o o o o o o

Roots repeated : odd odd odd even odd even even odd odd times

The relevant wavy curve of the functions $F(x) > 0$ and $F(x) < 0$ is :

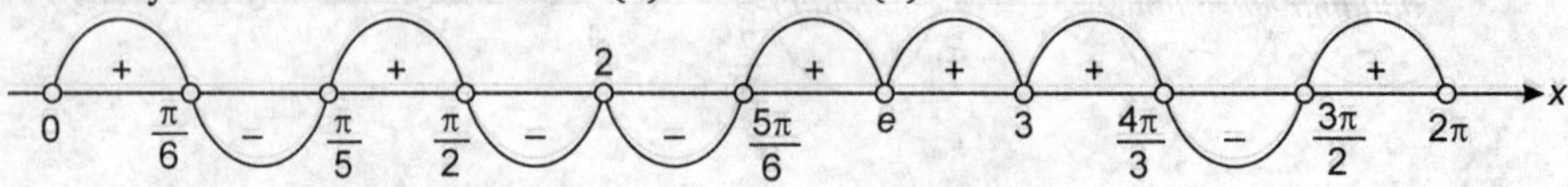

For $F(x) > 0$:

$$x \in \left(0, \frac{\pi}{6}\right) \cup \left(\frac{\pi}{3}, \frac{\pi}{2}\right) \cup \left(\frac{5\pi}{6}, e\right) \cup (e, 3) \cup \left(3, \frac{4\pi}{3}\right) \cup \left(\frac{3\pi}{2}, 2\pi\right)$$

or $x \in \left(0, \frac{\pi}{6}\right) \cup \left(\frac{\pi}{3}, \frac{\pi}{2}\right) \cup \left(\frac{5\pi}{6}, \frac{4\pi}{3}\right) \cup \left(\frac{3\pi}{2}, 2\pi\right) - \{e, 3\}$

For $F(x) < 0$:

$$x \in \left(\frac{\pi}{6}, \frac{\pi}{3}\right) \cup \left(\frac{\pi}{2}, 2\right) \cup \left(2, \frac{5\pi}{6}\right) \cup \left(\frac{4\pi}{3}, \frac{3\pi}{2}\right)$$

or $x \in \left(\frac{\pi}{6}, \frac{\pi}{3}\right) \cup \left(\frac{\pi}{2}, \frac{5\pi}{6}\right) \cup \left(\frac{4\pi}{3}, \frac{3\pi}{2}\right) - \{2\}$

Illustration 3. If $\frac{(1-x)(2-x)^2(3-x)^3}{(4-x)^4(5-x)^5} \geq 0$, then find the values of x.

Sol : Let $F(x) = \frac{(1-x)(2-x)^2(3-x)^3}{(4-x)^4(5-x)^5}$

$$= -\frac{(x-1)(x-2)^2(x-3)^3}{(x-4)^4(x-5)^5}$$

Here critical points are	:	1	2	3	4	5
Circles	:	•	•	•	o	o
Roots repeated	:	odd	even	odd	even	odd times

The relevant wavy curve of the function $F(x) \geq 0$ is

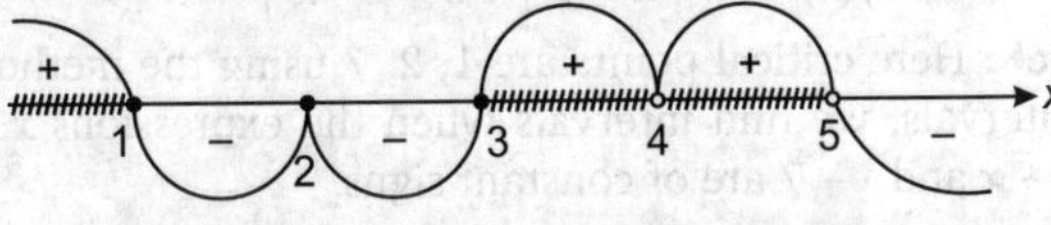

$\therefore \quad x \in (-\infty, 1] \cup [3, 4) \cup (4, 5)$

Illustration 4. If $(x-5)^5(x-77)^{77}(x-111)^{111}(x-2211)^{2211} < 0$, then find the values of x and also find the number of integeral solutions.

Sol : Let $F(x)$

$$= (x-5)^5(x-77)^{77}(x-111)^{111}(x-2211)^{2211}$$

Here critical points are	:	5	77	111	2211
Circles	:	o	o	o	o

and all roots repeated odd times.

The relevant wavy curve of the function $F(x) < 0$ is :

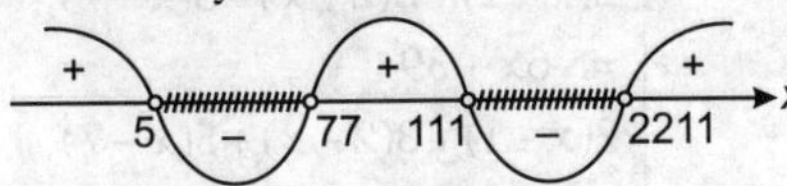

$\therefore \quad x \in (5, 77) \cup (111, 2211)$

$\Rightarrow x = 6, 7, 8, 9, ..., 76, 112, 113, ..., 2210$

$\therefore$ Number of integral solutions are

$$71 + 2099 = 2170$$

Illustration 5. Let $F(x) = \frac{(x-2)^2(x-6)^6(x-10)^{10}}{(x-4)^4(x-8)^8}$.

Solve the following inequalities :

(i) $F(x) \geq 0$ (ii) $F(x) \leq 0$
(iii) $F(x) > 0$ (iv) $F(x) < 0$

Sol : Here critical points are :

2 4 6 8 10

and all roots repeated even times.

(i) $F(x) \geq 0$ and (ii) $F(x) \leq 0$:

Critical points : 2 4 6 8 10

Circles : • o • o •

The relevant wavy curve of the functions $F(x) \geq 0$ and $F(x) \leq 0$ is :

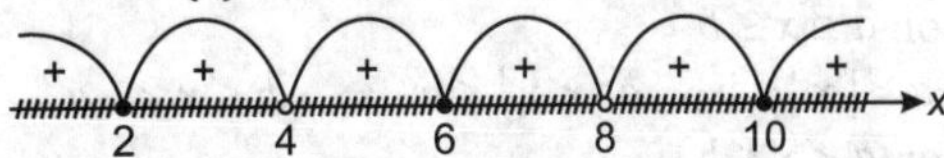

For $F(x) \geq 0$: $x \in (-\infty, 2] \cup [2, 4) \cup (4, 6] \cup [6, 8) \cup (8, 10] \cup [10, \infty)$

or $x \in (-\infty, 4) \cup (4, 8) \cup (8, \infty)$ or $x \in R \sim \{4, 8\}$

For $F(x) \leq 0$: $x = \{2, 6, 10\}$

(iii) For $F(x) > 0$ and (iv) For $F(x) < 0$:

Critical points : 2 4 6 8 10

Circles : o o o o o

The relevant wavy curve of the functions $F(x) > 0$ and $F(x) < 0$ is :

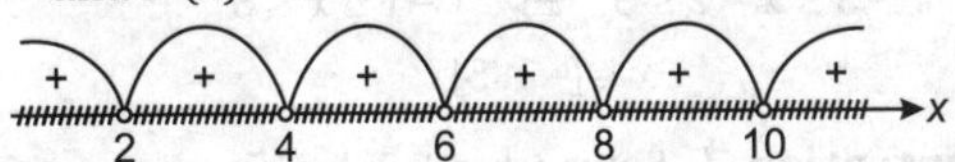

For $F(x) > 0$: $x \in (-\infty, 2) \cup (2, 4) \cup (4, 6) \cup (6, 8) \cup (8, 10) \cup (10, \infty)$

or $x \in R \sim \{2, 4, 6, 8, 10\}$

For $F(x) < 0$: $x \in \phi$

Modulus or Absolute Value

The magnitude of a real number x is called it's modulus or absolute value and is, in general, denoted by $|x|$ and is defined as

$$|x| = \begin{cases} x, & x \geq 0 \\ -x, & x < 0 \end{cases}$$

Or Any point x lying on the real number line will have it's co-ordinate as $(x, 0)$. Thus its distance from the origin is $\sqrt{x^2}$. Hence $|x| = \sqrt{x^2}$ or $|x| = \max.(x, -x)$

e.g., If $x = -3.9$, then $|x| = 3.9$, if $x = 2.8$, then $|x| = 2.8$.

Graph of $y = |x| = \sqrt{x^2}$:

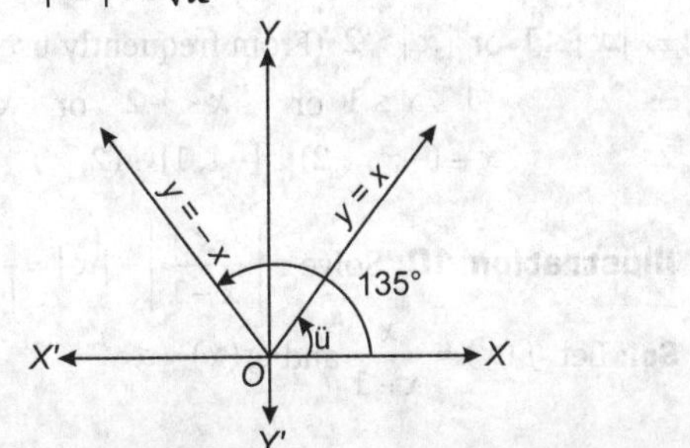

Properties of | x | :

(i) $x^2 = |x|^2 = |x^2|$ (ii) $|xy| = |x||y|$

(iii) $\left|\frac{x}{y}\right| = \left|\frac{x}{y}\right|$, provided $y \neq 0$

(iv) $|x \pm y| \leq |x| + |y|$ and if $xy \geq 0$, then $|x + y| = |x| + |y|$

(v) $|x \pm y| \geq ||x| - |y||$

(vi) $|x| \leq a$ or $x^2 \leq a^2 \Leftrightarrow -a \leq x \leq a\ (a > 0)$

(vii) $|x| < a$ or $x^2 < a^2 \Leftrightarrow -a < x < a\ (a > 0)$

(viii) $|x| \geq a$ or $x^2 \geq a^2 \Leftrightarrow x \leq -a$ or $x \geq a\ (a > 0)$

(ix) $|x| > a$ or $x^2 > a^2 \Leftrightarrow x < -a$ or $x > a\ (a > 0)$

(x) $a \leq |x| \leq b$ or $a^2 \leq x^2 \leq b^2 \Leftrightarrow -b \leq x \leq -a$ or $a \leq x \leq b$

(xi) $a < |x| \leq b$ or $a^2 < x^2 \leq b^2 \Leftrightarrow -b \leq x < -a$ or $a < x \leq b$

(xii) $a \leq |x| < b$ or $a^2 \leq x^2 < b^2 \Leftrightarrow -b < x \leq -a$ or $a \leq x < b$

(xiii) $a < |x| < b$ or $a^2 < x^2 < b^2 \Leftrightarrow -b < x < -a$ or $a < x < b$

(xiv) If $a < 0$, then $|x| \geq a \Rightarrow x \in R$ and $|x| \leq a \Rightarrow x \in \phi$

Illustration 6. Solve : $|x - 2| \leq 3$

Sol : We have $|x - 2| \leq 3$

$\Rightarrow \quad -3 \leq x - 2 \leq 3 \quad \Rightarrow \quad -1 \leq x \leq 5$

$\therefore \quad x \in [-1, 5]$

Illustration 7. Solve : $|x + 1| > 5$

Sol : We have $|x + 1| > 5$

$\Rightarrow \quad x + 1 < -5$ or $x + 1 > 5$

$\Rightarrow \quad x < -6$ or $x > 4$

$\therefore \quad x \in (-\infty, -6) \cup (4, \infty)$

Illustration 8. Solve : $0 < |x - 1| \leq 3$

Sol : We have $0 < |x - 1| \leq 3$

$\Rightarrow \quad -3 \leq x - 1 < 0$ or $0 < x - 1 \leq 3$

$\Rightarrow \quad -2 \leq x < 1$ or $1 < x \leq 4$

$\therefore \quad x \in [-2,) \cup (1, 4)$

Illustration 9. Solve : $\frac{1 - |x|}{2 - |x|} \geq 0$

Sol : We have $\frac{1 - |x|}{2 - |x|} \geq 0$

$\Rightarrow \quad \frac{|x| - 1}{|x| - 2} \geq 0$

$\Rightarrow |x| \leq 1$ or $|x| > 2$ (From frequently used inequality)

$\Rightarrow \quad -1 \leq x \leq 1$ or $x < -2$ or $x > 2$

$\therefore \quad x \in (-\infty, -2) \cup [-1, 1] \cup (2, \infty)$

Illustration 10. Solve : $\left|\frac{x}{x-1}\right| + |x| = \frac{x^2}{|x-1|}$

Sol: Let $f(x) = \frac{x}{x-1}$ and $g(x) = x$

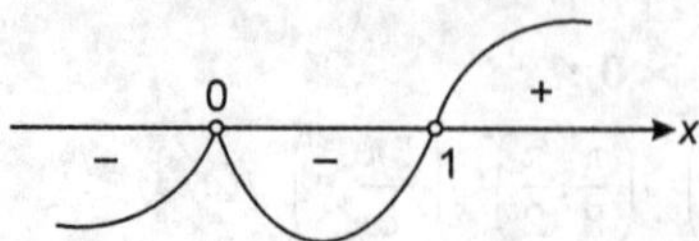

Then $\quad f(x) + g(x) = \frac{x}{x-1} + x = \frac{x^2}{x-1}$

$\therefore$ The given equation can be reduced in the form

$$|f(x)| + |g(x)| = |f(x) + g(x)|$$

Hence $\quad f(x) \cdot g(x) \geq 0 \Rightarrow \frac{x^2}{(x-1)} \geq 0$

From wave curve method, we get

$$x \in (1, \infty) \cup \{0\}$$

Function of the Form

$F(x) = |f_1(x)| + |f_2(x)| + |f_3(x)| + \ldots + |f_n(x)|$ ...(i)

Functions of this form are solved by *method of intervals*. We first find all critical points of $f_1(x), f_2(x), \ldots, f_n(x)$ if coefficient of x is positive, then graph starts with positive sign [Fig. (a)] and if coefficient of x is negative, then graph starts with negative sign [Fig (b)].

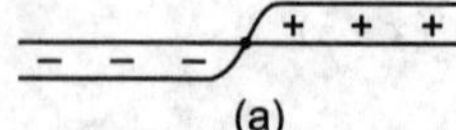

(a)

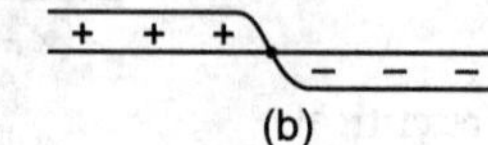

(b)

Then using the definition of the absolute value, we pass from function (i) to a collection of systems which do not contain the absolute value symbols.

Illustration 11. Solve the function

$$f(x) = 2|x - 1| - 3|2 - x| + 5|x - 7|$$

Sol : Here critical points are 1, 2, 7 using the method of intervals, we find intervals when the expressions $x - 1$, $2 - x$ and $x - 7$ are of constant signs.

i.e., $x < 1$, $1 < x < 2$, $2 < x < 7$, $x > 7$

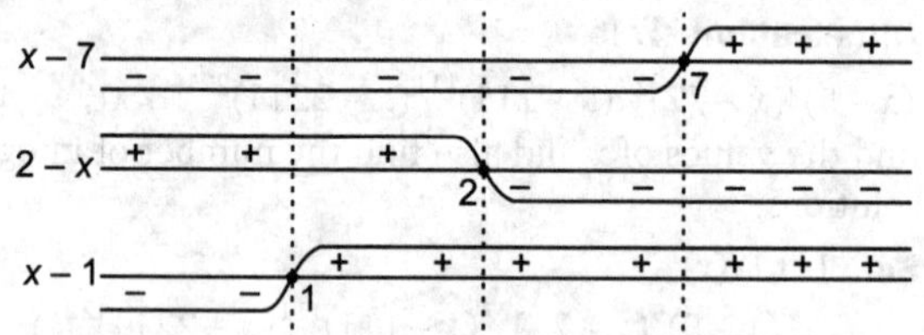

$$\text{Thus, } f(x) = \begin{cases} -2(x-1) - 3(2-x) - 5(x-7), & x < 1 \\ = -4x + 31 & \\ 2(x-1) - 3(2-x) - 5(x-7), & 1 \leq x < 2 \\ = 27 & \\ 2(x-1) + 3(2-x) - 5(x-7), & 2 \leq x < 7 \\ = -6x + 39 & \\ 2(x-1) + 3(2-x) + 5(x-7), & x \geq 7 \\ = 4x - 31 & \end{cases}$$

$$\text{Hence, } f(x) = \begin{cases} -4x + 31, & x < 1 \\ 27, & 1 \leq x < 2 \\ -6x + 39, & 2 \leq x < 7 \\ 4x - 31, & x \geq 7 \end{cases}$$

Illustration 12. If $a < b < c$, then find the least value of $f(x)$, where

$$f(x) = |x-a| + |x-b| + |x-c|$$

Sol : Here critical points are a, b, c using the method of intervals, we find intervals when the expressions $x-a$, $x-b$ and $x-c$ are of constant signs.

i.e., $x < a, a < x < b, b < x < c, x > c$

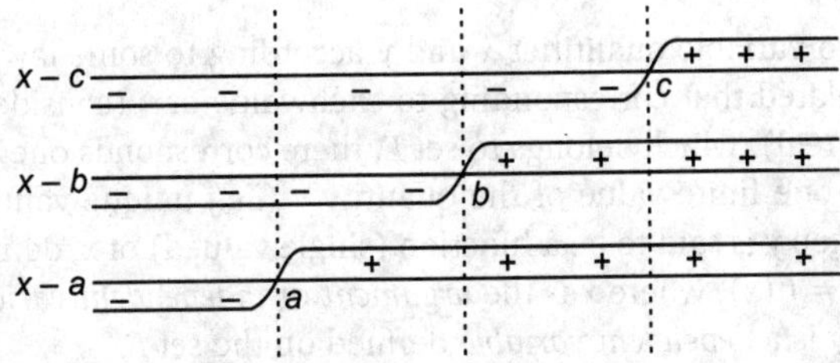

$$f(x) = \begin{cases} -(x-a)-(x-b)-(x-c), & x < a \\ = a+b+c-3x & \\ (x-a)-(x-b)-(x-c), & a \le x < b \\ = b+c-a-x & \\ (x-a)+(x-b)-(x-c), & b \le x < c \\ = c-a-b+x & \\ (x-a)+(x-b)+(x-c), & x \ge c \\ = 3x-a-b-c & \end{cases}$$

$$\Rightarrow \quad f(x) = \begin{cases} a+b+c-3x, & x < a \\ b+c-a-x, & a \le x < b \\ c-a-b+x, & b \le x < c \\ 3x-a-b-c, & x \ge c \end{cases}$$

Graph of given function is as shown

Clearly minimum value of $f(x)$ will occur at $x = b$ which is

$$f(b) = c - a - b + b$$
$$= c - a.$$

x = a x = b x = c

Illustration 13. Sketch the graph of the function $f(x) = 2|x-2| - |x+1| - x$.

Sol : Critical points of $f(x)$ are -1 and 2.

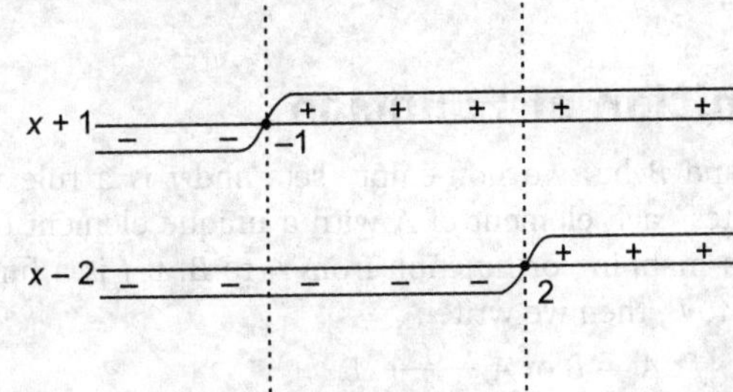

$$\therefore \quad f(x) = \begin{cases} -2(x-2)+(x+1)-x, & x < -1 \\ -2(x-2)-(x+1)-x, & -1 \le x < 2 \\ 2(x-2)-(x+1)-x, & x \ge 2 \end{cases}$$

$$\Rightarrow \quad f(x) = \begin{cases} 5-2x, & x < -1 \\ 3-4x, & -1 \le x < 2 \\ -5, & x \ge 2 \end{cases}$$

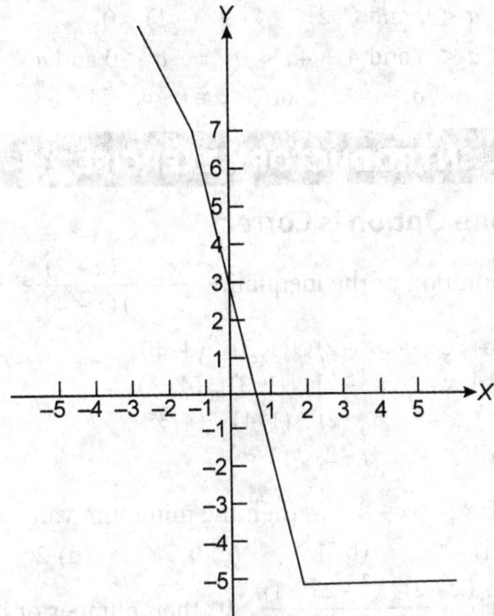

Illustration 14. Sketch the graph of the function $f(x) = x + \sqrt{x^2}$ or $f(x) = x + |x|$.

Sol : The given function can be written as :

$$f(x) = \begin{cases} 2x, & x \ge 0 \\ 0, & x < 0 \end{cases}$$

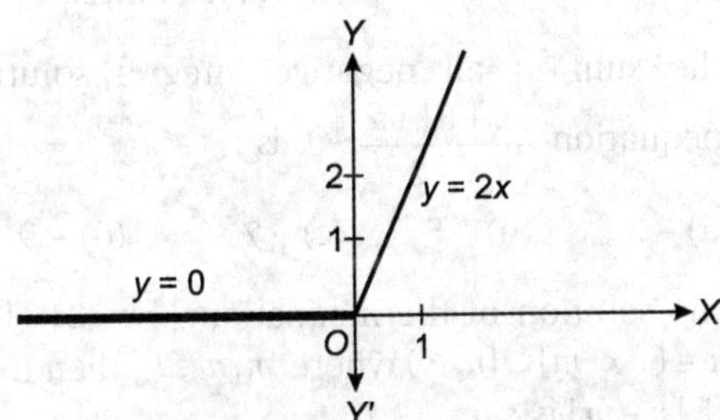

Quadratic Inequalities

Let $f(x) = ax^2 + bx + c, a \ne 0$ and $D = b^2 - 4ac$, then :

(i) $f(x) \ge 0, \forall x \in R \Rightarrow a > 0$ and $D \le 0$

(ii) $f(x) \le 0, \forall x \in R \Rightarrow a < 0$ and $D \le 0$

(iii) $f(x) > 0, \forall x \in R \Rightarrow a > 0$ and $D < 0$

(iv) $f(x) < 0, \forall x \in R \Rightarrow a < 0$ and $D < 0$

Illustration 15. Find a for which $x^2 + ax + 9 > 0, \forall x \in R$.

Sol : Here $x^2 + ax + 9 > 0$

$\therefore \quad a^2 - 4 \cdot 1 \cdot 9 < 0$

$\Rightarrow \quad a^2 < 36$

or $\quad -6 < a < 6$

or $\quad a \in (-6, 6)$

Illustration 16. Find a for which $ax^2 - 2x - 1 < 0, \forall x \in R$.

Sol : Here $ax^2 - 2x - 1 < 0$

$\Rightarrow$ $a < 0$ and $D < 0$

$\Rightarrow$ $a < 0$ and $(-2)^2 - 4 \cdot 9 \cdot (-1) < 0$

$\Rightarrow$ $a < 0$ and $4 + 4a < 0$ $\Rightarrow$ $a < 0$ and $a < -1$

$\therefore$ $a < -1$ or $a \in (-\infty, -1)$

INTRODUCTORY EXERCISE 2.2

Only One Option is Correct

1. Solution of the inequality $\dfrac{(x-1)(x+2)}{(x-3)(x-4)} \geq 0$ is :

(a) $x \in (-\infty, -2) \cup (1, 3) \cup (4, \infty)$
(b) $x \in (-\infty, -2] \cup [1, 3) \cup (4, \infty)$
(c) $x \in (-\infty, -2) \cup (1, 4) \cup (4, \infty)$
(d) $x \in (-\infty, -2) \cup [1, \infty)$

2. If $x^3 - 3x + 2 \geq 0$, then the minimum value of x is :

(a) – 2 (b) 1 (c) 0 (d) 2

3. If $\dfrac{(x+2)(x^2-2x+1)}{(4+3x-x^2)} \geq 0$, then number of positive integral solutions are :

(a) 3 (b) 4 (c) 5 (d) 6

4. The minimum value of $5^x + 3^x + e^x + 5^{-x} + 3^{-x} + e^{-x}$ is :

(a) 4 (b) 6
(c) 12 (d) None of these

5. The sum of all negative integral solutions of inequation $\dfrac{|x+3|+x}{x+2} > 1$ is :

(a) – 3 (b) – 5 (c) – 7 (d) – 9

6. The solution of the inequation $||x-2|-1| \geq 3$ is $x \in (-\infty, m] \cup [n, \infty)$ where $m, n \in I$, then the value of $|n - m|$ is :

(a) 8 (b) 12 (c) 16 (d) 20

7. The number of elements in the set $\{x \in R : (|x| - 3)\,|x + 4| = 6\}$ is equal to

[JEE Main 16 March 2021 (MS)]

(a) 3 (b) 2 (c) 4 (d) 1

8. The solution of the inequation $|x-1| + 2|x+1| + |x-2| \leq 8$ is $x \in [a, b]$, then the value of $|b-a| + \dfrac{1}{|b+a|}$ is :

(a) 4 (b) 6 (c) 8 (d) 10

9. The solution of the equation $|x^2 + 6x + 7| = |x^2 + 4x + 4| + |2x + 3|$ is :

(a) $x \in \left[-\dfrac{1}{2}, \infty\right)$ (b) $x \in (-1, \infty)$

(c) $x \in \left[-\dfrac{3}{2}, \infty\right)$ (d) $x \in (-2, \infty)$

10. The maximum value of a for which $ax^2 + x - 1 \leq 0, \forall x \in R$, is :

(a) – 7/2 (b) – 5/2
(c) – 3/4 (d) – 1/4

Function

Introduction

If two variable quantities x and y according to some law are so related that corresponding to each value of x (considered only real), which belongs to set E, there corresponds one and only one finite value of the quantity y (*i.e.*, unique value of y), then y is said to be a function (single valued) of x, defined by $y = f(x)$, where x is the *argument* or *independent variable* and y is a *dependent variable* defined on the set E.

For example : If r be the radius of the sphere and V its volume, the r and V are related by :

$$V = \frac{4}{3}\pi r^3 \quad \text{or} \quad V = f(r)$$

Then we say that the volume V of the sphere is a function of the radius r.

Graphically :

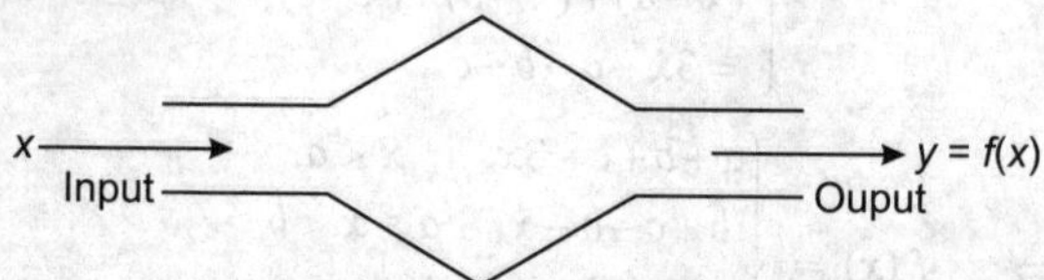

where y is the image of x and x is the pre-image of y under f.

Note :

(i) If to each value of x, which belongs to set E there corresponds one or more than one values of the quantity y. Then y is called the multiple valued function of x defined on the set E.

(ii) The word FUNCTION is used only for single valued function.

For example : $y = \sqrt{x}$ is single valued function but $y^2 = x$ is a multiple valued function.

$\because$ $y^2 = x \Rightarrow y = \pm\sqrt{x}$ for one value of x, y gives two values.

Definition of Function

Let A and B be two non-empty sets and f is a rule which associates each element of A with a unique element of B is called a mapping or function from A to B. If f is a function from A to B, then we write :

$$f : A \to B \text{ or } A \xrightarrow{f} B$$

which is read as f is a mapping from A to B.

Every function from $A \to B$ satisfied the following conditions :

(i) $f \subset A \times B$
(ii) $\forall\, a \in A \Rightarrow (a, f(a)) \in f$
(iii) $(a, b) \in f$ and $(a, c) \in f \Rightarrow b = c$

Definition of Function (Graphically)

If any line parallel to y-axis cuts the graph of the function at most one point, then it is a single valued function and if it cuts at more than one point then it is a multiple function or it does not form a function.

Illustration 17. Prove that $y = |\sin|x||$ is a function.

Sol : Since any line parallel to y-axis cuts $y = |\sin|x||$ at one and only one point. Thus, $y = |\sin|x||$ is a function.

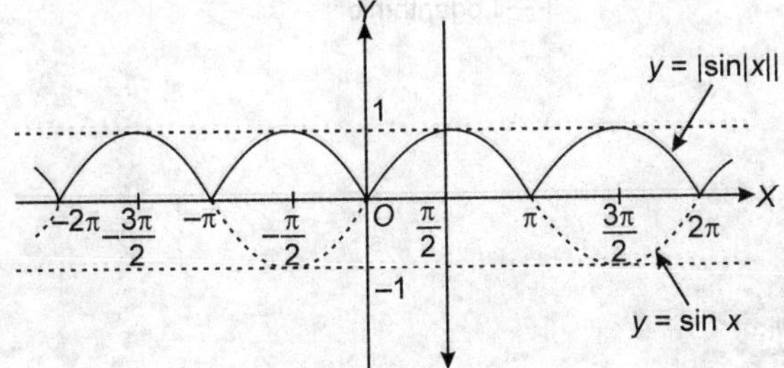

Illustration 18. Prove that $|y| = |\ln|x||$ is not a function.

Sol : Since any line parallel to y-axis meets the curve $|y| = |\ln|x||$ at two points. Thus, $|y| = |\ln|x||$ is not a function.

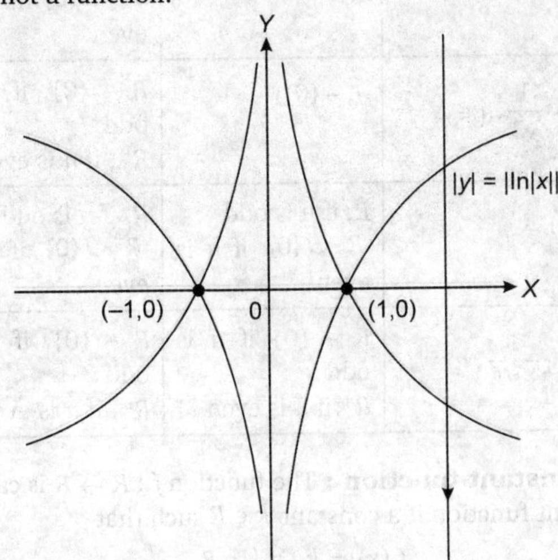

Domain, Co-domain and Range of a Function

Let $f: A \to B$ is a function from A to B, then the set A is called the ***domain*** of the function f (denoted by D_f) and the set B is called the ***co-domain*** of the function f (denoted by C_f). The set of all those elements of B which are the images of the elements of set A is called the ***range*** of the function f (denoted by R_f).

Domain of $f = D_f = \{a : a \in A, (a, f(a)) \in f\}$

Range of $f = R_f = \{f(a) : a \in A, f(a) \in B\}$

It should be noted that the range of f is always a subset of co-domain B.

i.e., $$f(A) \subset B$$

Illustration 19. Find the domain, co-domain and range of the function f which is denoted by the figure.

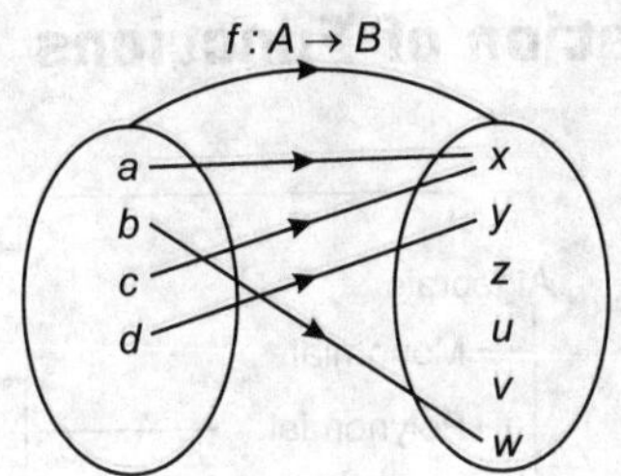

Sol : It is clear from the figure

Domain of $f = D_f$ = All elements of set $A = \{a, b, c, d\}$

Co-domain of $f = C_f$ = All elements of set $B = \{x, y, z, u, v, w\}$

Range of $f = R_f$ = All elements of B which are the images of the elements of set $A = \{x, y, w\}$

Also $\{x, y, w\} \subset \{x, y, z, u, v, w\}$

i.e., Range $\subset$ co-domain

Domain

The domain of $y = f(x)$ is the set of all real x for which y is defined.

Rules for the Domain of a Function

(i) Domain of $(f(x) \pm g(x))$ = domain of $f(x) \cap$ Domain of $g(x)$

i.e., $D_{f \pm g} = D_f \cap D_g$

(ii) Domain of $(f(x) \cdot g(x))$ = domain of $f(x) \cap$ Domain of $g(x)$

i.e., $D_{fg} = D_f \cap D_g$

(iii) Domain of $\left(\frac{f(x)}{g(x)}\right)$ = Domain of $f(x) \cap$ Domain of $g(x) \cap \{x : g(x) \neq 0\}$

i.e., $D_{f/g} = D_f \cap D_g \cap \{x : g(x) \neq 0\}$

(iv) Domain of $\sqrt{f(x)}$ = Domain of $f(x) \cap \{x : f(x) \geq 0\}$

i.e., $D_{\sqrt{f}} = D_f \cap \{x : f \geq 0\}$

(v) Domain of $\log_a f(x)$ = Domain of $f(x) \cap \{x : f(x) > 0\}$

i.e., $D_{\log_a f} = D_f \cap \{x : f > 0\}$

(vi) Domain of $(fog)\, x$ = Domain of $g(x)$

i.e., $D_{fog} = D_g$ [where $(fog)\, x = f\{g(x)\}$]

Range

The range of $y = f(x)$ is the set of all real y for which x is defined.

How to find range : First put $y = f(x)$ by suitable substitution, find x in terms of y. Then find all such y for which x is defined *i.e.,* in the domain. Set of these values of y is the range of $f(x)$.

Classification of Functions

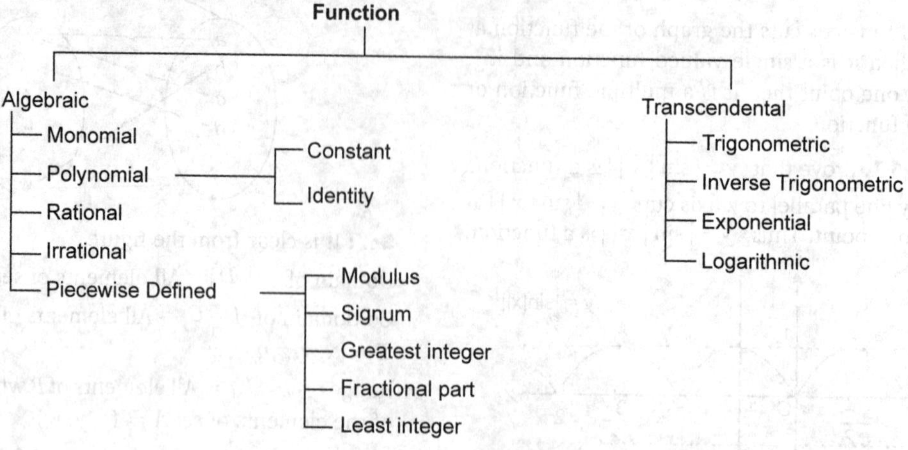

(A) Algebraic Functions

An algebraic function consists of a finite number of terms involving powers and roots of the variable x and the operations : addition, subtraction, multiplication and division.

e.g., $7x^3 + 2x^2 - 3x + 2$,

$$5x^2 + 2x^{1/2} + 7, \frac{3x^2 - 2x + 5}{7x^2 + 4x - 1} \text{ etc.}$$

(a) Monomial functions : Any function of the form $f(x) = kx^n$ where k is a constant and $n \in W$ is known as monomial function.

e.g., $f(x) = 5x^3$

> **Note :** The function $f(x) = ax$, $a \neq 0$ is called on odd linear function.

(b) Polynomial functions : Any function which is obtained by addition or subtraction of several monomials is called a polynomial whose domain and co-domain both is the set of real numbers. Clearly in this

$$f(x) = a_0x^n + a_1x^{n-1} + a_2x^{n-2} + ... + a_{n-1} + a_n (a_0 \neq 0)$$

where $a_0, a_1, a_2, ..., a_{n-1}, a_n \in R$ and $n \in N$.

'n' is called the degree of polynomial which is maximum value of the exponents of x.

	Polynomial	Degree
(i)	$f(x) = 125x^{150} + 22x^{99} + x^{27}$	150
(ii)	$f(x) = 3x^3 + 2x^2 + 3x + 7$	3
(iii)	$f(x) = 2x^2 + 4x - 7$	2
(iv)	$f(x) = 5x + 8$	1
(v)	$f(x) = 4$	0
(vi)	$f(x) = 0$	undefined

Note :

Functions ($y = f(x)$)	Domain (*i.e.,* values taken by x)	Range (*i.e.,* values taken by $f(x)$)
(i) $x^n, n \in N$	R	R, if n is odd $R^+ \cup \{0\}$ if n is even
(ii) $\frac{1}{x^n}, n \in N$	$R - \{0\}$	$R - \{0\}$, if n is odd R^+, if n is even
(iii) $x^{1/n}, n \in N$	R, if n is odd $R^+ \cup \{0\}$ if n is even	R, if n is odd $R^+ \cup \{0\}$ if n is even
(iv) $\frac{1}{x^{1/n}}, n \in N$	$R - \{0\}$ if n is odd R^+ if n is even	$R - \{0\}$, if n is odd R^+, if n is even

(i) Constant function : The function $f : R \to R$ is called a constant function if a constant $k \in R$ such that

$$f(x) = k, \forall k \in R$$

$\therefore$ Domain of $f(x) = D_f = R$

and Range of $f(x) = R_f = k$

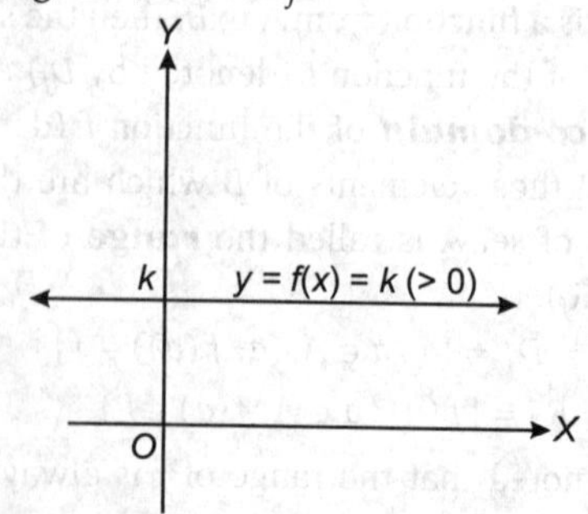

(ii) Identity function : The function $f : R \to R$ is called an identity function if

$$f(x) = x \ \forall \ x \in R$$

$$\therefore \quad D_f = R_f = R$$

(*i.e.*, output is same as the input)

(c) Rational function : A function defined by the quotient of two polynomial functions is called a rational function. Thus $f(x) = \dfrac{P(x)}{Q(x)}$,

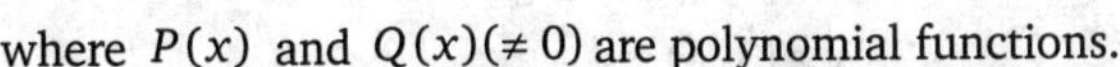

where $P(x)$ and $Q(x)(\neq 0)$ are polynomial functions.

	Rational Functions	Domain
(i)	$\dfrac{a_0x^n + a_1x^{n-1} + a_2x^{n-2} + ... + a_n}{b_0x^n + b_1x^{n-1} + b_2x^{n-2} + ... + b_n}$	All real numbers except the real roots of the equation $b_0x^n + b_1x^{n-1} + b_2x^{n-2} + ... + b_n = 0$
(ii)	$\dfrac{5x^2 - 7x + 1}{x^3 - 6x^2 + 11x - 6}$	$R - \{1, 2, 3\}$
(iii)	$\dfrac{7x^2 - 4x + 5}{x^2 - 3x + 2}$	$R - \{1, 2\}$
(iv)	$\dfrac{1}{x^3}$	$R - \{0\}$

(d) Irrational function : An algebraic function or rational function containing one or more radicals (non-integral rational powers of x) is called irrational function.

e.g., $\dfrac{3x^2 + 2\sqrt{x}}{3x - 2}$, $2x^3 - 7x^{1/2} + 5x - 2$, $\dfrac{(x^3 - 2x^2 + 7)^{2/3}}{\sqrt{(x+2)}}$

then domains are $x > 0$, $x \neq 2/3$; $x > 0$; $x > -2$

i.e., $(0, \infty) - \left\{\dfrac{2}{3}\right\}$; $(0, \infty)$; $(-2, \infty)$ respectively.

(e) Piecewise defined functions :

(i) Modulus function (or absolute value function)

It is defined as $y = f(x) = |x| = \begin{cases} x; & x \geq 0 \\ -x; & x < 0 \end{cases}$

$$D_f = R, R_f = [0, \infty)$$

Properties of modulus function : (For basic properties see page no. 98)

(i) max. $\{x, y\} = \dfrac{x+y}{2} + \left|\dfrac{x-y}{2}\right|$

(ii) min. $\{x, y\} = \dfrac{x+y}{2} - \left|\dfrac{x-y}{2}\right|$

Replacing y by $-x$, then

(a) max. $\{x, -x\} = |x|$

(b) min. $\{x, -x\} = -|x|$

Note :

If $f(x) = \dfrac{1}{|x|}$ then $D_f = R - \{0\}, R_f = R^+$

Illustration 20. Find the domain of definition of the function

$$f(x) = \sqrt{\left(\frac{1-|x|}{2-|x|}\right)}$$

Sol : Since $f(x) = \sqrt{\left(\dfrac{1-|x|}{2-|x|}\right)}$

$\Rightarrow f(x)$ is defined when

$$\frac{1-|x|}{2-|x|} \geq 0 \Rightarrow \frac{|x|-1}{|x|-2} \geq 0$$

$\therefore$ Critical points of $|x|$ are 1 and 2.

$\Rightarrow$ From wavy curve method, we get

$$|x| \leq 1 \text{ or } |x| > 2$$

Then $\quad -1 \leq x \leq 1$ or $x < -2$ or $x > 2$

$\Rightarrow \quad x \in (-\infty, -2) \cup [-1, 1] \cup (2, \infty)$

$\therefore \quad D_f = (-\infty, -2) \cup [-1, 1] \cup (2, \infty)$

(ii) Signum function : The signum function f is defined as

$$y = f(x) = \text{sgn}(x) = \begin{cases} \dfrac{|x|}{x} \text{ or } \dfrac{x}{|x|}; & x \neq 0 \\ 0 \quad ; & x = 0 \end{cases}$$

$$= \begin{cases} 1; & x > 0 \\ 0; & x = 0 \\ -1; & x < 0 \end{cases}$$

$D_f = R, R_f = \{-1, 0, 1\}$

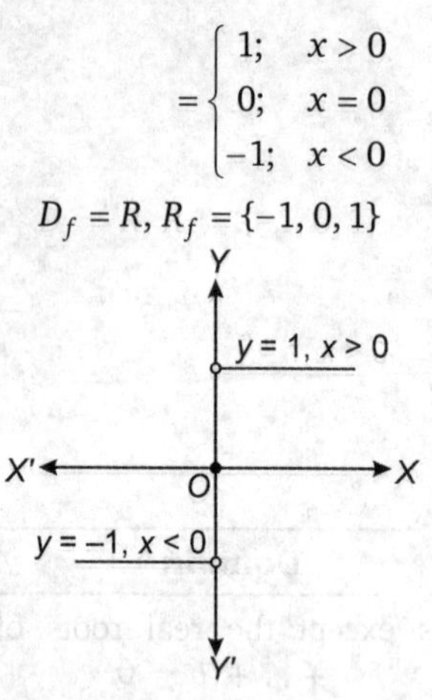

Properties of signum function :

(i) sgn (kx) = sgn (x), $k \in N$
(ii) x sgn $(x) = |x|$
(iii) $|x|$ sgn $(x) = x$
(iv) x (sgn x) (sgn x) $= x$

Illustration 21. Find the domain and range of the function $f(x)$ = sgn $(x^2 + 1)$

Sol : We have

$f(x)$ = sgn $(x^2 + 1)$

$$= \begin{cases} 1, & x^2+1>0 \\ 0, & x^2+1=0 \quad \text{(Impossible)} \\ -1, & x^2+1<0 \quad \text{(Impossible)} \end{cases}$$

$= 1, x^2 + 1 > 0$ or $f(x) = 1, x \in R$

Hence $D_f = R$ and $R_f = 1$

(iii) Greatest Integer or Step up function : $[x]$ denotes the greatest integer less than or equal to x *i.e.*, $[x] \le x$. It is also known **as *floor of x*.**

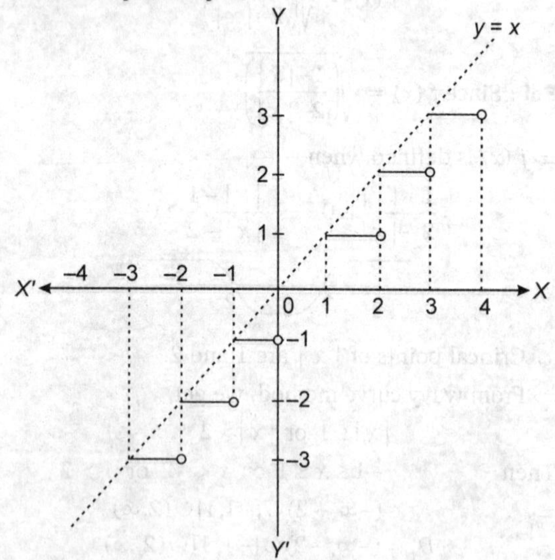

Thus, $[3.5779] = 3$, $[0.89] = 0$, $[3] = 3$,
$[-8.7285] = -9$, $[-0.6] = -1$

In general if n is an integer and x is any real number between n and $(n + 1)$.

i.e., $n \le x < n+1$, then $[x] = n$

$\therefore$ $y = f(x) = [x]$

For $x \in [-3, -2)$; $y = -3$
For $x \in [-2, -1)$; $y = -2$
For $x \in [-1, 0)$; $y = -1$
For $x \in [0, 1)$; $y = 0$
For $x \in [1, 2)$; $y = 1$
For $x \in [2, 3)$; $y = 2$
For $x \in [3, 4)$; $y = 3$

$D_f = R, R_f = I$

Properties of Greatest Integer Function :

(i) $[[x]] = [x]$
(ii) $[x] \le x < [x]+1$ and $x-1 < [x] \le x$, $0 \le x-[x] < 1$
(iii) $[x \pm n] = [x] \pm n, n \in I$
(iv) $[-x] = \begin{cases} -[x] = -x, & x \in I \\ -1-[x], & x \notin I \end{cases}$
(v) $[x]+[-x] = \begin{cases} 0, & x \in I \\ -1, & x \notin I \end{cases}$
(vi) $[x]-[-x] = \begin{cases} 2x, & x \in I \\ 2[x]+1, & x \notin I \end{cases}$
(vii) $[x]+[y] \le [x+y] \le [x]+[y]+1$
(viii) $\left[\frac{[x]}{n}\right] = \left[\frac{x}{n}\right], n \in N$
(ix) $[x] = \left[\frac{x}{2}\right] + \left[\frac{x+1}{2}\right]$
(x) $\left[\frac{n+1}{2}\right] + \left[\frac{n+2}{4}\right] + \left[\frac{n+4}{8}\right] + ... = n, n \in N$
(xi) $[x] + \left[x+\frac{1}{n}\right] + \left[x+\frac{2}{n}\right] + ... + \left[x+\frac{n-1}{n}\right] = [nx], n \in N$
(xii) $[x] \ge n \Leftrightarrow x \ge n, n \in I$
(xiii) $[x] > n \Leftrightarrow x \ge n+1, n \in I$
(xiv) $[x] \le n \Leftrightarrow x < n+1, n \in I$
(xv) $[x] < n \Leftrightarrow x < n, n \in I$
(xvi) $n_1 \le [x] \le n_2 \Rightarrow n_1 \le x < n_2+1$, $n_1, n_2 \in I$

Note :
If $f(x) = \frac{1}{[x]}$, then
$D_f = R - [0, 1)$, $R_f = \frac{1}{n}, n \in I - \{0\}$

Illustration 22. Let $[a]$ denote the larger integer not exceeding the real number a. If x and y satisfy the equations $y = 2[x] + 3$ and $y = 3[x-2]$ simultaneously, determine $[x + y]$.

Sol : We have

$y = 2[x] + 3 = 3[x-2]$

$\Rightarrow$ $2[x]+3 = 3[x]-6 \Rightarrow [x] = 9$

and $y = 2[x] + 3 = 2 \times 9 + 3 = 21$

$\therefore [x+y] = [x+21] = [x] + 21 = 9 + 21 = 30$

Illustration 23. If $[x]$ denotes the integral part of x for real x, then find the value of

$$\left[\frac{1}{4}\right]+\left[\frac{1}{4}+\frac{1}{200}\right]+\left[\frac{1}{4}+\frac{1}{100}\right]+\left[\frac{1}{4}+\frac{3}{200}\right]+\ldots+\left[\frac{1}{4}+\frac{199}{200}\right]$$

Sol : The given expression can be written as

$$\left[\frac{1}{4}\right]+\left[\frac{1}{4}+\frac{1}{200}\right]+\left[\frac{1}{4}+\frac{2}{200}\right]+\left[\frac{1}{4}+\frac{3}{200}\right]+\ldots+\left[\frac{1}{4}+\frac{199}{200}\right]$$

$$=\sum_{r=0}^{199}\left[\frac{1}{4}+\frac{r}{200}\right]=\sum_{r=0}^{149}\left[\frac{1}{4}+\frac{r}{200}\right]+\sum_{r=150}^{199}\left[\frac{1}{4}+\frac{r}{200}\right]$$

$= 0 + 50 = 50$

Alternative Method : (By prop. no (xi))

Here $x=\frac{1}{4}$ and $n = 200$

$$\therefore \left[\frac{1}{4}\right]+\left[\frac{1}{4}+\frac{1}{200}\right]+\left[\frac{1}{4}+\frac{2}{200}\right]+\ldots.+\left[\frac{1}{4}+\frac{199}{200}\right]$$

$$=\left[200\times\frac{1}{4}\right]=[50]=50$$

Illustration 24. Find the domain of definition of the function $f(x)=\sqrt{([x]-1)}+\sqrt{(4-[x])}$, where $[\cdot]$ denotes the greatest integer function.

Sol : Given $f(x)=\sqrt{([x]-1)}+\sqrt{(4-[x])}$

$\therefore f(x)$ is defined when $[x]-1\geq 0$ and $4-[x]\geq 0$

$\Rightarrow \quad 1\leq[x]\leq 4$ or $1\leq x<5$

Hence $\quad D_f=[1,5)$

Illustration 25. Find the domain of definition of the function $f(x)=[x]\sin\left(\frac{\pi}{[x+1]}\right)$, $[\cdot]$ denotes the greatest integer function.

Sol : Since $f(x)=[x]\sin\left(\frac{\pi}{[x+1]}\right)$

$\Rightarrow f(x)$ is defined when $[x+1]\neq 0$ or $x\notin[-1,0)$

Hence Domain of $f(x)$ is $R\sim[-1,0)$

or $\quad D_f=R-[-1,0)$

Illustration 26. Find the domain of definition of the function $f(x)=\frac{1}{\sqrt{|[|x|-1]|-5}}$, $[\cdot]$ denotes the greatest integer function.

Sol : Since $f(x)=\frac{1}{\sqrt{|[|x|-1]|-5}}$

$\Rightarrow \quad f(x)$ is defined when $|[|x|-1]|-5>0$

then $\quad [|x|-1]>5$

and $\quad [|x|-1]<-5$

$\Rightarrow \quad |x|-1\geq 6$

and $\quad |x|-1<-5$

$\Rightarrow \quad |x|\geq 7$ and $|x|-4$ (impossible)

$\Rightarrow \quad x\leq -7$ or $x\geq 7$

$\therefore \quad D_f=(-\infty,-7]\cup[7,\infty)$

(iv) Fractional Part function : It is denoted as $f(x) = \{x\}$ and defined as :

(i) $\{x\}=f$, if $x=n+f$, where $n\in I$ and $0\leq f<1$

(ii) $\{x\}=x-[x]$

e.g., $\quad \{2.7\}=2.7-2=0.7,$

$\{3\}=3-3=0$

and $\quad \{-3.6\}=\{-4+0.4\}=0.4$

$\therefore \quad y=f(x)=\{x\}=x-[x]$

For $x\in[-2,-1)$; $\quad y=x+2$

For $x\in[-1,0)$; $\quad y=x+1$

For $x\in[0,1)$; $\quad y=x$

For $x\in[1,2)$; $\quad y=x-1$

For $x\in[2,3)$; $\quad y=x-2$

... $\quad$ ... $\quad$ ...

$D_f=R,\ R_f=[0,1)$

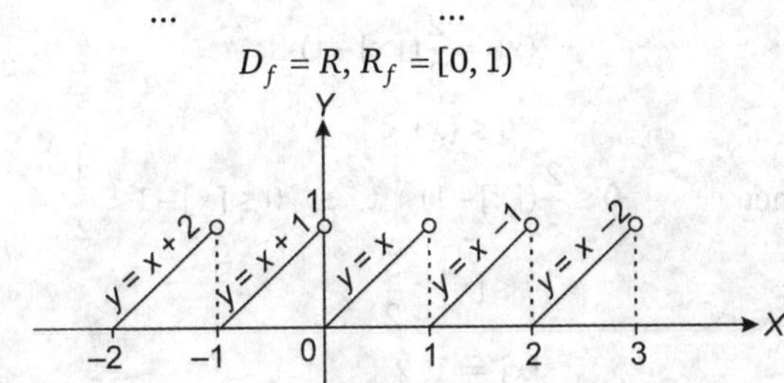

Properties of fractional part function :

(i) For proper fraction $0<\{x\}<1$

(ii) $\{x\pm n\}=\{x\},\ n\in I$

(iii) $\{x\}=0$, if $x\in I$

(iv) $\{x\}=x$, if $x\in[0,1)$

(v) $\{-x\}=1-\{x\}$, if $x\notin I$

(vi) $\{x\}+\{-x\}=\begin{cases}0, & x\in I\\ 1, & x\notin I\end{cases}$

(vii) $\{x\}-\{-x\}=\begin{cases}0, & x\in I\\ 2\{x\}-1, & x\notin I\end{cases}$

(viii) $[x+y]=\begin{cases}[x]+[y], & \{x\}+\{y\}<1\\ [x]+[y]+1, & \{x\}+\{y\}\geq 1\end{cases}$

Note : If $f(x)=\frac{1}{\{x\}}$, then $D_f=R-I,\ R_f=(1,\infty)$

Illustration 27. Let $\{x\}$ and $[x]$ denote the fractional and integral parts of a real number x respectively. Then solve the equation

$$4\{x\}=x+[x]$$

Sol : $\because \{x\}=x+[x]$ ...(i)

Substituting the value of x from (i) in the given equation, then

$$4\{x\}=[x]+\{x\}+[x]$$

or $\quad 3\{x\}=2[x]$

$\Rightarrow \quad \{x\}=\frac{2}{3}[x]$ ...(ii)

but $0 \le \{x\} < 1 \Rightarrow 0 \le \frac{2}{3}[x] < 1$ [From (ii)]

$\Rightarrow$ $0 \le [x] < \frac{3}{2}$ $\therefore$ $[x] = 0, 1$

From (ii), $\{x\} = 0, \frac{2}{3}$

Hence x from (i), $x = 0 + 0 = 0$ and $x = 1 + \frac{2}{3} = \frac{5}{3}$

Hence x = 0, 5/3

Illustration 28. Solve : $\{x + 1\} + 2x = 4[x + 1] - 6$, when $[\cdot]$ and $\{\cdot\}$ denote the greatest integer and fractional part functions respectively.

Sol : $\because$ $\{x + 1\} = \{x\}$ and $x = [x] + \{x\}$

Then given equation can be written as

$$\{x\} + 2([x] + \{x\}) = 4([x] + 1) - 6$$

$$\Rightarrow \quad 3\{x\} = 2[x] - 2$$

$$\Rightarrow \quad \{x\} = \frac{2}{3}([x] - 1) \quad \text{...(i)}$$

$\because$ $0 \le \{x\} < 1$

then $0 \le \frac{2}{3}([x] - 1) < 1 \Rightarrow 0 \le [x] - 1 < \frac{3}{2}$

$\Rightarrow$ $1 \le [x] < \frac{5}{2}$

$\therefore$ $[x] = 1, 2$

From (i), $\{x\} = 0, \frac{2}{3}$

Hence, $x = [x] + \{x\} = 1, 8/3$

Illustration 29. Solve : $|2x - 1| = 3[x] + 2\{x\}$, when $[\cdot]$ and $\{\cdot\}$ denote the greatest integer and fractional part functions respectively.

Sol :

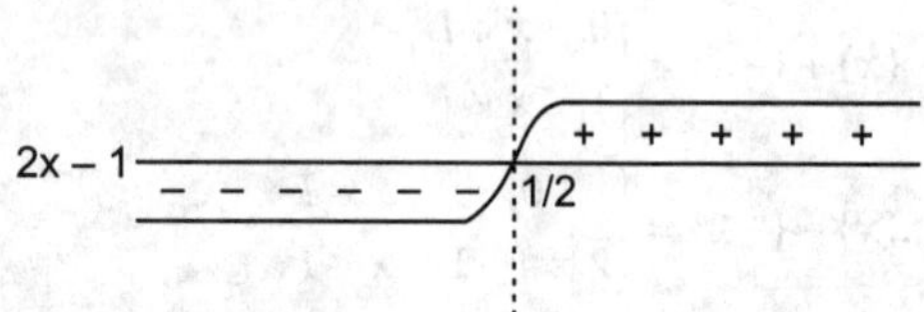

Case I. For $x < \frac{1}{2}$, $|2x - 1| = 1 - 2x$ and $[x] = x - \{x\}$

then given equation can be written as

$$1 - 2x = 3(x - \{x\}) + 2\{x\}$$

or $\{x\} = 5x - 1$ ...(i)

Now $0 \le \{x\} < 1$

$\Rightarrow$ $0 \le 5x - 1 < 1$ [from (i)]

$\Rightarrow$ $\frac{1}{5} \le x < \frac{2}{5} \Rightarrow [x] = 0 \Rightarrow x = \{x\}$

From (i),

$x = 5x - 1 \Rightarrow x = \frac{1}{4}$ which is a solution.

Case II. For $x \ge \frac{1}{2}$, $|2x - 1| = 2x - 1$ and $[x] = x - \{x\}$

then given equation can be written as

$2x - 1 = 3(x - \{x\}) + 2\{x\}$ or $\{x\} = x + 1$ ...(ii)

$\because$ $0 \le \{x\} < 1$

$\Rightarrow$ $0 \le x + 1 < 1$ [from (ii)]

$\Rightarrow$ $-\le x < 0$

Which is not possible since $x \ge \frac{1}{2}$

Hence $x = \frac{1}{4}$ is the only solution.

Illustration 30. Find the domain of definition of the function $f(x) = \sqrt{2\{x\}^2 - 3\{x\} + 1}$, $x \in [-1, 1]$, where $\{\cdot\}$ denotes the fractional part function.

Sol : $f(x)$ is defined for $2\{x\}^2 - 3\{x\} + 1 \ge 0$

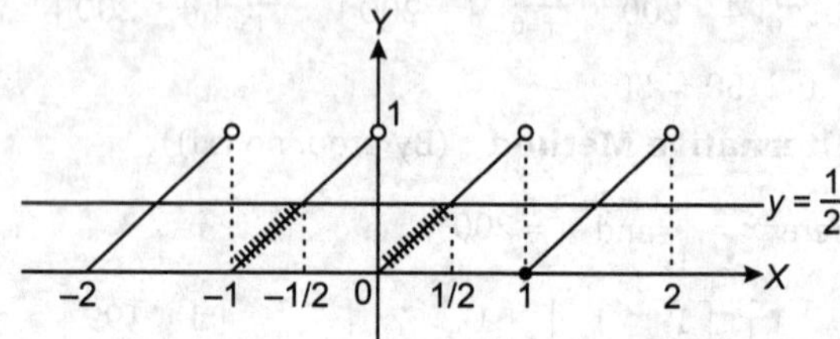

$\Rightarrow$ $(2\{x\} - 1)(\{x\} - 1) \ge 0$

$\therefore$ $\{x\} \le \frac{1}{2}$ or $\{x\} \ge 1$ (not possible) $\Rightarrow$ $\{x\} \le \frac{1}{2}$

Now we draw $y = \{x\}$ and $y = \frac{1}{2}$

From the graph

$$D_f = \left[-1, -\frac{1}{2}\right] \cup \left[0, \frac{1}{2}\right] \cup \{1\} \quad (\because x \in [-1, 1])$$

(v) Least Integer Function : (x) or $\lceil x \rceil$ denotes the least integer function which is greater than or equal to x. It is also known as ***ceiling of x***.

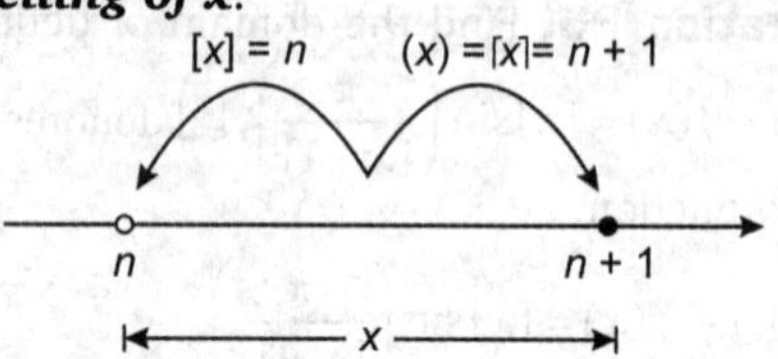

Thus, $(3.578) = 4$, $(0.87) = 1$, $(4) = 4$,

$(-8.239] = -8$, $[-0.7] = 0$

In general if n is an integer and x is any real number between n and $(n + 1)$

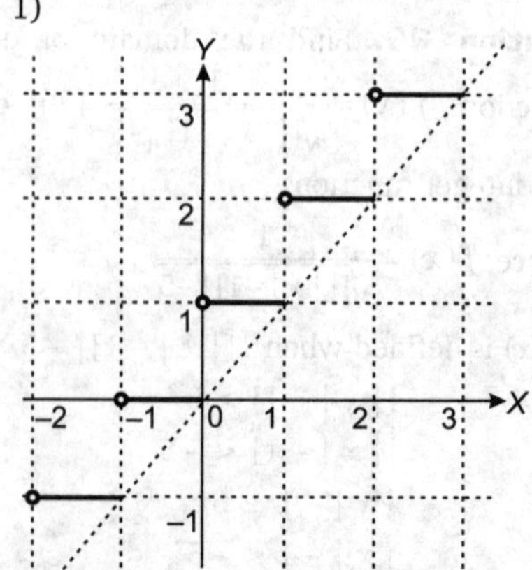

i.e., $n < x \le n + 1$, then $(x) = n + 1$

$\therefore \quad y = f(x) = (x) = \lceil x \rceil$

For $x \in (-2, -1]$; $y = -1$

For $x \in (-1, 0]$; $y = 0$

For $x \in (0, 1]$; $y = 1$

For $x \in (1, 2]$; $y = 2$

For $x \in (2, 3]$; $y = 3$

... ...

$$D_f = R,\ R_f = n+1,\ n \in I$$

Properties of Least Integer function :

(i) $((x)) = (x)$

(ii) $(x) - 1 < x \le (x)$ and $x \le (x) < x+1$

(iii) $(x \pm n) = (x) \pm n, n \in I$

(iv) $(x) = \begin{cases} [x] = x, & x \in I \\ [x]+1, & x \notin I \end{cases}$

(v) $(-x) = \begin{cases} -(x) = -x, & x \in I \\ 1-(x), & x \notin I \end{cases}$

(vi) $(x) + (-x) = \begin{cases} 0, & x \in I \\ 1, & x \notin I \end{cases}$

(vii) $(x) - (-x) = \begin{cases} 2x, & x \in I \\ 2(x)-1, & x \notin I \end{cases}$

(viii) $(x) + (y) - 1 \le (x+y) \le (x) + (y)$

(ix) $\left(\frac{(x)}{n}\right) = \left(\frac{x}{n}\right), n \in N$

(x) $(x) = \left(\frac{x}{2}\right) + \left(\frac{x-1}{2}\right)$

(xi) $\left(\frac{n+1}{2}\right) + \left(\frac{n+2}{4}\right) + \left(\frac{n+4}{8}\right) + \left(\frac{n+8}{16}\right) + \ldots + \left(\frac{n+2^{n-1}}{2^n}\right) = 2n-1, n \in N$

(xii) $(x) + \left(x + \frac{1}{n}\right) + \left(x + \frac{2}{n}\right) + \ldots + \left(x + \frac{n-1}{n}\right) = (nx) + n - 1,\ n \in N$

(xiii) $(x) \ge n \Leftrightarrow x > n-1, n \in I$

(xiv) $(x) > n \Leftrightarrow x > n, n \in I$

(xv) $(x) \le n \Leftrightarrow x < n, n \in I$

(xvi) $(x) < n \Leftrightarrow x < n-1, n \in I$

(xvii) $n_1 \le (x) \le n_2 \Rightarrow n_1 - 1 < x < n_2; n_1, n_2 \in I$

Note :

If $f(x) = \frac{1}{(x)}$, then $D_f = R - (-1, 0]$

and $R_f = \frac{1}{n}, n \in I - \{0\}$

Illustration 31. Solve the equation : $(x)^2 = [x]^2 + 2x$, where $[x]$ and (x) are the integers just less than or equal to x and just greater than or equal to x respectively.

Sol : Given $(x)^2 = [x]^2 + 2x$ ...(i)

Case I. If $x \in I$, then $(x) = [x] = x$

From (i), $x^2 = x^2 + 2x$

$\therefore \quad 2x = 0$ or $x = 0$

Case II. If $x \notin I$, then $(x) = [x]+1$

From (i), $[x]^2 + 1 + 2[x] = [x]^2 + 2x \Rightarrow x - [x] = 1/2$

or $\{x\} = 1/2$

$\therefore \quad x = [x] + \{x\} = n + \frac{1}{2},\ n \in I$

Hence, solutions of the original equation are

$$x = 0, n + 1/2;\ n \in I$$

Illustration 32. Find the solution set of $(x)^2 + (x+1)^2 = 25$, where (x) is the least integer greater than or equal to x.

Sol : We have $(x)^2 + (x+1)^2 = 25$

$\Rightarrow \quad (x)^2 + [(x)+1]^2 = 25$ (By property)

$\Rightarrow \quad (x)^2 + (x)^2 + 2(x) + 1 = 25$

or $\quad (x)^2 + (x) - 12 = 0$ or $((x)+4)((x)-3) = 0$

$\therefore \quad (x) = -4$ or $(x) = 3$

$\Rightarrow \quad -5 < x \le -4$ or $2 < x \le 3$

Hence, $x \in (-5, -4] \cup (2, 3]$

Illustration 33. Find the value of $\left(\frac{9}{2}\right) + \left(\frac{5}{2}\right) + \left(\frac{3}{2}\right) + \ldots$, where (x) denotes least integer greater than or equal to x.

Sol : The given expression can be written as :

$$\left(\frac{8+1}{2}\right) + \left(\frac{8+2}{4}\right) + \left(\frac{8+4}{8}\right) + \ldots + \left(\frac{8+2^7}{2^8}\right)$$

$$= \sum_{n=1}^{8} \left(\frac{8+2^{n-1}}{2^n}\right) = \sum_{n=1}^{8} \left(\frac{1}{2^{n-3}} + 0.5\right)$$

$$= \sum_{n=1}^{3} \left(\frac{1}{2^{n-3}} + 0.5\right) + \sum_{n=4}^{8} \left(\frac{1}{2^{n-3}} + 0.5\right)$$

$= (4.5) + (2.5) + (1.5) + \{1 + 1 + 1 + 1 + 1\}$

$= 5 + 3 + 2 + 5 = 15$

Alternative Method :

Here $n = 8$, then by property

$$\left(\frac{9}{2}\right) + \left(\frac{5}{2}\right) + \left(\frac{3}{2}\right) + \ldots$$

$$= \left(\frac{8+1}{2}\right) + \left(\frac{8+2}{4}\right) + \left(\frac{8+4}{8}\right) + \ldots + \left(\frac{8+2^7}{2^8}\right)$$

$= 2n - 1 = 2 \times 8 - 1 = 15$

Illustration 34. Find the domain of definition of the function $f(x) = \sqrt{\frac{2-(x)}{3-(x)}}$, where (x) denotes the least integer greater than or equal to x.

Sol : Given $f(x) = \sqrt{\left(\frac{2-(x)}{3-(x)}\right)}$ $\because f(x)$ is defined when

$$\frac{2-(x)}{3-(x)} \geq 0 \Rightarrow \frac{(x)-2}{(x)-3} \geq 0$$

$\therefore \quad (x) \leq 2 \quad$ or $\quad (x) > 3$

then $\quad x < 2 \quad$ or $x > 3$ (By property)

$\therefore \quad x \in (-\infty, 2) \cup (3, \infty)$

INTRODUCTORY EXERCISE 2.3

Only One Option is Correct

1. Which of the following graphs is the graph of a function?

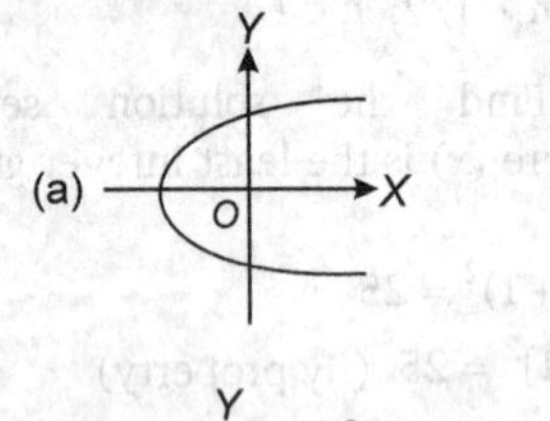

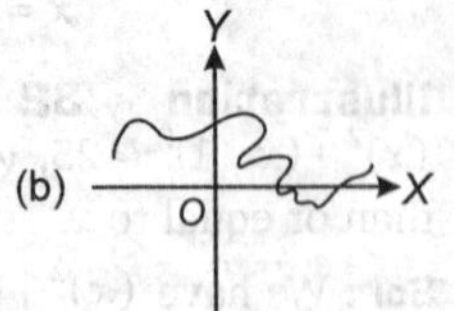

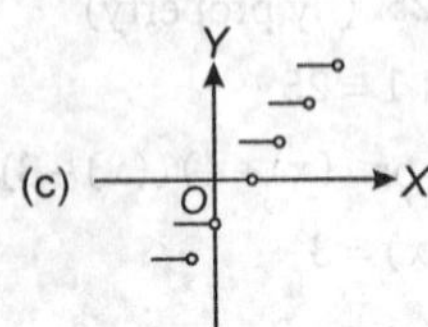

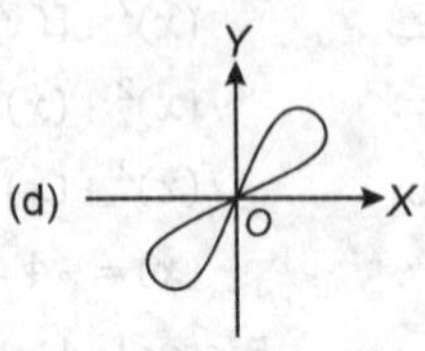

2. Which of the following graphs is the graph of a function?

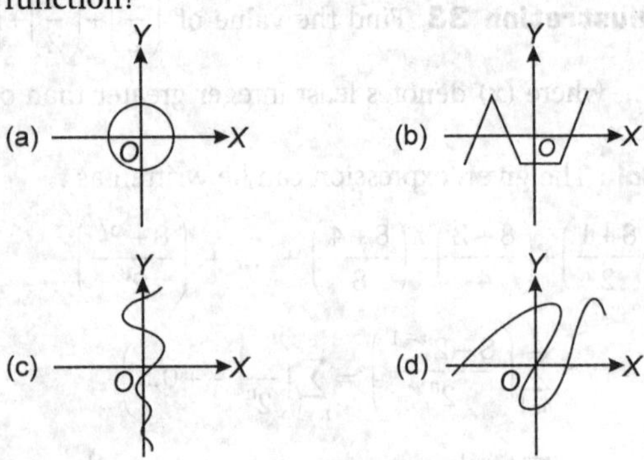

3. If $y = 3[x] + 1 = 4[x + 1] - 10$, where $[\cdot]$ denotes the greatest integer function, then the value of $[x + 2y]$ is :

(a) 61 (b) 51 (c) 39 (d) 29

4. The value of $\left[\frac{3}{4}\right] + \left[\frac{3}{4} + \frac{1}{100}\right] + \left[\frac{3}{4} + \frac{1}{50}\right] + \ldots + \left[\frac{3}{4} + \frac{99}{100}\right]$, where $[\cdot]$ denotes the greatest integer function, is :

(a) 25 (b) 50

(c) 75 (d) 100

5. The domain of definition of the function $f(x) = \frac{1}{\sqrt{x - [x]}}$, where $[\cdot]$ denotes the greatest integer function is :

(a) R (b) I (c) $R - I$ (d) N

6. The domain of the function $f(x) = \frac{1}{\sqrt{[x]^2 - 3[x] - 10}}$ is (where $[x]$ denotes the greatest integer less than or equal to x) **[JEE Main 11 April 2023 (ES)]**

(a) $(-\infty, -3] \cup (5, \infty)$ (b) $(-\infty, -2) \cup [6, \infty)$

(c) $(-\infty, -2) \cup (5, \infty)$ (c) $(-\infty, -3] \cup [6, \infty)$

7. If $f(x) = \left(\frac{x}{1 - |x|}\right)^{1/2001}$, then D_f is :

(a) $(-\infty, -1) \cup (1, \infty)$ (b) $(-1, 1)$

(c) R (d) $R - \{-1, 1\}$

8. For $f(x) = \sqrt{77^x - 777^x}$, D_f is :

(a) $(-\infty, -1]$ (b) $(-\infty, 0]$

(c) $(0, 1)$ (d) $(77, 777)$

9. If the functions are defined as $f(x) = \sqrt{x}$ and $g(x) = \sqrt{(1 - x)}$, then what is the common domain of the following functions : $f + g, f - g, f/g, g/f, g - f$, where $(f \pm g)(x) = f(x) \pm g(x)$, $\left(\frac{f}{g}\right)(x) = \frac{f(x)}{g(x)}$

[JEE Main 18 March 2021 (MS)]

(a) $0 \leq x \leq 1$ (b) $0 \leq x < 1$

(c) $0 < x < 1$ (d) $0 < x \leq 1$

10. Solution of $2[x] = x + \{x\}$, where $[\cdot]$ and $\{\cdot\}$ denotes the greatest integer and fractional part functions respectively are :

(a) 0, 1/2 (b) 0, 3/2 (c) 1/2, 3/2 (d) 1/2, 5/2

11. If $f(x) = \frac{x - [x]}{1 - [x] + x}$, where $[\cdot]$ denotes the greatest integer function, then $f(x) \in$:

(a) $[0, 1/2)$ (b) $[1/2, 1)$ (c) $[0, 1)$ (d) $[1/2, 3/2)$

12. If $f(x) = \text{sgn}(x^2 - 2x + 3)$, then the value of $f(x)$ is :

(a) 0 (b) 1 (c) 2 (d) 3

13. If $x(\text{sgn}\, x)(\text{sgn}\, x) + |x|(\text{sgn}\, x)^3 = \lambda x^\mu$, then the value of $\lambda + \mu$ is :

(a) 3 (b) 4 (c) 5 (d) 6

14. If $f(x) = \frac{1}{\sqrt{(\lceil x \rceil - 1)(3 - \lceil x \rceil)}}$, where $\lceil \cdot \rceil$ denotes the least integer function, then D_f is :

(a) $(0, 3]$ (b) $(1, 2]$ (c) $(1, 3)$ (d) $[2, 3)$

15. The solution of the equation $2[x] + \lceil x \rceil = 10$, where $[\cdot]$ and $\lceil \cdot \rceil$ denotes the greatest integer and least integer functions respectively, is :

(a) $(1, 2)$ (b) $(2, 3)$

(c) $(3, 4)$ (d) None of these

16. Given $y = 2\lceil x \rceil + 3$ and $y = 3\lceil x - 2 \rceil + 5$, where $\lceil \cdot \rceil$ denotes the least integer function, then the value of $\lceil x + 3y \rceil$ is :

(a) 15 (b) 17 (c) 25 (d) 37

(B) Transcendental Functions

All functions which are not algebraic are called transcendental functions.

(a) Trigonometric Functions

(i) Sine Function : $f(x) = \sin x$, then domain and range of $f(x)$ are R and $[-1, 1]$ respectively.

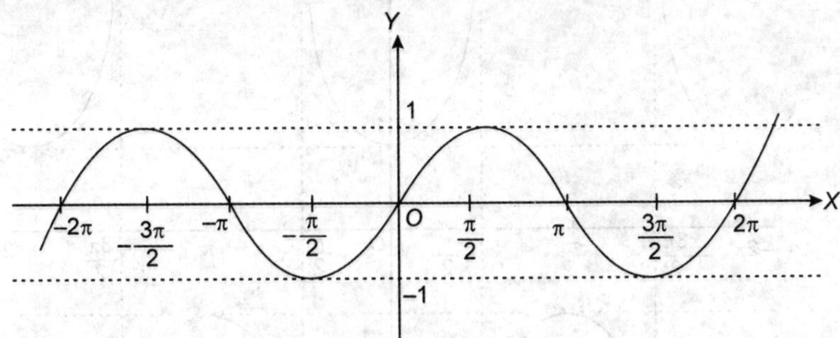

(ii) Cosine Function : $f(x) = \cos x$, then domain and range of $f(x)$ are R and $[-1, 1]$ respectively.

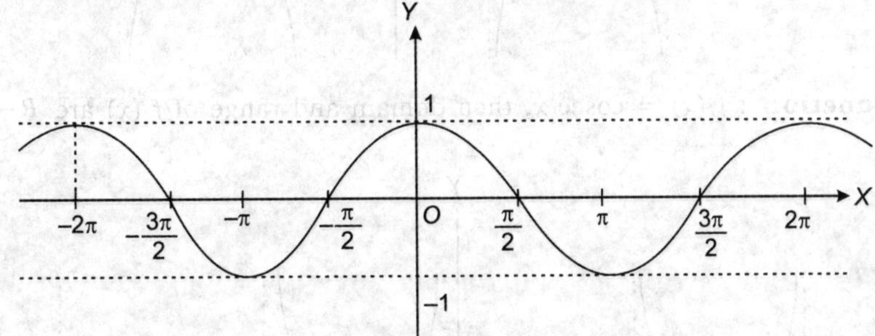

(iii) Tangent Function : $f(x) = \tan x$, then domain and range of $f(x)$ are $R - \left\{n\pi + \frac{\pi}{2}, n \in I\right\}$ and R respectively.

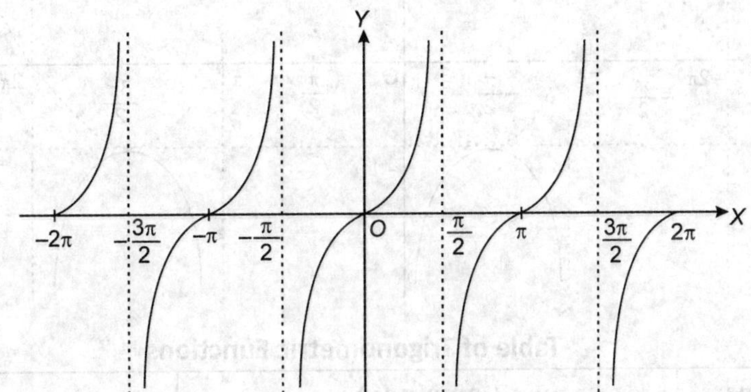

(iv) Cotangent Function : $f(x) = \cot x$, then domain and range of $f(x)$ are $R - \{n\pi, n \in I\}$ and R respectively.

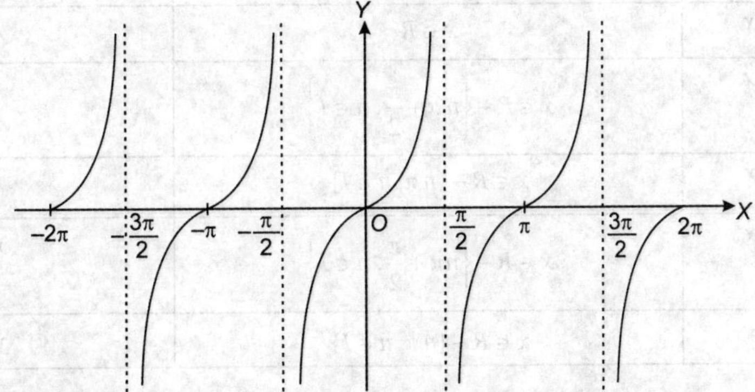

(v) Secant Function : $f(x) = \sec x$, then domain and range of $f(x)$ are $R - \left\{n\pi + \frac{\pi}{2}, n \in I\right\}$ and $R - (-1, 1)$ respectively.

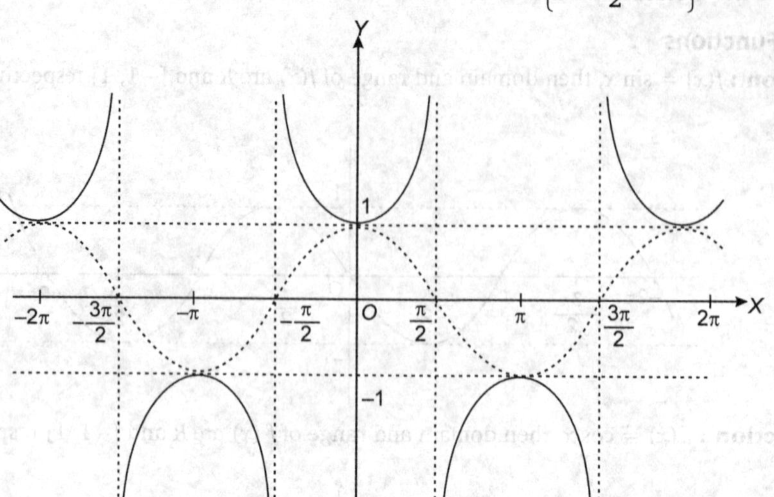

(vi) Cosecant Function : $f(x) = \operatorname{cosec} x$, then domain and range of $f(x)$ are $R - \{n\pi, n \in I\}$ and $R - (-1, 1)$ respectively.

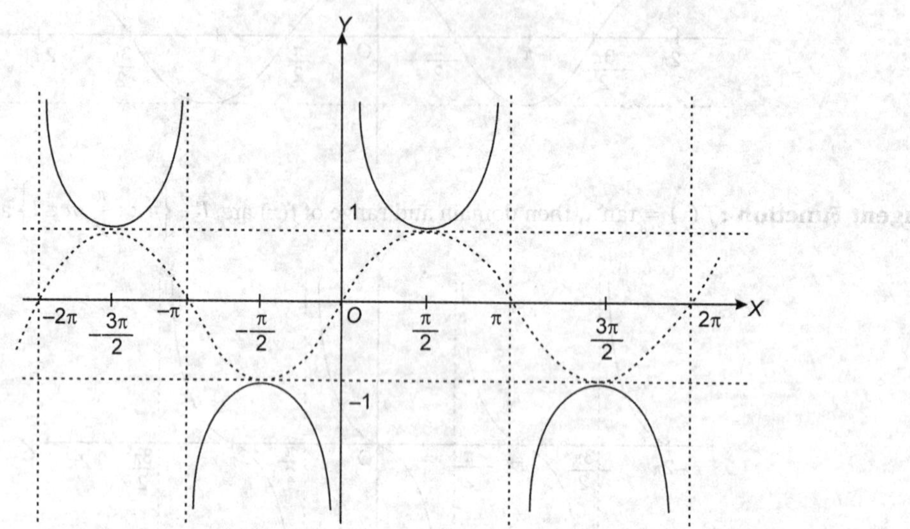

Table of Trigonometric Functions

S.No.	Functions	Domain (x)	Range (y)
1.	$y = \sin x$	$x \in R$	$y \in [-1, 1]$
2.	$y = \cos x$	$x \in R$	$y \in [-1, 1]$
3.	$y = \tan x$	$x \in R - \left\{n\pi + \frac{\pi}{2}, n \in I\right\}$	$y \in R$
4.	$y = \cot x$	$x \in R - \{n\pi, n \in I\}$	$y \in R$
5.	$y = \sec x$	$x \in R - \left\{n\pi + \frac{\pi}{2}, n \in I\right\}$	$y \in R - (-1, 1)$
6.	$y = \operatorname{cosec} x$	$x \in R - \{n\pi, n \in I\}$	$y \in R - (-1, 1)$

To Solve the Inequation with the help of the Graphs:

Suppose we have to solve

$$f(x) > k \quad \text{...(i)}$$

and $$f(x) < k \quad \text{...(ii)}$$

When solving the inequation, we often use the graphs of the functions $y = f(x)$ and $y = k$. Then the solutions of the inequation $f(x) > k$ and $f(x) < k$ are the values of x for which $y = f(x)$ lies above and below the straight line $y = k$ respectively.

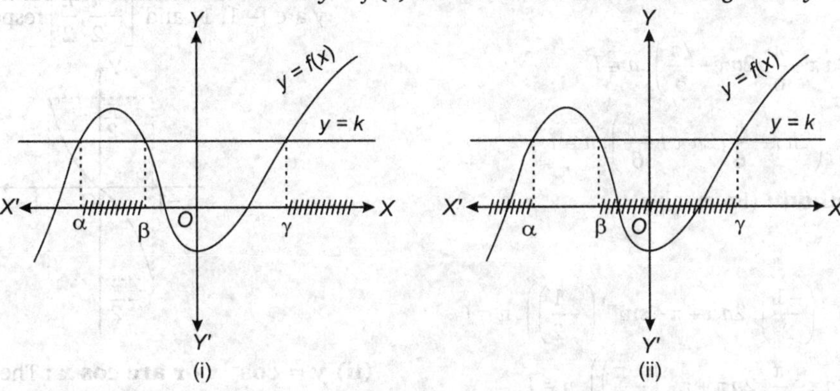

Therefore, we can easily see from the graphs, the solution set for $f(x) > k$ is given by (figure (i)) $x \in (\alpha, \beta) \cup (\gamma, \infty)$ and the solution set for $f(x) < k$ is given by (figure (ii)) $x \in (-\infty, \alpha) \cup (\beta, \gamma)$.

Inequalities Containing Trigonometric Functions :

Solutions of Elementary Inequalities

S.No.	Inequalities	Solution set of inequality, $n \in I$
1.	$\sin x > k, -1 < k < 1$	$x \in (2n\pi + \sin^{-1} k, 2n\pi + \pi - \sin^{-1} k)$
2.	$\sin x < k, -1 < k < 1$	$x \in (2n\pi - \pi - \sin^{-1} k, 2n\pi + \sin^{-1} k)$
3.	$\cos x > k, -1 < k < 1$	$x \in (2n\pi - \cos^{-1} k, 2n\pi + \cos^{-1} k)$
4.	$\cos x < k, -1 < k < 1$	$x \in (2n\pi + \cos^{-1} k, 2n\pi + 2\pi - \cos^{-1} k)$
5.	$\tan x > k, k \in R$	$x \in (n\pi + \tan^{-1} k, n\pi + \pi/2)$
6.	$\tan x < k, k \in R$	$x \in (n\pi - \pi/2, n\pi + \tan^{-1} k)$

Illustration 35. Find the domain of definition of the following functions :

(i) $f(x) = \sqrt{\cos(\sin x)}$

(ii) $f(x) = \sqrt{\sin(\cos x)}$

Sol : (i) We have $f(x) = \sqrt{\cos(\sin x)}$

$f(x)$ is defined when $\cos(\sin x) \geq 0$

$\therefore \quad 2n\pi - \frac{\pi}{2} \leq \sin x \leq 2n\pi + \frac{\pi}{2}, \ n \in I$

but $-1 \leq \sin x \leq 1 \ \forall \ x \in R$

Hence $\quad -1 \leq \sin x \leq 1$

then $\quad x \in R$

$\therefore \quad D_f = R$

(ii) We have $f(x) = \sqrt{\sin(\cos x)}$

$f(x)$ is defined when $\sin(\cos x) \geq 0$

$\therefore \quad 2n\pi \leq \cos x \leq 2n\pi + \pi, n \in I$

but $-1 \leq \cos x \leq 1$

Hence $\quad 0 \leq \cos x \leq 1$

$\therefore \quad x \in \left[2n\pi - \frac{\pi}{2}, 2n\pi + \frac{\pi}{2}\right], \ n \in I$

$\Rightarrow \quad D_f = \left[2n\pi - \frac{\pi}{2}, 2n\pi + \frac{\pi}{2}\right], n \in I$

Illustration 36. Solve $\sin x > -\frac{1}{2}$ or find the domain of definition of $f(x) = \frac{1}{\sqrt{(1 + 2\sin x)}}$.

Sol : We have $f(x) = \frac{1}{\sqrt{(1 + 2\sin x)}}$

$f(x)$ is defined when $1 + 2\sin x > 0$ or $\sin x > -\frac{1}{2}$

Now we construct the graph of $y = \sin x$ and $y = -\frac{1}{2}$.

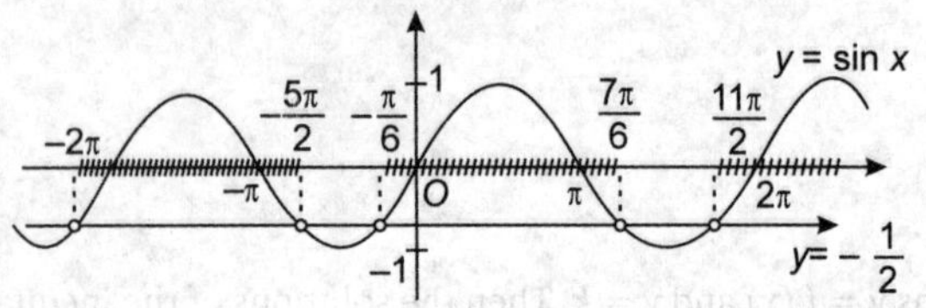

As $\sin x > -\dfrac{1}{2}$

$\therefore \quad x \in \left(2n\pi - \dfrac{\pi}{6}, 2n\pi + \dfrac{7\pi}{6}\right), n \in I$

or $\quad x = \bigcup_{n \in I} \left(2n\pi - \dfrac{\pi}{6}, 2n\pi + \dfrac{7\pi}{6}\right), n \in I$

Alternative Method : (From table)

Here $k = -\dfrac{1}{2}$

$\therefore \; x \in \left(2n\pi + \sin^{-1}\left(\dfrac{-1}{2}\right), 2n\pi + \pi - \sin^{-1}\left(-\dfrac{1}{2}\right)\right), n \in I$

or $\quad x \in \left(2n\pi - \dfrac{\pi}{6}, 2n\pi + \pi - \left(-\dfrac{\pi}{6}\right)\right), n \in I$

or $\quad x \in \left(2n\pi - \dfrac{\pi}{6}, 2n\pi + \dfrac{7\pi}{6}\right), n \in I$

or $\quad x = \bigcup_{n \in I} \left(2n\pi - \dfrac{\pi}{6}, 2n\pi + \dfrac{7\pi}{6}\right)$

Illustration 37. Find the domain of definition of the function $f(x) = \sqrt{\left(1 - \dfrac{x^2}{2} - \cos x\right)}$.

Sol : We have $f(x) = \sqrt{\left(1 - \dfrac{x^2}{2} - \cos x\right)}$

$f(x)$ is defined, when $1 - \dfrac{x^2}{2} - \cos x \geq 0$

or $1 - \dfrac{x^2}{2} \geq \cos x$

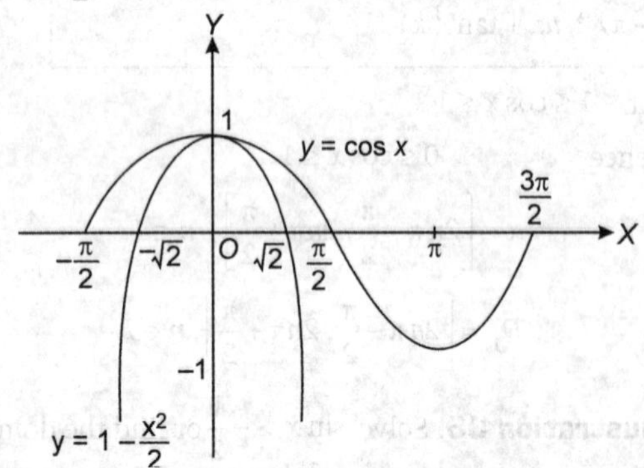

Now we construct the graph of $y = 1 - \dfrac{x^2}{2}$

i.e., $\quad x^2 = -2(y-1)$

and $\quad y = \cos x$

It is clear from the graph $1 - \dfrac{x^2}{2} \ngtr \cos x$

But $1 - \dfrac{x^2}{2} = \cos x$ only at $x = 0$

Thus, $\quad 1 - \dfrac{x^2}{2} \geq \cos x \Rightarrow x = 0$

Hence, $\quad D_f = \{0\}$

(b) Inverse Trigonometric Functions

(i) $y = \sin^{-1} x$ or arc sin x : The domain and range of y are $[-1, 1]$ and $\left[-\dfrac{\pi}{2}, \dfrac{\pi}{2}\right]$ respectively.

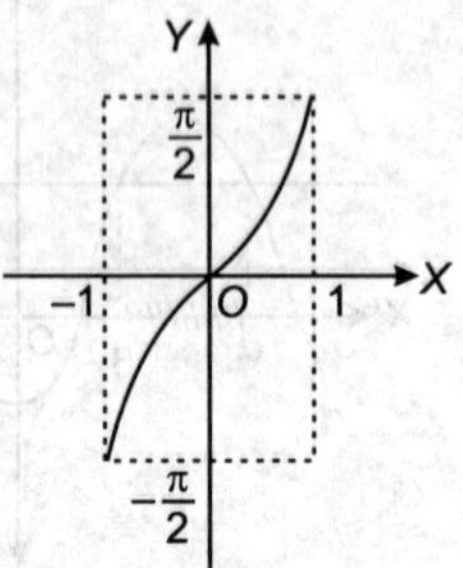

(ii) $y = \cos^{-1} x$ or arc cos x : The domain and range of y are $[-1, 1]$ and $[0, \pi]$ respectively.

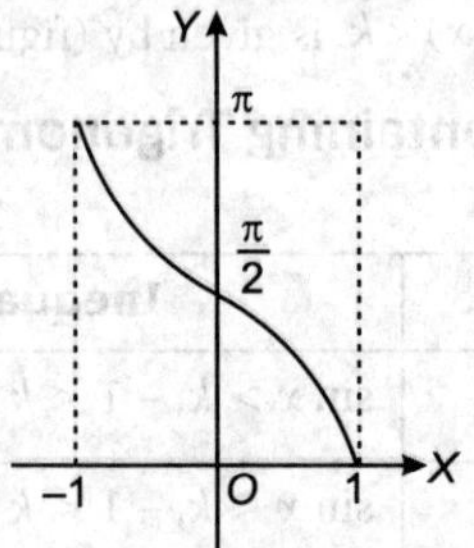

(iii) $y = \tan^{-1} x$ or arc tan x : The domain and range of y are R and $\left(-\dfrac{\pi}{2}, \dfrac{\pi}{2}\right)$ respectively.

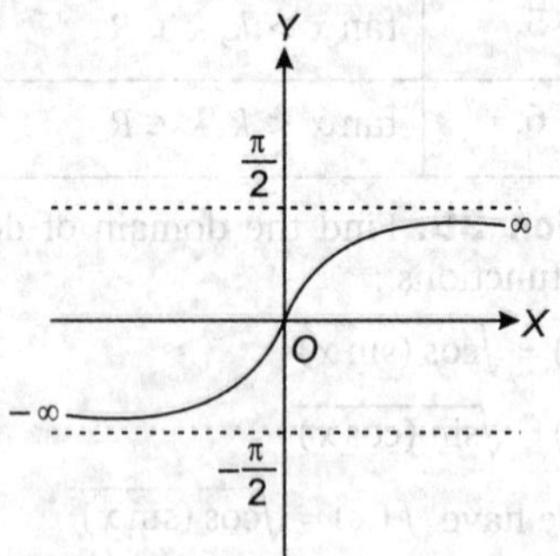

(iv) $y = \cot^{-1} x$ or arc cot x : The domain and range of y are R and $(0, \pi)$ respectively.

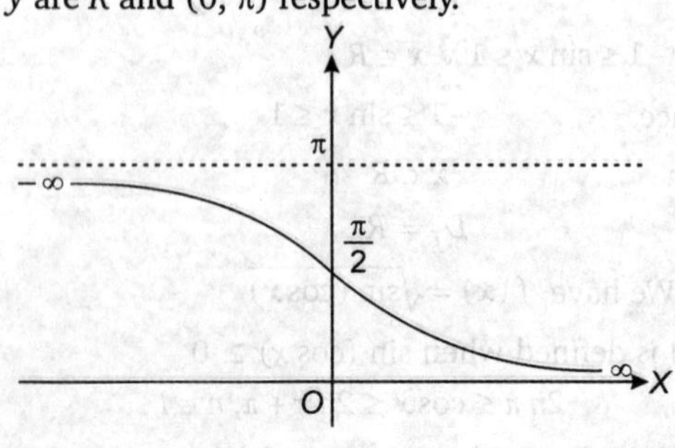

(v) $y = \sec^{-1} x$ or arc sec x : The domain and rage of y are $R-(-1, 1)$ and $[0, \pi] \sim \{\pi/2\}$ respectively.

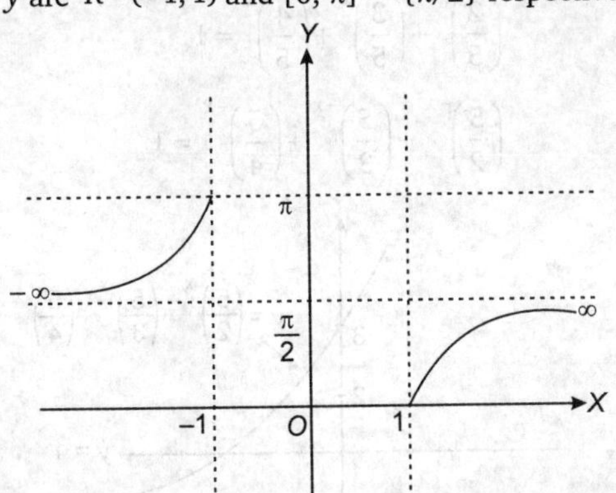

(vi) $y = \text{cosec}^{-1} x$ or arc cosec x : The domain and range of y are $R - (-1, 1)$ and $\left[-\frac{\pi}{2}, \frac{\pi}{2}\right] - \{0\}$ respectively.

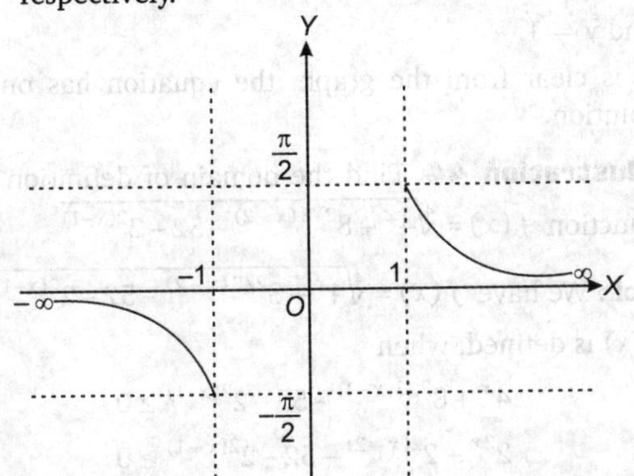

Table of Inverse Trigonometric Functions

S.No.	Functions	Domain (x)	Range (y)
1.	$y = \sin^{-1} x$	$[-1, 1]$	$\left[-\frac{\pi}{2}, \frac{\pi}{2}\right]$
2.	$y = \cos^{-1} x$	$[-1, 1]$	$[0, \pi]$
3.	$y = \tan^{-1} x$	R	$\left(-\frac{\pi}{2}, \frac{\pi}{2}\right)$
4.	$y = \cot^{-1} x$	R	$(0, \pi)$
5.	$y = \sec^{-1} x$	$R-(-1, 1)$	$[0, \pi]-\left\{\frac{\pi}{2}\right\}$
6.	$y = \text{cosec}^{-1} x$	$R-(-1, 1)$	$\left[-\frac{\pi}{2}, \frac{\pi}{2}\right]-\{0\}$

Inequalities Containing Inverse Trigonometric Functions : Solutions of Elementary Inequalities

S.No.	Inequalities	Solution set of inequality
1.	$\sin^{-1} x > k; -\frac{\pi}{2} < k < \frac{\pi}{2}$	$x \in (\sin k, 1]$
2.	$\sin^{-1} x < k; -\frac{\pi}{2} \le k \le \frac{\pi}{2}$	$x \in [-1, \sin k)$
3.	$\cos^{-1} x > k; 0 < k < \pi$	$x \in [-1, \cos k)$
4.	$\cos^{-1} x < k, 0 < k < \pi$	$x \in (\cos k, 1]$
5.	$\tan^{-1} x > k, -\frac{\pi}{2} < k < \frac{\pi}{2}$	$x \in (\tan k, \infty)$
6.	$\tan^{-1} x < k, -\frac{\pi}{2} < k < \frac{\pi}{2}$	$x \in (-\infty, \tan k)$
7.	$\cot^{-1} x > k, 0 < k < \pi$	$x \in (-\infty, \cot k)$
8.	$\cot^{-1} x < k, 0 < k < \pi$	$x \in (\cot k, \infty)$

Illustration 38. Solve the inequality
$$2(\sin^{-1} x)^2 - (\sin^{-1} x) - 6 \ge 0.$$

Sol : We have
$$2(\sin^{-1} x)^2 - (\sin^{-1} x) - 6 \ge 0$$
$$\Rightarrow \quad (2\sin^{-1} x + 3)(\sin^{-1} x - 2) \ge 0$$
$$\therefore \quad \sin^{-1} x \le -\frac{3}{2} \text{ or } \sin^{-1} x \ge 2 \left(\text{Impossible } \because 2 > \frac{\pi}{2}\right)$$
$$\therefore \quad \sin^{-1} x \le -\frac{3}{2} \quad \left(-\frac{\pi}{2} < -\frac{3}{2} < \frac{\pi}{2}\right)$$
$$\therefore \quad x \in \left[-1, \sin\left(-\frac{3}{2}\right)\right]$$

Illustration 39. Find the domain of definition of the function $f(x) = \sin^{-1}\left(\frac{x^2}{2}\right)$.

Sol : We have $f(x) = \sin^{-1}\left(\frac{x^2}{2}\right)$

$f(x)$ is defined, when
$$-1 \le \frac{x^2}{2} \le 1$$
or
$$-2 \le x^2 \le 2$$
$$\Rightarrow \quad 0 \le x^2 \le 2$$
$$\therefore \quad -\sqrt{2} \le x \le \sqrt{2}$$

Therefore $D_f = [-\sqrt{2}, \sqrt{2}]$

Illustration 40. Find the domain of definition of the function $f(x) = \sqrt{(\cos^{-1} x - \sin^{-1} x)}$.

Sol : We have $f(x) = \sqrt{(\cos^{-1} x - \sin^{-1} x)}$

$f(x)$ is defined, when
$$\cos^{-1} x - \sin^{-1} x \ge 0$$
$$\Rightarrow \quad \frac{\pi}{2} - \sin^{-1} x - \sin^{-1} x \ge 0 \Rightarrow \frac{\pi}{4} \ge \sin^{-1} x$$
or $\quad \sin^{-1} x \le \frac{\pi}{4} \quad$ (Here $k = \pi/4 < \pi/2$)
$$\therefore \quad x \in \left[-1, \frac{1}{\sqrt{2}}\right]$$

Illustration 41. Find the domain of definition of the function $f(x) = \sin^{-1}\left(\frac{1+x^2}{2x}\right)$ or $\cos^{-1}\left(\frac{1+x^2}{2x}\right)$.

Sol : We have $f(x) = \sin^{-1}\left(\frac{1+x^2}{2x}\right)$ or $\cos^{-1}\left(\frac{1+x^2}{2x}\right)$

$f(x)$ is defined, when

$$-1 \le \frac{1+x^2}{2x} \le 1$$

$$\Rightarrow \quad \left|\frac{1+x^2}{2x}\right| \le 1$$

$$\Rightarrow \quad \frac{|1+x^2|}{2|x|} \le 1 \Rightarrow \frac{1+x^2}{2|x|} \le 1$$

$$\Rightarrow \quad \frac{1+|x|^2}{2|x|} - 1 \le 0$$

$$\Rightarrow \quad \frac{(|x|-1)^2}{2|x|} \le 0$$

or $\quad \frac{(|x|-1)^2}{2|x|} = 0 \quad \left(\because \frac{(|x|-1)^2}{2|x|} \not< 0\right)$

or $\quad (|x|-1)^2 = 0$

or $\quad |x|-1 = 0$

$\therefore \quad x = \pm 1$

$\therefore \quad D_f = \{-1, 1\}$

Illustration 42. Find the domain of definition of the function $f(x) = \sin^{-1}[2-3x^2]$, where $[\cdot]$ denotes the greatest integer function.

Sol : We have $f(x) = \sin^{-1}[2-3x^2]$

$f(x)$ is defined, when $-1 \le [2-3x^2] \le 1$

$\Rightarrow \quad -1 \le 2-3x^2 < 2$

or $\quad -3 \le -3x^2 < 0$

or $\quad 1 \ge x^2 > 0 \Rightarrow 0 < x^2 \le 1$

$\therefore \quad x \in [-1, 1] - \{0\}$

Hence $\quad D_f = [-1, 1] - \{0\}$

(c) Exponential Functions

Let $a \ne 1, a > 0$ be a real number, then $f : R \to R$ defined by $f(x) = a^x$ is called exponential function. Its domain and range are R and $(0, \infty)$ respectively.

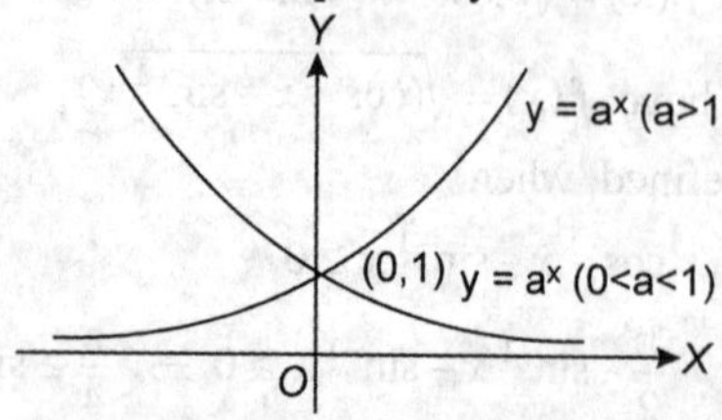

Illustration 43. Find the number of solutions of $2^x + 3^x + 4^x - 5^x = 0$.

Sol : We have $2^x + 3^x + 4^x - 5^x = 0$

$$\Rightarrow \quad 2^x + 3^x + 4^x = 5^x$$

$$\Rightarrow \quad \left(\frac{2}{5}\right)^x + \left(\frac{3}{5}\right)^x + \left(\frac{4}{5}\right)^x = 1$$

or $\quad \left(\frac{5}{2}\right)^{-x} + \left(\frac{5}{3}\right)^{-x} + \left(\frac{5}{4}\right)^{-x} = 1$

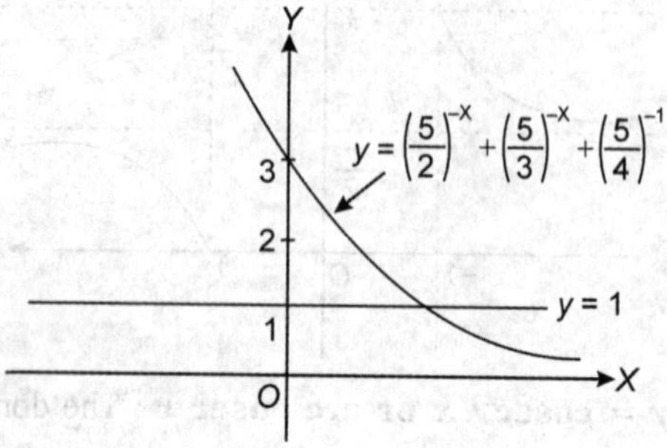

We construct the graphs of $y = \left(\frac{5}{2}\right)^{-x} + \left(\frac{5}{3}\right)^{-x} + \left(\frac{5}{4}\right)^{-x}$ and $y = 1$.

It is clear from the graph, the equation has only one solution.

Illustration 44. Find the domain of definition of the function $f(x) = \sqrt[4]{4^x + 8^{2/3(x-2)} - 52 - 2^{2(x-1)}}$.

Sol : We have $f(x) = \sqrt[4]{4^x + 8^{2/3(x-2)} - 52 - 2^{2(x-1)}}$

$f(x)$ is defined, when

$$4^x + 8^{2/3(x-2)} - 52 - 2^{2(x-1)} \ge 0$$

$$\Rightarrow \quad 2^{2x} + 2^{2(x-2)} - 52 - 2^{2(x-1)} \ge 0$$

$$\Rightarrow \quad 2^{2x}(1 + 2^{-4} - 2^{-2}) \ge 52$$

$$\Rightarrow \quad 2^{2x}\left(\frac{13}{16}\right) \ge 52$$

$$\Rightarrow \quad 2^{2x} \ge 64$$

or $\quad 2^{2x} \ge 2^6 \Rightarrow 2x \ge 6$

$\Rightarrow \quad x \ge 3$

$\therefore \quad D_f = [3, \infty)$

Illustration 45. Find the domain of definition of the function $f(x) = \sqrt[6]{\left(\frac{1-5^x}{7^{-x}-7}\right)}$.

Sol : We have $f(x) = \sqrt[6]{\left(\frac{1-5^x}{7^{-x}-7}\right)}$

$f(x)$ is defined, when $\frac{1-5^x}{7^{-x}-7} \ge 0$

or $\quad \frac{5^x-1}{7^{-x}-7} \le 0$

Case I. $\quad 5^x - 1 \ge 0$ and $7^{-x} - 7 < 0$

$\Rightarrow \quad 5^x \ge 1$ and $7^{-x} < 7^1$

$\Rightarrow \quad 5^x \ge 5^0$ and $7^{-x} < 7^1$

$\therefore \quad x \ge 0$ and $-x < 1$

$\Rightarrow \quad x \ge 0 \quad$ and $\quad x > -1$

Hence, $\quad x \ge 0 \quad$ or $\quad x \in [0, \infty)$

Case II. $5^x - 1 \le 0 \quad$ and $\quad 7^{-x} - 7 > 0$

$\Rightarrow \quad 5^x \le 1 \quad$ and $\quad 7^{-x} > 7$

$\Rightarrow \quad 5^x \le 5^0 \quad$ and $\quad 7^{-x} > 7^1$

$\therefore \quad x \le 0 \quad$ and $\quad -x > 1$

$\Rightarrow \quad x \le 0 \quad$ and $\quad x < -1$

Hence, $x < -1$ or $x \in (-\infty, -1)$

Combining both cases, we get :

$$x \in (-\infty, -1) \cup [0, \infty)$$

Hence, $\quad D_f = (-\infty, -1) \cup [0, \infty)$

Illustration 46. Find the domain of definition of the function $f(x) = \sqrt{(0.625)^{4-3x} - (1.6)^{x(x+8)}}$.

Sol : We have $f(x) = \sqrt{(0.625)^{4-3x} - (1.6)^{x(x+8)}}$

$f(x)$ is defined, when

$$(0.625)^{4-3x} - (1.6)^{x(x+8)} \ge 0$$

$$\Rightarrow \quad \left(\frac{625}{1000}\right)^{4-3x} - \left(\frac{16}{10}\right)^{x(x+8)} \ge 0$$

$$\Rightarrow \quad \left(\frac{5}{8}\right)^{4-3x} \ge \left(\frac{8}{5}\right)^{x(x+8)}$$

$$\Rightarrow \quad \left(\frac{8}{5}\right)^{3x-4} \ge \left(\frac{8}{5}\right)^{x(x+8)}$$

$$\therefore \quad 3x - 4 \ge x(x+8)$$

$$\Rightarrow \quad x^2 + 5x + 4 \le 0$$

$$\Rightarrow \quad (x+1)(x+4) \le 0$$

or $\quad -4 \le x \le -1$

or $\quad x \in [-4, -1]$

Hence, $\quad D_f = [-4, -1]$

(d) Logarithmic Functions

Let $a \ne 1, a > 0$ be a real number. Then $f : (0, \infty) \to R$ defined by $f(x) = \log_a x$ is called logarithmic function. Its domain and range are $(0, \infty)$ and R respectively.

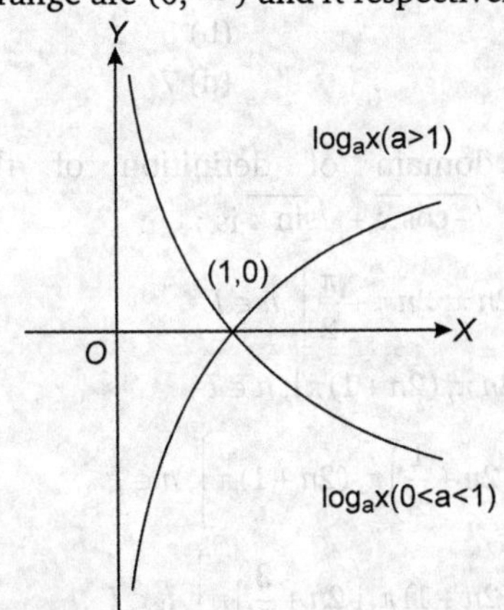

Properties of Logarithmic Functions :

f and g are two functions and $a > 0$, $a \ne 1$, then if $f > 0$, $g > 0$ we have

(i) $\log_a (f \cdot g) = \log_a f + \log_a g$

(ii) $\log_a (f/g) = \log_a f - \log_a g$

(iii) $\log_a f^{2\alpha} = 2\alpha \log_a |f| \qquad (\alpha \in R)$

(iv) $\log_{a^\beta} f^\alpha = \dfrac{\alpha}{\beta} \log_a f \qquad (\alpha \in R$ and $\beta \in R - \{0\})$

(v) $f^{\log_a g} = g^{\log_a f}$

(vi) $a^{\log_a f} = f$

(vii) If $\log_a f > \log_a g \Rightarrow \begin{cases} f > g, & \text{if } a > 1 \\ f < g, & \text{if } 0 < a < 1 \end{cases}$

(viii) If $\log_a f = g \Rightarrow f = a^g$

(ix) If $\log_a f > g \Rightarrow \begin{cases} f > a^g, & \text{if } a > 1 \\ f < a^g, & \text{if } 0 < a < 1 \end{cases}$

(x) If $\log_a f < g \Rightarrow \begin{cases} f < a^g, & \text{if } a > 1 \\ f > a^g, & \text{if } 0 < a < 1 \end{cases}$

(xi) $\log_g f = \dfrac{\log_a f}{\log_a g}, a > 0, a \ne 1 = \dfrac{1}{\log_f g}$

$\left(\begin{array}{l}\text{If } f = g = b > 0, \ne 1 \\ \text{then } \log_b b = 1\end{array}\right)$

(xii) $\log_{fg} a = \dfrac{1}{\log_a f + \log_a g}$

Illustration 47. Solve $\log_x (x^2 - 1) \le 0$.

Sol : Given $\log_x (x^2 - 1) \le 0$

Case I. If $x > 1$, then

$$x^2 - 1 \le x^0$$

$$\Rightarrow \quad x^2 - 1 \le 1$$

$$\Rightarrow \quad x^2 \le 2$$

$$\therefore \quad -\sqrt{2} \le x \le \sqrt{2}$$

but $x > 1$

$$\therefore \quad 1 < x \le \sqrt{2}$$

$$\Rightarrow \quad x \in (1, \sqrt{2})$$

Case II. If $0 < x < 1$, then

$$x^2 - 1 \ge x^0$$

$$\Rightarrow \quad x^2 - 1 \ge 1$$

or $\quad x^2 \ge 2$

$$\therefore \quad x \in (-\infty, -\sqrt{2}] \cup [\sqrt{2}, \infty)$$

but $\quad 0 < x < 1$

$$\therefore \quad x \in \phi$$

Combining both cases, we get

$$x \in (1, \sqrt{2}]$$

Hence, $\quad D_f = (1, \sqrt{2}]$

Illustration 48. Find the domain of definition of the function $f(x) = \sin^{-1}\left(\log_9\left(\frac{x^2}{4}\right)\right)$.

Sol : We have $f(x) = \sin^{-1}\left(\log_9\left(\frac{x^2}{4}\right)\right)$

$f(x)$ is defined, when

$$-\leq \log_9\left(\frac{x^2}{4}\right) \leq 1 \quad \text{and } x^2 > 0$$

$$\Rightarrow \quad 9^{-1} \leq \frac{x^2}{4} \leq 9^1 \quad \text{and } x^2 > 0$$

or $$\frac{4}{9} \leq x^2 \leq 36 \quad \text{and } x^2 > 0$$

$$\Rightarrow \quad \left(\frac{2}{3}\right)^2 \leq x^2 \leq 6^2$$

$$\therefore \quad x \in \left[-6, -\frac{2}{3}\right] \cup \left[\frac{2}{3}, 6\right]$$

Hence, $$D_f = \left[-6, -\frac{2}{3}\right] \cup \left[\frac{2}{3}, 6\right]$$

Illustration 49. Find the domain of definition of the function $f(x) = \sqrt{\log_{0.5}\left(\frac{x-2}{x+7}\right)}$.

Sol : We have $f(x) = \sqrt{\log_{0.5}\left(\frac{x-2}{x+7}\right)}$

$f(x)$ is defined, when

$$\log_{0.5}\left(\frac{x-2}{x+7}\right) \geq 0 \quad \text{and } \left(\frac{x-2}{x+7}\right) > 0$$

$$\Rightarrow \quad \left(\frac{x-2}{x+7}\right) \leq (0.5)^0 \quad \text{and } \left(\frac{x-2}{x+7}\right) > 0$$

$$\Rightarrow \quad \left(\frac{x-2}{x+7}\right) \leq 1 \quad \text{and } \left(\frac{x-2}{x+7}\right) > 0$$

$$\Rightarrow \quad \left(\frac{x-2}{x+7} - 1\right) \leq 0 \quad \text{and } \left(\frac{x-2}{x+7}\right) > 0$$

$$\Rightarrow \quad \frac{-9}{(x+7)} \leq 0 \quad \text{and } \left(\frac{x-2}{x+7}\right) > 0$$

$$\Rightarrow \quad (x+7) > 0 \quad \text{and } \left(\frac{x-2}{x+7}\right) > 0$$

$$\Rightarrow \quad (x+7) > 0 \quad \text{and } (x-2) > 0$$

$$\Rightarrow \quad x > -7 \quad \text{and } x > 2$$

$$\therefore \quad x > 2$$

or $$x \in (2, \infty)$$

Hence, $$D_f = (2, \infty).$$

Illustration 50. Find the domain of definition of the function $f(x) = \log_{10}\{1 - \log_{10}(x^2 - 5x + 16)\}$.

Sol : We have $f(x) = \log_{10}\{1 - \log_{10}(x^2 - 5x + 16)\}$

$f(x)$ is defined, when

$$1 - \log_{10}(x^2 - 5x + 16) > 0 \text{ and } x^2 - 5x + 16 > 0$$

$$\Rightarrow \quad \log_{10}(x^2 - 5x + 16) < 1 \text{ and } x^2 - 5x + 16 > 0$$

$$\Rightarrow \quad x^2 - 5x + 16 < 10^1 \text{ and } \left(x - \frac{5}{2}\right)^2 + \frac{39}{4} > 0$$

$$\Rightarrow \quad x^2 - 5x + 6 < 0 \text{ and } \left(x - \frac{5}{2}\right)^2 + \frac{39}{4} > 0$$

$$\Rightarrow \quad (x-2)(x-3) < 0 \text{ and } x \in R$$

$$\Rightarrow \quad x \in (2, 3) \text{ and } x \in R$$

$$\Rightarrow \quad x \in (2, 3)$$

Hence, $D_f = (2, 3)$

Illustration 51. Find the domain of definition of the function $f(x) = \log_2 \log_3 \log_{4/\pi}(\tan^{-1} x)^{-1}$.

Sol : We have $f(x) = \log_2 \log_3 \log_{4/\pi}(\tan^{-1} x)^{-1}$

$f(x)$ is defined, when

$$\log_3 \log_{4/\pi}(\tan^{-1} x)^{-1} > 0 \text{ and } (\tan^{-1} x)^{-1} > 0$$

$$\Rightarrow \quad \log_{4/\pi}(\tan^{-1} x)^{-1} > 3^0 \text{ and } \frac{1}{\tan^{-1} x} > 0$$

$$\Rightarrow \quad \log_{4/\pi}(\tan^{-1} x)^{-1} > 1 \text{ and } \tan^{-1} x > 0$$

$$\Rightarrow \quad (\tan^{-1} x)^{-1} > \left(\frac{4}{\pi}\right)^1 \text{ and } \tan^{-1} x > 0$$

$$\Rightarrow \quad \tan^{-1} x < \frac{\pi}{4} \text{ and } \tan^{-1} x > 0$$

$$\Rightarrow \quad 0 < \tan^{-1} x < \frac{\pi}{4}$$

or $$0 < x < 1 \Rightarrow \quad x \in (0, 1)$$

Hence, $$D_f = (0, 1)$$

INTRODUCTORY EXERCISE 2.4

Only One Option is Correct

1. The number of solutions of the equation $10 \sin x = x$ are :

(a) 3 (b) 5
(c) 6 (d) 7

2. The domain of definition of the function $f(x) = \sqrt{-\cos x} + \sqrt{\sin x}$ is :

(a) $\left[2n\pi, 2n\pi + \frac{\pi}{2}\right], n \in I$

(b) $[2n\pi, (2n+1)\pi], n \in I$

(c) $\left[\left(2n + \frac{1}{2}\right)\pi, (2n+1)\pi\right], n \in I$

(d) $\left[(2n+1)\pi, \left(2n + \frac{3}{2}\right)\pi\right], n \in I$

3. The domain of definition of the function $f(x)=\dfrac{1}{\sqrt{|\cos x|+\cos x}}$ is :

(a) $[-2n\pi, 2n\pi], n \in I$
(b) $(2n\pi, (2n+1)\pi), n \in I$
(c) $\left(\dfrac{(4n+1)\pi}{2}, \dfrac{(4n+3)\pi}{2}\right), n \in I$
(d) $\left(\dfrac{(4n-1)\pi}{2}, \dfrac{(4n+1)\pi}{2}\right), n \in I$

4. The domain of definition of the function $f(x)=\dfrac{1}{\sqrt{\sin x}}+\sqrt[3]{\sin x}$ is :

(a) $(2n\pi, (2n+1)\pi), n \in I$
(b) $(0, \pi)$
(c) $(\pi, 2\pi)$
(d) $((2n+1)\pi, (2n+2)\pi), n \in I$

5. If $\sin\left(\dfrac{\cos^{-1} x}{5}\right)=1$, then x equals :

(a) 0 (b) $\dfrac{\pi}{2}$
(c) $\dfrac{5\pi}{2}$ (d) None of these

6. The domain of definition of the function $f(x)=\sin^{-1}(|x-1|-2)$ is :

(a) $[-2, 1]\cup[2, 4]$ (b) $[-2, 0] \cup [2, 4]$
(c) $[-2, 1] \cup [2, 3]$ (d) $[-2, 0] \cup [2, 3]$

7. If $[\cdot]$ stands for the greatest integer function, domain of definition of the function $f(x)=\text{cosec}^{-1}[2x-1]$ is :

(a) $\left(-\infty, \dfrac{1}{2}\right)\cup\left[\dfrac{3}{2}, \infty\right)$ (b) $\left(-\infty, \dfrac{1}{2}\right)\cup(1, \infty)$
(c) $\left(-\infty, -\dfrac{1}{2}\right)\cup\left[\dfrac{3}{2}, \infty\right)$ (d) None of these

8. The domain of definition of the function $f(x)=\cos^{-1}\left(\dfrac{6-3x}{4}\right)+\text{cosec}^{-1}\left(\dfrac{x-1}{2}\right)$ is :

(a) $\left[\dfrac{2}{3}, \dfrac{10}{3}\right]$ (b) $(-\infty, -1]\cup[3, \infty)$
(c) $\left[3, \dfrac{10}{3}\right]$ (d) $\left[\dfrac{10}{3}, \infty\right)$

9. The domain of the function,

$$f(x)=\sin^{-1}\left(\frac{3x^2+x-1}{(x-1)^2}\right)+\cos^{-1}\left(\frac{x-1}{x+1}\right) \text{ is :}$$

[JEE Main 26 August 2021 (ES)]

(a) $\left[0, \dfrac{1}{2}\right]$ (b) $\left[0, \dfrac{1}{4}\right]$
(c) $\left[\dfrac{1}{4}, \dfrac{1}{2}\right]\cup\{0\}$ (d) $[-2, 0]\cup\left[\dfrac{1}{4}, \dfrac{1}{2}\right]$

10. Number of solutions of the equation $3^x \cdot 8^{x/(x+1)}=36$ are :

(a) 2 (b) 3 (c) 4 (d) 5

11. The domain of definition of the function $y(x)$ given by $2^x+2^y=2$ is :

(a) $(0, 1]$ (b) $[0, 1]$ (c) $(-\infty, 0]$ (d) $(-\infty, 1)$

12. The domain of definition of the function $f(x)=\sqrt{(27\sqrt{3}-3^{x^2-17x+63.5})}$ is :

(a) $[-\infty, 5]$ (b) $[12, \infty]$
(c) $[5, 12]$ (d) None of these

13. The domain of definition of the function $f(x)=\dfrac{1}{\sqrt{(0.2)^{(2x-3)/(x-2)}-5}}$ is :

(a) $(1, 5/3)$ (b) $(5/3, 2)$
(c) $(2, 7/3)$ (d) $(7/3, 3)$

14. Number of positive integral solutions of inequation $\log_{0.5}\left(\log_6\left(\dfrac{x^2+x}{x+4}\right)\right)>0$ are :

(a) 3 (b) 5 (c) 7 (d) 9

15. The domain of definition of the function $f(x)=\sqrt{\log_{10}\left\{\dfrac{\log_{10} x}{2(3-\log_{10} x)}\right\}}$ is :

(a) $(10^2, 10^3)$ (b) $[10^2, 10^3)$
(c) $(10^2, 10^3]$ (d) $[10^2, 10^3]$

16. The domain of $f(x)=\dfrac{\log_{(x+1)}(x-2)}{e^{2\log_e x}-(2x+3)}$, $x \in R$ is

[JEE Main 29 Jan 2023 (MS)]

(a) $R-\{-1, 3\}$ (b) $(2, \infty)-\{3\}$
(c) $(-1, \infty)-\{3\}$ (d) $R-\{3\}$

17. The domain of definition of 1the function $f(x)=\dfrac{1}{\sqrt{\log_{1/2}(x^2-7x+13)}}$ is :

(a) $(-\infty, 3)$ (b) $(3, 4)$ (c) $(4, 7)$ (d) $(7, \infty)$

18. If $\{x\}$ and $[x]$ stand for fractional part of x and greatest integer $x \le x$ respectively, then domain of the function $f(x)=\log_{\{1-x^2\}}[1-x^2]$ is :

(a) $(-1, 1)$ (b) $[-1, 1]$
(c) $R-(-1, 1)$ (d) None of these

Range

Range of a function $y=f(x)$ is the set of all possible values of y or collection of all outputs $\{f(x)\}$ corresponding to values of x in the domain.

i.e., R_f = Range of $y=f(x)$ is the set of all outputs $\{f(x)\}$, $\forall\, x \in D_f$.

Its study is little but complicated. It needs through investigation of a curve and as we move further knowing many aspects of calculus. We will get more ideas clear some methods are discussed below :

(i) Find the domain of $y = f(x)$, first of all.

(ii) Range of a polynomial function with odd degree is "R".

(iii) Range of $a\cos\theta + b\sin\theta + c$ is $[c - \sqrt{(a^2+b^2)},\ c + \sqrt{(a^2+b^2)}]$, when a, b, c are real constants and $\theta \in R$.

(iv) Sum of a positive number and its reciprocal is greater than or equal to two.

i.e., $a + \frac{1}{a} \geq 2$ if $a > 0$.

(v) To find the range of a function whose domain is $(-\infty, \infty)$ proceed as follows :

Let $y = f(x)$

Expressing x in terms of y. Let is becomes $x = g(y)$

Now, find 'y' for which x defined.

(vi) To find the range of a function whose domain is $(-\infty, \infty) - \{(a, b, c)\}$; proceed as follows :

Let $y = f(x)$

Expressing x in terms of y. Let it becomes $x = g(y)$.

Now, find 'y' for which x defined and cast-off the limiting values of $f(x)$ at $x \to a, x \to b, x \to c$.

(vii) To find the range of a continuous and differentiable function whose domain is a finite interval $[a, b]$ proceed as follows. Find the sub-interval in which $f(x)$ is increasing or decreasing and then calculate the least and greatest values among them. To find increasing and decreasing function, we use derivatives *i.e.*, $f(x)$ increases if $f'(x) > 0$ and $f(x)$ decreases if $f'(x) < 0$.

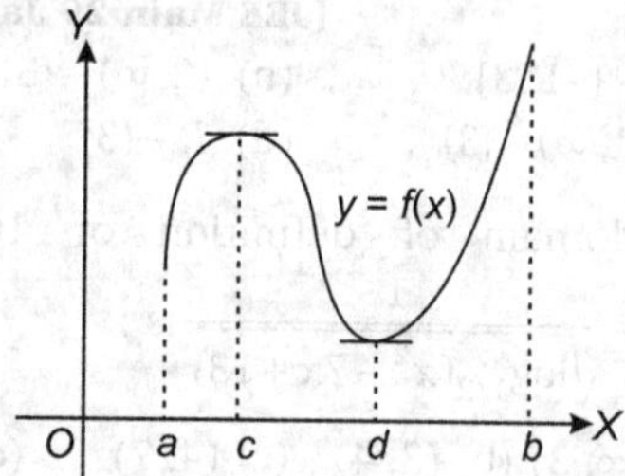

In the above figure function $f(x)$ increases in (a, c) and in (d, b) and decreases in $(c, d]$ clearly greatest and least values are $f(b)$ and $f(d)$ respectively. So $R_f = [f(d), f(b)]$.

(viii) To find the range of some trigonometric function we may proceed as follows :

If $f(x) = \sin^4 x + \cos^4 x$

$$= (\sin^2 x + \cos^2 x)^2 - 2\sin^2 x\cos^2 x = 1 - \frac{\sin^2 2x}{2}$$

Now, $0 \leq \sin^2 2x \leq 1$

$$\Rightarrow \quad 0 \leq \frac{\sin^2 2x}{2} \leq \frac{1}{2}$$

$$\Rightarrow \quad 0 \geq -\frac{\sin^2 2x}{2} \geq -\frac{1}{2}$$

$$\Rightarrow \quad 1 \geq 1 - \frac{\sin^2 2x}{2} \geq 1 - \frac{1}{2} \text{ or } 1 \geq f(x) \geq \frac{1}{2}$$

$$\therefore \quad R_f = \left[\frac{1}{2}, 1\right]$$

and if $g(x) = \sin^6 x + \cos^6 x$

$$= (\sin^2 x + \cos^2 x)^3 - 3\sin^2 x\cos^2 x(\sin^2 x + \cos^2 x)$$

$$= (1)^3 - 3\sin^2 x\cos^2 x(1) = 1 - \frac{3}{4}\sin^2 2x$$

Now, $0 \leq \sin^2 2x \leq 1 \Rightarrow 0 \leq \frac{3}{4}\sin^2 2x \leq \frac{3}{4}$

$$\Rightarrow \quad 0 \geq -\frac{3}{4}\sin^2 2x \geq -\frac{3}{4}$$

$$\Rightarrow \quad 1 \geq 1 - \frac{3}{4}\sin^2 2x \geq 1 - \frac{3}{4} \text{ or } 1 \geq g(x) \geq \frac{1}{4}$$

$$\therefore \quad R_g = \left[\frac{1}{4}, 1\right]$$

Note :

(i) For a quadratic function $f(x) = ax^2 + bx + c$, where $a, b, c \in R$ and $a \neq 0$.

Case I. If $a > 0$, then the parabola open upwards and $D > 0$ or $= 0$ or < 0

$$R_f = \left[-\frac{D}{4a}, \infty\right); \text{ where } D = b^2 - 4ac$$

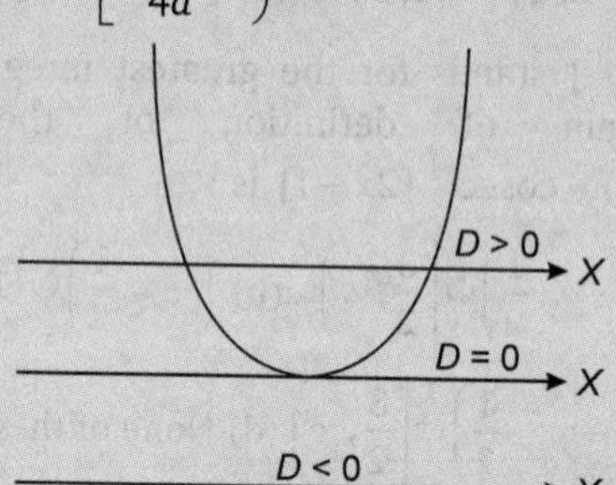

Case II. If $a < 0$, then the parabola open downwards and $D < 0$ or $= 0$ or > 0

$$R_f = \left(-\infty, -\frac{D}{4a}\right]; \text{ where } D = b^2 - 4ac$$

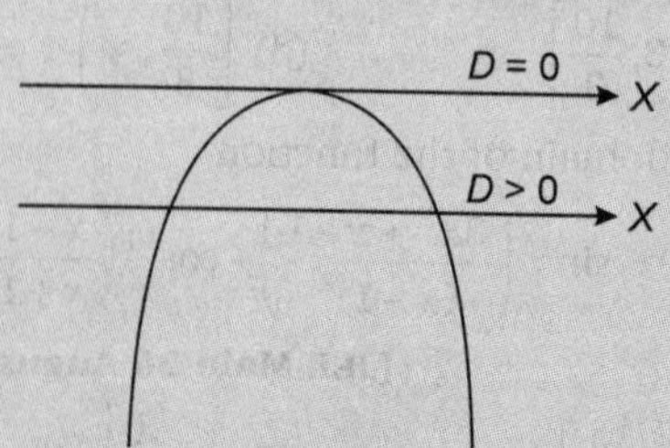

(ii) For a quadratic function $f(x) = ax^2 + bx + c$, where $a, b, c \in R$ and $a \neq 0$, where $x \in [x_1, x_2]$.

Case I. If $-\frac{b}{2a} \in (x_1, x_2)$, then

$$R_f = \left[\min.\left\{f(x_1), f(x_2), \frac{-D}{4a}\right\},\right.$$

$$\left.\max.\left\{f(x_1), f(x_2), \frac{-D}{4a}\right\}\right]$$

Case II. If $-\frac{b}{2a} \notin (x_1, x_2)$, then

$$R_f = [\min.\{f(x_1), f(x_2)\}, \max.\{f(x_1), f(x_2)\}]$$

(iii) The range of the function $f(x) = \frac{ax^2+bx+c}{px^2+qx+r}$, where a, b, c, p, q, r are real constants.

Case I. If $px^2+qx+r \neq 0$

Let $y = f(x) = \frac{ax^2+bx+c}{px^2+qx+r}$

Find the value $\frac{-B \pm \sqrt{(B^2-4AC)}}{2A}$ say y_1, y_2; $y_1 < y_2$

where $A = q^2 - 4pr$,

$B = 4ar + 4pc - 2bq$

and $C = b^2 - 4ac$

If $A < 0$, then $R_f = [y_1, y_2]$

If $A > 0$, then $R_f = R - (y_1, y_2)$

Case II. If $ax^2+bx+c = a(x-\alpha)(x-\beta)$

and $px^2+qx+r = p(x-\alpha)(x-\gamma)$

$$\because \quad y = f(x) = \frac{a(x-\alpha)(x-\beta)}{p(x-\alpha)(x-\gamma)} = \frac{a(x-\beta)}{p(x-\gamma)}, \; x \neq \alpha$$

$$\Rightarrow \quad x = \frac{py\gamma - a\beta}{py - a}$$

Clearly x is defined if $py - a \neq 0$ i.e., $y \neq \frac{a}{p}$

Also when $x = \alpha$, $y = \frac{a(\alpha - p)}{p(\alpha - \gamma)}$

$$\therefore \quad R_f = R - \left\{\frac{a}{p}, \frac{a(\alpha-\beta)}{p(\alpha-\gamma)}\right\}$$

Illustration 52. Find the range of the function $f(x) = 1 - x - x^2$.

Sol : We have

$$f(x) = -(x^2+x-1)$$

$$= -\left(\left(x+\frac{1}{2}\right)^2 - \frac{1}{2^2} - 1\right) = \frac{5}{4} - \left(x+\frac{1}{2}\right)^2 \leq \frac{5}{4}$$

$$\therefore \quad R_f = \left(-\infty, \frac{5}{4}\right]$$

Alternative Method : Here $a = -1 < 0$, $b = -1, c = 1$

$\therefore$ $D = b^2 - 4ac = 1 - 4(-1)(1) = 5$

Then from note (i), $R_f = \left(-\infty, -\frac{D}{4a}\right] = \left(-\infty, \frac{5}{4}\right]$

Illustration 53. Find the range of the function $f(x) = \frac{1}{\sqrt{(x^2-4x+6)}}$.

Sol : Since $x^2 - 4x + 6 = (x-2)^2 + 2 \geq 2$

$$\Rightarrow \quad x^2 - 4x + 6 \geq 2$$

$$\sqrt{(x^2-4x+6)} \geq \sqrt{2}$$

or $$\frac{1}{\sqrt{(x^2-4x+6)}} \leq \frac{1}{\sqrt{2}}$$

but $f(x) > 0$

$$\therefore \quad 0 < \frac{1}{\sqrt{x^2-4x+6}} \leq \frac{1}{\sqrt{2}}$$

Hence, $$R_f = \left(0, \frac{1}{\sqrt{2}}\right]$$

Illustration 54. Find the range of the function $f(x) = \sin^4 x(1+\sin^2 x) + \cos^4 x(1+\cos^2 x)$.

Sol : We have

$$f(x) = \sin^4 x(1+\sin^2 x) + \cos^4 x(1+\cos^2 x)$$

$$= \sin^4 x + \cos^4 x + \sin^6 x + \cos^6 x$$

$$= (\sin^2 x + \cos^2 x)^2 - 2\sin^2 x\cos^2 x + (\sin^2 x + \cos^2 x)^3 - 3\sin^2 x\cos^2 x(\sin^2 x + \cos^2 x)$$

$$= 1 - 2\sin^2 x\cos^2 x + 1 - 3\sin^2 x\cos^2 x = 2 - \frac{5}{4}\sin^2 2x$$

Now, $$0 \leq \sin^2 2x \leq 1$$

$$\Rightarrow \quad 0 \geq -\frac{5}{4}\sin^2 2x \geq -\frac{5}{4}$$

$$\Rightarrow \quad 2 \geq 2 - \frac{5}{4}\sin^2 2x \geq 2 - \frac{5}{4}$$

or $$2 \geq f(x) \geq \frac{3}{4}$$

$$\therefore \quad R_f = \left[\frac{3}{4}, 2\right]$$

Illustration 55. Find the range of the function $f(x) = 3\sin\sqrt{\left(\frac{\pi^2}{16} - x^2\right)}$.

Sol : At $x = 0$

$f(x)$ has maximum value.

max. $f(x) = 3\sin\frac{\pi}{4} = \frac{3}{\sqrt{2}}$ and at $x = \pm\frac{\pi}{4}$, $f(x)$ has minimum value.

min. $f(x) = 0$

$$\therefore \quad R_f = \left[0, \frac{3}{\sqrt{2}}\right]$$

Illustration 56. Find the range of the function $f(x) = 3|\sin x| - 4|\cos x|$.

Sol : $\because$ $0 \leq |\sin x| \leq 1$ and $0 \leq |\cos x| \leq 1$

For maximum value of $f(x)$

Put $|\sin x| = 1$ and $|\cos x| = 0$

$\therefore$ max. $f(x) = 3$ and for minimum value of $f(x)$

put $|\sin x| = 0$ and $|\cos x| = 1$

$\therefore$ min. $f(x) = -4$

Hence, $R_f = (\text{min. } f(x), \text{max. } f(x)] = [-4, 3]$

Illustration 57. Find the range of the function $f(x) = \dfrac{1}{2-\cos 5x}$.

Sol : $\because$ $-1 \le \cos 5x \le 1$

$\Rightarrow$ $1 \ge -\cos 5x \ge -1$

or $2+1 \ge 2-\cos 5x \ge 2-1$

or $3 \ge 2-\cos 5x \ge 1$

or $\dfrac{1}{3} \le \dfrac{1}{2-\cos 5x} \le 1$

or $\dfrac{1}{3} \le f(x) \le 1$

$\therefore$ $R_f = \left[\dfrac{1}{3}, 1\right]$

Illustration 58. Find the range of the function $f(x) = \left[\dfrac{1}{\sin\{x\}}\right]$, where [·] and { · } denotes the greatest integer and fractional part functions.

Sol : Since $0 < \{x\} < 1$

$\therefore$ $0 < \sin\{x\} < \sin 1$

or $\dfrac{1}{\sin\{x\}} > \dfrac{1}{\sin 1} > 1.15$

($\because \sin 1 \approx \sin 57° < \sin 60°$) or $(\sin 1 < 0.866)$

or $\left[\dfrac{1}{\sin\{x\}}\right] = 1, 2, 3, 4, \ldots$

Hence $R_f = N$

Illustration 59. Find the range of the function $f(x) = \ln\left\{\dfrac{1}{\sqrt{[\cos x] - [\sin x]}}\right\}$, $[x] \le x$.

Sol : Since $f(x)$ is defined when

$[\cos x] > [\sin x]$

which is possible only when

$[\cos x] = 1$ and $[\sin x] = 0$

$\therefore$ $f(x) = \ln\{1\} = 0$

Hence, $R_f = \{0\}$

Illustration 60. Find the range of the function $f(x) = \cot^{-1}(2x - x^2)$.

Sol : Let $y = f(x) = \cot^{-1}(2x - x^2)$.

$\because$ $y \in (0, \pi)$ ...(i)

and $\cot y = 2x - x^2$

$\Rightarrow$ $(x-1)^2 = 1 - \cot y \ge 0$

$\Rightarrow$ $\cot y \le 1$

$y \ge \dfrac{\pi}{4}$ ...(ii)

Hence, range of $f(x)$ from (i) and (ii) is $\left[\dfrac{\pi}{4}, \pi\right)$.

$\Rightarrow$ $R_f = \left[\dfrac{\pi}{4}, \pi\right)$

Illustration 61. Find the range of the function $f(x) = \cos^{-1}\left(\dfrac{\sqrt{(2x^2+1)}}{x^2+1}\right)$.

Sol : Given $f(x) = \cos^{-1}\left(\dfrac{\sqrt{(2x^2+1)}}{x^2+1}\right)$

Let $y = f(x) = \cos^{-1}\left(\dfrac{\sqrt{(2x^2+1)}}{x^2+1}\right)$

$\therefore$ $y \in [0, \pi]$ ...(i)

Also, $\cos y = \dfrac{\sqrt{(2x^2+1)}}{(x^2+1)} > 0$

$\Rightarrow$ $\cos y > 0$, $y < \dfrac{\pi}{2}$ ...(ii)

From (i) and (ii), $y \in \left[0, \dfrac{\pi}{2}\right)$

Hence, range of $f(x) = R_f = \left[0, \dfrac{\pi}{2}\right)$

Illustration 62. Find the range of the function $f(x) = \cos^{-1}\sqrt{\log_{[x]}\dfrac{|x|}{x}}$, where $[x] \le x$.

Sol : $f(x)$ is defined for $[x] > 0$, $[x] \ne 1$ and $\dfrac{|x|}{x} > 0$

$\Rightarrow x \ge 1, x \notin [1, 2)$ and $x > 0$

$\therefore$ $x \in [2, \infty) \Rightarrow \dfrac{|x|}{x} = 1$

$\therefore$ $f(x) = \cos^{-1}\sqrt{\cos_{[x]} 1} = \cos^{-1}\sqrt{0} = \cos^{-1} 0 = \pi/2$

Hence, $R_f = \{\pi/2\}$.

Illustration 63. Find the range of the function $f(x) = 4^x - 2^x + 1$.

Sol : Given $f(x) = 4^x - 2^x + 1 = y$ (let)

$\therefore$ $D_f = R \Rightarrow y = (2^x)^2 - 2^x + 1$

$\Rightarrow$ $y = \left(2^x - \dfrac{1}{2}\right)^2 + \dfrac{3}{4} \ge \dfrac{3}{4}$

$\therefore$ $y \ge \dfrac{3}{4}$

Hence, $R_f = \left[\dfrac{3}{4}, \infty\right)$

Illustration 64. Find the range of the function $f(x)=\log_2(\sqrt{(x-4)}+\sqrt{(6-x)})$.

Sol : Given : $f(x)=\log_2(\sqrt{(x-4)}+\sqrt{(6-x)})$

Let $\quad y=f(x)=\log_2(\sqrt{x-4}+\sqrt{(6-x)})$

$\Rightarrow \quad 2^y=\sqrt{(x-4)}+\sqrt{(6-x)}$

On squaring, $2^{2y}=x-4+6-x+2\sqrt{(x-4)}\sqrt{(6-x)}$

$\Rightarrow \quad 2^{2y}-2=2\sqrt{(x-4)(6-x)}$

$\Rightarrow \quad 2^{2y-1}-1=\sqrt{(x-4)(6-x)}\geq 0$

$\therefore \quad 2^{2y-1}-\geq 0$ ...(i)

Also $(2^{2y-1}-1)^2=-x^2+10x-24$

$\Rightarrow -(2^{2y-1}-1)^2=x^2-10x+24=(x-5)^2-1$

or $\quad 1-(2^{2y-1}-1)^2=(x-5)^2\geq 0$

$\Rightarrow \quad (2^{2y-1}-1)^2\leq 1 \Rightarrow -1\leq 2^{2y-1}-1\leq 1$ ...(ii)

$\Rightarrow \quad -1\leq 2^{2y-1}-1\leq 1$

From (i) and (ii), $0\leq 2^{2y-1}-1\leq 1$

or $\quad 1\leq 2^{2y-1}\leq 2 \Rightarrow 2^0\leq 2^{2y-1}\leq 2^1 \Rightarrow 0\leq 2y-1\leq 1$

$\Rightarrow \quad 1\leq 2y\leq 2 \therefore \frac{1}{2}\leq y\leq 1$

Hence, range of $f(x)=R_f=\left[\frac{1}{2},1\right]$

Illustration 65. Find the range of the function $f(x)=5\cos x+3\cos(x+\pi/3)+3$.

Sol : We have $f(x)=5\cos x+3\cos(x+\pi/3)+3$

$$=5\cos x+3\left\{\cos x\cdot\frac{1}{2}-\sin x\cdot\frac{\sqrt{3}}{2}\right\}+3$$

$$=\frac{13}{2}\cos x-\frac{3\sqrt{3}}{2}\sin x+3$$

$$\Rightarrow \quad 3-\sqrt{\left(\frac{13}{2}\right)^2+\left(\frac{-3\sqrt{3}}{2}\right)^2}$$

$$\leq f(x)\leq 3+\sqrt{\left(\frac{13}{2}\right)^2+\left(\frac{-3\sqrt{3}}{2}\right)^2}$$

$\Rightarrow \quad 3-7\leq f(x)\leq 3+7 \Rightarrow -4\leq f(x)\leq 10$

$\therefore \quad R_f=[-4,10]$

Illustration 66. Find the range of the function $f(x)=\log_2\left(\frac{\sin x-\cos x+3\sqrt{2}}{\sqrt{2}}\right)$.

Sol : $\because \frac{\sin x-\cos x+3\sqrt{2}}{\sqrt{2}}=\frac{1}{\sqrt{2}}\sin x-\frac{1}{\sqrt{2}}\cos x+3$.

Now, $3-\sqrt{\left(\frac{1}{\sqrt{2}}\right)^2+\left(\frac{-1}{\sqrt{2}}\right)^2}\leq\frac{1}{\sqrt{2}}\sin x-\frac{1}{\sqrt{2}}\cos x+3$

$$\leq 3+\sqrt{\left(\frac{1}{\sqrt{2}}\right)^2+\left(\frac{-1}{\sqrt{2}}\right)^2}$$

$\Rightarrow \quad 3-1\leq\frac{\sin x-\cos x+3\sqrt{2}}{\sqrt{2}}\leq 3+1$

$\Rightarrow \quad 2\leq\left(\frac{\sin x-\cos x+3\sqrt{2}}{\sqrt{2}}\right)\leq 2^2$

$\therefore \quad \log_2 2\leq\log_2\left(\frac{\sin x-\cos x+3\sqrt{2}}{\sqrt{2}}\right)\leq 2\log_2 2$

or $\quad 1\leq f(x)\leq 2$

Hence, $\quad R_f=[1,2]$

Illustration 67. Find the range of the function $f(x)=\frac{(1+x+x^2)(1+x^4)}{x^3}$, when $x>0$.

Sol : $\because \; f(x)=\frac{1+x^4+x+x^5+x^2+x^6}{x^3}$

$$=\left(\frac{1}{x^3}+x^3\right)+\left(x+\frac{1}{x}\right)+\left(x^2+\frac{1}{x^2}\right)$$

$$\geq 2\sqrt{\frac{1}{x^3}\cdot x^3}+2\cdot\sqrt{x\cdot\frac{1}{x}}+2\sqrt{x^2\cdot\frac{1}{x^2}}=6$$

$$\left(\text{use } a+\frac{1}{a}\geq 2, a>0\right)$$

$\Rightarrow \quad f(x)\geq 6$

$\therefore \quad R_f=[6,\infty)$

Illustration 68. Find the range of the function $f(x)=(\tan^{-1}x)^2+\frac{2}{\sqrt{(1+x^2)}}, \forall x>0$.

Sol : We have $f(x)=(\tan^{-1}x)^2+\frac{2}{\sqrt{(1+x^2)}}$

$$\therefore \quad f'(x)=\frac{2\tan^{-1}x}{(1+x^2)}-\frac{2x}{(1+x^2)^{3/2}}$$

$$=\frac{2}{(1+x^2)}\left\{\tan^{-1}x-\frac{x}{\sqrt{(1+x^2)}}\right\}$$

Let $\quad g(x)=\tan^{-1}x-\frac{x}{\sqrt{(1+x^2)}}$

$$\therefore \; g'(x)=\frac{1}{(1+x^2)}-\left\{\frac{\sqrt{(1+x^2)}\cdot 1-x\cdot\frac{1}{2\sqrt{(1+x^2)}}\cdot 2x}{(1+x^2)}\right\}$$

$$=\frac{1}{(1+x^2)}\left\{1-\frac{1}{\sqrt{(1+x^2)}}\right\}>0, \forall x\in R$$

$g'(x)>0$

$\therefore$ $g(x)$ is increasing function, then $f'(x)>0, \;\forall x\in R$.

So, range of $f(x)$ is $\left(f(0)\lim_{x\to\infty}f(x)\right)$

$$=\left(2,\lim_{x\to\infty}(\tan^{-1}x)^2+\frac{2}{\sqrt{(1+x^2)}}\right)=\left(2,\left(\frac{\pi}{2}\right)^2\right)=\left(2,\frac{\pi^2}{4}\right)$$

Illustration 69. Find the range of the function $f(x)=\dfrac{x^2+14x+9}{x^2+2x+3}$, where $x \in R$.

Sol : Here $a=1$, $b=14$, $c=9$, $p=1$, $q=2$, $r=3$

$\therefore \quad A=2^2-4\cdot1\cdot3=-8, B=4\cdot1\cdot3+4\cdot1\cdot9-2\cdot14\cdot2$

$=12+36-56=-8$

and $C=14^2-4\cdot1\cdot9=196-36=160$

$\therefore \quad \dfrac{-B\pm\sqrt{B^2-4AC}}{2A}=\dfrac{8\pm\sqrt{64-4\cdot(-8)(160)}}{2\times-8}$

$=\dfrac{1\pm\sqrt{1+80}}{-2}=\dfrac{1\pm9}{-2}=-5, 4$

Here $A<0$ and $y_1=-5, y_2=4$

$\therefore \quad R_f=[-5,4]$

Illustration 70. Find the range of the function $f(x)=\dfrac{x^2-3x+2}{x^2+x-6}$

Sol : Let $y=f(x)=\dfrac{x^2-3x+2}{x^2+x-6}$

$=\dfrac{(x-1)(x-2)}{(x+3)(x-2)}=\dfrac{x-1}{x+3}, x\neq-3, 2$

$\Rightarrow \quad xy+3y=x-1$

$\Rightarrow \quad x=\dfrac{3y+1}{1-y}$

also as $x\to2$, $y\to1/5$

$\therefore \quad R_f=R-\left\{1,\dfrac{1}{5}\right\}$

Illustration 71. For what real values of a does the range of $f(x)=\dfrac{x+1}{a+x^2}$ contain the interval [0, 1]?

Sol : Let $y=f(x)=\dfrac{x+1}{a+x^2}$

$\Rightarrow \quad y(a+x^2)=x+1$

$\Rightarrow \quad yx^2-x+ay-1=0$ has real roots for every $y\in[0,1]$

$\therefore \quad 1-4y(ay-1)\geq0$

$\Rightarrow \quad 1-4ay^2+4y\geq0 \qquad$ ...(i)

Holds for $0\leq y\leq1$

Case I. If $y=0$

Here $y=0$ is assumed at $x=-1$

For any $a\neq-1$ for $a=-1$, $y=\dfrac{x+1}{x^2-1}$ is undefined for $x=-1$

$\therefore$ Value $y=0$ is not assumed.

Case II. If $0<y\leq1$

put $\quad y=\dfrac{1}{z}$, then $z\geq1$

Inequation (i) hold

$1-\dfrac{4a}{z^2}+\dfrac{4}{z}\geq0$

i.e., $\quad z^2+4z-4a\geq0$

$(z+2)^2-4-4a\geq0$

$\Rightarrow \quad (1+2)^2-4-4a\geq0 \qquad (\because z\geq1)$

or $\quad a\leq\dfrac{5}{4}, a\neq-1$

$\therefore \quad a\in\left(-\infty,\dfrac{5}{4}\right]-\{-1\}$

or $\quad a\in(-\infty,-1)\cup\left(-1,\dfrac{5}{4}\right]$

Miscellaneous Examples on Domain and Range

Illustration 72. Find the domain of definition of the function $f(x)=e^x+\sin^{-1}\left(\dfrac{x}{2}-1\right)+\ln\sqrt{(x-[x])}$, where [·] denotes the greatest integer function.

Sol : Given $f(x)=e^x+\sin^{-1}\left(\dfrac{x}{2}-1\right)+\ln\sqrt{(x-[x])}$

$=f_1(x)+f_2(x)+f_3(x) \qquad$ (say)

To find the domain of $f_1(x)$:

$\because \quad f_1(x)=e^x \therefore x\in R$

$D_{f_1}=R \qquad$ ...(i)

To find the domain of $f_2(x)$:

$\because \quad f_2(x)=\sin^{-1}\left(\dfrac{x}{2}-1\right)\Rightarrow f_2(x)$ is defined

If $\quad -1\leq\dfrac{x}{2}-1\leq1$ or $0\leq x\leq4$

$D_{f_2}=[0,4] \qquad$ ...(ii)

To find the domain of $f_3(x)$:

$\therefore \quad f_3(x)=\ln\sqrt{(x-[x])}\Rightarrow f_3(x)$ is defined

If $x-[x]>0$ or $x>[x], x\notin I \qquad$ ... (iii)

$D_{f_3}=R-I$

Now from (i), (ii) and (iii),

Required Domain $D_f=D_{f_1}\cap D_{f_2}\cap D_{f_3}$

$=R\cap[0,4]\cap R-I$

$=(0,4)-\{1,2,3\}$

Illustration 73. Find the domain of definition of the function $f(x)=\dfrac{\sqrt{(\cos x-1/2)}}{\sqrt{(6+35x-6x^2)}}$

Sol : Given $f(x)=\dfrac{\sqrt{(\cos x-1/2)}}{\sqrt{(6+35x-6x^2)}}$

For $f(x)$ to be defined the following two inequalties must be satisfied.

(I) $\cos x - \frac{1}{2} \ge 0 \Rightarrow \cos x \ge \frac{1}{2} = \cos\frac{\pi}{3}$

$\therefore \cos x \ge \cos\frac{\pi}{3}$

$2n\pi - \frac{\pi}{3} \le x \le 2n\pi + \frac{\pi}{3},\ n \in I$

$\therefore x \in \ldots \cup \left[-\frac{\pi}{3}, \frac{\pi}{3}\right] \cup \left[\frac{5\pi}{3}, \frac{7\pi}{3}\right] \cup \ldots$

(II) $6 + 35x - 6x^2 > 0 \Rightarrow (6x+1)(x-6) < 0$

$\therefore -\frac{1}{6} < x < 6$

Hence, domain of $f(x)$ is D_f = Domain of I $\cap$ Domain of II

$$= \left(-\frac{1}{6}, \frac{\pi}{3}\right] \cup \left[\frac{5\pi}{3}, 6\right)$$

Illustration 74. (i) Find the range of the function $f(x) = \frac{e^{-x}}{1+[x]}$, $[\cdot]$ denotes the greatest integer function.

(ii) Find the range of the function $f(x) = \frac{e^x}{1+[x]}$, $x \ge 0$ and $[\cdot]$ denotes the greatest integer function.

Sol : (i) Let $f(x) = \frac{e^{-x}}{1+[x]}$

or $y(1+[x]) = e^{-x} > 0$

or $y(1+[x]) > 0$

If $1+[x] > 0$, then $y > 0$

and if $1+[x] < 0$, then $y < 0$

Hence, $R_f = R - \{0\}$

(ii) $f(x)$ is an increasing function in $[0, \infty)$.

$\therefore R_f = [f(0), f(\infty)) = [1, \infty)$

Illustration 75. Find the range of the function $f(x) = \frac{e^x - e^{|x|}}{e^x + e^{|x|}}$.

Sol : We have $f(x) = \frac{e^x - e^{|x|}}{e^x + e^{|x|}} = \begin{cases} 0, & x \ge 0 \\ \frac{e^x - e^{-x}}{e^x + e^{-x}}, & x < 0 \end{cases}$

$R_f = \{0\}$ for $x \ge 0$ ...(i)

and let $y = f(x) = \frac{e^x - e^{-x}}{e^x + e^{-x}}$ for $x < 0$

or $\frac{y+1}{1-y} = \frac{2e^x}{2e^{-x}} = e^{2x}$

(From componendo and dividendo rule)

$\Rightarrow 0 < \frac{y+1}{1-y} < 1 \quad \left(\because x < 0 \Rightarrow e^{2x} < 1 \quad \therefore 0 < e^{2x} < 1\right)$

$\Rightarrow 0 > \frac{y+1}{y-1} > -1$

$\Rightarrow 0 > \frac{y+1}{y-1}$ and $\frac{y+1}{y-1} > -1$

$\Rightarrow \frac{y+1}{y-1} < 0$ and $\frac{2y}{y-1} > 0$

$\Rightarrow y \in (-1, 1)$ and $y \in (-\infty, 0) \cup (1, \infty)$

Hence, $y \in (-1, 0)$ ...(ii)

From (i) and (ii), we get $R_f = (-1, 0]$

Illustration 76. Find the range of the function $f(x) = \ln((\cos x)^{\cos x} + 1)$, $x \in (0, \pi/2)$

Sol : As $0 < x < \frac{\pi}{2}$

$\Rightarrow 0 < \cos x < 1$

$\therefore$ Range of more $\ln((\cos x)^{\cos x} + 1)$ in $(0, \pi/2)$

= Range of $\ln(x^x + 1)$ in $(0, 1)$

Let $g(x) = x^x + 1$

$\therefore g'(x) = x^x(1 + \ln x)$

$g'(x) < 0$ in $\left(0, \frac{1}{e}\right)$

$\therefore R_f = (\ln(g(1/e), \ln(g(0))$

$$= \left(\ln\left(\left(\frac{1}{e}\right)^{\frac{1}{e}} + 1\right), \ln\left(\lim_{x \to 0} x^x + 1\right)\right)$$

$$= \left(\ln\left(\left(\frac{1}{e}\right)^{\frac{1}{e}} + 1\right), \ln\left(e^{\lim_{x \to 0} x \ln x} + 1\right)\right)$$

$$= \left(\ln\left(\left(\frac{1}{e}\right)^{\frac{1}{e}} + 1\right), \ln\left(e^{\lim_{x \to 0} \frac{\ln x}{1/x}} + 1\right)\right)$$

$$= \left(\ln\left(\left(\frac{1}{e}\right)^{\frac{1}{e}} + 1\right), \ln\left(e^{\lim_{x \to 0} \frac{\frac{1}{x}}{-1/x^2}} + 1\right)\right),$$

(Using L' Hospital rule)

$$= \left(\ln\left(\left(\frac{1}{e}\right)^{\frac{1}{e}} + 1\right), \ln(1+1)\right)$$

$$= \left(\ln\left(\left(\frac{1}{e}\right)^{\frac{1}{e}} + 1\right), \ln 2\right)$$

and $g'(x) > 0$ in $\left(\frac{1}{e}, 1\right)$

$\therefore R_f = (\ln(g(1/e), \ln(g(1))$

$$= \left(\ln\left(\left(\frac{1}{e}\right)^{\frac{1}{e}} + 1\right), \ln 2\right)$$

Equal or Identical Functions

Two functions f and g are said to be identical if

(i) The domain of f = the domain of g

i.e., $D_f = D_g$

(ii) The range of f = the range of g i.e., $R_f = R_g$

(iii) $f(x) = g(x)\ \forall\ x \in D_f$ or $x \in D_g$

e.g., $f(x) = x$ and $g(x) = \sqrt{x^2}$ are not identical.

$\because \quad D_f = D_g = R, R_f = R, R_g = [0, \infty)$

$\Rightarrow \quad R_f \neq R_g$

and also $f(x) = \dfrac{1}{x}$ and $g(x) = \dfrac{x}{x^2}$ are identical.

$\because \quad D_f = D_g = R - \{0\}$

Also $\quad R_f = R_g = R - \{0\}$

Also $\quad f(x) = g(x)\ \forall\ x \in R - \{0\}$

Illustration 77. Is each of the following functions identical?

(i) $f(x) = \ln x^2$ and $g(x) = 2\ln x$

(ii) $f(x) = \ln e^x$ and $g(x) = e^{\ln x}$

(iii) $f(x) = \log_x e$ and $g(x) = \dfrac{1}{\log_e x}$

(iv) $f(x) = \sqrt{x^2}$ and $g(x) = (\sqrt{x})^2$

(v) $f(x) = \text{sgn}\,(x^2 + 1)$ and $g(x) = \sin^2 x + \cos^2 x$

(vi) $f(x) = x\,|x|$ and $g(x) = x^2 \,\text{sgn}\, x$

(vii) $f(x) = \sin(\cos^{-1} x)$ and $g(x) = \cos(\sin^{-1} x)$

Sol : (i) $\because\ f(x) = \ln x^2$ and $g(x) = 2\ln x$

$\therefore \quad D_f = R - \{0\}$ and $D_g = (0, \infty) \Rightarrow D_f \neq D_g$

Hence, $f(x)$ and $g(x)$ are not identical.

(ii) $\because\ f(x) = \ln e^x$ and $g(x) = e^{\ln x}$

$\therefore \quad D_f = R$ and $D_g = (0, \infty) \Rightarrow D_f \neq D_g$

Hence, $f(x)$ and $g(x)$ are not identical.

(iii) $\because\ f(x) = \log_x e$ and $g(x) = \dfrac{1}{\log_e x}$

$\therefore \quad D_f = (0, \infty) - \{1\}$ and $D_g = (0, \infty) - \{1\}$

$\Rightarrow \quad D_f = D_g$

Also for this domain

$$\log_e x = \frac{1}{\log_x e}$$

$\therefore \quad R_f = R_g$

Hence, $f(x)$ and $g(x)$ are identical.

(iv) $\because\ f(x) = \sqrt{x^2}$ and $g(x) = (\sqrt{x})^2$

$\therefore \quad D_f = R$ and $D_g = [0, \infty) \Rightarrow D_f \neq D_g$

Hence, $f(x)$ and $g(x)$ are not identical.

(v) $\because\ f(x) = \text{sgn}\,(x^2 + 1)$ and $g(x) = \sin^2 x + \cos^2 x$

$\therefore \quad D_f = R$ and $D_g = R \Rightarrow D_f = D_g$

and $\quad f(x) = 1$ and $g(x) = 1$

$\therefore \quad R_f = R_g$

Hence, $f(x)$ and $g(x)$ and identical.

(vi) $\because\ f(x) = x\,|x|$ and $g(x) = x^2 \,\text{sgn}\, x$

$\therefore \quad D_f = R$ and $D_g = R$

$\therefore \quad D_f = D_g$

and $\quad f(x) = x\,|x|$ and $g(x) = x^2 \cdot \dfrac{|x|}{x} = x\,|x|$

$\Rightarrow \quad R_f = R_g$

Hence, $f(x)$ and $g(x)$ are identical.

(vii) $\because\ f(x) = \sin(\cos^{-1} x)$ and $g(x) = \cos(\sin^{-1} x)$

$\therefore \quad D_f = [-1, 1]$ and $D_g = [-1, 1]$

and $\quad R_f = [-1, 1]$ and $R_g = [-1, 1]$

$\Rightarrow \quad D_f = D_g$ and $R_f = R_g$

Also $g(x) = \cos(\sin^{-1} x) = \cos\left(\dfrac{\pi}{2} - \cos^{-1} x\right)$

$= \sin(\cos^{-1} x) = f(x)$

Hence, $f(x)$ and $g(x)$ are identical function.

INTRODUCTORY EXERCISE 2.5

Only One Option is Correct

1. The range of the function $f(x) = x^{2n+1} + x^{2n-1} + 1$, $n \in N$ is :

(a) $(-\infty, \infty)$ (b) $(0, \infty)$ (c) $(1, \infty)$ (d) $(0, 1)$

2. The range of the function $f(x) = x \ln |x|$, is $(x \neq 0)$:

(a) $(0, 1)$ (b) $(1, e)$ (c) $\left[-\dfrac{1}{e}, \infty\right)$ (d) $[1, \infty)$

3. Range of the function $f(x) = \dfrac{x-1}{x+3}$ is :

(a) R (b) $R - \{-1\}$ (c) $R - \{1\}$ (d) $R - \{-3\}$

4. Let $f : R \to R$ be defined by $f(x) = \dfrac{x}{1+x^2}$, $x \in R$. Then the range of f is :

[JEE Main 11 Jan 2019 (MS)]

(a) $\left[-\dfrac{1}{2}, \dfrac{1}{2}\right]$ (b) $R - [-1, 1]$

(c) $R - \left[-\dfrac{1}{2}, \dfrac{1}{2}\right]$ (d) $(-1, 1) - \{0\}$

5. Range of the function $f(x) = \sin^{-1}\left(\dfrac{x}{1+x^2}\right)$ is:

(a) $\left[-\dfrac{\pi}{6}, \dfrac{\pi}{6}\right]$ (b) $\left[-\dfrac{\pi}{4}, \dfrac{\pi}{4}\right]$

(c) $\left[-\dfrac{\pi}{3}, \dfrac{\pi}{3}\right]$ (d) $\left[-\dfrac{\pi}{2}, \dfrac{\pi}{2}\right]$

6. The range of $f(x) = 4\sin^{-1}\left(\frac{x^2}{x^2+1}\right)$ is :

[JEE Main 13 Apr 2023 (ES)]

(a) $[0, 2\pi]$ (b) $[0, \pi]$ (c) $[0, 2\pi)$ (d) $[0, \pi)$

7. The range of the function $f(x) = |x-4|-|x+4|$ is given by :

(a) $(-\infty, -8)$ (b) $[-8, 8]$
(c) $[8, \infty)$ (d) R

8. Range of the function $f(x) = \cos^{-1}(|1-x^2|)$ is :

(a) $\left[0, \frac{\pi}{2}\right]$ (b) $\left[0, \frac{\pi}{3}\right]$ (c) $(0, \pi)$ (d) $\left(\frac{\pi}{2}, \pi\right)$

9. Range of the function
$$f(x) = \frac{1}{\cos\{\sin^{-1}(\sin x + \cos x)\}}$$ is :

(a) $[-1, 1] - \{0\}$ (b) $(-\infty, -1] \cup [1, \infty)$
(c) $(0, 1]$ (d) $[1, \infty)$

10. Range of the function $f(x) = \frac{1}{|\sin x|} + \frac{1}{|\cos x|}$ is :

(a) $[2\sqrt{2}, \infty)$ (b) $(\sqrt{2}, 2\sqrt{2})$
(c) $(0, 2\sqrt{2})$ (d) $(2\sqrt{2}, 4)$

11. Range of the function $f(x) = \min. (\sin x, \cos 2x)$ is :

(a) $[-1, 1]$ (b) $\left[\frac{1}{2}, 1\right]$ (c) $\left[-1, \frac{1}{2}\right]$ (d) $\left[-1, -\frac{1}{2}\right]$

12. Range of the function $f(x) = \sin^2 x - 5\sin x - 6$ is :

(a) $[-10, 0]$ (b) $[-1, 1]$ (c) $[0, \pi]$ (d) $\left[-\frac{49}{4}, 0\right]$

13. Range of the function $f(x) = \log_e\left\{\cos\sqrt{\left(\frac{\pi^2}{9} - x^2\right)}\right\}$ is :

(a) $[0, \log_e 2]$ (b) $[-\log_e 2, 0]$
(c) $[-\log_e 2, \log_e 2]$ (d) None of these

14. Range of the function $f(x) = \log_{3/4}(x^2 - 7x + 13)$ is

(a) $[1, \infty)$ (b) $(-\infty, 1]$
(c) $(0, 1]$ (d) None of these

15. Range of the function $f(x) = e^{\sin(x-[x])\pi}$, where $[\cdot]$ denotes the greatest integer function, is :

(a) $[0, e]$ (b) $[1, e]$
(c) $\left[\frac{1}{e}, e\right]$ (d) None of these

16. Range of the function
$$f(x) = \sqrt{\sin^{-1}|\sin x| - \cos^{-1}|\cos x|}$$ is :

(a) $\{0\}$ (b) $\left[0, \sqrt{\frac{\pi}{2}}\right]$
(c) $[0, \sqrt{\pi}]$ (d) None of these

17. Range of the function $f(x) = [x^2 + x + 1]$, where $[\cdot]$ denotes the greatest integer function, is :

(a) all positive integers (b) all negative integers
(c) $[0, \infty)$ (d) $[1, \infty)$

18. Range of the function $f(x) = \sqrt{\ln(\cos(\sin x))}$ is :

(a) $[0, \ln(\text{cosec } 1))$
(b) $[0, \ln(\sec 1))$
(c) $(\ln(\text{cosec } 1), \ln(\sec 1))$
(d) $\{0\}$

19. Range of the function
$$f(x) = \frac{\sin x}{\sqrt{(1+\tan^2 x)}} - \frac{\cos x}{\sqrt{(1+\cot^2 x)}}$$ is :

(a) $[-1, 0]$ (b) $[0, 1]$
(c) $[-1, 1]$ (d) None of these

20. If $[5\sin x] + [\cos x] + 6 = 0$, where $[\cdot]$ denotes the greatest integer function, then the range of the function $f(x) = \sin x + \sqrt{3}\cos x$, corresponding to the solution set of the given equation is :

(a) $[-2, 0]$ (b) $[-\sqrt{3}, 0]$
(c) $[-2, \sqrt{3})$ (d) $[-2, -\sqrt{3})$

21. Range of the function $f(x) = [|\sin x| + |\cos x|]$, where $[\cdot]$ denotes the greatest integer function, is :

(a) $\{0\}$ (b) $\{0, 1\}$
(c) $\{1\}$ (d) None of these

22. Range of the function
$$f(x) = 2x^3 - 15x^2 + 36x + \sin^{-1} x$$ is :

(a) $\left[\frac{-106-\pi}{2}, \frac{106+\pi}{2}\right]$

(b) $\left[\frac{-106-\pi}{2}, \frac{46+\pi}{2}\right]$

(c) $\left[\frac{\pi}{2}, \frac{46+\pi}{2}\right]$

(d) None of the above

23. Range of the function
$$f(x) = \cos^{-1}\left(\frac{1+x^2}{2x}\right) + \sqrt{(2-x^2)}$$ is :

(a) $\left\{0, 1+\frac{\pi}{2}\right\}$ (b) $\{0, 1+\pi\}$
(c) $\{1, 1+\pi\}$ (d) $\left\{1, \frac{1+\pi}{2}\right\}$

24. Range of the function
$$f(x) = \frac{1}{\pi}(\sin^{-1} x + \tan^{-1} x) + \frac{(x+1)}{x^2+2x+5}$$ is :

(a) $\left[-\frac{3}{4}, \frac{1}{5}\right]$ (b) $\left[-\frac{3}{4}, \frac{5}{4}\right]$
(c) $\left[-\frac{3}{4}, \frac{5}{4}\right]$ (d) $\left[-\frac{3}{4}, 1\right]$

25. Range of the functions

$f(x) = \sin^{-1}\left[x^2 + \frac{1}{2}\right] + \cos^{-1}\left[x^2 - \frac{1}{2}\right]$, where [·] denotes the greatest integer function is :

(a) $\left\{\frac{\pi}{2}, \pi\right\}$ (b) $\{\pi\}$

(c) $\left\{\frac{\pi}{2}\right\}$ (d) None of these

26. Which pair of functions is identical?

(a) $e^{\ln \sec^{-1} x}$ and $\sec^{-1} x$

(b) $\ln (x-1)^2$ and $2\ln (x-1)$

(c) sgn x and sgn (sgn x)

(d) $\sin^{-1}(\sin x)$ and $\sin (\sin^{-1} x)$

27. If $f(x) = \sin^{-1}\left(\frac{2x}{1+x^2}\right)$ and $g(x) = \cos^{-1}\left(\frac{1-x^2}{1+x^2}\right)$, then $f(x) = g(x)$ holds, then the interval for x is :

(a) $[0, \infty)$ (b) $(-\infty, 1]$

(c) $[0, 1]$ (d) None of these

Value of a Function

The value of a function $f(x)$ at $x = a$ is denoted by $f(a)$ and is obtained by putting a in place of x in $f(x)$.

Illustration 78. If $f(x) = \frac{x^3 - 8}{x-2}$, then find $f(2)$.

Sol : We have $f(x) = \frac{x^3-8}{x-2}$

$\therefore \quad f(2) = \frac{2^3 - 8}{2-2} = \frac{8-8}{2-2} = \frac{0}{0}$ = does not exist

Wrong Method :

We have $f(x) = \frac{x^3-8}{x-2} = \frac{x^3 - 2^3}{x-2}$

$= \frac{(x-2)(x^2+2x+2^2)}{(x-2)}$

$= x^2 + 2x + 4$

$\therefore \quad f(2) = 4 + 4 + 4 = 12$

Which is wrong.

Only when $f(x) = x^2 + 2x + 4$ for $x \neq 2$

Illustration 79. If $f(x) = x^2$, find the value of $\frac{f(11.1) - f(11)}{11.1 - 11}$.

Sol : Given $f(x) = x^2$

$\therefore \frac{f(11.1) - f(11)}{11.1 - 11}$

$= \frac{(11.1)^2 - (11)^2}{0.1} = \frac{(11.1+11)(11.1-11)}{(0.1)}$

$= \frac{(22.1)(0.1)}{(0.1)} = 22.1$

Illustration 80. If $y = f(x) = \frac{1-x}{1+x}$, prove that $x = f(y)$.

Sol : $f(y) = \frac{1-y}{1+y} \qquad \left(\because f(x) = \frac{1-x}{1+x}\right)$

$$= \frac{1 - \frac{1-x}{1+x}}{1 + \frac{1-x}{1+x}} = \frac{(1+x)-(1-x)}{(1+x)+(1-x)} = \frac{2x}{2} = x$$

Illustration 81. If $f(x) = \cos(\log x)$, show that $2f(x)f(y) = f(xy) + f\left(\frac{x}{y}\right)$.

Sol : R.H.S. $= f(xy) + f\left(\frac{x}{y}\right)$

$= \cos(\log(xy)) + \cos\left(\log\left(\frac{x}{y}\right)\right)$

$= \cos(\log x + \log y) + \cos(\log x - \log y)$

$= 2\cos\left(\frac{\log x + \log y + \log x - \log y}{2}\right)$

$\cos\left(\frac{\log x + \log y - \log x + \log y}{2}\right)$

$= 2\cos(\log x)\cos(\log y) = f(x)f(y)$ = L.H.S.

Illustration 82. If $f(x) = \frac{|x|}{x}, x \neq 0$, then find the value of $|f(\alpha) - f(-\alpha)|, \alpha \neq 0$.

Sol : Case I. If $\alpha > 0, -\alpha < 0$

$\therefore \quad f(\alpha) = \frac{|\alpha|}{\alpha} = \frac{\alpha}{\alpha} = 1$

and $\quad f(-\alpha) = \frac{|-\alpha|}{-\alpha} = \frac{-(-\alpha)}{\alpha} = -1$

$\Rightarrow \quad |f(\alpha) - f(-\alpha)| = |1-(-1)| = 2$

Case II. If $\alpha < 0, -\alpha > 0$

$\therefore \; f(\alpha) = \frac{|\alpha|}{\alpha} = \frac{-\alpha}{\alpha} = -1$ and $f(-\alpha) = \frac{|-\alpha|}{-\alpha} = \frac{-\alpha}{-\alpha} = 1$

$\Rightarrow \quad |f(\alpha) - f(-\alpha)| = |-1-1| = |-2| = 2$

If Domain of *f*(*x*) is [*a*, *b*], then Domain of *f*(*g*(*x*)) is [$g^{-1}(a)$, $g^{-1}(b)$]

Proof : Given Domain of $f(x)$ is $[a, b]$.

For domain of $f(g(x))$.

$$a \le g(x) \le b \Rightarrow g^{-1}(a) \le x \le g^{-1}(b)$$

Hence, Domain of $f(g(x))$ is $[g^{-1}(a), g^{-1}(b)]$.

Illustration 83. The function $f(x)$ is defined on the interval [0, 1], then find the domain of definition of the functions :

(i) $f(\ln x)$ (ii) $f(\cos x)$ (iii) $f(2x + 3)$

Sol : (i) $\because \; 0 \le \ln x \le 1 \Rightarrow e^0 \le x \le e^1$ or $1 \le x \le e$

Hence, Domain of $f(\ln x)$ is $[1, e]$.

(ii) $\because\ 0 \le \cos x \le 1 \Rightarrow 2n\pi - \frac{\pi}{2} \le x \le 2n\pi + \frac{\pi}{2}, n \in I$

Hence, Domain of $f(\cos x)$ is $\left[2n\pi - \frac{\pi}{2}, 2n\pi + \frac{\pi}{2}\right], n \in I.$

(iii) $\because\ 0 \le 2x+3 \le 1 \Rightarrow -3 \le 2x \le -2$ or $-\frac{3}{2} \le x \le -1$

Hence, Domain of $f(2x+3)$ is $\left[-\frac{3}{2}, -1\right]$.

If $f(x) \cdot f(1/x) = f(x) + f(1/x)\ \forall\ x \in R - \{0\}$ where $f(x)$ be a Polynomial Function of Degree $n \in N$, then $f(x) = 1 \pm x^n$

Proof : Let $f(x) = a_0 + a_1x + a_2x^2 + \ldots + a_nx^n$, so

$$f(x) \cdot f\left(\frac{1}{x}\right) = f(x) + f\left(\frac{1}{x}\right)$$

$$\Rightarrow (a_0 + a_1x + a_2x^2 + \ldots + a_nx^n)\left(a_0 + \frac{a_1}{x} + \frac{a_2}{x^2} + \ldots + \frac{a_n}{x^n}\right)$$

$$= (a_0 + a_1x + a_2x^2 + \ldots + a_nx^n) + \left(a_0 + \frac{a_1}{x} + \frac{a_2}{x^2} + \ldots + \frac{a_n}{x^n}\right)$$

Multiplying both sides by x^n, we get

$$(a_0 + a_1x + a_2x^2 + \ldots + a_nx^n)(a_0x^n + a_1x^{n-1} + a_2x^{n-2} + \ldots + a_n)$$
$$= a_0x^n + a_1x^{n+1} + a_2x^{n+2} + \ldots + a_nx^{2n} + a_0x^n + a_1x^{n-1} + a_2x^{n-2} + \ldots + a_n$$

Comparing the coefficients of $x^{2n}, x^{2n-1}, \ldots, x^{n+1}$ on both sides, we get

$$a_0a_n = a_n \Rightarrow a_0 = 1 \qquad (\because a_n \ne 0)$$

$$a_na_1 + a_{n-1}a_0 = a_{n-1} \Rightarrow a_1 = 0,\ a_n \ne 0$$

Similarly, we get

$$a_2 = a_3 = \ldots = a_{n-1} = 0$$

Now comparing the coefficient of x^n on both sides, we get

$$2a_0 = a_0^2 + a_1^2 + a_2^2 + \ldots + a_{n-1}^2 + a_n^2$$

$$\Rightarrow \quad a_n^2 = 1 \Rightarrow a_n = \pm 1$$

Hence, $\quad f(x) = 1 \pm x^n$.

Alternative Method :

$$\because \quad f(x) \cdot f\left(\frac{1}{x}\right) = f(x) + f\left(\frac{1}{x}\right)$$

$$\Rightarrow \quad f(x)\left\{f\left(\frac{1}{x}\right) - 1\right\} = f\left(\frac{1}{x}\right)$$

$$\therefore \quad f(x) = \frac{f\left(\frac{1}{x}\right)}{f\left(\frac{1}{x}\right) - 1} \qquad \ldots(i)$$

Also $\quad f\left(\frac{1}{x}\right)\{f(x) - 1\} = f(x)$

$$\therefore \quad f\left(\frac{1}{x}\right) = \frac{f(x)}{f(x) - 1} \qquad \ldots(ii)$$

$$\therefore \quad f(x) \cdot f\left(\frac{1}{x}\right) = \frac{f\left(\frac{1}{x}\right)}{f\left(\frac{1}{x}\right) - 1} \cdot \frac{f(x)}{f(x) - 1}$$

[From (i) and (ii)]

$$\text{or} \quad f(x) \cdot f\left(\frac{1}{x}\right)\left\{\left(f\left(\frac{1}{x}\right) - 1\right)(f(x) - 1) - 1\right\} = 0$$

but $\quad f(x) \ne 0,\ f\left(\frac{1}{x}\right) \ne 0, \qquad (\because x \ne 0)$

$$\therefore \quad \left(f\left(\frac{1}{x}\right) - 1\right)(f(x) - 1) = 1 \qquad \ldots(iii)$$

So, $f\left(\frac{1}{x}\right) - 1$ and $f(x) - 1$ are reciprocal to each other.

Also, $x, \frac{1}{x}$ are reciprocal to each other.

Thus (i) can hold only, when

$$f(x) - 1 = \pm x^n, n \in N$$

Hence, $\quad f(x) = 1 \pm x^n$

Illustration 84. If $f(x)$ be a function satisfying $f(x) \cdot f\left(\frac{1}{x}\right) = f(x) + f\left(\frac{1}{x}\right)$ and $f(4) = 65$. Then find $f(x)$ and $f(3)$.

Sol : $\because\ f(x) \cdot f\left(\frac{1}{x}\right) = f(x) + f\left(\frac{1}{x}\right)$

$\therefore\ f(x) = 1 \pm x^n \Rightarrow f(4) = 1 \pm 4^n = 65$ or $\pm 4^n = 64 = 4^3$

Taking positive sign

$\Rightarrow \quad 4^n = 4^3 \quad \Rightarrow n = 3$

then $\quad f(x) = 1 + x^3$ and $f(3) = 1 + 3^3 = 28$

Illustration 85. If $g(x)$ is a polynomial satisfying $g(x)\,g(y) = g(x) + g(y) + g(xy) - 2 \quad \forall\ x, y \in R$ and $g(2) = 5$ then, find the value of $g(3)$.

Sol : Given $g(x)g(y) = g(x) + g(y) + g(xy) - 2 \qquad \ldots(i)$

Putting $x = 2, y = 1$, we have

$$g(2)g(1) = g(2) + g(1) + g(2) - 2$$

$$\Rightarrow \quad 5g(1) = 5 + g(1) + 5 - 2, \qquad (\because g(2) = 5)$$

$$\therefore \quad g(1) = 2$$

Again putting $y = \frac{1}{x}$ in (i) $(x \ne 0)$, we get

$$g(x)\,g\left(\frac{1}{x}\right) = g(x) + g\left(\frac{1}{x}\right) + g(1) - 2$$
$$= g(x) + g\left(\frac{1}{x}\right) + 2 - 2$$
$$= g(x) + g\left(\frac{1}{x}\right) \qquad (\because g(1) = 2)$$

$\therefore\ g(x) = 1 \pm x^n \Rightarrow g(2) = 1 \pm 2^n = 5 \Rightarrow \pm 2^n = 4$

Taking +ve sign, $2^n = 2^2$

$\therefore$ $n = 2$

then, $g(x) = 1 + x^2$

$\therefore$ $g(3) = 1 + 3^2 = 10$

If $f(x) = \dfrac{a^x}{a^x + \sqrt{a}} (a > 0)$, then $f(x) + f(1 - x) = 1$

and (I) $\sum_{r=1}^{2n-1} f\left(\dfrac{r}{2n}\right) = n - \dfrac{1}{2}$ and (II) $\sum_{r=1}^{2n} f\left(\dfrac{r}{2n+1}\right) = n$

Proof : We have $f(x) = \dfrac{a^x}{a^x + \sqrt{a}}$

Now, $f(x) + f(1-x) = \dfrac{a^x}{a^x + \sqrt{a}} + \dfrac{a^{1-x}}{a^{1-x} + \sqrt{a}}$

$$= \frac{a^x}{a^x + \sqrt{2}} + \frac{a}{a + a^x\sqrt{a}} = \frac{a^x}{a^x + \sqrt{a}} + \frac{\sqrt{a}}{\sqrt{a} + a^x} = 1 \quad \text{...(i)}$$

(i) Let $S = \sum_{r=1}^{2n-1} f\left(\dfrac{r}{2n}\right)$ ...(ii)

Replacing r by $(2n - r)$, then

$$S = \sum_{r=1}^{2n-1} f\left(\frac{2n-r}{2n}\right) \quad \text{...(iii)}$$

Adding (ii) and (iii), we get

$$2S = \sum_{r=1}^{2n-1} f\left(\frac{r}{2n}\right) + f\left(\frac{2n-r}{2n}\right)$$

$$= \sum_{r=1}^{2n-1} (1) = (2n-1) \quad \text{[from (i)]}$$

$$\therefore \quad S = n - \frac{1}{2}$$

(ii) Let $S = \sum_{r=1}^{2n} f\left(\dfrac{r}{2n+1}\right)$ ...(iv)

Replacing r by $2n + 1 - r$, then

$$S = \sum_{r=1}^{2n} f\left(\frac{2n+1-r}{2n+1}\right) \quad \text{...(v)}$$

Adding (iv) and (v), we get

$$2S = \sum_{r=1}^{2n} f\left(\frac{r}{2n+1}\right) + f\left(\frac{2n+1-r}{2n+1}\right) = \sum_{r=1}^{2n} (1) \quad \text{[from (i)]}$$

$$= 2n$$

$$\therefore \quad S = n$$

Illustration 86. Let $f(x) = \dfrac{9^x}{9^x + 3}$, evaluate the sum

$$f\left(\frac{1}{1996}\right) + f\left(\frac{2}{1996}\right) + f\left(\frac{3}{1996}\right) + \ldots + f\left(\frac{1995}{1996}\right).$$

Sol : $\because\ f(x) = \dfrac{9^x}{9^x + 3}$

$$\therefore \quad f(1-x) = \frac{9^{1-x}}{9^{1-x} + 3} = \frac{9}{9 + 3\cdot 9^x} = \frac{3}{3 + 9^x}$$

$\Rightarrow \quad f(x) + f(1-x) = 1$

Let $S = f\left(\dfrac{1}{1996}\right) + f\left(\dfrac{2}{1996}\right) + \ldots + f\left(\dfrac{1995}{1996}\right)$

or $S = f\left(\dfrac{1995}{1996}\right) + f\left(\dfrac{1994}{1996}\right) + \ldots + f\left(\dfrac{1}{1996}\right)$ (writing in reverse order)

$$\Rightarrow \quad 2S = \left(f\left(\frac{1}{1996}\right) + f\left(\frac{1995}{1996}\right)\right)$$

$$+ \left(f\left(\frac{2}{1996}\right) + f\left(\frac{1994}{1996}\right)\right) + \ldots + \left(f\left(\frac{1995}{1996}\right) + f\left(\frac{1}{1996}\right)\right)$$

$$= 1 + 1 + \ldots + 1 = 1995$$

Hence $S = 997.5$

Alternative Method : The given series can be written as :

$$\sum_{r=1}^{1995} f\left(\frac{r}{1996}\right) = \frac{1995}{2} = 997.5 \quad (\text{Here } n = 1995)$$

INTRODUCTORY EXERCISE 2.6

Only One Option is Correct

1. Let f be a linear function for which $f(6) - f(2) = 12$. The value of $f(12) - f(2)$ is equal to :

(a) 12 (b) 18 (c) 24 (d) 30

2. Let $f(x) = ax^2 + bx + c$ be such that $f(1) = 3$, $f(-2) = \lambda$ and $f(3) = 4$. If $f(0) + f(1) + f(-2) + f(3) = 14$, then λ is equal to **[JEE Main 28 Jul 2022 (ES)]**

(a) – 4 (b) 13/2 (c) 23/2 (d) 4

3. Suppose f is a real function satisfying $f(x + f(x)) = 4f(x)$ and $f(1) = 4$, then the value of $f(21)$ is :

(a) 16 (b) 21 (c) 64 (d) 105

4. Let f be a function satisfying $f(xy) = \dfrac{f(x)}{y}$ for all positive real numbers x and y. If $f(30) = 20$, then the value of $f(40)$ is :

(a) 15 (b) 20 (c) 40 (d) 60

5. If $f(x) = 2x^n + a$, if $f(2) = 26$ and $f(4) = 138$, then $f(3)$ is equal to :

(a) 56 (b) 82 (c) 64 (d) 122

6. If $P(x) + P(2x) = 5x^2 - 18$, then the value of $P(4)$ is :

(a) 0 (b) 7 (c) 9 (d) 16

7. If $f\left(2x^2 + \dfrac{y^2}{8}, 2x^2 - \dfrac{y^2}{8}\right) = xy$, then $f(5,4) + f(10,6)$ is equal to :

(a) 7 (b) 11 (c) 17 (d) 25

8. If function $f(x)$ is defined in $[-2, 2]$, then domain of definition of $f(|x|+1)$ is :

(a) $[-2, 2]$ (b) $[-3, 3]$ (c) $[-1, 1]$ (d) $[-3, 1]$

9. If $[\cdot]$ stands for integral part of x and $f(x)$ has domain $\left[-\frac{3}{2}, 3\right]$, then domain of $f([|x|])$ is :

(a) $(-2, 2)$ (b) $(-3, 3)$ (c) $(-4, 4)$ (d) $\{0, 1, 2, 3\}$

10. If the domain of definition of $f(x)$ is $(0, 1)$, then domain of definition of $f(\ln x)+f(\cos x)$ is :

(a) $(0, e)$ (b) $\left(1, \frac{\pi}{2}\right)$

(c) $\left(-\infty, \frac{\pi}{2}\right)$ (d) None of these

11. If the domain of definition of $f(x)$ is $(0, 1)$, then domain of definition of $f(c^x)+f(\log_{10}|x|)$ is $(c > 1)$:

(a) $(-\infty, 0)$ (b) $(0, 1)$
(c) $(0, \infty)$ (d) None of these

12. If $f(x)$ be a polynomial satisfying $f(x)\cdot f\left(\frac{1}{x}\right) = f(x)+f\left(\frac{1}{x}\right)$ and $f(4)=65$, then the value of $f(6)$ is :

(a) 220 (b) 217 (c) 214 (d) 37

13. If $g(x)$ be a polynomial satisfying
$g(x)\cdot g\left(\frac{1}{x}\right) = g(x)+g\left(\frac{1}{x}\right)$ and $g(3)=-26$, then the value of $g(4)$ is :

(a) – 126 (b) – 65 (c) – 63 (d) – 124

14. If $f(x)=\frac{4^x}{4^x+2}$ and given that $f(x)+f(1-x)=1$, then :

$f\left(\frac{1}{1997}\right)+f\left(\frac{2}{1997}\right)+f\left(\frac{3}{1997}\right)+\ldots+f\left(\frac{1996}{1997}\right)$ is :

(a) 992 (b) 994 (c) 996 (d) 998

15. If $f(x)=\frac{9^x}{9^x+3}$, then the value of the sum $f\left(\frac{1}{2006}\right)+f\left(\frac{2}{2006}\right)+f\left(\frac{3}{2006}\right)+\ldots+f\left(\frac{2005}{2006}\right)$ is :

(a) 1001.5 (b) 1002.5 (c) 1003.5 (d) 1004.5

16. A function $f(x)$ is given by $f(x)=\frac{5^x}{5^x+5}$, then the sum of the series

$f\left(\frac{1}{20}\right)+f\left(\frac{2}{20}\right)+f\left(\frac{3}{20}\right)+\ldots+f\left(\frac{39}{20}\right)$ is equal to

[JEE Main 25 Feb 2021 (ES)]

(a) 19/2 (b) 49/2 (c) 39/2 (d) 29/2

17. Let $f(x)=\frac{7^{1-x}}{7^{1-x}+\sqrt{7}}$, then $f(x)+f(1-x)$ is :

(a) – 1 (b) 1
(c) 2 (d) None of these

Different Types of Mappings

(1) One-One Mapping

The mapping $f: A \to B$ is called one-one mapping if different elements in A have different f images in B. Such a mapping is also known an *injective mapping* or *an injection* or *monomorphism*.

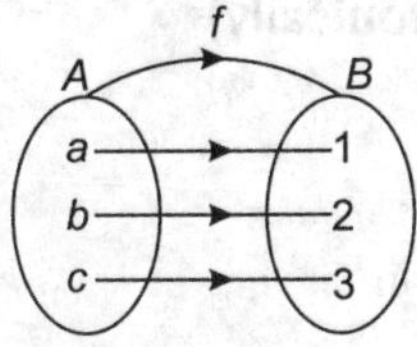

$f: A \to B$ is one-one

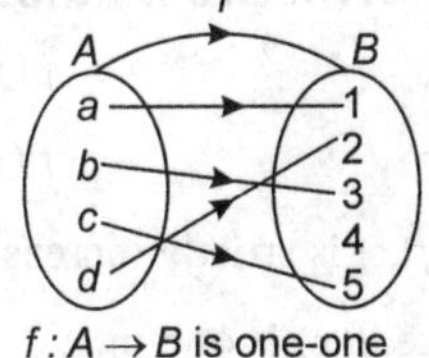

$f: A \to B$ is one-one

Methods to Test one-one :

(a) Analytically : If $x_1, x_2 \in A$, then $f(x_1)=f(x_2) \Rightarrow x_1 = x_2$ or Equivalently

$$x_1 \neq x_2 \Rightarrow f(x_1) \neq f(x_2)$$

(b) Graphically : If any line parallel to x-axis cuts the graph of the function at most at one point, then the function is one-one.

(c) Monotonically : Any function, which is entirely increasing or decreasing in whole domain, then $f(x)$ is one-one.

Note : To find one-one or injective mapping we mainly use Graphically or Monotonically method.

Illustration 87. Prove that the mapping $f: A \to B$ given by $f(x)=2x+5$ is one-one.

Sol : Let us consider set $A=\{-1, 2, 4\}$ and then $B=\{3, 9, 13\}$

i.e., Two different values of x do not have same values of $f(x)$. Hence, $f(x)$ is one-one.

I. Alternative Method (Analytically)

Let $x_1, x_2 \in A$

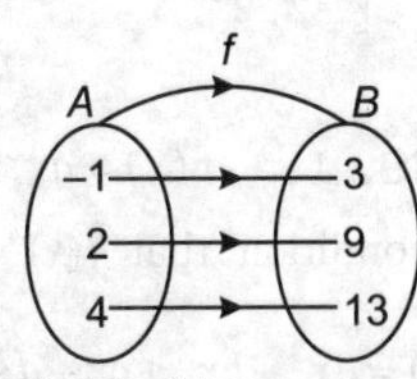

$f: A \to B$ is one-one

then $f(x_1)=f(x_2) \Rightarrow 2x_1+5=2x_2+5$

$\Rightarrow 2x_1 = 2x_2 \Rightarrow x_1 = x_2$

or if $x_1 \neq x_2 \Rightarrow 2x_1 \neq 2x_2$

$\Rightarrow 2x_1+5 \neq 2x_2+5 \Rightarrow f(x_1) \neq f(x_2)$

Hence, $f(x)$ is one-one.

II. Alternative Method (Graphically)

First draw the graph of $y = 2x + 5$

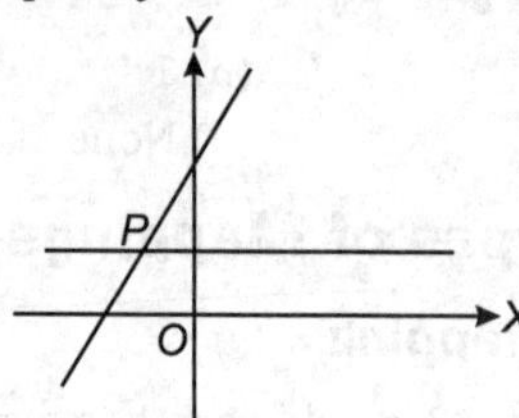

Since any line parallel to x-axis cuts the graph at one point P hence $f(x)$ is one-one.

III. Alternative Method (Monotonically)

Since $\quad f(x) = 2x + 5$

$\therefore \quad f'(x) = 2 > 0$

$\Rightarrow$ $f(x)$ is strictly increasing function.

Hence, $f(x)$ is one-one.

Illustration 88. Prove that functions $f(x) = \sqrt{x}, x \geq 0$ and $f(x) = \sin x \; \forall \; x \in \left[-\frac{\pi}{2}, \frac{\pi}{2}\right]$ are one-one.

Sol : $f(x) = \sqrt{x}, x \geq 0$

$\because$ Graph of $f(x) = \sqrt{x}, x \geq 0$ is

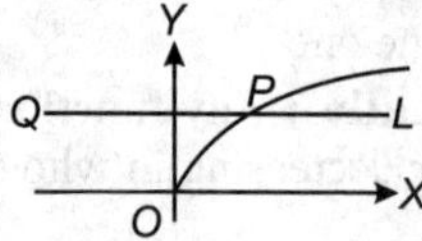

Since any line QL parallel to x-axis cuts the graph of $f(x) = \sqrt{x}$ at one point P, hence $f(x)$ is one-one.

Also graph of $f(x) = \sin x, x \in \left[-\frac{\pi}{2}, \frac{\pi}{2}\right]$ is :

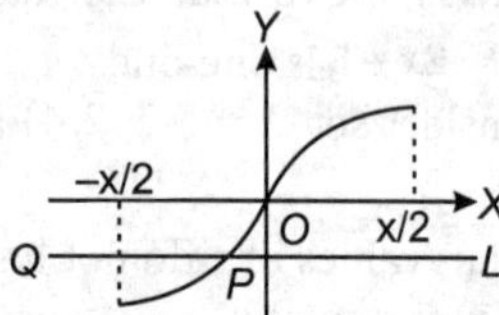

Since any line QL, parallel to x-axis cuts the graph of $f(x) = \sin x \; \forall \; x \in \left[-\frac{\pi}{2}, \frac{\pi}{2}\right]$ at one point P, hence $f(x)$ is one-one.

Illustration 89. Let $f(x) = ax^3 + bx^2 + cx + d \sin x$, then find the condition that $f(x)$ is always one-one function.

Sol : Given $f(x) = ax^3 + bx^2 + cx + d \sin x$.

Now for $f(x)$ to be one-one either f is entirely increasing or entirely decreasing.

$\therefore \; f'(x) = 3ax^2 + 2bx + c + d \cos x \geq 3ax^2 + 2bx + c - |d|$

for $f'(x) > 0 \; \forall \; x \in R$

$a > 0$ and $4b^2 - 4 \cdot 3a(c - |d|) < 0 \Rightarrow b^2 < 3a(c - |d|)$

and $f'(x) = 3ax^2 + 2bx + c + d \cos x \leq 3ax^2 + 2bx + c + |d|$

for $\quad f'(x) < 0 \; \forall \; x \in R$

$a < 0$ and $4b^2 - 12a(c + |d|) < 0$

$\Rightarrow \quad b^2 < 3a(c + |d|)$

Hence, required condition is $b^2 < 3a(c - |d|)$

Number of One-one Mappings

Let $f : A \to B$ be a maping such that A and B are finite sets having m and n elements respectively, (where $n > m$).

Let $A = \{x_1, x_2, ..., x_m\}$ and $B = \{y_1, y_2, ..., y_n\}$.

It is clear for one-one mapping.

Number of possible images for $x_1 = n$

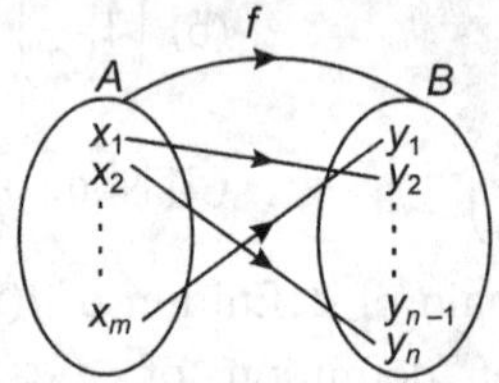

After x_1 number of possible images for $x_2 = (n - 1)$

After x_2 number of possible images for $x_3 = (n - 2)$

---- ---- ---- ----

---- ---- ---- ----

After x_{m-1} number of possible images for

$x_m = (n - m + 1)$

Hence, number of one-one mappings

$= n(n-1)(n-2) \dots (n-m+1)$

$$= \begin{cases} {}^nP_m, & n \geq m \\ 0, & n < m \end{cases}$$

$$= \begin{cases} {}^nC_m \cdot m!, & n \geq m \\ 0, & n < m \end{cases}$$

Illustration 90. A mapping is selected at random from the set of all mappings of the set $A = \{1, 2, 3, ..., n\}$ into itself. Then find the number of one-one mappings.

Sol : The number of one-one mappings $= {}^nP_n = n!$

Total Number of Mappings from *A* to *B*

Let $f : A \to B$ be a mapping such that A and B are finite sets having m and n elements respectively (where $n > m$).

Let $\quad A = \{x_1, x_2, ..., x_m\}$

and $\quad B = \{y_1, y_2, ..., y_n\}$

Number of possible mappings for $x_1 = n$

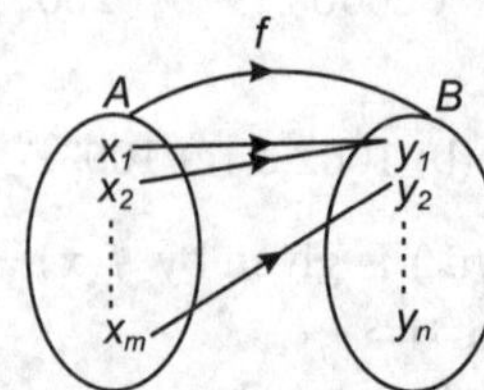

After x_1 number of possible mappings for $x_2 = n$

After x_2 number of possible mappings for $x_3 = n$

---- ---- ---- ----
---- ---- ---- ----

After x_{m-1} number of possible mappings for $x_m = n$

Hence, total number of mappings from A to B

$$= \underbrace{n \times n \times n \times \ldots \times n}_{m\text{–times}} = n^m$$

Illustration 91. A mapping is selected at random from the set of all mappings of the set $A = \{1, 2, 3, \ldots, n\}$ into itself. Then find the number of mappings from A to A.

Sol : Hence, total number of functions from A to B

$$= \underbrace{n \times n \times n \times \ldots \times n}_{m\text{–times}} = n^n$$

(2) Many-One Mapping

The mapping $f : A \to B$ is called *many-one* mapping if two or more than two different elements in A have the same image in B.

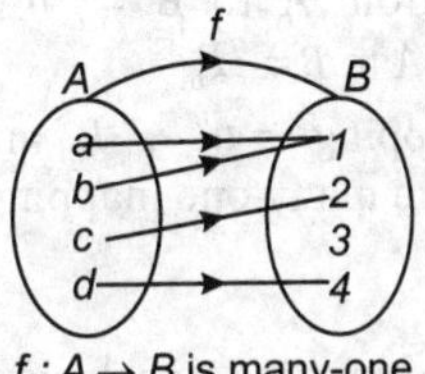

$f : A \to B$ is many-one

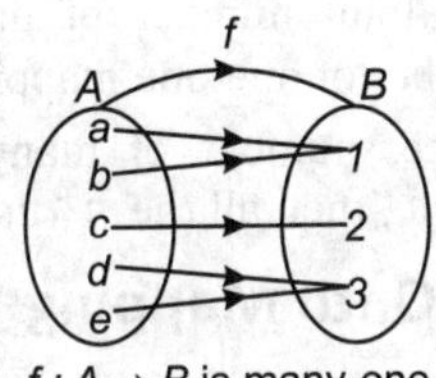

$f : A \to B$ is many-one

Methods to Test Many-One

(a) Analytically : If $x_1, x_2 \in A$, then $f(x_1) = f(x_2) \Rightarrow x_1 \neq x_2$.

(b) Graphically : If any line parallel to x-axis cuts the graph of the function at least two points, then the function is many-one.

(c) Monotonically : Any function which is neither increasing nor decreasing in whole domain, then $f(x)$ is many-one.

Or

Any continuous functions $f(x)$, which has at least one local maxima or local minima, is many-one.

Illustration 92. Prove that the mapping $f : A \to B$ given by $f(x) = x^2 + x + 1 \; \forall \; x \in R$ is many-one.

Sol : Let us consider set $A = \{-1, 0, 2\}$ and then $B = \{1, 7\}$ *i.e.*, two different values of x, $f(x)$ has same value. Hence $f(x)$ is many-one.

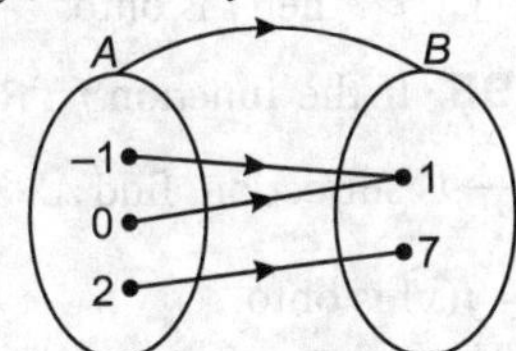

I. Alternative Method (Analytically) :

Let $\quad x_1, x_2 \in A$

Then $\quad f(x_1) = f(x_2)$

$\Rightarrow \quad x_1^2 + x_1 + 1 = x_2^2 + x_2 + 1$

$\Rightarrow \quad x_1^2 - x_2^2 + x_1 - x_2 = 0$

$\Rightarrow \quad (x_1 - x_2)(x_1 + x_2 + 1) = 0$

$\Rightarrow \quad x_1 \neq x_2 \qquad \{\because \text{ if } x_1 = -1, x_2 = 0\}$

Hence, $f(x)$ is many-one.

II. Alternative Method (Graphically) :

Let $\quad y = f(x) = x^2 + x + 1$

$\Rightarrow \quad y = \left(x + \frac{1}{2}\right)^2 + \frac{3}{4}$ or $\left(x + \frac{1}{2}\right)^2 = \left(y - \frac{3}{4}\right)$

Which is a parabola whose vertex is at $\left(-\frac{1}{2}, \frac{3}{4}\right)$.

Cuts the y-axis at (0, 1) and neither cuts nor touch the x-axis.

Graph of $\; y = x^2 + x + 1$ is as shown in graph.

Since any line QL parallel to x-axis cuts the graph at two points R and S. Hence, $f(x)$ is many-one.

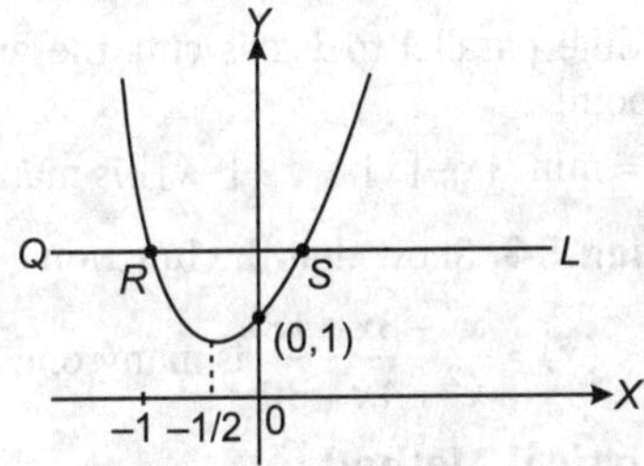

III. Alternative Method (Monotonically) :

Since $\quad f(x) = x^2 + x + 1$

$\therefore \quad f'(x) = 2x + 1$

$\Rightarrow \quad f'(x) > 0$ if $x > -1/2$

and $\quad f'(x) < 0$ if $x < -1/2$

Therefore $f(x)$ is neither increasing nor decreasing and continuous everywhere which has minimum value 3/4 at $x = -1/2$.

Hence, $f(x)$ is many-one.

Illustration 93. Prove that the functions

$$|y| = |\sin|x||, \; y = (x - [x])^2$$

and y = min. $\{x - [x], -x - [-x]\}$ are many-one where $x \in R$ and $[\cdot]$ denotes the greatest integer function.

Sol : Graph of $|y| = |\sin|x||$

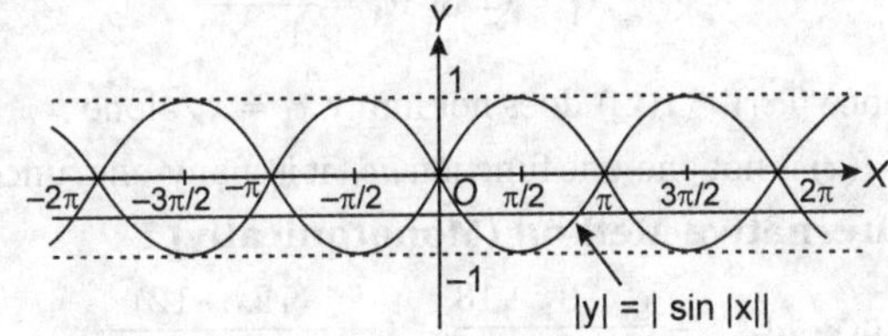

Since any line parallel to x-axis cuts the graph at more than one point.

Hence, $|y| = |\sin|x||$ is many-one.

Graph of $y = (x - [x])^2$

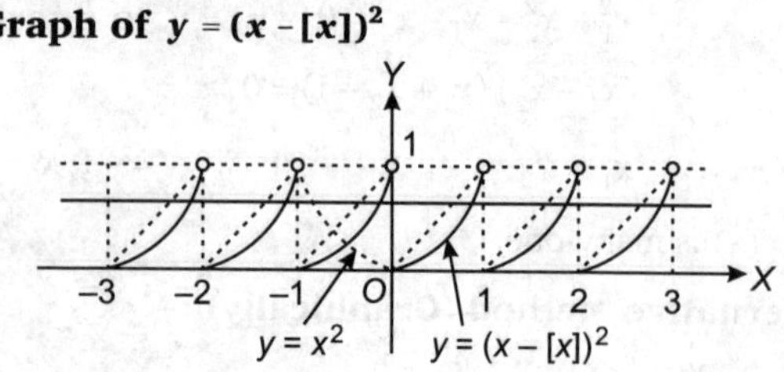

Since any line parallel to x-axis cuts the graph at more than one point.

Hence, $y = (x - [x])^2$ is many-one.

Graph of $y = \min.\{x - [x], -x - [x]\}$

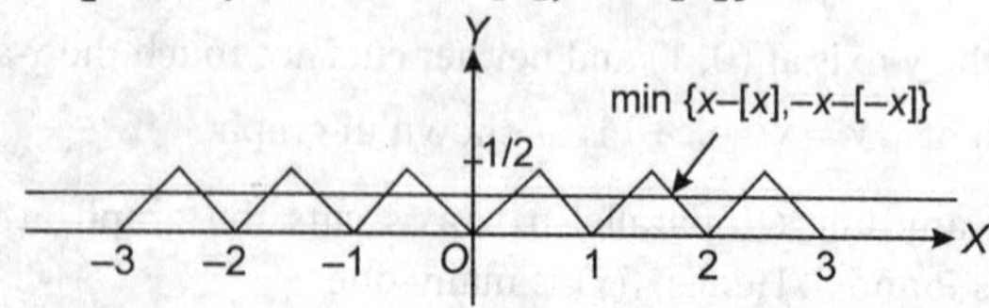

Since any line parallel to x-axis cuts the graph at more than one point.

Hence, $y = \min.\{x - [x], -x - [-x]\}$ is many-one.

Illustration 94. Show that the function

$$f(x) = \frac{x^2 - 8x + 18}{x^2 + 4x + 30} \text{ is many-one.}$$

Sol : Analytical Method :

Let $\quad x_1, x_2 \in D_f$

$\therefore \quad f(x_1) = f(x_2)$

$$\Rightarrow \frac{x_1^2 - 8x_1 + 18}{x_1^2 + 4x_1 + 30} = \frac{x_2^2 - 8x_2 + 18}{x_2^2 + 4x_2 + 30}$$

$$\Rightarrow (x_1^2 - 8x_1 + 18)(x_2^2 + 4x_2 + 30) = (x_1^2 + 4x_1 + 30)(x_2^2 - 8x_2 + 18)$$

$$\Rightarrow x_1^2x_2^2 + 4x_1^2x_2 + 30x_1^2 - 8x_1x_2^2 - 32x_1x_2 - 240x_1 + 18x_2^2 + 72x_2 + 540$$

$$= x_1^2x_2^2 - 8x_1^2x_2 + 18x_1^2 + 4x_1x_2^2 - 32x_1x_2 + 72x_1 + 30x_2^2 - 240x_2 + 540$$

$$\Rightarrow 12x_1^2x_2 - 12x_1x_2^2 + 12x_1^2 - 12x_2^2 - 312x_1 + 312x_2 = 0$$

$$\Rightarrow (x_1 - x_2)(12x_1x_2 + 12(x_1 + x_2) - 312) = 0$$

$$\Rightarrow (x_1 - x_2)(x_1x_2 + x_1 + x_2 - 26) = 0$$

$$\Rightarrow \quad x_1 = x_2 \text{ or } x_1 = \frac{26 - x_2}{1 + x_2}$$

Since $f(x_1) = f(x_2)$ does not imply $x_1 = x_2$ alone

$\therefore f(x)$ is not one-one function *i.e.*, it is many-one function.

Alternative Method (Monotonically) :

Since $f(x) = \dfrac{x^2 - 8x + 18}{x^2 + 4x + 30} = 1 - \dfrac{(12x + 12)}{(x^2 + 4x + 30)}$

Differentiating w.r.t. x, we get

$$f'(x) = 0 - \left\{\frac{(x^2 + 4x + 30)\cdot 12 - (12x + 12)(2x + 4)}{(x^2 + 4x + 30)^2}\right\}$$

$$= \frac{12(x^2 + 2x - 26)}{(x^2 + 4x + 30)^2}$$

$$f'(x) = 0 \quad \Rightarrow \quad x^2 + 2x - 26 = 0$$

$$\therefore \quad x = \frac{-2 \pm \sqrt{4 + 4 \times 26}}{2} = -1 \pm 3\sqrt{3}$$

$$\therefore \quad f'(x) = \frac{12(x + 1 + 3\sqrt{3})(x + 1 - 3\sqrt{3})}{(x^2 + 4x + 30)^2}$$

$$f'(x) > 0 \Rightarrow \quad x \in (-\infty, -1 - 3\sqrt{3}) \cup (-1 + 3\sqrt{3}, \infty)$$

$$f'(x) < 0 \Rightarrow \quad x \in (-1 - 3\sqrt{3}, -1 + 3\sqrt{3})$$

Since $f(x)$ is increasing as well as decreasing.

Hence, $f(x)$ is many-one function.

Number of Many-One Mappings

Let $f : A \to B$ be a mapping such that A and B are finite sets having m and n elements respectively (where $n > m$).

Let $A = \{x_1, x_2, x_3, ..., x_m\}$ and $B = \{y_1, y_2, y_3, ..., y_n\}$

$\because$ Total number of mappings from A to B = n^m and number of one-one mappings from A to $B = {}^nP_m$

Hence, number of many-one mapping $= n^m - {}^nP_m$ and if $n < m$, then all the n^m mappings are many-one mappings.

(3) Onto Mapping

The mapping $f : A \to B$ is called onto mapping if each element in B (co-domain) is the f image of at least one element in A, then we say that f is a function of ***A* 'onto' *B***. Thus, $f : A \to B$ such that $f(A) = B$ *i.e.*, Range = Co-domain such a mapping is also known as ***surjective*** or ***surjection***.

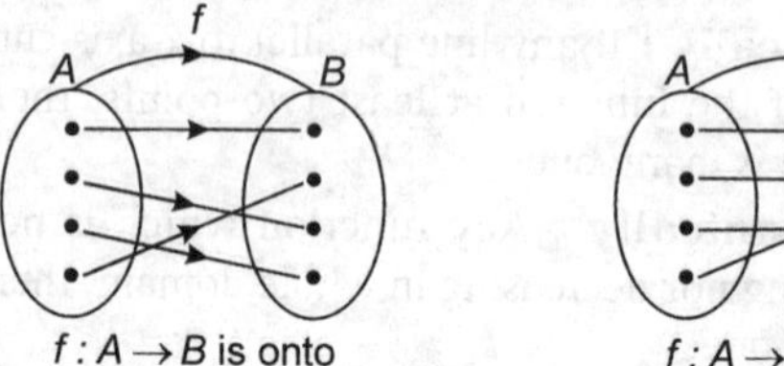

$f : A \to B$ is onto

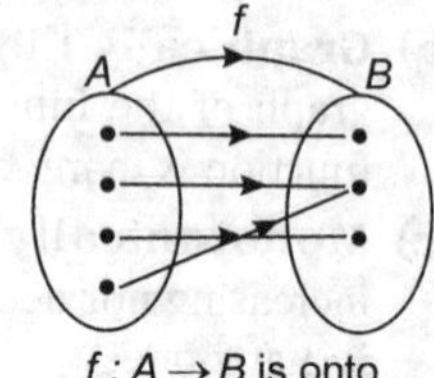

$f : A \to B$ is onto

Note :

Every polynomial function $f : R \to R$ of degree odd is ONTO

Method to Test Onto Mapping

Let $f : A \to B$ be a mapping, let y be an arbitrary element in B and then $y = f(x)$, where $x \in A$. Then express x in terms of y. Now if $x \in A \;\forall\; y \in B$, then f is onto.

Illustration 95. If the function $f : R \to A$ is given by $f(x) = \dfrac{e^x - e^{-|x|}}{e^x + e^{|x|}}$ is surjection, find A.

Sol : Since, $y = f(x)$ is onto.

$$\therefore \quad R_f = C_f = A$$

We have $\quad f(x) = \dfrac{e^x - e^{-|x|}}{e^x + e^{|x|}}$

Case I. If $x \geq 0$, then $f(x) = \dfrac{e^x - e^{-x}}{e^x + e^x} = \dfrac{1}{2} - \dfrac{1}{2}e^{-2x}$

but $e^{-2x} > 0$ and $\because\ x \geq 0 \Rightarrow -2x \leq 0$

$\Rightarrow\quad e^{-2x} \leq 1 \Rightarrow 0 < e^{-2x} \leq 1$

$\Rightarrow\quad 0 > -\frac{1}{2}e^{-2x} \geq -\frac{1}{2}$

$\Rightarrow\quad \frac{1}{2} > \frac{1}{2} - \frac{1}{2}e^{-2x} \geq 0$

$\therefore\quad 0 \leq f(x) < \frac{1}{2}$

$\therefore\quad R_f = \left[0, \frac{1}{2}\right)$ for $x \geq 0$

Case II. If $x < 0$, then $f(x) = \frac{e^x - e^x}{e^x + e^{-x}} = 0$

$\therefore\quad R_f = \{0\}$ for $x < 0$

Finally, $\quad R_f = \left[0, \frac{1}{2}\right) = A$

Hence, $\quad A = \left[0, \frac{1}{2}\right)$

Illustration 96. If the function $f : R \to A$ is given by $f(x) = \frac{x^2}{x^2+1}$ is surjection, then find A.

Sol : Since $y = f(x)$ is onto.

$\therefore\quad R_f = C_f = A$

We have $\quad y = f(x) = \frac{x^2}{x^2+1} \Rightarrow x^2y + y = x^2$

$\Rightarrow\quad y = x^2(1-y)$

$\Rightarrow\quad x^2 = \frac{y}{1-y}$

$\Rightarrow\quad x = \pm\sqrt{\left(\frac{y}{1-y}\right)}$ exists if

$\Rightarrow\quad \frac{y}{1-y} \geq 0 \Rightarrow \frac{y}{y-1} \leq 0$

$\therefore\quad 0 \leq y < 1 \Rightarrow y \in [0, 1)$

$\therefore\quad R_f = [0, 1)$

Hence, $\quad A = [0, 1)$

Illustration 97. Let $f(x) = \frac{ax^2+2x+1}{2x^2-2x+1}$. Find the values of a if : $f : R \to [-1, 2]$ is onto.

Sol : Since $y = f(x)$ is onto.

$\therefore\quad R_f = C_f = [-1, 2]$

We have $\quad y = f(x) = \frac{ax^2+2x+1}{2x^2-2x+1}$

$\Rightarrow\quad y(2x^2 - 2x + 1) = ax^2 + 2x + 1$

$\Rightarrow\quad (2y - a)x^2 - 2(y+1)x + (y-1) = 0$

$\because\quad x \in R$

$\therefore\quad 4(y+1)^2 - 4(2y-a)(y-1) \geq 0$

$\Rightarrow\quad (y+1)^2 - (2y-a)(y-1) \geq 0$

At $y = -1$, $0 - (-2-a)(-1-1) \geq 0$

$\Rightarrow\quad -2(2+a) \geq 0 \Rightarrow 2 + a \leq 0$

$\therefore\quad a \leq -2 \quad$ ...(i)

and at $\quad y = 2, 9 - (4-a)(1) \geq 0$

$\Rightarrow\quad 5 + a \geq 0$

$\therefore\quad a \geq -5 \quad$ ...(ii)

From (i) and (ii), we get

$a \in [-5, -2]$

Number of Onto Mappings

Let $f : A \to B$ be a mapping such that A and B are finite sets having m and n elements respectively. Then number of onto (surjective) mappings from A to B.

$$= \begin{cases} \sum_{r=1}^{n} (-1)^{n-r} \cdot {}^nC_r \cdot r^m, & \text{if } 1 \leq n \leq m \\ 0, & \text{if } n > m \end{cases}$$

Alternative Method :

Number of onto (surjective) mappings from A to B

$$= \begin{cases} \sum_{r=0}^{n-1} (-1)^r \cdot {}^nC_r \cdot (n-r)^m, & \text{if } n < m \\ n!, & \text{if } n = m \\ 0, & \text{if } n > m \end{cases}$$

OR

Coefficient of x^m in $m!(e^x - 1)^n$

Note :

Coefficient of x^r in $e^{px} = \frac{p^r}{n!}$

Illustration 98. Find the number of surjections from $A = \{1, 2, 3, ..., n\}$, $n \geq 2$ to $B = \{a, b, c\}$.

Sol : Number of surjections from A to B is :

$$= \sum_{r=1}^{3} (-1)^{3-r} \cdot {}^3C_r \cdot r^n$$

$$= (-1)^2 \cdot {}^3C_1 \cdot (1)^n + (-1)^1 \cdot {}^3C_2 \cdot 2^n + (-1)^0 \cdot {}^3C_3 \cdot 3^n$$

$$= 3 - 3 \cdot 2^n + 3^n = 3^n - 3(2^n - 1)$$

I. Alternative Method :

The number of surjections from A to B is the coefficient of x^n in $n!(e^x - 1)^3$.

or coefficient of x^n in $n!(e^{3x} - {}^3C_1e^{2x} + {}^3C_2e^x - 1)$

or coefficient of x^n in $n!(e^{3x} - 3e^{2x} + 3e^x - 1)$ is

$$n!\left(\frac{3^n}{n!} - 3\cdot\frac{2^n}{n!} + 3\cdot\frac{1^n}{n!}\right) = 3^n - 3(2^n - 1)$$

II. Alternative Method :

The total number of mappings from A to B is n^m *i.e.*, 3^n of which three functions $f_1(x) = a$, $f_2(x) = b$ and $f_3(x) = c\ \forall\ x \in A$ are not surjections.

$\therefore$ No. of not surjections $= {}^3C_1 2^n - {}^3C_2(1)^n = 3(2^n - 1)$

Hence, no. of surjections from A to B is $3^n - 3(2^n - 1)$.

(4) Into Mapping

If $f : A \to B$ is such that there exists at least one element in co-domain which is not the image of any element in domain, then $f(x)$ is into.

Thus, $f : A \to B$ such that $f(A) \subset B$

i.e., Range $\subset$ co-domain

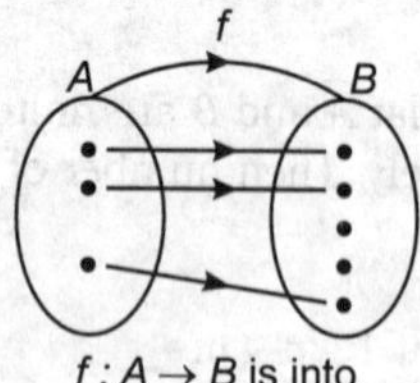

$f : A \to B$ is into

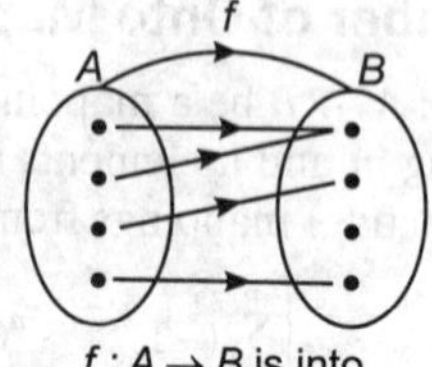

$f : A \to B$ is into

Note :

Every polynomial function $f : R \to R$ of degree even is INTO.

Method to Test Into Mapping :

Let $f : A \to B$ be a mapping, let y be an arbitrary element in B and then $y = f(x)$ and express x in terms of y. Now if $x \notin A \ \forall \ y \in B$, then f is into.

Illustration 99. If $f : R \to R$ be a function such that $f(x) = \sin x$, then show that f is into and if $f : R \to [-1, 1]$ then f is into.

Sol : $\because$ $\quad -1 \le \sin x \le 1$

$\therefore$ $\quad R_f = [-1, 1]$

For first part $R_f \subset C_f \Rightarrow f$ is into.

and for second part $R_f = C_f \Rightarrow f$ is into.

Illustration 100. Let $f : N \to I$ be a function defined as $f(x) = x - 100$, show that f is an into.

Sol : Let $\quad y = f(x) = x - 100$

$\Rightarrow \quad x = y + 100 = g(y)$ (say)

$\therefore \quad g(y) = y + 100$

It is celar that $g(y)$ is defined $\forall \ y \in I$

but $g(y) \notin N$ for $y \le -100$

Hence, f is into.

Illustration 101. If $f : R \to R$ be a function defined as

$$f(x) = \begin{cases} x|x| - 9, & x \in Q \\ x|x| - \sqrt{5}, & x \notin Q \end{cases}$$, show that f is an into.

Sol : Here co-domain $= R$

and $\quad f(x) \ne \sqrt{5} \ \forall \ x \in R$

$\therefore \quad$ Range $= R \sim \{\sqrt{5}\}$

Range $\subset$ co-domain. Hence f is into.

Number of Into Mappings

Let $f : A \to B$ be a mapping such that A and B are finite sets having m and n elements respectively.

Then number of into mapping from A to B

= Total mappings – Number of onto mappings

$$= \begin{cases} \sum_{r=1}^{n-1} (-1)^{r-1} \cdot {}^nC_r \cdot (n-r)^m, & \text{if } n \le m \\ n^m, & \text{if } n > m \end{cases}$$

Note :

Total number of constant function from A to B are n.

Illustration 102. Find the number of into mapping from $A = \{1, 2, 3, 4, 5\}$ to $B = \{a, b, c, d\}$

Sol : Number of into mappings from A to B is :

$$\sum_{r=1}^{4-1} (-1)^{r-1} \cdot {}^4C_r \cdot (4-r)^5 = \sum_{r=1}^{3} (-1)^{r-1} \cdot {}^4C_r \cdot (4-r)^5$$

$$= {}^4C_1 \cdot (4-1)^5 - {}^4C_2 \cdot (4-2)^5 + {}^4C_3(4-3)^5$$

$$= 4 \cdot 3^5 - 6 \cdot 2^5 + 4 \cdot 1^5$$

$$= 972 - 192 + 4 = 784$$

Explanation :

Total number of mappings from A to $B = 4^5 = 1024$ and number of surjections from A to B is the coefficient of x^5 in $5!(e^x - 1)^4$.

or coefficient of x^5 in $5!(e^{4x} - 4e^{3x} + 6e^{2x} - 4e^x + 1)$

is $5!\left(\frac{4^5}{5!} - 4 \cdot \frac{3^5}{5!} + 6 \cdot \frac{2^5}{5!} - 4 \cdot \frac{1^5}{5!} + 0\right)$

$= 4^5 - 4 \cdot 3^5 + 6 \cdot 2^5 - 4 = 1024 - 972 + 192 - 4 = 240$

Hence, number of into mappings

= Number of total mappings from A to B – Number of surjection from A to B = 1024 – 240 = 784

From Above Classification, we Conclude that the Function is of Four Types

(a) One-One Onto (Bijective) Mapping :

A mapping $f : A \to B$ is said to be one-one onto or bijective if and only if

(i) It is one-one i.e., $f(x_1) = f(x_2)$

$\Rightarrow \quad x_1 = x_2 \ \forall \ x_1, x_2 \in A$

(ii) It is onto i.e., $\forall \ y \in B$, there exists $x \in A$ such that $f(x) = y$.

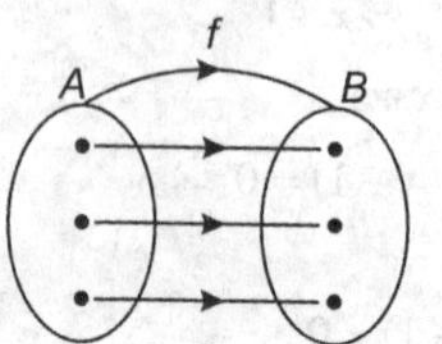

$f : A \to B$ is one-one onto

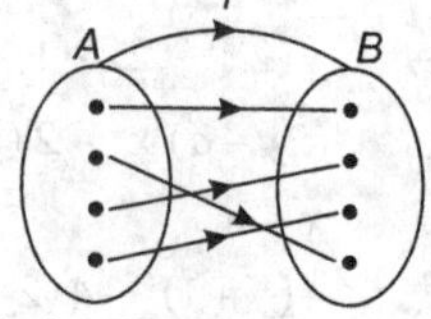

$f : A \to B$ is one-one onto

Number of One-One Onto (Bijective) Mappings

Let $f : A \to B$ be a mapping such that A and B are finite sets having the same number of elements. If A has n elements, then B has also n elements.

Let $\quad A = \{x_1, x_2, x_3, ..., x_n\}$

and $\quad B = \{y_1, y_2, y_3, ..., y_n\}$

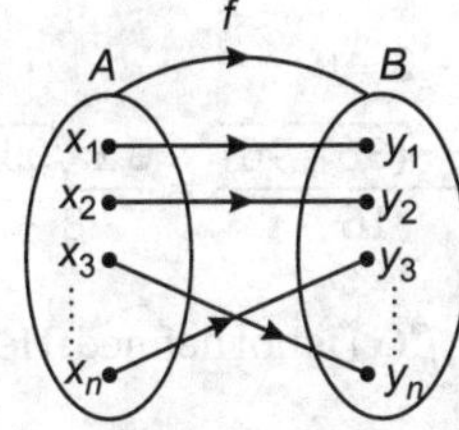

It is clear for bijective mapping.

Number of possible images for $x_1 = n$

After x_1 number of possible images for $x_2 = (n-1)$

After x_2 number of possible images for $x_3 = (n-2)$

---- ---- ---- ----

---- ---- ---- ----

After x_{n-1} number of possible images for $x_n = 1$

Hence, number of bijective mappings

$$= n(n-1)(n-2)(n-3) \ldots 2 \cdot 1 = n!$$

(b) One-One into Mapping :

A mapping $f : A \to B$ is said to be one-one into if and only if

(i) It is one-one

i.e., $f(x_1) = f(x_2) \Rightarrow x_1 = x_2 \ \forall \ x_1, x_2 \in A$

(ii) It is into *i.e.*, $\forall \ y \in B$ there exists $x \notin A$

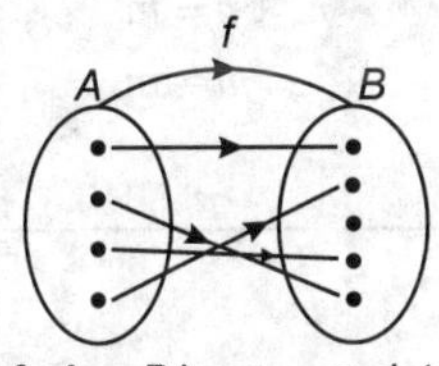

$f : A \to B$ is one-one into

(c) Many-One Onto Mapping :

A mapping $f : A \to B$ is said to be many-one onto if and only if :

(i) It is many-one

i.e., $x_1 \neq x_2 \Rightarrow f(x_1) = f(x_2) \ \forall \ x_1, x_2 \in A$

(ii) It is onto *i.e.*, $\forall \ y \in B$ there exists $x \in A$ such that $f(x) = y$

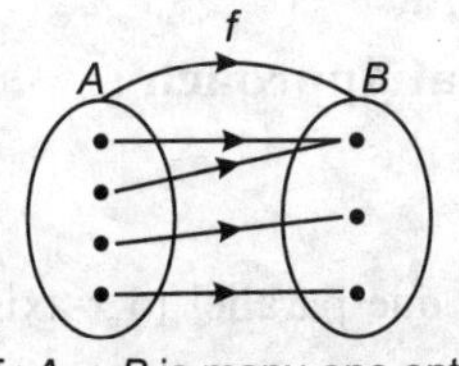

$f : A \to B$ is many-one onto

(d) Many-one Into Mapping :

A mapping $f : A \to B$ is said to be many-one into if and only if

(i) It is many-one *i.e.*,

$$x_1 \neq x_2$$
$$\Rightarrow \quad f(x_1) = f(x_2) \ \forall \ x_1, x_2 \in A$$

(ii) It is into *i.e.*, $\forall \ y \in B$ there exists $x \notin A$

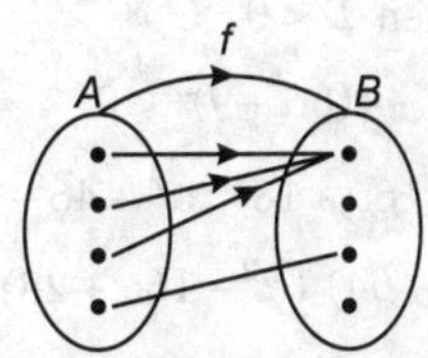

$f : A \to B$ is many-one into

Illustration 103. Let $A = R - \{3\}$, $B = R - \{1\}$ and let $f : A \to B$ be defined by $f(x) = \dfrac{x-2}{x-3}$. Is f bijective? Given reasons.

Sol : To test whether *f* is one-one (injective) :

Let $\quad x_1, x_2 \in A$

Then $\quad f(x_1) = f(x_2)$

$$\Rightarrow \quad \frac{x_1 - 2}{x_1 - 3} = \frac{x_2 - 2}{x_2 - 3}$$

$$\Rightarrow \quad x_1x_2 - 3x_1 - 2x_2 + 6 = x_1x_2 - 2x_1 - 3x_2 + 6$$

$$\Rightarrow \quad x_1 = x_2$$

Hence, f is injective.

To test whether *f* is onto (surjective) :

Let $\quad y = f(x) = \dfrac{x-2}{x-3} \ \forall \ y \in B$

$$\Rightarrow \quad xy - 3y = x - 2$$

$$\Rightarrow \quad x(y-1) = 3y - 2$$

$$\Rightarrow \quad x = \frac{3y-2}{y-1}$$

$\therefore \quad$ Range of $f = R - \{1\} = B$

Hence, f is onto.

Finally, f is one-one onto *i.e.*, bijective.

Illustration 104. A function $f : R \to R$ is defined by $f(x) = \dfrac{\alpha x^2 + 6x - 8}{\alpha + 6x - 8x^2}$. Find the interval of values of α for which f is onto. Is the function one-to-one for $\alpha = 3$? Justify your answer.

Sol : As $f : R \to R$, the function will be onto if $= \dfrac{\alpha x^2 + 6x - 8}{\alpha + 6x - 8x^2}$ can assume any real value $\forall \ x \in R$.

Let $\quad y = \dfrac{\alpha x^2 + 6x - 8}{\alpha + 6x - 8x^2}$

or $\quad \alpha y+6xy-8x^2y=\alpha x^2+6x-8$

$\Rightarrow \quad (8y+\alpha)x^2-6x(y-1)-(\alpha y+8)=0$

$\therefore \quad B^2-4AC>0\ \forall\ x\in R$

$\Rightarrow \quad 36(y-1)^2+4\cdot(8y+\alpha)\cdot(\alpha y+8)>0$

$\Rightarrow \quad 9(y^2-2y+1)+(8\alpha y^2+64y+\alpha^2y+8\alpha)>0$

$\Rightarrow \quad (8\alpha+9)y^2+(\alpha^2+46)y+(8\alpha+9)>0$

If $8\alpha+9>0$, then $D<0$

$\Rightarrow \quad (\alpha^2+46)^2-4(8\alpha+9)^2<0$

$\Rightarrow \quad (\alpha^2+46+16\alpha+18)\ (\alpha^2+46-16\alpha-18)<0$

$\Rightarrow \quad (\alpha^2+16\alpha+64)\ (\alpha^2-16\alpha+28)<0$

$\Rightarrow \quad (\alpha+8)^2\ (\alpha-14)\ (\alpha-2)<0 \quad (\because (\alpha+8)^2>0)$

$\therefore \quad (\alpha-14)(\alpha-2)<0 \Rightarrow \quad \alpha\in(2,14)$

Now, when $\alpha=3$, then $f(x)=\dfrac{3x^2+6x-8}{3+6x-8x^2}$.

$\therefore \quad f'(x)=\dfrac{66\left\{\left(x-\frac{5}{6}\right)^2+\frac{11}{36}\right\}}{(3+6x-8x^2)^2}$ if $3+6x-8x^2=0$

or $\quad 8x^2-6x-3=0$

or $\quad x=\dfrac{6\pm\sqrt{(36+96)}}{16}=\dfrac{3\pm\sqrt{33}}{8}$

For $x=\dfrac{3\pm\sqrt{33}}{8}$, $f'(x)$ is not defined. Hence $\forall\ x\in R$, $f(x)$ is neither increasing nor decreasing. Hence $f(x)$ is not one-one.

Number of Functions (Mappings) at One Place in a Table

Let $f: A\rightarrow B$ be a mapping such that A and B are finite sets having m and n elements respectively, then :

	Description of Mappings	Number of functions (Mappings)
1.	Total number of mappings from A to B	n^m
2.	Total number of one-one mappings from A to B	$\begin{cases} {}^nP_m, & n\ge m \\ 0, & n<m \end{cases}$
3.	Total number of many-one mappings from A to B	$\begin{cases} n^m-{}^nP_m, & n\ge m \\ n^m, & n<m \end{cases}$
4.	Total number of onto (surjective) mappings from A to B	$\begin{cases} \sum_{r=0}^{n-1}(-1)^r\cdot {}^nC_r\cdot(n-r)^m, & n<m \\ n!, & n=m \\ 0, & n>m \end{cases}$
5.	Total number of one-one onto (bijective) mappings from A to B	$n!, n=m$
6.	Total number of into mappings from A to B	$\begin{cases} \sum_{r=0}^{n-1}(-1)^{r-1}\cdot {}^nC_r\cdot(n-r)^m, & n\le m \\ n^m, & n>m \end{cases}$
7.	Total number of constant mappings from A to B	n

Mixed Illustration on Mappings

Illustration 105. Let $A=\{x:-1\le x\le 1\}=B$. For each of the following functions from A to B, find whether it is injective, surjective or bijective.

(i) $f(x)=\dfrac{x}{2}$

(ii) $g(x)=x\,\text{sgn}\,x$

(iii) $h(x)=x\,|\,x\,|$

(iv) $i(x)=x^3$

(v) $J(x)=\lceil x\rceil, \lceil x\rceil\ge x$

(vi) $k(x)=\sin\pi x$

Sol : Graphical Approach :

(i) $f(x)=\dfrac{x}{2}$

Here any line parallel to x-axis cuts the graph at one point.

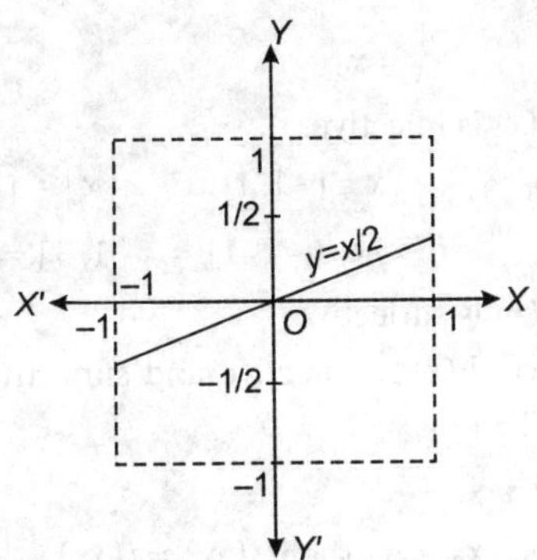

$\therefore f(x)$ is injective

and from graph, $R_f = \left[-\dfrac{1}{2}, \dfrac{1}{2}\right] \subset [-1, 1] = C_f$

i.e., $R_f \subset C_f$

$\therefore f$ is not surjective.

i.e., into mapping.

(ii) $g(x) = x \operatorname{sgn} x$

$$= x\begin{cases} 1, & x > 0 \\ -1, & x < 0 \\ 0, & x = 0 \end{cases} = \begin{cases} x, & x > 0 \\ -x, & x < 0 \\ 0, & x = 0 \end{cases} = |x|$$

Here any line parallel to x-axis cuts the graph at two points.

$\therefore g(x)$ is not injective

i.e., many one mapping and from graph,

$$R_g = [0, 1] \subset [-1, 1] = C_g$$

i.e., $R_g \subset C_g$

$\therefore g$ is not surjective

i.e., into mapping.

(iii) $h(x) = x\,|x| = \begin{cases} x^2, & x \geq 0 \\ -x^2, & x < 0 \end{cases}$

Here any line parallel to x-axis cuts the graph at one point.

$\therefore h(x)$ is injective.

and from graph, $R_h = [-1, 1] = C_h$

$\therefore h(x)$ is surjective.

Hence, $h(x)$ is bijective.

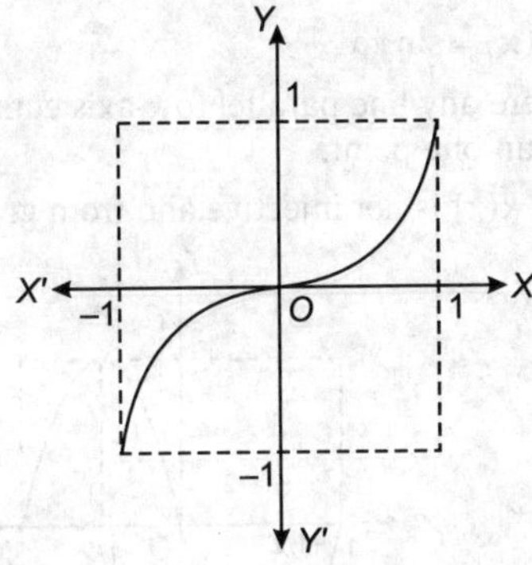

(iv) $i(x) = x^3$

Here any line parallel to x-axis cuts the graph at one point.

$\therefore i(x)$ is injective.

And from graph, $R_i = [-1, 1] = C_i$

$\therefore i(x)$ is surjective

Hence, $i(x)$ is bijective.

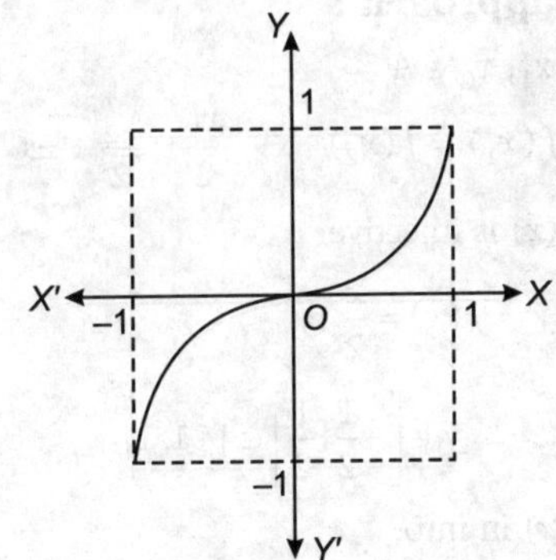

(v) $J(x) = \lceil x \rceil = \begin{cases} -1, & x = -1 \\ 0, & -1 < x \leq 0 \\ 1, & 0 < x \leq 1 \end{cases}$

Here any line parallel to x-axis cuts the graph at more than one point.

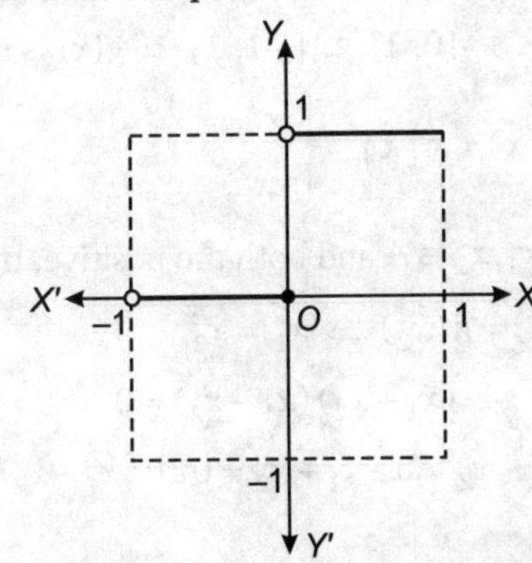

$\therefore J(x)$ is not injective and from graph,

$$R_j = \{-1, 0, 1\} \subset [-1, 1] = C_j$$

i.e., $R_j \subset C_j$

$\therefore J$ is not surjective *i.e.*, into mapping.

(vi) $k(x) = \sin \pi x$

Here any line parallel to x-axis cuts the graph more than one point.

$\therefore$ $k(x)$ is not injective and from graph,

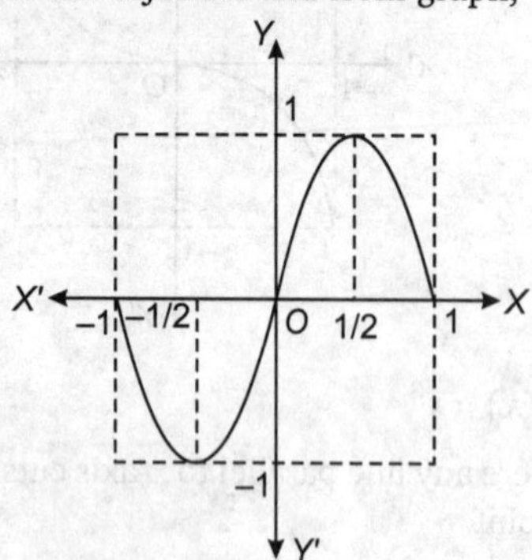

$$R_k = [-1, 1] = C_k$$

$\therefore$ k is surjective

i.e., onto mapping.

Analytical approach :

(i) Let $x_1, x_2 \in A$

$\therefore \quad f(x_1) = f(x_2) \Rightarrow \dfrac{x_1}{2} = \dfrac{x_2}{2} \Rightarrow x_1 = x_2$

$\therefore$ $f(x)$ is injective.

Let $y = f(x) = \dfrac{x}{2}$ $\qquad \because x \in [-1, 1]$

$\therefore \quad y \in \left[-\dfrac{1}{2}, \dfrac{1}{2}\right] \subset [-1, 1]$

$\therefore$ $f(x)$ in into.

(ii) Let $x_1, x_2 \in A$

$\therefore \quad g(x_1) = g(x_2) \Rightarrow x_1 \operatorname{sgn} x_1 = x_2 \operatorname{sgn} x_2$

$\Rightarrow \quad |x_1| = |x_2| \Rightarrow \quad x_1 = \pm x_2$

$\therefore \quad g(x)$ is many-one *i.e.*, not injective.

Also $g(x) = x \operatorname{sgn} x = |x| \in [0, 1]$

$\therefore \quad [0, 1] \subset [-1, 1] \Rightarrow g(x)$ is into.

(iii) $\because h(x) = x|x| = \begin{cases} x^2, & x \geq 0 \\ -x^2, & x < 0 \end{cases}$

Let $x_1, x_2 \in A$ and both are positive, then

$h(x_1) = h(x_2) \Rightarrow x_1^2 = x_2^2$

$\Rightarrow \quad (x_1 + x_2)(x_1 - x_2) = 0$

$\Rightarrow x_1 - x_2 = 0, \; x_1 + x_2 \neq 0 \;\; (\because x_1 + x_2 > 0)$

$\Rightarrow \quad x_1 = x_2$

and if x_1, x_2 both are negative, then

$h(x_1) = h(x_2) \Rightarrow -x_1^2 = -x_2^2 \Rightarrow x_1^2 - x_2^2 = 0$

$\Rightarrow \quad (x_1 + x_2)(x_1 - x_2) = 0$

$\Rightarrow x_1 - x_2 = 0, \; x_1 + x_2 \neq 0 \;\; (\because x_1 + x_2 < 0)$

$\Rightarrow \quad x_1 = x_2$

$\therefore$ $h(x)$ is injective.

Also, $\because \quad x \in [-1, 1] \;\; \therefore \;\; x|x| \in [-1, 1]$

or $\quad h(x) \in [-1, 1] = [-1, 1] = B$

$\therefore$ $h(x)$ is surjective.

Hence, $h(x)$ is injective and surjective *i.e.*, $h(x)$ is bijective.

(iv) $i(x) = x^3$

Let $x_1, x_2 \in A$, then $i(x_1) = i(x_2)$

$\Rightarrow \quad x_1^3 = x_2^3 \Rightarrow x_1^3 - x_2^3 = 0$

$\Rightarrow \quad (x_1 - x_2)(x - \omega x_2)(x_1 - \omega^2 x_2) = 0$

$\Rightarrow x_1 - x_2 = 0 \qquad (\because x_1 - \omega x_2 \neq 0, x_1 - \omega^2 x_2 \neq 0)$

$\Rightarrow \quad x_1 = x_2$

$\Rightarrow$ $i(x)$ is injective.

Now, $\because \;\; x \in [-1, 1]$

$\therefore \quad x^3 \in [-1, 1]$ or $i(x) \in [-1, 1] = B$

$\therefore$ $i(x)$ is surjective.

Hence, $i(x)$ is injective and surjective *i.e.*, $i(x)$ is bijective

(v) $J(x) = \lceil x \rceil$ $\qquad \because 0.8 \neq 0.9$

$\Rightarrow \quad \lceil 0.8 \rceil = \lceil 0.9 \rceil \Rightarrow 1 = 1$

$\therefore$ $f(x)$ is many one.

$$R_J = \{-1, 0, 1\} \subset [-1, 1] = B$$

$\Rightarrow R_J \subset B$

Hence, $J(x)$ is into.

(vi) $k(x) = \sin \pi x$

Let $x_1, x_2 \in A$, then $k(x_1) = k(x_2)$

$\Rightarrow \quad \sin \pi x_1 = \sin \pi x_2$

$\Rightarrow \quad \pi x_1 = n\pi + (-1)^n \pi x_2, \; n \in I$

$\Rightarrow \quad x_1 = n + (-1)^n x_2$

$\Rightarrow \quad x_1 \neq x_2$ for all $n \in I$

$\therefore$ $k(x)$ is many-one.

$\therefore \;\; \sin \pi x \in [-1, 1]$ or $k(x) \in [-1, 1] = B$

Hence, $k(x)$ is surjective.

Illustration 106. If $A = \{1, 2, 3\}$ and $B = \{a, b, c, d\}$ for $f : A \to B$, find the total number of :

(i) mappings

(ii) one-one mappings

(iii) many-one mappings

(iv) onto mappings

(v) into mappings

(vi) constant mappings

Sol : Here $m = 3$ and $n = 4 \; (n > m)$

(i) Total number of mappings $= n^m = 4^3 = 64$

(ii) Total number of one-one mappings

$= {}^nP_m = {}^4P_3 = 4\cdot3\cdot2 = 24$

(iii) Total number of many-one mappings

$= n^m - {}^nP_m = 64 - 24 = 40$

(iv) Total number of onto mappings = 0 (Here $n > m$)

(v) Total number of into mappings $= n^m = 4^3 = 64$ (Here $n > m$)

(vi) Total number of constant mappings $= n = 4$

Illustration 107. If $A = \{1, 2, 3, 4\}$ and $B = \{a, b, c\}$ for $f : A \to B$, find the total number of :

(i) mappings
(ii) one-one mappings
(iii) many-one mappings
(iv) onto mappings
(v) into mappings
(vi) constant mappings

Sol : Here $m = 4$ and $n = 3$ $(n < m)$

(i) Total number of mappings $= n^m = 3^4 = 81$

(ii) Total number of one-one mappings $= 0$ $(\because n < m)$

(iii) Total number of many-one mappings

$= n^m = 3^4 = 81 \qquad (\because n < m)$

(iv) Total number of onto mappings

$$= \sum_{r=0}^{2} (-1)^r \cdot {}^3C_r \cdot (3-r)^4$$

$$= {}^3C_0(3-0)^4 - {}^3C_1\cdot(3-1)^4 + {}^3C_2\cdot(3-2)^4$$

$$= 81 - 48 + 3 = 36$$

(v) Total number of into mappings

$$= \sum_{r=1}^{2} (-1)^{r-1} \cdot {}^3C_r \cdot (3-r)^4$$

$$= {}^3C_1\cdot(3-1)^4 - {}^3C_2\cdot(3-2)^4 = 48 - 3 = 45$$

(vi) Total number of constant mappings $= n = 3$

Composition of Functions

Let A, B and C be three non-empty sets. Let $f : A \to B$ and $g : B \to C$ be two functions, then $gof : A \to C$. This function is called the product or composite of f and g, given by

$$(gof)\, x = g\{f(x)\} \; \forall\, x \in A$$

gof : A→C

Thus the image of every $x \in A$ under the function gof is the g-image of the f-image of x.

Note :

1. The gof is defined only if $\forall\, x \in A$, $f(x)$ is an element of the domain of g so that we can take its g-image.
2. The range of f must be *a* **subset** of the domain of g in gof.
3. (i) $(fog)x = f\{g(x)\}$ (ii) $(fof)x = f\{f(x)\}$
   (iii) $(gog)x = g\{g(x)\}$
   (iv) $(fg)x = f(x)\cdot g(x)$
   (v) $(f \pm g)x = f(x) \pm g(x)$
   (vi) $\left(\dfrac{f}{g}\right)x = \dfrac{f(x)}{g(x)};\; g(x) \neq 0$

Properties of Composition of Functions

(1) The composition of functions is not commutative

i.e., $fog \neq gof$

(2) The composition of functions is associative *i.e.,* if f, g, h are three functions such that $(fog)\, oh$ and $fo\,(goh)$ exists, then $(fog)oh = fo(goh)$

Proof : Let $h : A \to B$, $g : B \to C$ and $f : C \to D$ be any three given functions. Then $fog : B \to D$ and $goh : A \to C$

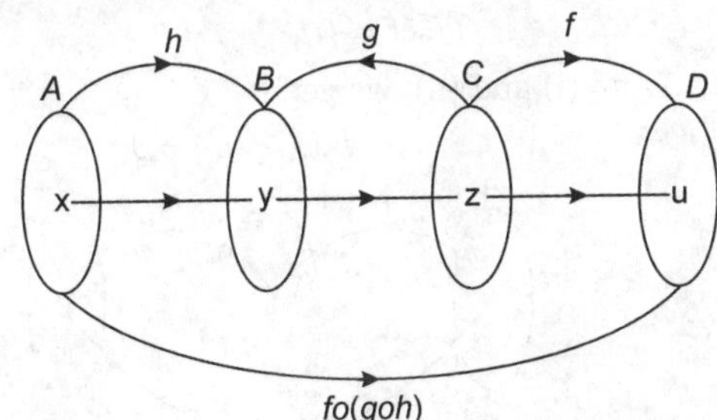

Therefore, $(fog)\, oh : A \to D$ and $fo\,(goh) : A \to D$

i.e., Domain of $\{(fog)\, oh\}$ = Domain of $\{fo\,(goh)\}$

further, $\{(fog)\, oh\}\,(x) = (fog)\,(h\,(x))$

$= (fog)\,(y) \qquad (\because h(x) = y)$

$= f(g(y))$

$= f(z) \qquad \ldots(i) \qquad (\because g(y) = z)$

and $\{fo\,(goh)\}\; x = f\{(goh)(x)\} = f\{g(h(x))\}$

$= f\{g(y)\} \qquad (\because h(x) = y)$

$= f(z) \qquad \ldots(ii) \qquad (\because g(y) = z)$

Therefore from (i) and (ii), we get

$$(fog)\, oh = fo\,(goh)$$

(3) Let $f : A \to B$, $g : B \to C$ be two functions, then

(i) f and g are injective $\Rightarrow$ gof is injective
(ii) f and g are surjective $\Rightarrow$ gof is surjective
(iii) f and g are bijective $\Rightarrow$ gof is bijective

(4) Let $f : A \to B$, $g : B \to C$ be two functions, then

(i) $gof : A \to C$ is injective $\Rightarrow f : A \to B$ is injective
(ii) $gof : A \to C$ is surjective $\Rightarrow g : B \to C$ is surjective

(iii) $gof : A \to C$ is injective and $g : B \to C$ is subjective $f : A \to B$ is injective

(iv) $gof : A \to C$ is surjective and $g : B \to C$ is injective $\Rightarrow f : A \to B$ is surjective

(5) An injective mapping from a finite set to itself is bijective.

(6) The composition of any function with the identity function is the function itself.

i.e., $f : A \to B$, then $foI_A = I_B of = f$

where I_A and I_B are the identity functions of A and B respectively.

Proof : Let $x \in A$ and $f(x) = y$, then

$$(foI_A)(x) = f(I_A(x)) = f(x)$$

$$\therefore \quad foI_A = f \quad \text{...(i)}$$

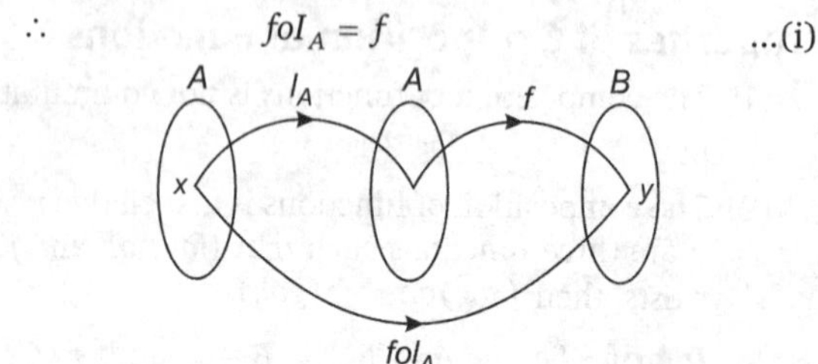

Again, $(I_B of)(x) = I_B(f(x)) = I_B(y) = y$

$= f(x) \quad (\because y = f(x))$

$$\therefore \quad I_B of = f \quad \text{...(ii)}$$

From (i) and (ii), we get

$$foI_A = I_B of = f$$

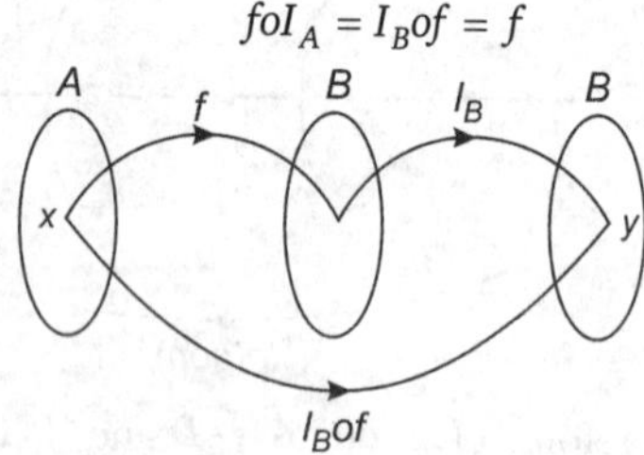

Illustration 108. If $f : R \to R$ and $g : R \to R$ be two mappings such that $f(x) = \sin x$ and $g(x) = x^2$, then :

(a) show that $fog \neq gof$,

(b) find the values of $(fog)\sqrt{\pi}/2$ and $(gof)\pi/3$.

Sol : (a) Let $x \in R$

$\therefore \quad (fog)x = f\{g(x)\} = f(x^2) \quad \{\because g(x) = x^2\}$

$= \sin x^2$

and $\quad (gof)x = g\{f(x)\} \quad \{\because f(x) = \sin x\}$ ...(i)

$= g\{\sin x\} \quad (\because f(x) = \sin x)$

$= \sin^2 x$ ...(ii)

From (i) and (ii), $(fog)x \neq (gof)x \; \forall \; x \in R$

Hence, $fog \neq gof$

(b) From (i), $(fog)x = \sin x^2$

$$\therefore \quad (fog)\frac{\sqrt{\pi}}{2} = \sin\frac{\pi}{4} = \frac{1}{\sqrt{2}}$$

and from (ii), $(gof)x = \sin^2 x$

$$\therefore \quad (gof)\pi/3 = \sin^2 \pi/3 = (\sqrt{3}/2)^2 = 3/4$$

Illustration 109. Let $f(x) = 1 + x^2$ and $f(g(x)) = 1 + x^2 - 2x^3 + x^4$, then find $g(x)$.

Sol : $\because \; f(g(x)) = 1 + x^2 - 2x^3 + x^4$

$\therefore \quad 1 + (g(x))^2 = 1 + x^2(1 - 2x + x^2) \; (\because f(x) = 1 + x^2)$

$= 1 + x^2(1 - x)^2 = 1 + \{x(1 - x)\}^2$

$\Rightarrow \quad (g(x))^2 = \{x(1 - x)\}^2$

$\Rightarrow \quad g(x) = \pm x(1 - x)$

Illustration 110. Let $f(x)$ and $g(x)$ be bijective functions where $f : (a, b, c, d) \to \{1, 2, 3, 4\}$ and $g : \{3, 4, 5, 6\} \to \{w, x, y, z\}$ respectively. Then find the number of elements in the range set of $g(f(x))$.

Sol : The range of $f(x)$ for which $g(f(x))$ is defined is $\{3, 4\}$. Hence, the domain of $g(f(x))$ has two elements.

$\therefore$ The range of $g(f(x))$ also has two elements.

Illustration 111. If the mappings f and g are given by $f = \{(1, 2), (3, 5), (4, 1)\}$ and $g = \{(2, 3), (5, 1), (1, 3)\}$ then write down pairs in the mappings fog and gof.

Sol : $\because \; f = \{(1, 2), (3, 5), (4, 1)\}$

$\therefore \quad f(1) = 2, f(3) = 5, f(4) = 1$ ...(i)

and $\quad g = \{(2, 3), (5, 1), (1, 3)\}$

$\therefore \quad g(2) = 3, g(5) = 1, g(1) = 3$ ...(ii)

Again $D_f = \{1, 3, 4\}$, $R_f = \{2, 5, 1\}$

and $\quad D_g = \{2, 5, 1\}$, $R_g = \{1, 3\}$

Then gof mapping is defined in the following way :

$$\{1,3,4\} \xrightarrow{f} \{2,5,1\} \xrightarrow{g} \{1,3\}$$

gof

$\therefore \quad (gof)(1) = g\{f(1)\} = g(2) = 3$

$(gof)(3) = g\{f(3)\} = g(5) = 1$

and $\quad (gof)(4) = g\{f(4)\} = g(1) = 3$

Hence, $gof = \{(1, 3), (3, 1), (4, 3)\}$

Now since range of $g \subset$ domain of f

$\therefore$ fog is defined.

The fog mapping is defined in the following way :

$$\{2,5,1\} \xrightarrow{g} \{1,3,4\} \xrightarrow{f} \{2,5,1\}$$

fog

$\therefore \quad (fog)(2) = f\{g(2)\} = f(3) = 5$

$(fog)(5) = f\{g(5)\} = f(1) = 2$

and $\quad (fog)(1) = f\{g(1)\} = f(3) = 5$

Hence, $\quad fog = \{(2, 5), (5, 2), (1, 5)\}$

Illustration 112. If $f(x) = \begin{cases} 2 + x, & \text{if } x \geq 0 \\ 2 - x, & \text{if } x < 0 \end{cases}$, then find $(fof)(x)$.

Sol : Given : $f(x) = \begin{cases} 2 + x, & \text{if } x \geq 0 \\ 2 - x, & \text{if } x < 0 \end{cases} = 2 + |x|$

$\therefore\ (fof)(x) = f(f(x)) = f(2+|x|) = 2+|2+|x||$

$$= 2+2+|x| = 4+|x|$$

$$= \begin{cases} 4+x, & \text{if } x \ge 0 \\ 4-x, & \text{if } x < 0 \end{cases}$$

Illustration 113. Let $f(x) = \begin{cases} |x+1|; & x < 1 \\ 1-x; & x \ge 1 \end{cases}$ and $g(x) = \begin{cases} x-2; & x < 0 \\ x+3; & x \ge 0 \end{cases}$ then find $(f+g)x$ and $(fg)x$.

Sol : We have

$$f(x) = \begin{cases} -x-1; & x < -1 \\ x+1; & -1 \le x < 1 \\ 1-x; & x \ge 1 \end{cases}$$

and $$g(x) = \begin{cases} x-2; & x < 0 \\ x+3; & x \ge 0 \end{cases}$$

Making the same domain of f and g

$$f(x) = \begin{cases} -x-1; & x < -1 \\ x+1; & -1 \le x < 0 \\ x+1; & 0 \le x < 1 \\ 1-x; & x \ge 1 \end{cases}$$

and $$g(x) = \begin{cases} x-2; & x < -1 \\ x-2; & -1 \le x < 0 \\ x+3; & 0 \le x < 1 \\ x+3; & x \ge 1 \end{cases}$$

Now, $$(f+g)x = \begin{cases} -x-1+x-2 = -3; & x < -1 \\ x+1+x-2 = 2x-1; & -1 \le x < 0 \\ x+1+x+3 = 2x+4; & 0 \le x < 1 \\ 1-x+x+3 = 4; & x \ge 1 \end{cases}$$

and $(fg)x$

$$= \begin{cases} (-x-1)(x-2) = -x^2+x+2; & x < -1 \\ (x+1)(x-2) = x^2-x-2; & -1 \le x < 0 \\ (x+1)(x+3) = x^2+4x+3; & 0 \le x < 1 \\ (1-x)(x+3) = -x^2-2x+3; & x \ge 1 \end{cases}$$

Illustration 114. Let two functions are defined as :

$$f(x) = \begin{cases} x+1, & x \le 1 \\ 2x+1, & 1 < x \le 2 \end{cases}$$

and $$g(x) = \begin{cases} x^2, & -1 \le x < 2 \\ x+2, & 2 \le x \le 3 \end{cases}$$

then find *fog* and *gof*.

Sol : Analytical Approach : We have

$$(fog)(x) = f(g(x)) = \begin{cases} g(x)+1, & g(x) \le 1 \\ 2g(x)+1, & 1 < g(x) \le 2 \end{cases}$$

$$= \begin{cases} x^2+1, & x^2 \le 1 \text{ and } -1 \le x < 2 \\ 2x^2+1, & 1 < x^2 \le 2 \text{ and } -1 \le x < 2 \\ (x+2)+1, & x+2 \le 1 \text{ and } 2 \le x \le 3 \\ 2(x+2)+2, & 2 \le x+2 \le 3 \text{ and } 2 \le x \le 3 \end{cases}$$

$$= \begin{cases} x^2+1, & -1 \le x \le 1 \text{ and } -1 \le x < 2 \\ 2x^2+1, & -\sqrt{2} \le x < -1 \text{ and } 1 < x \le \sqrt{2} \\ & \text{and } -1 \le x < 2 \\ x+3, & x \le -1 \text{ and } 2 \le x \le 3 \\ 2x+6, & 0 \le x \le 1 \text{ and } 2 \le x \le 3 \end{cases}$$

$$= \begin{cases} x^2+1, & -1 \le x \le 1 \\ 2x^2+1, & 1 < x \le \sqrt{2} \\ x+3, & x \in \phi \\ 2x+6, & x \in \phi \end{cases} = \begin{cases} x^2+1, & -1 \le x \le 1 \\ 2x^2+1, & 1 < x \le \sqrt{2} \end{cases}$$

$$(gof)(x) = g(f(x)) = \begin{cases} f^2(x), & -1 \le f(x) < 2 \\ f(x)+2, & 2 \le f(x) \le 3 \end{cases}$$

$$= \begin{cases} (x+1)^2, & -1 \le x+1 < 2 \text{ and } x \le 1 \\ (x+1)+2, & 2 \le x+1 \le 3 \text{ and } x \le 1 \\ (2x+1)^2, & -2 \le 2x+1 < 2 \text{ and } 1 < x \le 2 \\ (2x+1)+2, & 2 \le 2x+1 \le 3 \text{ and } 1 < x \le 2 \end{cases}$$

$$= \begin{cases} (x+1)^2, & -2 \le x < 1 \text{ and } x \le 1 \\ x+3, & -3 \le x \le 2 \text{ and } x \le 1 \\ (2x+1)^2, & -1 \le x < \frac{1}{2} \text{ and } 1 < x \le 2 \\ 2x+3, & \frac{1}{2} \le x \le 1 \text{ and } 1 < x \le 2 \end{cases}$$

$$= \begin{cases} (x+1)^2, & -2 \le x < 1 \\ x+3, & -3 \le x \le 1 \\ (2x+1)^2, & x \in \phi \\ 2x+3, & x \in \phi \end{cases} = \begin{cases} (x+1)^2, & -2 \le x < 1 \\ x+3, & x = 1 \end{cases}$$

$$= (x+1)^2, -2 \le x \le 1$$

Graphical Approach :

We have $f(g(x)) = \begin{cases} g(x)+1, & g(x) \le 1 \\ 2g(x)+1, & 1 < g(x) \le 2 \end{cases}$

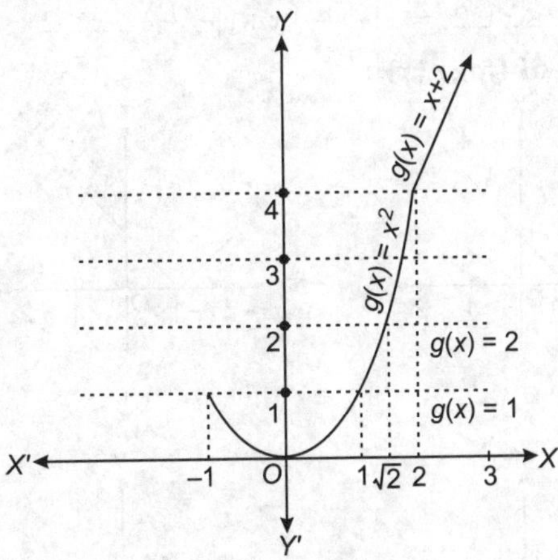

It is clear that $g(x) \le 1\ \forall\ x \in [-1, 1]$

and $\quad 1 < g(x) \le 2 \quad \forall\ x \in (1, \sqrt{2}]$

$$\Rightarrow \quad (fog)(x) = f(g(x)) = \begin{cases} x^2+1, & -1 \le x \le 1 \\ 2x^2+1, & 1 < x \le \sqrt{2} \end{cases}$$

and we have

$$(gof)(x) = g(f(x)) = \begin{cases} (f(x))^2, & -1 \le f(x) < 2 \\ f(x)+2, & 2 \le f(x) \le 3 \end{cases}$$

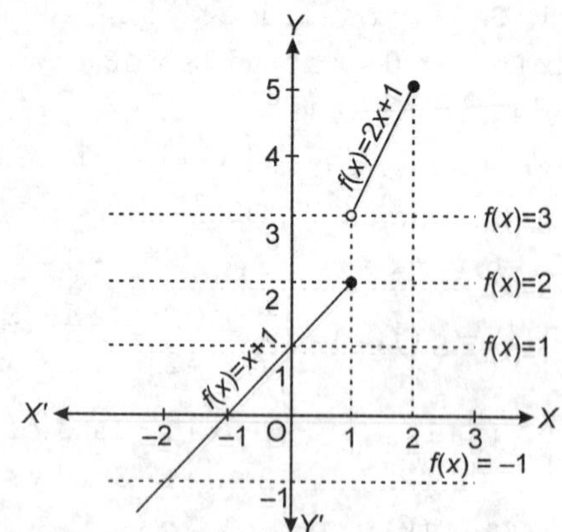

It is clear that

$$-1 \le f(x) < 2 \quad \forall\, x \in [-2, 1)$$

and $\quad 2 \le f(x) \le 3 \quad \forall\, x \in \{1\}$

$$\therefore \quad (gof)(x) = g(f(x)) = \begin{cases} (x+1)^2, & -2 \le x < 1 \\ x+3, & x = 1 \end{cases}$$

$$= (x+1)^2, -2 \le x \le 1$$

Illustration 115. If $f(x) = -1 + |x-1|;\ -1 \le x \le 3$ and $g(x) = 2 - |x+1|;\ -2 \le x \le 2$, then calculate $(fog)(x)$ and $(gof)(x)$ and also draw their graphs.

Sol : $(fog)(x) = f\{g(x)\}$

$= f\{2 - |x+1|\};\ -2 \le x \le 2$

$= -1 + |2 - |x+1| - 1|;\ -2 \le x \le 2$

and $-1 \le 2 - |x+1| \le 3$

$= -1 + |1 - |x+1||;\ -2 \le x \le 2$

and $-4 \le x \le 2$ $= -1 + |1 - |x+1||;\ -2 \le x \le 2$

Graph of (*fog*) (*x*) :

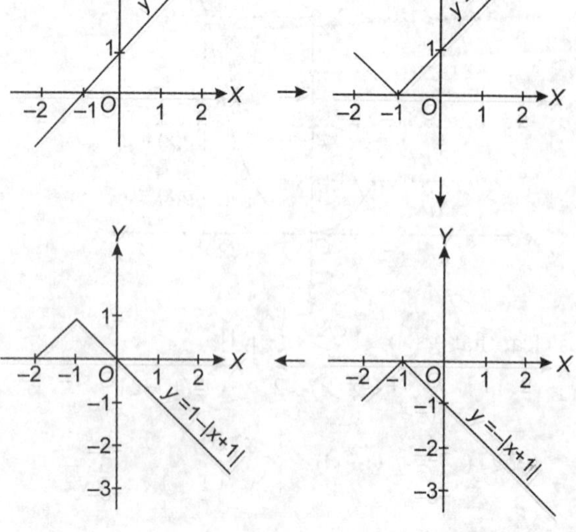

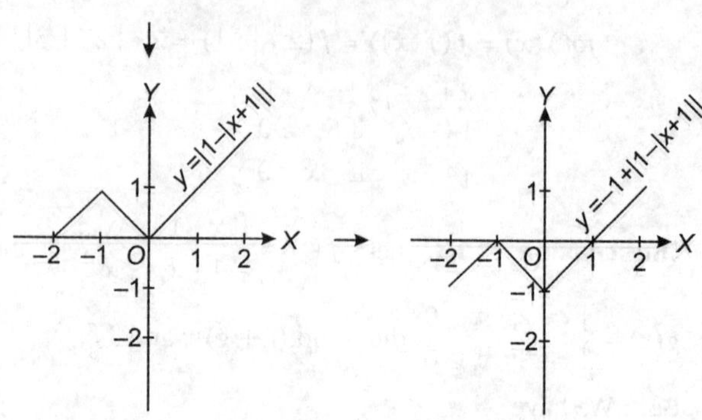

and $(gof)(x) = g(f(x))$

$= g\{-1 + |x-1|\};\ -1 \le x \le 3$

$= 2 - |-1 + |x-1| + 11;\ -1 \le x \le 3$

and $-2 \le -1 + |x-1| \le 2$

$= 2 - |x-1|;\ -1 \le x \le 3$

and $-2 \le x \le 4$

$= 2 - |x-1|;\ -1 \le x \le 3$

Graph of (*gof*) (*x*) :

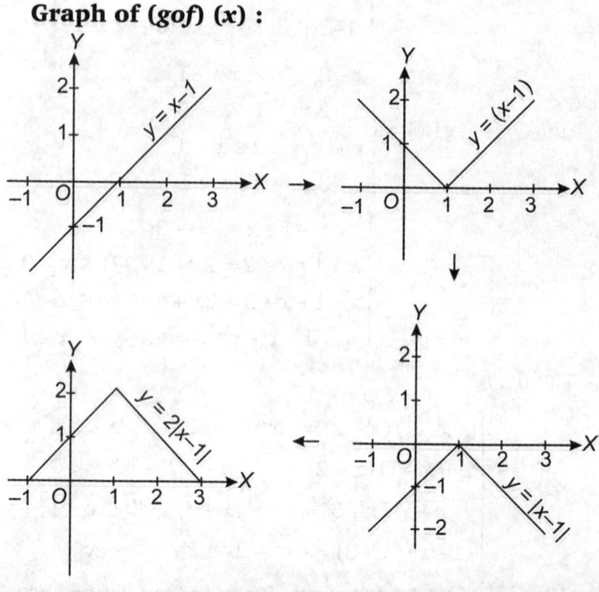

INTRODUCTORY EXERCISE 2.7

Only One Option is Correct

1. Let function $f : R \to R$ be defined by $f(x) = 2x + \sin x \ \forall\, x \in R$, then f is :

(a) one-one onto
(b) one-one, but not onto
(c) onto but not one-one
(d) neither one-one not onto

2. The function $f : R \to R$ defined by $f(x) = 6^x + 6^{|x|}$, then f is :

(a) one-one onto (b) one-one into
(c) many-one onto (d) many-one into

3. The function $f : R \to R$ defined by $f(x) = \dfrac{x^2 - 4}{x^2 + 1}$, then f is :
(a) one-one onto (b) one-one into
(c) many-one onto (d) many-one into

4. Let $f : R \to R$ be a function such that $f(x) = \dfrac{x^2 + 2x + 1}{x^2 + 1}$. Then **[JEE Main 29 Jan 2023 (MS)]**
(a) $f(x)$ is many one in $(-\infty, -1)$
(b) $f(x)$ is many one in $(1, \infty)$
(c) $f(x)$ is one-one in $[1, \infty)$ but not in $(-\infty, \infty)$
(d) $f(x)$ is one-one in $(-\infty, \infty)$

5. The function $f : N \to N$ defined by $f(x) = x - (-1)^x$, then f is :
(a) one-one onto (b) one-one into
(c) many-one onto (d) many-one into

6. Function $f(x) = ax^3 + bx^2 + cx + d$ $(a \neq 0)$ is one-one onto function if :
(a) $b^2 - 4ac < 0$ (b) $b^2 - 3ac < 0$
(c) $b^2 - 4ac > 0$ (d) $b^2 - 3ac > 0$

7. If $f : R \to R$ is a function such that $f(x) = x^3 + 3x^2 + 5x + \cos x$, then f is :
(a) one-one onto (b) one-one into
(c) many-one onto (d) many-one into

8. If $f : R \to R$ is a function such that $f(x) = x^3 + 3x^2 + 12x - 2\sin x$, then f is :
(a) one-one onto (b) one-one into
(c) many-one onto (d) many-one into

9. If the function $f : R - \{1, -1\} \to A$ defined by $f(x) = \dfrac{x^2}{1 - x^2}$ is surjective, then A is equal to : **[JEE Main 9 Apr 2019 (MS)]**
(a) $[0, \infty)$ (b) $R - \{-1\}$
(c) $R - [-1, 0)$ (d) $R - (-1, 0)$

10. If $f : R \to R$, where $f(x) = ax + \cos x$ and $f(x)$ is bijective, then :
(a) $a \in R$ (b) $a \in R^+$
(c) $a \in R^-$ (d) $a \in R - (-1, 1)$

11. Let $f : R \to R$ be a function such that $f(x) = x^3 - 6x^2 + 11x - 6$, $f(x)$ is :
(a) one-one onto (b) one-one into
(c) many-one onto (d) many-one into

12. The total number of functions $f : \{1, 2, 3, 4\} \to \{1, 2, 3, 4, 5, 6\}$ such that $f(1) + f(2) = f(3)$, is equal to **[JEE Main 25 July 2022 (MS)]**
(a) 60 (b) 90 (c) 108 (d) 126

13. Let $A = \{a_1, a_2, a_3, a_4, a_5, a_6)$ and $B = \{b_1, b_2, b_3\}$. The number of functions of $f : A \to B$ such that it is onto and there are exactly three elements in A such that $f(A) = b_1$ is :
(a) 75 (b) 90 (c) 100 (d) 120

14. Let $A = \{a_1, a_2, a_3, a_4, a_5\}$ and $B = \{b_1, b_2, b_3, b_4, b_5\}$, then the number of one-one functions of $f : A \to B$ such that $f(a_i) \neq b_i$, $i = 1, 2, 3, 4, 5$ is :
(a) 40 (b) 44 (c) 24 (d) 60

15. The number of bijective functions of $f : A \to A$, where $A = \{1, 2, 3\}$ such that
$f(1) \neq 3,$, $f(2) \neq 1$, $f(3) \neq 2$ is :
(a) 1 (b) 2
(c) 9 (d) None of these

16. Suppose that $g(x) = 1 + \sqrt{x}$ and $f(g(x)) = 3 + 2\sqrt{x} + x$, then $f(x)$ is :
(a) $1 + 2x^2$ (b) $2 + x^2$ (c) $1 + x$ (d) $2 + x$

17. If $g(x) = x^2 + x - 2$ and $(gof)(x) = 4x^2 - 10x + 4$, then $f(x)$ is :
(a) $2x - 3$ (b) $2x + 3$
(c) $2x^2 + 3x + 1$ (d) $2x^2 - 3x - 1$

18. For $x \in R$, two real valued functions $f(x)$ and $g(x)$ are such that, $g(x) = \sqrt{x} + 1$ and $(fog)(x) = x + 3 - \sqrt{x}$. Then $f(0)$ is equal to : **[JEE Main 13 Apr 2023 (MS)]**
(a) 1 (b) 5 (c) 0 (d) – 3

19. Let $f(x) = \dfrac{x - 1}{x + 1}$, $x \in R - \{0, -1, 1\}$. If $f^{n+1}(x) = f(f^n(x))$ for all $n \in N$, then $f^6(6) + f^7(7)$ is equal to : **[JEE Main 26 Jun 2022 (MS)]**
(a) 7/6 (b) – 3/2 (c) 7/12 (d) –11/12

20. Let $f : R \to R$, $g : R \to R$ be two given functions such that f is injective and g is surjective, then which of the following is injective?
(a) gof (b) fog
(c) gog (d) None of these

21. If f and g are one-one functions, then :
(a) $f + g$ is one-one (b) fg is one-one
(c) fog is one-one (d) None of these

22. If $f(x) = \dfrac{1}{x}$, $g(x) = \dfrac{1}{x^2}$ and $h(x) = x^2$, then :
(a) $(fog)(x) = x^2$, $x \,^1\, 0$, $h(g(x)) = \dfrac{1}{x^2}$
(b) $h(g(x)) = \dfrac{1}{x^2}$, $x \neq 0$, $(fog)(x) = x^2$
(c) $(fog)(x) = \dfrac{1}{x^2}$, $x \neq 0$, $h(g(x)) = (g(x))^2$, $x \neq 0$
(d) None of the above

23. For $x \in \left(0, \frac{3}{2}\right)$, let $f(x) = \sqrt{x}$, $g(x) = \tan x$ and $h(x) = \frac{1-x^2}{1+x^2}$. If $\phi(x) = (hofog)(x)$, then $\phi\left(\frac{\pi}{3}\right)$ is equal to : **[JEE Main 12 Apr 2019 (MS)]**

(a) $\tan\left(\frac{\pi}{12}\right)$ (b) $\tan\left(\frac{5\pi}{12}\right)$

(c) $\tan\left(\frac{7\pi}{12}\right)$ (d) $\tan\left(\frac{11\pi}{12}\right)$

24. If $f(g(x)) = |\sin x|$ and $g(f(x)) = \sin^2 \sqrt{x}$, then :

(a) $f(x) = \sqrt{x}, g(x) = \sin^2 x$

(b) $f(x) = \sin\sqrt{x}, g(x) = x^2$

(c) $f(x) = \sqrt{\sin x}, g(x) = x^2$

(d) $f(x) = |x|, g(x) = \sin^2 x$

25. Let $f(x) = \log_e (\sin x)$, $0 < x < \pi$ and $g(x) = \sin^{-1} (e^{-x})$, $(x \geq 0)$. If α is a positive real number such that $a = (fog)'(\alpha)$ and $b = (fog)(\alpha)$, then : **[JEE Main 10 Apr 2019 (ES)]**

(a) $a\alpha^2 + b\alpha + a = 0$ (b) $a\alpha^2 + b\alpha - a = -2\alpha$

(c) $a\alpha^2 - b\alpha - a = 0$ (d) $a\alpha^2 - b\alpha - a = 1$

26. If $f(x) = \begin{cases} x & \text{when } x \text{ is rational} \\ 1-x & \text{when } x \text{ is irrational} \end{cases}$, then $(fof)(x)$ is given as :

(a) 1 (b) x

(c) $1 + x$ (d) None of these

27. If $f(x) = \begin{cases} x^2, & x \geq 0 \\ x, & x < 0 \end{cases}$, then $(fof)(x)$ is given by :

(a) $\begin{cases} x^2, & x \geq 0 \\ x, & x < 0 \end{cases}$ (b) $\begin{cases} x^4, & x \geq 0 \\ x^2, & x < 0 \end{cases}$

(c) $\begin{cases} x^4, & x \geq 0 \\ -x^2, & x < 0 \end{cases}$ (d) $\begin{cases} x^4, & x \geq 0 \\ x, & x < 0 \end{cases}$

Even and Odd Functions

Even Function :

A function $y = f(x)$ is said to be an even function if

$$f(-x) = f(x) \ \forall \ x \in D_f.$$

e.g., $f(x) = \cos 2x$ is even because

$$f(-x) = \cos(-2x) = \cos 2x = f(x)$$

The graph of an even function $y = f(x)$ is **symmetrical about the *y*-axis**. *i.e.*, if point (x, y) lies on the graph, then $(-x, y)$ will also lie on the graph or we say that if any point lies on the graph, then its image about y-axis also lies on the graph *e.g.*,

(i)

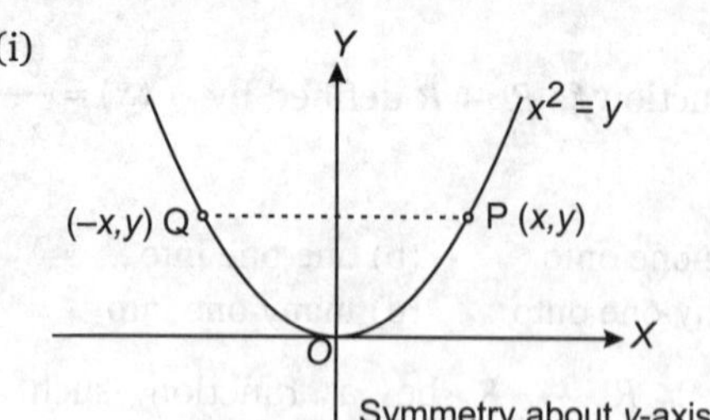

Symmetry about y-axis

(ii)

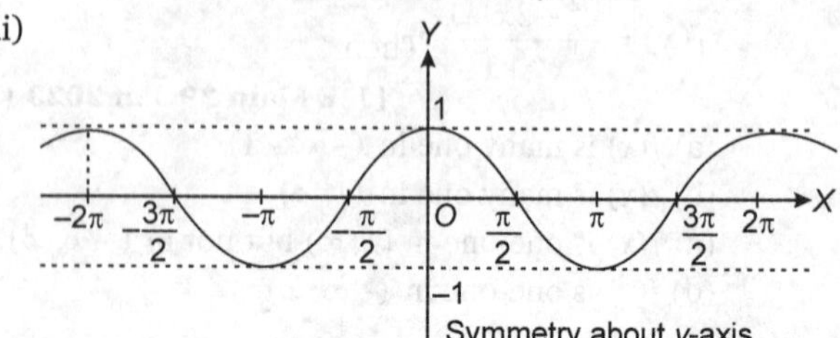

Symmetry about y-axis

(iii)

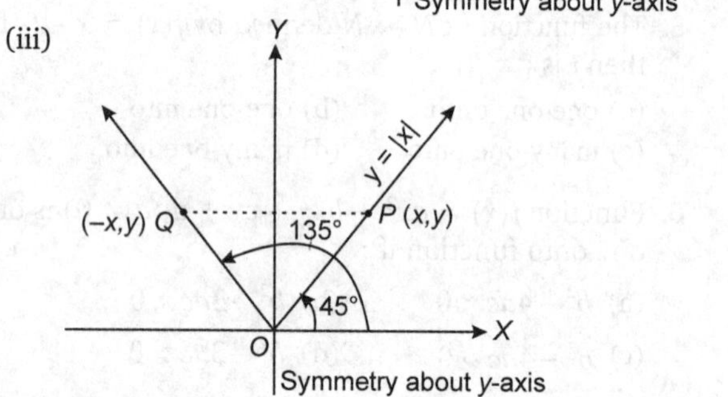

Symmetry about y-axis

Odd Function :

A function $y = f(x)$ is said to be an odd function if

$$f(-x) = -f(x) \ \forall \ x \in D_f$$

e.g., $f(x) = x^3 + x$ is odd because $f(-x) = -x^3 - x = -f(x)$

The graph of an odd function $y = f(x)$ is ***symmetrical in opposite quadrants***. *i.e.*, if point (x, y) lies on the graph, then $(-x, -y)$ will also lies on the graph or we say that if any point lies on the graph, then its image about origin also lies on the graph. *e.g.*,

(i)

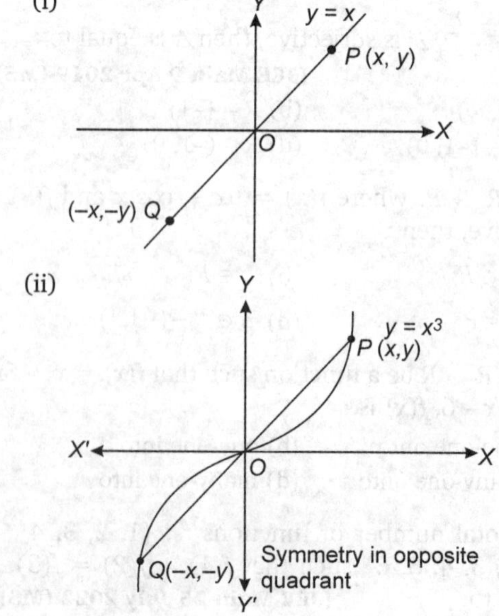

(ii)

Symmetry in opposite quadrant

(iii)

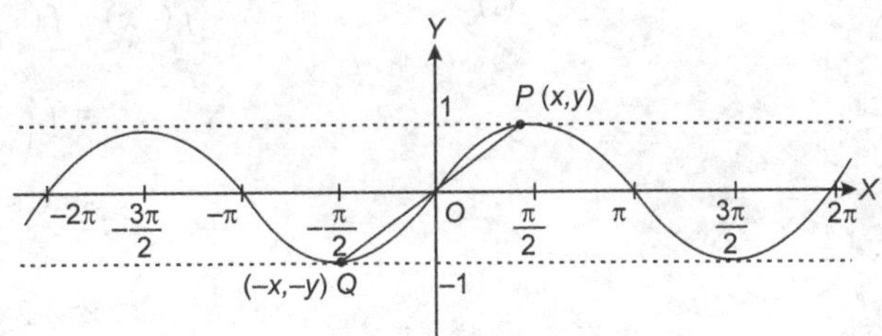

Symmetry in opposite quadrant

Note :

1. Inverse of an even function is not defined.
2. Every function can be expressed as the sum of an even and an odd function.
   i.e.,
   $$f(x)=\frac{1}{2}\{f(x)+f(-x)\}+\frac{1}{2}\{f(x)-f(-x)\}=\{\text{Even function}\}+\{\text{Odd function}\}$$
3. If $f(x) - f(-x) = 0$, then $f(x)$ is an even function and if $f(x) + f(-x) = 0$, then $f(x)$ is an odd function.
4. It is not essential that every function is even or odd. It is possible to have same function which are neither even nor odd function.
5. $f(x) = 0$ the only function which is defined on the entire number line is even and odd at the same time.
6. Every odd continuous function passes through origin.
7. Every even function $y = f(x)$ is many-one $\forall x \in D_f$.
8. The first derivative of an even function is an odd function and vice-versa.
9. The square of an even or an odd function is always an even function.
10. If $x = 0 \in D_f$, then for odd function $f(x)$ which is continuous at $x = 0, f(0) = 0$, *i.e.*, if for a function $f(0) \neq 0$, then that function can not be odd. It follows that for a differentiable even function $f'(0) = 0$, *i.e.*, if for a differentiable function $f'(0) \neq 0$, then the function $f(x)$ can not be even.

Table of Two Functions which are Attached :

$f(x)$	$g(x)$	$f(x) + g(x)$	$f(x) - g(x)$	$f(x)\,g(x)$	$f(x)/g(x)$	$(fog)x$
Even	Even	Even	Even	Even	Even	Even
Even	Odd	Neither even nor odd	Neither even nor odd	Odd	Odd	Even
Odd	Even	Neither even nor odd	Neither even nor odd	Odd	Odd	Even
Odd	Odd	Odd	Odd	Even	Even	Odd

Illustration 116. Prove that $f(x)=\ln(x+\sqrt{(1+x^2)})$ is an odd function.

Sol : $\because\ f(x)=\ln(x+\sqrt{(1+x^2)})$

$\therefore\quad f(-x)=\ln(-x+\sqrt{1+(-x)^2})$

$=\ln(-x+\sqrt{1+x^2})$

Now $f(x)+f(-x)$

$=\ln(\sqrt{(1+x^2)}+x)+\ln(\sqrt{(1+x^2)}-x)$

$=\ln(1+x^2-x^2)=\ln 1=0$

$\Rightarrow\quad f(-x)=-f(x)$

Hence, $f(x)$ is an odd function.

Alternative Method :

$\because\quad f(x)=\ln(x+\sqrt{1+x^2})$

$\therefore\quad f'(x)=\frac{1}{\sqrt{(1+x^2)}}=f'(-x)$

$$\left\{\because \int\frac{dx}{\sqrt{(1+x^2)}}=\ln\{x+\sqrt{1+x^2}\}\right\}$$

$\Rightarrow\quad f'(x)$ is an even function.

Hence, $f(x)$ is an odd function.

Illustration 117. If $f(t)=\frac{t}{e^t-1}+\frac{t}{2}+1$, show that $f(t)$ is an even function.

Sol : Since $f(t)=\dfrac{t}{e^t-1}+\dfrac{t}{2}+1$ ...(i)

$\therefore \quad f(-t)=\dfrac{-t}{e^{-t}-1}-\dfrac{t}{2}+1$

$=\dfrac{te^t}{e^t-1}-\dfrac{t}{2}+1$ ...(ii)

Subtracting (ii) from (i), we get

$$f(t)-f(-t)=\frac{t}{(e^t-1)}(1-e^t)+t=-t+t=0$$

$\Rightarrow \quad f(-t)=f(t)$

Hence $f(t)$ is an even function.

Illustration 118. Prove that $f(x)$ given by $f(x+y)=f(x)+f(y)\ \forall\ x,y\in R$ is an odd function.

Sol : Given $f(x+y)=f(x)+f(y)$ ...(i)

Replace y by $-x$, we have

$f(0)=f(x)+f(-x)$ ...(ii)

Now put $x=y=0$ in (i), we have

$$f(0)=f(0)+f(0)\Rightarrow f(0)=0$$

Then, from (ii), we get

$$0=f(x)+f(-x)\Rightarrow f(-x)=-f(x)$$

Hence, $f(x)$ is an odd function.

Illustration 119. Express ae^x as the sum of an even and an odd function.

Sol : Let $f(x)=ae^x$, then $f(-x)=ae^{-x}$

Now $f(x)=\dfrac{1}{2}\{f(x)+f(-x)\}+\dfrac{1}{2}\{f(x)-f(-x)\}$

$=\dfrac{1}{2}(ae^x+ae^{-x})+\dfrac{1}{2}(ae^x-ae^{-x})=$

(Even function) + (odd function)

Illustration 120. Prove that $x^2-\text{arc cos } y=\pi$ is an explicit function of x. Also find the domain of definition of the explicit function.

Sol : Since $x^2-\text{arc cos } y=\pi$

$\Rightarrow \quad x^2-\cos^{-1}y=\pi$ or $\cos^{-1}y=x^2-\pi$

$\Rightarrow \quad y=\cos(x^2-\pi)=\cos(\pi-x^2)\Rightarrow y=-\cos x^2$

Hence y is an explicit function of x.

For domain of explicit function $0\le\cos^{-1}y\le\pi$

$\Rightarrow \quad 0\le x^2-\pi\le\pi\Rightarrow\pi\le x^2\le 2\pi$

or $\quad \sqrt{\pi}\le|x|\le\sqrt{2\pi}$

Hence $x\in[-\sqrt{2\pi},-\sqrt{\pi}]\cup[\sqrt{\pi},\sqrt{2\pi}]$.

Illustration 121. If 'f' is an even function defined on (–5, 5) then find the four real values of x satisfying the equation $f(x)=f\left(\dfrac{x+1}{x+2}\right)$.

Sol : Since $f(x)$ is an even function

$\therefore \quad f(-x)=f(x)$ ...(i)

$=f\left(\dfrac{x+1}{x+2}\right)\quad\left(\because f(x)=f\left(\dfrac{x+1}{x+2}\right)\right)$

$\Rightarrow \quad -x=\dfrac{x+1}{x+2}\Rightarrow -x^2-2x=x+1$

or $\quad x^2+3x+1=0$

$\therefore \quad x=\dfrac{-3\pm\sqrt{9-4}}{2}=\dfrac{-3\pm\sqrt{5}}{2}$

and from (i),

$f(-x)=f(x)$

$\Rightarrow \quad f\left(\dfrac{-x+1}{-x+2}\right)=f(x)\quad\left(\because f(x)=f\left(\dfrac{x+1}{x+2}\right)\right)$

$\Rightarrow \quad \dfrac{-x+1}{-x+2}=x$

$\Rightarrow \quad -x+1=-x^2+2x$

$\Rightarrow \quad x^2-3x+1=0$

$\therefore \quad x=\dfrac{3\pm\sqrt{9-4}}{2}=\dfrac{3\pm\sqrt{5}}{2}$

Hence four real values of x are

$$\frac{-3\pm\sqrt{5}}{2},\ \frac{3\pm\sqrt{5}}{2}$$

Extension of a Function

Let a function $f(x)$ is defined on the interval $[0, a]$, it can be extend on $[-a, a]$, so that $f(x)$ is either even or odd function on the interval $[-a, a]$.

(i) Even Extension : As $f(x)$ is defined on the interval $[0, a]$, then

$$0\le x\le a\Rightarrow -a\le -x\le 0$$

$\therefore \quad -x\in[-a,0]$

We define $f(x)$ in the interval $[-a, 0]$ such that $f(x)=f(-x)$. Let $F_e(x)$ be the even extension, then

$$F_e(x)=\begin{cases}f(x); & x\in[0,a]\\ f(-x); & x\in[-a,0]\end{cases}$$

Graphically : If the graph of $y=f(x)$, $x\in[0,a]$ is as follows :

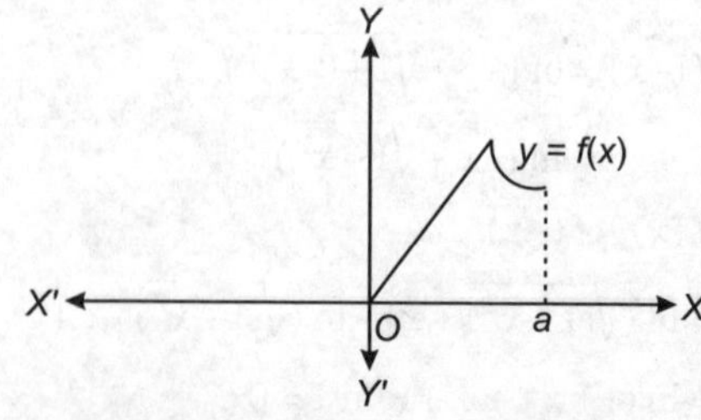

Even extension function of $y=f(x)$ is obtained as a function which is symmetrical about y-axis is as follows :

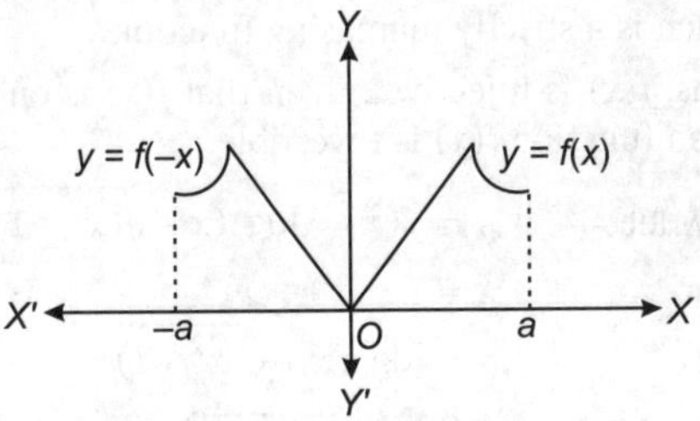

(ii) Odd Extension : As $f(x)$ is defined on the interval $[0, a]$, then $0 \le x \le a$

$\Rightarrow \quad -a \le -x \le 0$

$\therefore \quad -x \in [-a, 0]$

we define $f(x)$ in the interval $[-a,0]$ such that $f(x) = -f(-x)$. Let $F_o(x)$ be the odd extension, then

$$F_o(x) = \begin{cases} f(x); & x \in [0, a] \\ -f(-x); & x \in [-a, 0] \end{cases}$$

Graphically : Odd extension function of $y = f(x)$ is obtained as a function which is symmetrical about origin is as follows :

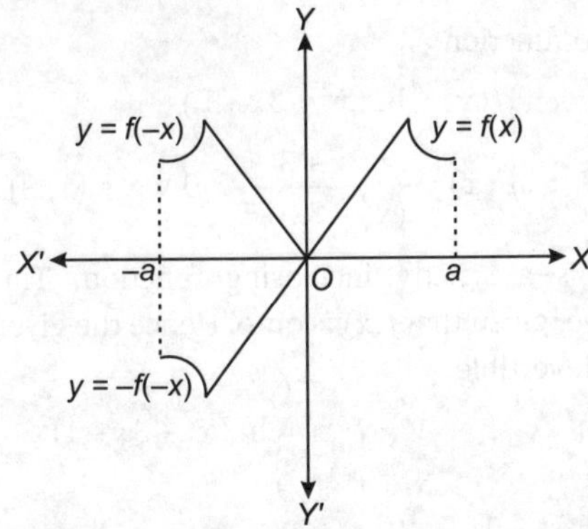

Illustration 122. Let $f(x) = e^x + \sin x$ be defined on the interval [–4, 0]. Find the odd and even extensions of $f(x)$ in the interval [– 4, 4].

Sol : Odd Extension : Let $F_o(x)$ be the odd extension of $f(x)$, then

$$F_o(x) = \begin{cases} f(x); & x \in [-4, 0] \\ -f(-x); & x \in [0, 4] \end{cases}$$

$$= \begin{cases} e^x + \sin x; & x \in [-4, 0] \\ -e^{-x} + \sin x; & x \in [0, 4] \end{cases}$$

Even Extension : Let $F_e(x)$ be the even extension of $f(x)$, then

$$F_e(x) = \begin{cases} f(x); & x \in [-4, 0] \\ f(-x); & x \in [0, 4] \end{cases}$$

$$= \begin{cases} e^x + \sin x; & x \in [-4, 0] \\ e^{-x} - \sin x; & x \in [0, 4] \end{cases}$$

Illustration 123. Let the function

$$f(x) = x^2 + x + \sin x - \cos x + \ln(1 + |x|)$$

be defined over the interval [0, 1]. Find the odd and even extensions of $f(x)$ in the interval [– 1, 1].

Sol : Odd Extension : Let $F_o(x)$ be the odd extension of $f(x)$, then

$$F_o(x) = \begin{cases} f(x); & x \in [0, 1] \\ -f(-x); & x \in [-1, 0] \end{cases}$$

$$= \begin{cases} x^2 + x + \sin x - \cos x + \ln(1 + |x|), & x \in [0, 1] \\ -x^2 + x + \sin x + \cos x - \ln(1 + |x|), & x \in [-1, 0] \end{cases}$$

Even Extension : Let $F_e(x)$ be the even extension of $f(x)$, then

$$F_e(x) = \begin{cases} f(x); & x \in [0, 1] \\ f(-x); & x \in [-1, 0] \end{cases}$$

$$= \begin{cases} x^2 + x + \sin x - \cos x + \ln(1 + |x|), & x \in [0, 1] \\ x^2 - x - \sin x - \cos x + \ln(1 + |x|), & x \in [-1, 0] \end{cases}$$

Illustration 124. Let $f(x) = \begin{cases} x^3 + x^2; & 0 \le x \le 2 \\ x + 2; & 2 < x \le 4 \end{cases}$

Then find even and odd extensions of $f(x)$.

Sol : Given that

$$f(x) = \begin{cases} x^3 + x^2; & 0 \le x \le 2 \\ x + 2; & 2 < x \le 4 \end{cases}$$

$$\Rightarrow \quad f(-x) = \begin{cases} (-x)^3 + (-x)^2; & 0 \le -x \le 2 \\ (-x) + 2; & 2 < -x \le 4 \end{cases}$$

$$= \begin{cases} -x^3 + x^2; & -2 \le x \le 0 \\ -x + 2; & -4 \le x < -2 \end{cases}$$

Even Extension : Let $F_e(x)$ be the even extension of $f(x)$, then

$$F_e(x) = \begin{cases} f(x); & x \in [0, 4] \\ f(-x); & x \in [-4, 0] \end{cases}$$

$$= \begin{cases} \begin{cases} x^3 + x^2; & 0 \le x \le 2 \\ x + 2; & 2 < x \le 4 \end{cases} \\ \begin{cases} -x^3 + x^2; & -2 \le x \le 0 \\ -x + 2; & -4 \le x < -2 \end{cases} \end{cases}$$

Odd Extension : Let $F_o(x)$ be the odd extension of $f(x)$, then

$$F_o(x) = \begin{cases} f(x); & x \in [0, 4] \\ -f(-x); & x \in [-4, 0] \end{cases}$$

$$= \begin{cases} \begin{cases} x^3 + x^2; & 0 \le x \le 2 \\ x + 2; & 2 < x \le 4 \end{cases} \\ \begin{cases} x^3 + x^2; & -2 \le x \le 0 \\ x - 2; & -4 \le x < -2 \end{cases} \end{cases}$$

Inverse of a Function

Let $f : A \to B$ be a function defined by $y = f(x)$ such that f is both one-one (injective) and onto (surjective) (*i.e.*, bijective), then there exists a unique function $g : B \to A$ such that

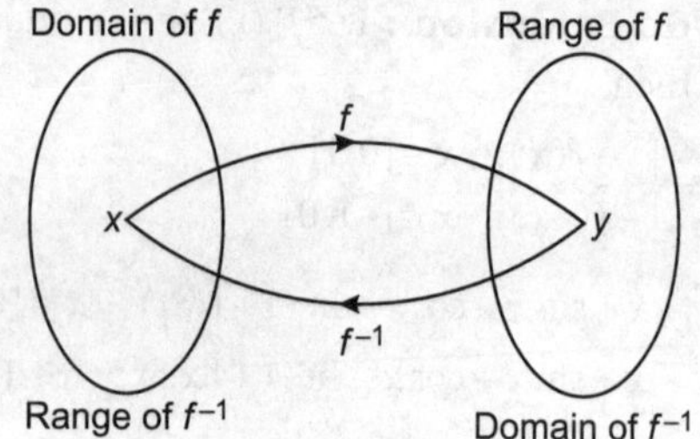

$$f(x) = y \Leftrightarrow g(y) = x \ \forall \ x \in A \text{ and } y \in B,$$

then g is said to be inverse of f.

Thus $g = f^{-1} : B \to A = \{(f(x), x) : (x, f(x)) \in f\}$

If f and g are inverse to each other, then

$$(fog)(x) = (gof)(x) = x$$

i.e., $f\{g(x)\} = g\{f(x)\} = x$

Consider

$$f = \{(a, \gamma), (b, \delta), (c, \alpha), (d, \beta)\}$$

$$g = \{(\alpha, c), (\beta, d), (\gamma, a), (\delta, b)\}$$

Here g is called the inverse of f *i.e.,* $g = f^{-1}$

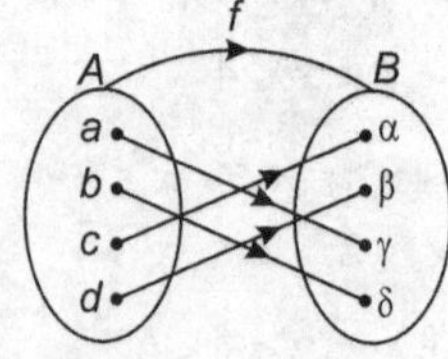

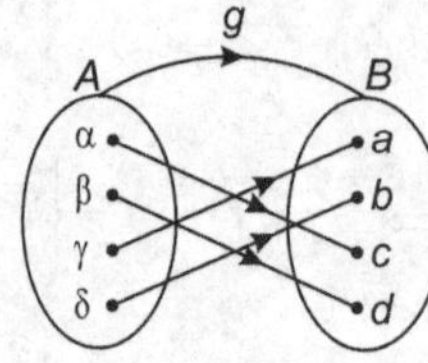

How to Find f^{-1} :

Given $y = f(x)$

(i) First of all test whether the given function $y = f(x)$ is invertible *i.e.,* bijective.

(ii) If $y = f(x)$ is invertible, then

$$x = f^{-1}(y) \qquad \text{...(i)}$$

and solve the equation $y = f(x)$

For x in terms of $y \Rightarrow x = \phi(y)$ (say) ...(ii)

from (i) and (ii), we get

$$f^{-1}(y) = \phi(y)$$

Replacing y by x, we get

$$f^{-1}(x) = \phi(x)$$

which is the required inverse of $f(x)$.

OR

If $y = f(x)$ is invertible, then use $(f^{-1} o f)(x) = fof^{-1}(x) = x$ and apply the formula of f on $f^{-1}(x)$ and solve for $f^{-1}(x)$.

Illustration 125. Find the inverse of the function $f(x) = \log_a(x + \sqrt{(x^2+1)})$; $a > 0, a \neq 1$ and assuming it to be an onto function.

Sol : Given $f(x) = \log_a(x + \sqrt{x^2+1})$

$$\therefore \quad f'(x) = \frac{\log_a e}{\sqrt{(1+x^2)}} > 0$$

which is a strictly increasing function.

Thus, $f(x)$ is injective, given that $f(x)$ is onto. Hence the given function $f(x)$ is invertible.

Now let $\quad y = f(x) = \log_a(x + \sqrt{(x^2+1)})$,

then $\quad x = f^{-1}(y) \qquad$ ...(i)

and $\quad y = \log_a(x + \sqrt{(x^2+1)})$

$\Rightarrow \quad a^y = x + \sqrt{(x^2+1)}$

and $\quad a^{-y} = \sqrt{(x^2+1)} - x$

$$\therefore \quad x = \frac{1}{2}(a^y - a^{-y}) \qquad \text{...(ii)}$$

From (i) and (ii), we get

$$f^{-1}(y) = \frac{1}{2}(a^y - a^{-y})$$

Hence $\quad f^{-1}(x) = \frac{1}{2}(a^x - a^{-x})$

Illustration 126. Find the inverse of the function $f(x) = \ln(x^2 + 3x + 1)$; $x \in [1, 3]$ and assuming it to be an onto function.

Sol : Given $f(x) = \ln(x^2 + 3x + 1)$

$$\therefore \quad f'(x) = \frac{2x+3}{(x^2+3x+1} > 0 \ \forall \ x \in [1, 3]$$

which is a strictly increasing function. Thus $f(x)$ is injective, given that $f(x)$ is onto. Hence the given function $f(x)$ is invertible.

Now let $\quad y = f(x) = \ln(x^2 + 3x + 1)$

then $\quad x = f^{-1}(y) \qquad$ ...(i)

and $\quad y = \ln(x^2 + 3x + 1)$

$\Rightarrow \quad e^y = x^2 + 3x + 1 \Rightarrow x^2 + 3x + 1 - e^y = 0$

$$\therefore \quad x = \frac{-3 \pm \sqrt{9 - 4 \cdot 1 \cdot (1 - e^y)}}{2} = \frac{-3 \pm \sqrt{(5 + 4e^y)}}{2}$$

$$= \frac{-3 + \sqrt{(5+4e^y)}}{2} \qquad (\because x \in [1, 3]) \text{ ...(ii)}$$

From (i) and (ii), we get

$$f^{-1}(y) = \frac{-3 + \sqrt{(5+4e^y)}}{2}$$

Hence $\quad f^{-1}(x) = \dfrac{-3 + \sqrt{(5+4e^x)}}{2}$

Illustration 127. Find the inverse of the function

$$f(x) = \text{arc } \sin\frac{x}{3}, x \in [-3, 3].$$

Sol : Given $f(x) = \text{arc } \sin\frac{x}{3} = \sin^{-1}\frac{x}{3}$

$$\because \quad -1 \leq \frac{x}{3} \leq 1 \quad \Rightarrow \quad -3 \leq x \leq 3$$

and $\quad y \in \left[-\frac{\pi}{2}, \frac{\pi}{2}\right]$.

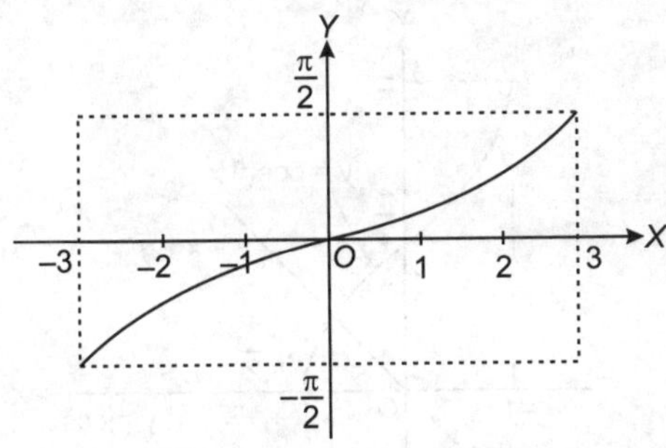

It is clear from the graph $f(x)$ is one-one and onto. Hence the given function is invertible.

Let $y = f(x) = \sin^{-1}\dfrac{x}{3}$,

then $x = f^{-1}(y)$

and $y = \sin^{-1}\dfrac{x}{3}$ ...(i)

$\Rightarrow$ $x = 3\sin y$ ...(ii)

From (i) and (ii), $f^{-1}(y) = 3\sin y$

Hence $f^{-1}(x) = 3\sin x$

Illustration 128. (i) Let a function $f: R \to R$ be defined by $f(x) = x - [x]$, (where $[x]$ is greatest integer $\le x$) for all $x \in R$. Is the function Invertible? (ii) Let $f: R \to R$ defined by $f(x) = \cos(5x+2)$. Is f invertible? Justify your answer.

Sol : (i) We know that any function $f: A \to B$ is invertible iff it is bijective.

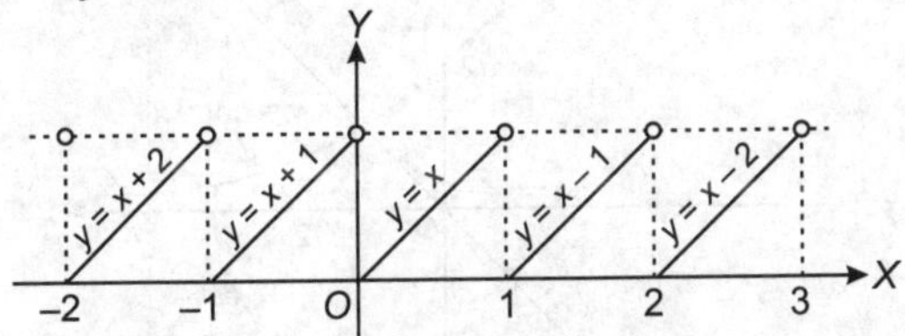

It is clear from the graph $f(x) = x - [x]$ is neither injective nor surjective.

Since any line parallel to x-axis cuts the graph more than one point

$\therefore$ $f(x) = x - [x]$ is many-one.

and Range of $f(x) = x - [x] = [0, 1) \subset R \Rightarrow f(x)$ is into

Thus f is not invertible.

> **Note :**
> All periodic functions are many-one. Hence they are not invertible when defined on whole of R. Each of these can be made invertible on a restricted domain.
>
> *e.g.*, $y = \sin x$ is invertible $\forall\, x \in \left[-\dfrac{\pi}{2}, \dfrac{\pi}{2}\right]$ and not invertible $\forall\, x \in R$.

(ii) Since period of $f(x) = \cos(5x+2)$ is $\dfrac{2\pi}{5}$.

$\therefore f(x)$ is periodic with period $\dfrac{2\pi}{5}$

$\Rightarrow$ $f(x)$ is not injective *i.e.*, many-one.

Also, $-1 \le \cos(5x+2) \le 1$

$\Rightarrow$ $-1 \le y \le 1$

$\therefore$ Range of $f(x) = R_f = [-1, 1] \subset R$

$\therefore f$ is not surjective.

Thus f is not invertible.

Illustration 129. If the function $f: [1, \infty) \to [1, \infty)$ is defined by $f(x) = 2^{x(x-1)}$. Is $f(x)$ invertible? If so, find its inverse.

Sol : Given $f(x) = 2^{x(x-1)}$

$\therefore$ $f'(x) = 2^{x(x-1)} \ln 2 \cdot (2x-1) > 0 \ \forall\, x \in [1, \infty)$

which is strictly increasing function.

Thus $f(x)$ is injective.

Also for $x \to 1$, $f(x) \to 1$ and for $x \to \infty$, $f(x) \to \infty$

Hence, $R_f = [1, \infty) = C_f$

So, $f(x)$ is onto

$\Rightarrow$ $f(x)$ is both one-one and onto.

$\therefore f(x)$ is invertible.

To Find f^{-1} :

Let f^{-1} be the inverse function of f, then by the rule of identity

$$fof^{-1}(x) = x$$

$\Rightarrow$ $f(f^{-1}(x)) = x$

$\Rightarrow$ $2^{f^{-1}(x)(f^{-1}(x)-1)} = x \quad (\because f(x) = 2^{x(x-1)})$

$\Rightarrow$ $f^{-1}(x)(f^{-1}(x)-1) = \log_2 x$

$\Rightarrow$ $(f^{-1}(x))^2 - f^{-1}(x) - \log_2 x = 0$

$\therefore$ $f^{-1}(x) = \dfrac{1 \pm \sqrt{1 + 4\log_2 x}}{2}$

as $f^{-1}(x) \in [1, \infty)$

$\therefore$ $f^{-1}(x) = \dfrac{1}{2}(1 + \sqrt{(1 + 4\log_2 x)})$

Graph of the Inverse of an Invertible Function

Let $f: A \to B$ be a bijective function and let $g: B \to A$ be the inverse of f. A point $P(x_1, y_1)$ lies on the graph $y = f(x)$, then $y_1 = f(x_1) \Leftrightarrow x_1 = g(y_1) \Leftrightarrow Q(y_1, x_1)$ lies on the graph $y = g(x)$. It is clear that point $P(x_1, y_1)$ and $Q(y_1, x_1)$ are reflections of each other w.r.t. the line $y = x$ *i.e.*, the line joining the points $P(x_1, y_1)$ and $Q(y_1, x_1)$ is perpendicular bisector by the line $y = x$.

OR

The graph of $f^{-1}(x)$ is obtained by taking the mirror image of $f(x)$ about the line $y = x$ and if the graph of a function $f(x)$ is symmetrical about the line $y = x$, then $f(x)$ and $f^{-1}(x)$ are identical functions.

e.g., $f(x) \equiv x^3 + y^3 - 3axy$

Here $f(x) = f^{-1}(x) = x^3 + y^3 - 3axy$

Also if curves $y = f(x)$ and $y = f^{-1}(x)$ intersect, then the solution of $f(x) = f^{-1}(x)$ is given by $y = x$. *i.e.*, $x = f(x)$

Illustration 130. Find the inverse of the function $f(x) = x^2 + 1, x \geq 0$ with the help of the graph $f(x)$.

Sol :

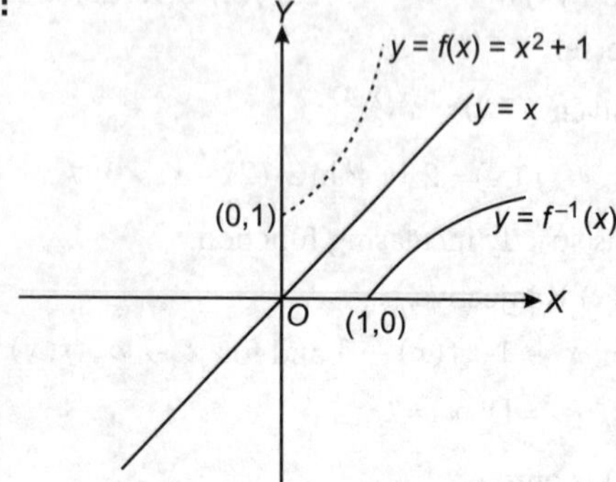

Illustration 131. Find the inverse of the function $f(x) = x^2 + 2ax + \frac{1}{16}$ with the help of the graph $f(x)$.

Sol :

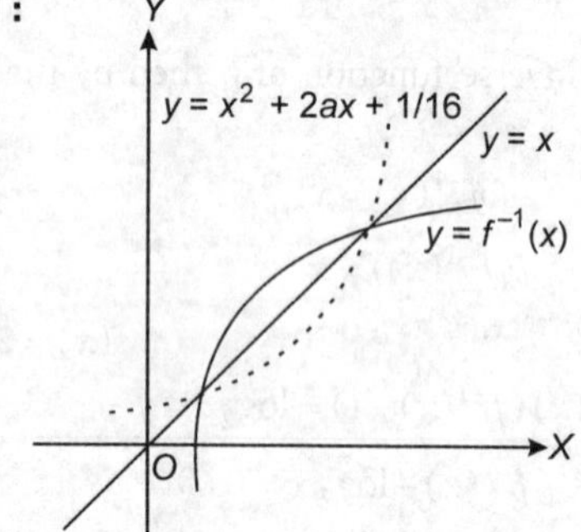

Symmetry of Graph of a Function about $y = x$ and its Inverse :

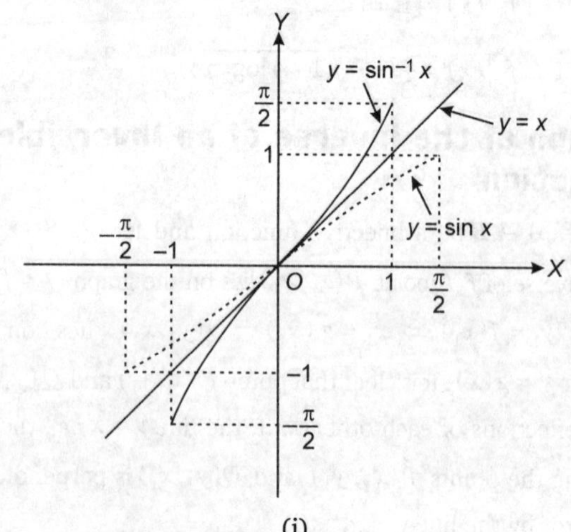

(i)

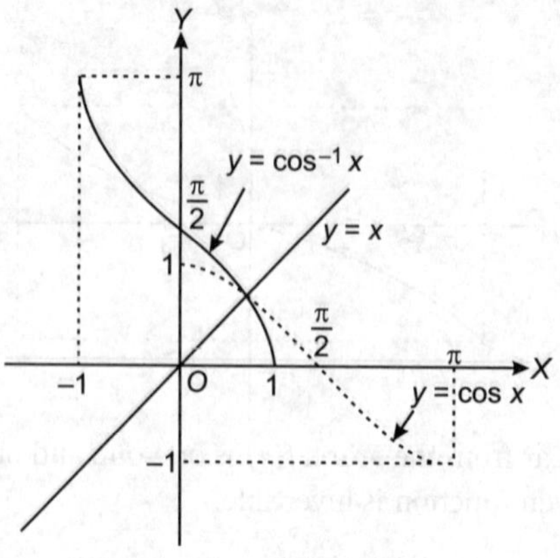

(ii)

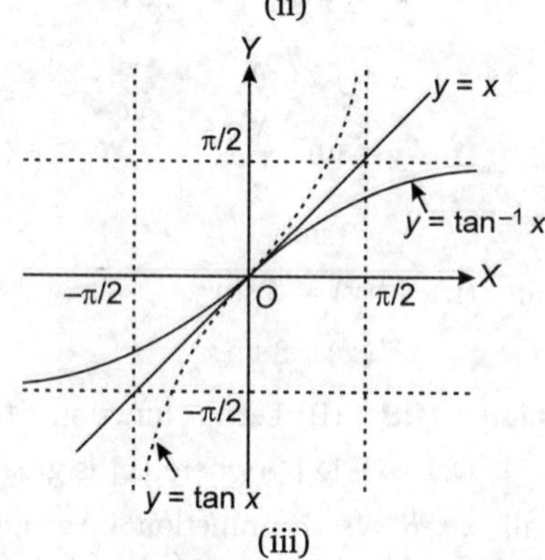

(iii)

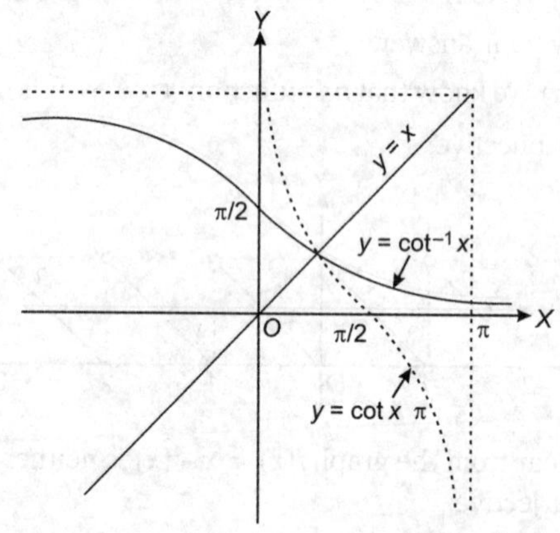

(iv)

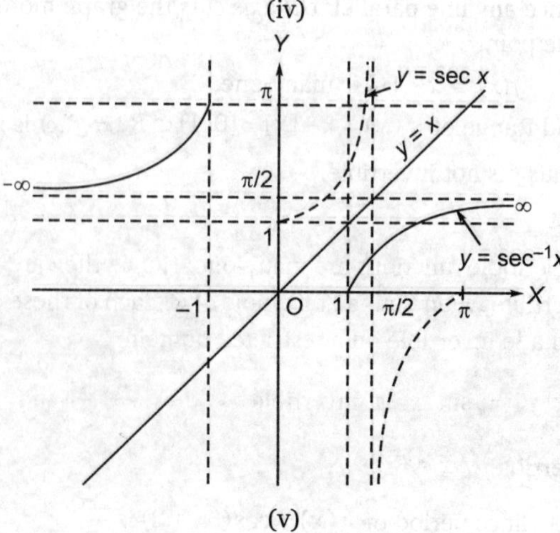

(v)

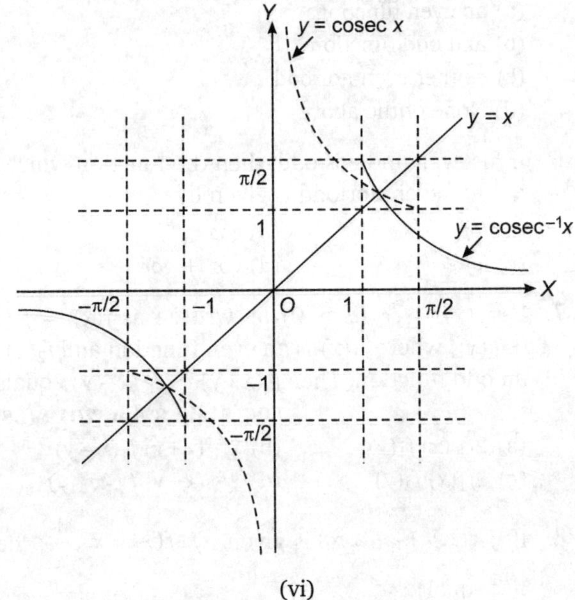

(vi)

Properties of Inverse Functions :

1. The inverse of bijective function is unique and bijective.
2. Let $f : A \to B$ be a function such that f is bijective and $g : B \to A$ is inverse of f, then $fog = I_B$ = Identity function of set B and $gof = I_A$ = Identity function of set A.
3. If $fog = gof$, then either $f^{-1} = g$ or $g^{-1} = f$ and $(fog)(x) = (gof)(x) = x$.
4. If f and g are two bijective functions such that $f : A \to B$ and $g : B \to C$.

   Also, $(gof)^{-1} = (f^{-1}og^{-1})$

In general :

$$(f_1 o f_2 o f_3 o f_4 o \ldots o f_n)^{-1} = f_n^{-1} o \ldots o f_4^{-1} o f_3^{-1} o f_2^{-1} o f_1^{-1}$$

Illustration 132. Show that the functions $f(x) = x^2 - x + 1$, $x \geq \frac{1}{2}$ and $g(x) = \frac{1}{2} + \sqrt{\left(x - \frac{3}{4}\right)}$ are mutually inverse and solve the equation $x^2 - x + 1 = \frac{1}{2} + \sqrt{\left(x - \frac{3}{4}\right)}$.

Sol : The function $y = f(x) = x^2 - x + 1 = \left(x - \frac{1}{2}\right)^2 + \frac{3}{4}$ increases in the interval $\left[\frac{1}{2}, \infty\right)$ and x varying in the indicated interval we have $y = f(x) \geq \frac{3}{4}$ *i.e.*, $y \in \left[\frac{3}{4}, \infty\right)$

$$\therefore \quad x^2 - x + 1 - y = 0$$

$$\Rightarrow \quad x = \frac{1 \pm \sqrt{1 - 4 \cdot 1 \cdot (1 - y)}}{2} = \frac{1 \pm \sqrt{(4y - 3)}}{2}$$

$$= \frac{1}{2} \pm \sqrt{\left(y - \frac{3}{4}\right)} = \frac{1}{2} + \sqrt{\left(y - \frac{3}{4}\right)} \quad \left(\because x \geq \frac{1}{2}\right)$$

$$= g(y)$$

$$\therefore \quad y = g^{-1}(x)$$

$$\Rightarrow \quad f(x) = g^{-1}(x)$$

Hence $f(x)$ and $g(x)$ are mutually inversion. Again it gives $x^2 - x + 1 = \frac{1}{2} + \sqrt{\left(x - \frac{3}{4}\right)}$

Since the graphs of the original and inverse functions can intersect only on the straight line $y = x$.

$$\therefore \quad x = f(x)$$

$$\Rightarrow \quad x = x^2 - x + 1$$

$$\Rightarrow \quad x^2 - 2x + 1 = 0$$

$$\Rightarrow \quad (x - 1)^2 = 0$$

$$\therefore \quad x = 1$$

Illustration 133. Find the real roots of the equation $x^2 + 2ax + \frac{1}{16} = -a + \sqrt{\left(a^2 + x - \frac{1}{16}\right)}$, where $0 < a < \frac{1}{4}$

Sol : We have $x^2 + 2ax + \frac{1}{16} = -a + \sqrt{\left(a^2 + x - \frac{1}{16}\right)}$

Let $$y = x^2 + 2ax + \frac{1}{16} \quad \ldots\text{(i)}$$

and $$y_1 = -a + \sqrt{\left(a^2 + x - \frac{1}{16}\right)} \quad \ldots\text{(ii)}$$

$$\because \quad y = y_1 \quad \ldots\text{(iii)}$$

From (ii), $$(y_1 + a)^2 = a^2 + x - \frac{1}{16}$$

$$\therefore \quad x = y_1^2 + 2ay_1 + \frac{1}{16}$$

$$x = y^2 + 2ay + \frac{1}{16} \quad \text{[From (iii)]} \quad \ldots\text{(iv)}$$

Equations (i) and (iv) are symmetrical about the line

$$y = x \quad \ldots\text{(v)}$$

$\therefore$ From (i) and (v), $x = x^2 + 2ax + \frac{1}{16}$

$$\Rightarrow \quad x^2 - (1 - 2a)x + \frac{1}{16} = 0$$

$$\therefore \quad x = \frac{(1 - 2a) \pm \sqrt{(1 - 2a)^2 - \frac{1}{4}}}{2}$$

For real roots, $1 - 2a > \frac{1}{2}$

$$\Rightarrow \qquad 2a < \frac{1}{2} \text{ or } a < \frac{1}{4}$$

Hence $\quad x_1 = \left(\frac{1-2a}{2}\right) + \sqrt{\left(\frac{1-2a}{2}\right)^2 - \frac{1}{16}}$

and $\quad x_2 = \left(\frac{1-2a}{2}\right) - \sqrt{\left(\frac{1-2a}{2}\right)^2 - \frac{1}{16}}$

are real roots and given facts, satisfy the original equation.

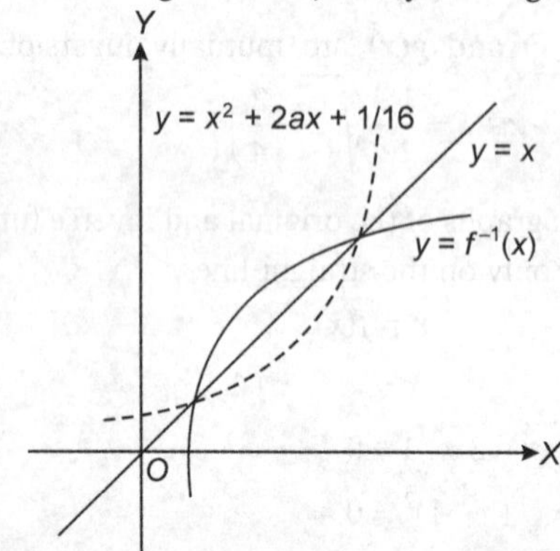

INTRODUCTORY EXERCISE 2.8

Only One Option is Correct

1. The function $f(x) = \sin(\ln(x + \sqrt{(1+x^2)}))$ is :
(a) an even function (b) and odd function
(c) neither even nor odd (d) None of these

2. If $f(x) = \sqrt[n]{x^m}, n \in N$, is an even function, then m is :
(a) an even integer (b) an odd integer
(c) any integer (d) $f(x)$ even is not possible

3. If f is even and g is odd, then fog is :
(a) even
(b) odd
(c) neither even nor odd
(d) Nothing can be said

4. If $f(x) = \begin{cases} x^2 \sin\left(\frac{\pi x}{2}\right), & |x| \le 1 \\ x|x|, & |x| \ge 1 \end{cases}$ then $f(x)$ is :
(a) an even function
(b) and odd function
(c) neither even nor odd
(d) None of these

5. $f(x) = \dfrac{\cos x}{\left[\frac{2x}{\pi}\right] + \frac{1}{2}}$, where x is not an integral multiple of π and $[\cdot]$ denotes the greatest integer function, is :
(a) an even function
(b) and odd function
(c) neither even nor odd
(d) None of the above

6. If f is even and g is odd, then the function which is neither even nor odd is given by :
(a) fof (b) gog
(c) $fof + fog$ (d) $gog + gof$

7. Let $f(x) = a^x$ $(a > 0)$ be written as $f(x) = f_1(x) + f_2(x)$, where $f_1(x)$ is an even function and $f_2(x)$ is an odd function. Then $f_1(x+y) + f_1(x-y)$ equals : **[JEE Main 8 Apr. 2019 (ES)]**
(a) $2f_1(x) f_1(y)$ (b) $2f_1(x+y) f_1(x-y)$
(c) $2f_1(x) f_2(y)$ (d) $2f_1(x+y) f_2(x-y)$

8. If $f : [1, \infty) \to [2, \infty)$ is given by $f(x) = x + \frac{1}{x}$, then f^{-1} equals :
(a) $\frac{x + \sqrt{x^2-4}}{2}$ (b) $\frac{x}{1+x^2}$
(c) $\frac{x - \sqrt{x^2-4}}{2}$ (d) $1 + \sqrt{(x^2-4)}$

9. If the function $f : [1, \infty) \to [1, \infty)$ is defined by $f(x) = 2^{x(x-1)}$, then $f^{-1}(x)$ is :
(a) $\left(\frac{1}{2}\right)^{x/(x-1)}$ (b) $\frac{1}{2}(1 + \sqrt{(1 + 4\log_2 x)})$
(c) $\frac{1}{2}(1 - \sqrt{(1 + 4\log_2 x)})$ (d) either (b) or (c)

10. The inverse of $y = 5^{\log x}$ is : **[JEE Main 17 Mar 2021 (MS)]**
(a) $x = 5^{\log y}$ (b) $x = y^{\log 5}$
(c) $y = x^{\frac{1}{\log 5}}$ (d) $x = 5^{\frac{1}{\log y}}$

11. Let $f : \left[-\frac{\pi}{3}, \frac{2\pi}{3}\right] \to [0, 4]$ be a function defined as $f(x) = \sqrt{3}\sin x - \cos x + 2$, then $f^{-1}(x)$ is given by :
(a) $\sin^{-1}\left(\frac{x-2}{2}\right) - \frac{\pi}{6}$ (b) $\sin^{-1}\left(\frac{x-2}{2}\right) + \frac{\pi}{6}$
(c) $\frac{2\pi}{3} + \cos^{-1}\left(\frac{x-2}{2}\right)$ (d) None of these

12. Which of the following functions is inverse to itself?
(a) $f(x) = \frac{1-x}{1+x}$
(b) $f(x) = 5^{\ln x}$
(c) $f(x) = 2^{x(x-1)}$
(d) None of the above

Periodic Function

A function $f(x)$ is said to be periodic function of x, if there exists a positive real number T such that $f(x+T)=f(x)$. The smallest value of T is called the period of the function. If positive value of T is independent of x, then $f(x)$ is periodic function and if the value of T depends upon x, then $f(x)$ is non-periodic.

Definition (Graphically) :

A function is said to be periodic if its graph repeat itself after a fixed interval and the width of that interval is called its period.

Illustration 134. If $\cos x$ is periodic, then find its period.

Sol : Let $f(x)=\cos x$

If T is period, then $f(x+T)=f(x)$

$\Rightarrow \quad \cos(x+T)=\cos x$

$\Rightarrow \quad x+T=2n\pi \pm x$

Put $x=0$, then $T=2n\pi$

$\therefore \quad T=2\pi, 4\pi, 6\pi, \ldots \quad (\because T>0)$

$\because$ Period of $f(x)$ is smallest positive real number. Thus, period of $f(x)=2\pi$

Graphically :

$\because$ Graph of $f(x)=\cos x$, repeats at an interval 2π.

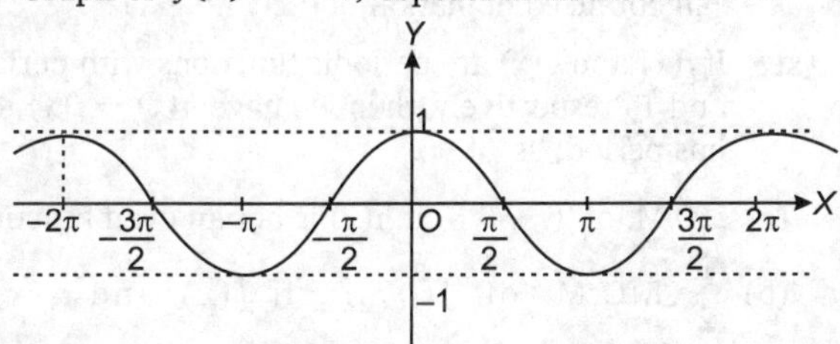

Thus, $f(x) = \cos x$ is periodic with period 2π.

Note :
Periods of $\sin x$, $\cos x$ and $\tan x$ are 2π, 2π and π respectively.

Illustration 135. If $\sqrt{(x-[x])}$ is periodic, then find its period, where [·] denotes the greatest integer function.

Sol : Let $f(x)=\sqrt{(x-[x])}$

If T is period, then $f(x+T)=f(x)$

$\Rightarrow \quad \sqrt{(x+T)-[x+T]}=\sqrt{x-[x]}$

On squaring, we get $x+T-[x+T]=x-[x]$

or $\quad [x+T]-[x]=T \; \forall \; x \in R$

$\Rightarrow \quad T=1, 2, 3, 4, \ldots$

The smallest value of T is 1.

Hence period of $f(x)$ is 1.

Graphically :

$\because$ Graph of $f(x)=\sqrt{x-[x]}$ repeats at an interval 1.

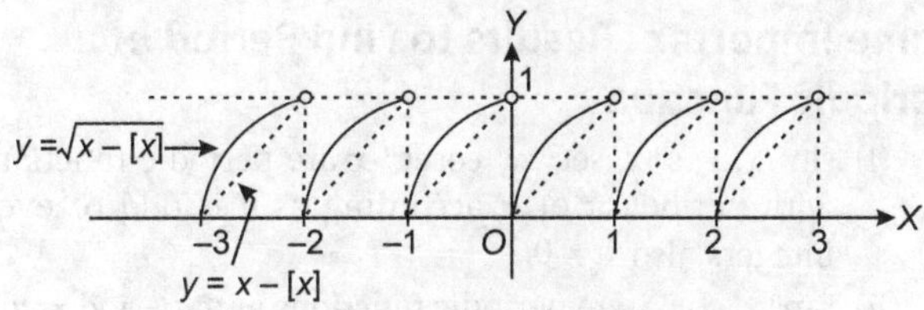

Thus, $f(x)=\sqrt{x-[x]}$ is periodic with period 1.

Illustration 136. Let f be a real valued function defined for all real numbers x such that for some fixed $a>0$.

$$f(x+a)=\frac{1}{2}+\sqrt{f(x)-\{f(x)\}^2} \text{ for all } x.$$

Show that the function $f(x)$ is periodic with period $2a$.

Sol : Given $f(x+a)=\dfrac{1}{2}+\sqrt{f(x)-\{f(x)\}^2}$ ...(i)

Replacing x by $x-a$ and $x+a$ in (i),

then $f(x)=\dfrac{1}{2}+\sqrt{f(x-a)-\{f(x-a)\}^2}>\dfrac{1}{2}$ ...(ii)

and $f(x+2a)=\dfrac{1}{2}+\sqrt{f(x+a)-\{f(x+a)\}^2}$

$=\dfrac{1}{2}+\sqrt{f(x+a)\{1-f(x+a)\}}$

$=\dfrac{1}{2}+\sqrt{\left(\dfrac{1}{2}+\sqrt{f(x)-\{f(x)\}^2}\right)\left(\dfrac{1}{2}-\sqrt{f(x)-\{f(x)\}^2}\right)}$ (from (i))

$=\dfrac{1}{2}+\sqrt{\dfrac{1}{4}-f(x)+\{f(x)\}^2}=\dfrac{1}{2}+\sqrt{\left(f(x)-\dfrac{1}{2}\right)^2}$

$=\dfrac{1}{2}+\left|f(x)-\dfrac{1}{2}\right|$

$=\dfrac{1}{2}+f(x)-\dfrac{1}{2}=f(x) \quad \left\{\because f(x)>\dfrac{1}{2}\right\}$

$\therefore \quad f(x+2a)=f(x)$

Hence $f(x)$ is periodic with period $2a$.

Illustration 137. Is $\cos\sqrt{t}$ a periodic function? If yes, then find the period. If not, then give reasons to explain your answer.

Sol : Let $f(t)=\cos\sqrt{t}$ be periodic with period $\lambda, \lambda \neq 0, \lambda>0$,

$\therefore \quad f(t+\lambda)=f(t)$

$\cos\sqrt{(t+\lambda)}=\cos\sqrt{t}$

$\Rightarrow \quad \sqrt{(t+\lambda)}=2n\pi \pm \sqrt{t}$ for all $t \geq 0$

Put $t=0$, then $\sqrt{\lambda}=2n\pi$ ...(i)

and put $t=\lambda$, then $\sqrt{2\lambda}=2n\pi \pm \sqrt{\lambda}$ ...(ii)

From (i) and (ii), $\sqrt{2\lambda}=\sqrt{\lambda} \pm \sqrt{\lambda}$

or $\quad \sqrt{2}=1 \pm 1$ which is impossible.

Hence $\cos\sqrt{t}$ is non-periodic function.

Some Important Results to Find Period of a Periodic Function

(i) $\sin^n x$, $\cos^n x$, $\sec^n x$, $\text{cosec}^n x$ are periodic functions with period 2π or π according as n is odd or even integers also $n \neq 0$.

(ii) $\tan^n x$, $\cot^n x$ are periodic functions with period x, n is odd or even integers also $n \neq 0$.

(iii) $\sin x^n$, $\cos x^n$, $\tan x^n$, $\cot x^n$, $\sec x^n$, $\text{cosec}\, x^n$ are non-periodic functions for $x \in R \sim \{0, 1\}$.

(iv) $\sin^{-1}(\sin x)$, $\cos^{-1}(\cos x)$, $\sec^{-1}(\sec x)$, $\text{cosec}^{-1}(\text{cosec}\, x)$ are periodic with period 2π and $\tan^{-1}(\tan x)$, $\cot^{-1}(\cot x)$ are periodic with period π.

(v) $\sin(\sin^{-1} x)$, $\cos(\cos^{-1} x)$, $\tan(\tan^{-1} x)$, $\cot(\cot^{-1} x)$, $\sec(\sec^{-1} x)$, $\text{cosec}(\text{cosec}^{-1} x)$ are non periodic.

(vi) $|\sin x|$, $|\cos x|$, $|\tan x|$, $|\cot x|$, $|\sec x|$, $|\text{cosec}\, x|$ are periodic functions with period π.

(vii) Algebraic functions $x^2, 3x^2 + 5x - 4, 2\sqrt{x} + 5, \ldots$ etc. are non-periodic.

(viii) Every constant function is always periodic with no fundamental period.

(ix) Inverse of a periodic function does not exist but in case of trigonometric function since domain and range is restricted and defined, then inverse exist.

(x) If $f(T) = f(0) = f(-T)$, then 'T' is period.

(xi) If $f(x)$ is periodic with period T, then $a\,f(x \pm b) \pm c$, where $a, b, c \in R\ (a \neq 0)$ is also periodic with period T.

(xii) If $f(x)$ is periodic with period T, then $f(ax + b)$, where $a, b \in R\ (a \neq 0)$ is also period with period $\frac{T}{|a|}$.

Proof : Consider $a > 0$

Let $\quad f(x + T) = f(x)$

and $\quad f(a(x + T') + b) = f(ax + b)$

$\Rightarrow \quad f(ax + b + aT') = f(ax + b)$

$\Rightarrow \quad f(y + aT') = f(y) = f(y + T) \quad (\because f(x + T) = f(x))$

$\Rightarrow \quad aT' = T$

or $\quad |a|T' = T \quad (\because a$ is positive$)$

$$\therefore \quad T' = \frac{T}{|a|}$$

e.g., $\sin 3x$ has period $2\pi/3$, $\cos 2x$ has a period $\frac{2\pi}{2} = \pi$ and $\tan(-7x + 8)$ has a period $\frac{\pi}{|-7|} = \frac{\pi}{7}$.

(xiii) Let $f(x)$ has period $T_1 = m/n$ ($m, n \in N$ and co-prime) and $g(x)$ has period $T_2 = r/s$ ($r, s \in N$ and co-prime), then $F(x) = k_1 f(x) \pm k_2 g(x)$ is a periodic function and its period is L.C.M. of T_1 and T_2, where k_1 and k_2 are constants.

i.e., Period of $F(x) =$ L.C.M. of $\{T_1, T_2\}$

$$= \text{L.C.M. of } \left\{\frac{m}{n}, \frac{r}{s}\right\} = \frac{\text{L.C.M. of } \{m, r\}}{\text{H.C.F. of } \{n, s\}}$$

Note :

(1) L.C.M. of T_1 and T_2 exists if T_1 and T_2 are rational quantities. If T_1 and T_2 are irrational, then L.C.M. of T_1 and T_2 does not exist unless they have same kind irrational and L.C.M. of rational and irrational is not possible.

e.g.,

(a) L.C.M. of $\left\{1, \frac{1}{2}, \frac{1}{3}, \ldots, \frac{1}{n}\right\}$

$$= \frac{\text{L.C.M. of } \{1, 1, 1, \ldots, 1\}}{\text{H.C.F. of } \{1, 2, 3, \ldots, n,\}} = \frac{1}{1} = 1$$

(b) L.C.M. of $\left\{2\pi, \frac{\pi}{3}\right\} = \frac{\text{L.C.M. of } [2\pi, \pi\}}{\text{H.C.F. of } \{1, 3\}}$

$$= \frac{2\pi}{1} = 2\pi$$

(c) L.C.M. of $[2\pi, e\}$ is not possible

($\because 2\pi$ and e are different kind irrational)

(d) L.C.M. of $\{2\pi, 7\}$ is not possible

($\because 2\pi \in$ irrational and $7 \in$ rational)

(2) If T be the period of $f(x) + g(x)$, provided there does not exist a positive number $k(<T)$ for which $f(x + k) + g(x + k) = f(x) + g(x)$, also k will be the period. The same rule is applicable for any other algebraic combination of $f(x)$ and $g(x)$.

(xiv) If $f(x)$ and $g(x)$ are periodic functions with periods T_1 and T_2 respectively, then we have $h(x) = f(x) + g(x)$ has period, as

(a) L.C.M. of $\{T_1, T_2\}$, if $h(x)$ is not an even function

(b) $\frac{1}{2}$ L.C.M. of $\{T_1, T_2\}$, if $f(x)$ and $g(x)$ are complementary pairwise comparable functions.

e.g., period of $\sin^4 x + \cos^4 x = \frac{1}{2}$ L.C.M. of $\{\pi, \pi\} = \frac{\pi}{2}$

($\because \sin^4 x$ and $\cos^4 x$ are complementary pairwise comparable functions) and period of $\sin^3 x + |\cos x|$ = L.C.M. of $\{\pi, \pi\} = \pi$ ($\because \sin^3 x + |\cos x|$ is not an even function)

Note :

Failure of L.C.M. Rule

The rule of L.C.M. is not applicable if $h(x) = f(x) + g(x)$ reduces to a constant or $f(x)$ and $g(x)$ are both even functions and having same exponent.

e.g., Let $\quad h(x) = \sin^2 x + \cos^2 x$

$\because$ period of $\sin^2 x = \pi$ and period of $\cos^2 x = \pi$ and $\sin^2 x$, $\cos^2 x$ are complementary pairwise comparable functions.

$\therefore$ period of $h(x) = \frac{1}{2}$ L.C.M. of $\{\pi, \pi\} = \frac{\pi}{2}$, which is not correct whereas $\sin^2 x + \cos^2 x = 1$, is a constant function. So, $h(x)$ is periodic with no fundamental period and let $h(x) = \sec^2 x - \tan^2 x$.

$\therefore$ period of $\sec^2 x = \pi$ and period of $\tan^2 x = \pi$ and $\sec^2 x, \tan^2 x$ are not complementary pairwise comparable functions.

$\therefore$ period of $h(x) =$ L.C.M. of $\{\pi, \pi\} = \pi$, which is not correct

whereas $\sec^2 x - \tan^2 x = 1$, is a constant functions. So, $h(x)$ is periodic with no fundamental period.

(xv) $a \mid \sin x \mid + b \mid \cos x \mid$, $a \mid \tan x \mid + b \mid \cot x \mid$, $a \mid \sec x \mid + b \mid \operatorname{cosec} x \mid$ are periodic with periods π and $\pi/2$ according as $a \neq b$ and $a = b$ respectively.

(xvi) $\mid \sin x \mid + \mid \tan x \mid$, $\mid \cos x \mid - \mid \sin x \mid$, $\mid \operatorname{cosec} x \mid - \mid \sec x \mid$, $\mid \tan x \mid + \mid \sec x \mid$, ... etc. are periodic functions with period π.

(xvii) A continuous periodic function is bounded.

(xviii) If $g(x)$ is periodic function with period T, then $f(g(x))$ will be periodic function, period of $f(g(x))$ may or may not be equal to T.

e.g., period of cos (sin x) is π, but period of sin x is 2π and period of sin (sin x) is 2π and period of sin x is 2π.

(xix) If $f(x)$ is periodic and $g(x)$ is strictly monotonic (other than linear), then $f(g(x))$ is non-periodic.

Illustration 138. Find the period of the function

$$f(x) = e^{x - [x] + |\cos \pi x| + |\cos 2\pi x| + \ldots + |\cos n\pi x|},$$

where [·] denotes the greatest integer function.

Sol : Given $f(x) = e^{x - [x] + |\cos \pi x| + |\cos 2\pi x| + \ldots + |\cos n\pi x|}$

$\because$ period of $x - [x] = 1$

period of $|\cos \pi x| = \frac{\pi}{\pi} = 1$

period of $|\cos 2\pi x| = \frac{\pi}{2\pi} = \frac{1}{2}$

... ...

... ...

period of $|\cos n\pi x| = \frac{\pi}{n\pi} = \frac{1}{n}$

So, period of $f(x) =$ L.C.M. of $\left\{1, \frac{1}{2}, \frac{1}{3}, \ldots, \frac{1}{n}\right\}$

$$= \frac{\text{L.C.M. of } \{1, 1, 1, \ldots, 1\}}{\text{H.C.F. of } \{1, 2, 3, \ldots, n,\}}$$

$$= \frac{1}{1} = 1$$

Illustration 139. Find the period of the function,

$f(x) = \sin^4 x + \cos^4 x$.

Sol : Since $f(x) = \sin^4 x + \cos^4 x$

$$= (\sin^2 x)^2 + (\cos^2 x)^2$$
$$= (\sin^2 x + \cos^2 x)^2 - 2\sin^2 x \cos^2 x$$
$$= 1 - \frac{1}{2}\sin^2 2x = 1 - \frac{1}{4}(1 - \cos 4x)$$
$$= \frac{3}{4} + \frac{1}{4}\cos 4x$$

Hence period of $f(x) = \frac{2\pi}{4} = \frac{\pi}{2}$

Illustration 140. Find the period of

$$f(x) = [\sin 3x] + |\cos 6x|,$$

where [·] denotes the greatest integer function.

Sol: Here, period of $[\sin 3x] = \frac{2\pi}{3}$ and period of $|\cos 6x| = \frac{\pi}{6}$.

Hence, period of $f(x) =$ L.C.M. of $\left\{\frac{2\pi}{3}, \frac{\pi}{6}\right\}$

$$= \frac{\text{L.C.M. of } \{2\pi, \pi\}}{\text{H.C.F. of } \{3, 6\}} = \frac{2\pi}{3}$$

Illustration 141. Find the period of the function

$$f(x) = 7\sin\left(\frac{\pi x}{3}\right) + 9\cos\left(\frac{\pi x}{4}\right) - 12\cot\left(\frac{\pi x}{6}\right)$$

Sol : Here, period of $7\sin\left(\frac{\pi x}{3}\right) = \frac{2\pi}{\pi/3} = 6$,

period of $9\cos\left(\frac{\pi x}{4}\right) = \frac{2\pi}{\pi/4} = 8$

and period of 12 $\cot\left(\frac{\pi x}{6}\right) = \frac{\pi}{\pi/6} = 6$

Hence, period of $f(x) =$ L.C.M. of $\{6, 8, 6\} = 24$

Illustration 142. Find the period of the function :

$$f(x) = \cos(\sin x) + \cos(\cos x)$$

Sol : Let T be the period of the function, then

$$f(x + T) = f(x)$$

$\Rightarrow$ $\cos(\sin(x + T)) + \cos(\cos(x + T))$

$$= \cos(\sin x) + \cos(\cos x)$$

Put $x = 0$, then

$\cos(\sin T) + \cos(\cos T) = \cos(0) + \cos(1)$

$$= \cos\left(\cos\frac{\pi}{2}\right) + \cos\left(\sin\frac{\pi}{2}\right)$$

On comparing, we get $T = \pi/2$.

Illustration 143. If $f(x) = \sin x + \cos ax$ is a periodic function, show that a is a rational number.

Sol : Given $f(x) = \sin x + \cos ax$

$\because$ Period of $\sin x = \frac{2\pi}{1}$ and period of $\cos ax = \frac{2\pi}{a}$

Hence, period of $f(x)$ = L.C.M. of $\left\{\frac{2\pi}{1}, \frac{2\pi}{a}\right\}$

$$= \frac{\text{L.C.M. of } \{2\pi, 2\pi\}}{\text{H.C.F. of } \{1, a\}} = \frac{2\pi}{k}$$

where k = H.C.F. of 1 and a

$\therefore \quad \frac{1}{k}$ = integer $= q$ (say), $(\neq 0)$

and $\quad \frac{a}{k}$ = integer $= p$ (say)

$\therefore \quad \frac{a/k}{1/k} = \frac{p}{q} \Rightarrow a = \frac{p}{q}$

$\therefore a$ = rational number

Illustration 144. Prove that $\sin(1/x), (x \neq 0)$ is a non-periodic function.

Sol : Let $f(x) = \sin(1/x)$ be periodic with period $T, T \neq 0, T > 0$

$\therefore \quad f(x+T) = f(x)$

$\Rightarrow \quad \sin\left(\frac{1}{x+T}\right) = \sin\left(\frac{1}{x}\right)$

$\Rightarrow \quad \frac{1}{x+T} = n\pi + (-1)^n \frac{1}{x}$ ...(i)

Put $x = T$ and $x = 2T$ in (i), then

$\frac{1}{2T} = n\pi + (-1)^n \frac{1}{T}$ ...(ii)

and $\quad \frac{1}{3T} = n\pi + (-1)^n \frac{1}{2T}$ ...(iii)

Subtracting (iii) from (ii), we get

$$\frac{1}{6T} = (-1)^n \cdot \frac{1}{2T}$$

or $\quad \frac{1}{3} = (-1)^n$ which is impossible.

Hence $\sin(1/x), (x \neq 0)$ is non-periodic function.

Functional Equations

An equation in which unknowns are functions is called a function equation.

Types of Functional Equations

(i) Functional Equations on Natural Numbers

Illustration 145. If $f(x+y) = f(x) + f(y) - xy - 1$ for all $x, y, \in N$ and $f(1) = 1$, then find the value of $\sum_{r=1}^{5} f(r)$.

Sol : For $x = y = 1$,

$f(1+1) = f(1) + f(1) - 1 - 1 \quad (\because f(1) = 1)$

$= 1 + 1 - 2 = 0 \quad \therefore f(2) = 0$

for $x = 2, y = 1$

$f(2+1) = f(2) + f(1) - 2 - 1$

$= 0 + 1 - 3 = -2 \quad (\because f(2) = 0)$

$\therefore \quad f(3) = -2$

for $x = y = 2$

$\therefore \quad f(2+2) = f(2) + f(2) - 4 - 1$

$= 0 + 0 - 5 = -5 \quad (\because f(2) = 0)$

$\therefore \quad f(4) = -5$ and for $x = 3, y = 2$

$f(3+2) = f(3) + f(2) - 6 - 1 = -2 + 0 - 7$

$(\because f(2) = 0, f(3) = -2)$

$\therefore \quad f(5) = -9$

$\Rightarrow \quad \sum_{r=1}^{5} f(r) = f(1) + f(2) + f(3) + f(4) + f(5)$

$= 1 + 0 - 2 - 5 - 9 = -15$

Illustration 146. If $f(x+y) = f(x) \cdot f(y)$ for all natural numbers x and y and $f(1) = 2$, then find $f(x)$.

Sol : For $x = y = 1$

$f(1+1) = f(1) \cdot f(1) = 2.2 \quad (\because f(1) = 2)$

$\therefore \quad f(2) = 2^2$ ...(i)

for $\quad x = 2, y = 1$

$f(2+1) = f(2) \cdot f(1) = 2^2 \cdot 2$ (from (i) and $f(1) = 2$)

$\therefore \quad f(3) = 2^3$

for $\quad x = y = 2$

$\therefore f(2+2) = f(2) \cdot f(2) = 2^2 \cdot 2^2$ (from (i))

$\therefore f(4) = 2^4$

....

similarly we get $f(x) = 2^x \quad \forall x \in N$.

(ii) Functional Equations on Real Line

Illustration 147. If $f(x) + 2f(1-x) = x^2 + 2 \ \forall x \in R$, then find $f(x)$.

Sol : Given $f(x) + 2f(1-x) = x^2 + 2$ ...(i)

Replacing x by $(1-x)$ in equation (i), then we get

$f(1-x) + 2f(x) = (1-x)^2 + 2$

or $\quad 2f(1-x) + 4f(x) = 2(1-x)^2 + 4$ ...(ii)

Now subtracting Eq. (i) from Eq. (ii), we get

$3f(x) = 2(1-x)^2 + 4 - x^2 - 2$

$= 2(1 - 2x + x^2) - x^2 + 2 = x^2 - 4x + 4 = (x-2)^2$

Hence, $\quad f(x) = \frac{(x-2)^2}{3}$

Illustration 148. Consider the real valued function satisfying $2f(\sin x) + f(\cos x) = x$, then find $f(x)$.

Sol : Given $2f(\sin x) + f(\cos x) = x$ ...(i)

Replacing x by $\left(\frac{\pi}{2} - x\right)$ in equation (i), we get

$$2f\left(\sin\left(\frac{\pi}{2} - x\right)\right) + f\left(\cos\left(\frac{\pi}{2} - x\right)\right) = \frac{\pi}{2} - x$$

or $\quad 2f(\cos x) + f(\sin x) = \frac{\pi}{2} - x \quad$ ...(ii)

Multiplying in equation (i) by 2, then

$$4f(\sin x) + 2f(\cos x) = 2x \quad \text{...(iii)}$$

Now subtracting Eq. (ii) from Eq. (iii), we get

$$3f(\sin x) = 2x - \left(\frac{\pi}{2} - x\right) = 3x - \frac{\pi}{2}$$

or $\quad f(\sin x) = x - \frac{\pi}{6}$ or $f(x) = \sin^{-1} x - \frac{\pi}{6}$

Hence, $f(x) = \sin^{-1} x - \frac{\pi}{6} \quad \forall\, x \in [-1, 1]$

Illustration 149. If $3f(x) + 2f\left(\frac{1}{x}\right) = x^2 + 2$ $\forall\, x \in R - \{0\}$, then find $f(x)$.

Sol : Given $3f(x) + 2f\left(\frac{1}{x}\right) = x^2 + 2 \quad$ ...(i)

Replacing x by $\frac{1}{x}$ in Eq. (i), we get

$$3f\left(\frac{1}{x}\right) + 2f(x) = \frac{1}{x^2} + 2 \quad \text{...(ii)}$$

Now multiplying Eq. (i) by 3 and Eq. (ii) by 2, then

$$9f(x) + 6f\left(\frac{1}{x}\right) = 3x^2 + 6 \quad \text{...(iii)}$$

and $\quad 4f(x) + 6f\left(\frac{1}{x}\right) = \frac{2}{x^2} + 4 \quad$ ...(iv)

Now subtracting Eq. (iv) from Eq. (iii), then we get

$$5f(x) = 3x^2 - \frac{2}{x^2} + 2 \text{ or } f(x) = \left(\frac{3x^4 + 2x^2 - 2}{5x^2}\right)$$

Illustration 150. If $f(x) + f\left(\frac{x-1}{x}\right) = 1 + x$ $\forall\, x \in R - \{0, 1\}$, then find $f(x)$.

Sol : Given $f(x) + f\left(1 - \frac{1}{x}\right) = 1 + x \quad$ ...(i)

Replacing x by $\left(1 - \frac{1}{x}\right)$ in Eq. (i), then we get

$$f\left(1 - \frac{1}{x}\right) + f\left(\frac{1}{1-x}\right) = 1 + 1 - \frac{1}{x} \quad \text{...(ii)}$$

and replacing x by $\left(\frac{1}{1-x}\right)$ in Eq. (i), then we get

$$f\left(\frac{1}{1-x}\right) + f(x) = 1 + \frac{1}{1-x} \quad \text{...(iii)}$$

Now adding Eq. (i) and Eq. (iii) and then subtracting Eq. (ii), then we get

$$2f(x) = 1 + x + 1 + \frac{1}{1-x} - 2 + \frac{1}{x}$$

$$= x + \frac{1}{1-x} + \frac{1}{x}$$

$$= \frac{x^2(1-x) + x + (1-x)}{x(1-x)} = \frac{(1 + x^2 - x^3)}{x(1-x)}$$

or $\quad f(x) = \frac{(1 + x^2 - x^3)}{2x(1-x)}$

(iii) Functional Equations of Difference Form

Let $f(x)$ be a function defined on the set of whole numbers 0, 1, 2, 3, ..., n is called a sequence.

(a) If $f(x+1) + pf(x) = q$, then

$$f(x) = \begin{cases} A + qx, & p = -1 \\ (-p)^x A + \frac{q[1 - (-p)^x]}{(1+p)}, & p \neq -1 \end{cases}$$

where $A = f(0)$.

(b) If $f(x+2) + pf(x+1) + qf(x) = 0$, then

$$f(x) = \begin{cases} A \cdot \alpha^x + B \cdot \beta^x, & \alpha \neq \beta \\ (A + Bx)\alpha^x, & \alpha = \beta \end{cases}$$

Where α, β are the roots of the equation $x^2 + px + q = 0$ and A, B are determined if $f(0)$ and $f(1)$ are given.

Illustration 151. If $f(x+1) - f(x) = 3$ and $f(0) = 1$, then find $f(x)$.

Sol : Comparing $f(x+1) - f(x) = 3$ with $f(x+1) + pf(x) = q$, we get

$$p = -1, q = 3$$

$\therefore \quad f(x) = A + qx = f(0) + qx = 1 + 3x$

Illustration 152. If $f(x+1) - 3f(x) = 0$ and $f(0) = 1$, then find $f(x)$.

Sol : Comparing $f(x+1) - 3f(x) = 0$ with $f(x+1) + pf(x) = q$, we get

$$p = -3, q = 0 \quad (\text{Here } p \neq -1)$$

$\therefore \quad f(x) = (-p)^x \cdot A + \frac{q[1 - (-p)^x]}{(1+p)}$

$$= 3^x \cdot f(0) + 0 = 3^x$$

Hence, $\quad f(x) = 3^x$

Illustration 153. If $f(x+2) - 5f(x+1) + 6f(x) = 0$ and $f(0) = 2, f(1) = 5$, then find $f(x)$.

Sol : Comparing $f(x+2) - 5f(x+1) + 6f(x) = 0$ with $f(x+2) + pf(x+1) + qf(x) = 0$, we get

$$p = -5, q = 6$$

Let α, β be the roots of the equation

$$x^2 - 5x + 6 = 0$$

$\therefore \quad \alpha = 2, \beta = 3$

As, $\alpha \neq \beta$

$\therefore \quad f(x) = A \cdot \alpha^x + B \cdot \beta^x = A(2)^x + b(3)^x \quad$ ...(i)

$\because \quad f(0) = 2 \Rightarrow A + B = 2 \quad$ ...(ii) and $f(1) = 5 \Rightarrow 2A + 3B = 5$ ...(iii)

solving Eq. (ii) and Eq. (iii), we get

$$A = B = 1$$

$$\therefore \quad f(x) = 2^x + 3^x$$

Illustration 154. If $f(x+2) - 6f(x+1) + 9f(x) = 0$ and $f(0) = 1, f(1) = 2$, then find $f(x)$.

Sol : Comparing $f(x+2) - 6f(x+1) + 9f(x) = 0$ with $f(x+2) + pf(x+1) + qf(x) = 0$, we get

$$p = -6, q = 9$$

Let α, β be the roots of the equation

$$x^2 - 6x + 9 = 0$$

$$\Rightarrow \quad (x-3)^2 = 0$$

$$\therefore \quad \alpha = \beta = 3$$

so $\quad \alpha = \beta$

$\therefore \quad f(x) = (A + Bx)\alpha^x = (A + Bx)3^x$ ...(i)

$\because \quad f(0) = 1 \Rightarrow A = 1$ ...(ii)

and $\quad f(1) = 2 \Rightarrow (A + B)\,3^1 = 2$ ...(iii)

Solving Eqs. (ii) and (iii), we get

$$A = 1, \; B = -\frac{1}{3}$$

Hence, $f(x) = \left(1 - \frac{x}{3}\right)3^x = (3 - x)3^{x-1}$ (from (i))

Illustration 155. If $f(x+1) + f(x-1) = \sqrt{3}f(x)$ and $f(2) = 4$, then find $f(x)$.

Sol : Given $f(x+1) + f(x-1) = \sqrt{3}f(x)$ ...(i)

Replacing x by $(x+1)$ in Eq. (i), then

$f(x+2) + f(x) = \sqrt{3}\,f(x+1)$ ...(ii)

Eq. (ii) can be rewritten as

$f(x+2) - \sqrt{3}f(x+1) + f(x) = 0$ ...(iii)

Comparing Eq. (iii) with

$f(x+2) + pf(x+1) + qf(x) = 0$, we get

$$p = -\sqrt{3}, q = 1$$

Let α, β be the roots of equation, $x^2 - \sqrt{3}x + 1 = 0$

or $$x = \frac{\sqrt{3} \pm \sqrt{(3-4)}}{2} = \frac{\sqrt{3} \pm i}{2} = \left(\cos\frac{\pi}{6} \pm i\sin\frac{\pi}{6}\right)$$

$$\therefore \quad \alpha = \cos\frac{\pi}{6} + i\sin\frac{\pi}{6},$$

$$\beta = \cos\frac{\pi}{6} - i\sin\frac{\pi}{6}, \; i = \sqrt{-1} \text{ as } \alpha \neq \beta$$

$$\therefore \; f(x) = A\cdot\alpha^x + B\cdot\beta^x$$

$= A\left(\cos\frac{\pi}{6} + i\sin\frac{\pi}{6}\right)^x + B\left(\cos\frac{\pi}{6} - i\sin\frac{\pi}{6}\right)^x$ ...(iv)

$\because \quad f(2) = 4$

$$\Rightarrow \quad A\left(\cos\frac{\pi}{6} + i\sin\frac{\pi}{6}\right)^2 + B\left(\cos\frac{\pi}{6} - i\sin\frac{\pi}{6}\right)^2 = 4$$

$$\Rightarrow \quad A\left(\cos\frac{\pi}{3} + i\sin\frac{\pi}{3}\right) + B\left(\cos\frac{\pi}{3} - i\sin\frac{\pi}{3}\right) = 4$$

$$\Rightarrow \quad A(1 + i\sqrt{3}) + B(1 - i\sqrt{3}) = 8$$

On comparing real and imaginary parts, we get

$$A + B = 8, \; A - B = 0$$

$$\Rightarrow \quad A = B = 4$$

$$\therefore f(x) = 4\left[\left(\cos\frac{\pi}{6} + i\sin\frac{\pi}{6}\right)^x + \left(\cos\frac{\pi}{6} - i\sin\frac{\pi}{6}\right)^x\right]$$

(from (iv))

$$= 4\left[\left(\cos\frac{\pi x}{6} + i\sin\frac{\pi x}{6}\right) + \left(\cos\frac{\pi x}{6} - i\sin\frac{\pi x}{6}\right)\right]$$

$$= 4\left(2\cos\frac{\pi x}{6}\right)$$

$$\therefore \quad f(x) = 8\cos\left(\frac{\pi x}{6}\right)$$

Note :

Period of $f(x) = 8\cos\left(\frac{\pi x}{6}\right)$ is $\frac{2\pi}{\pi/6} = 12$

Hence, If $f(x+1) + f(x-1) = \sqrt{3}\,f(x)$, then $f(x)$ is periodic with period 12.

INTRODUCTORY EXERCISE 2.9

Only One Option is Correct

1. The function $f(x) = |\cos 2x| + \frac{2x}{n} - \left[\frac{2x}{\pi}\right]$ is periodic with period ([·] is the greatest integer function) :
   (a) π (b) 2π
   (c) $\frac{\pi}{2}$ (d) None of these
2. $f(x) = x^a + \sin x - ax$ $(x > 0)$ is periodic function for :
   (a) one value of a (b) two values of a
   (c) infinite values of a (d) no value of a
3. The smallest positive period of $\cos^6 3x + \sin^6 3x$ is :
   (a) $\frac{\pi}{6}$ (b) $\frac{\pi}{3}$ (c) $\frac{\pi}{2}$ (d) π
4. Period of the function $f(x) = 4\cos\left(\frac{x}{4}\right) + 4\sin\left(\frac{x}{5}\right)$ is :
   (a) 40π (b) 20π
   (c) 2π (d) π
5. Period of the function
   $f(x) = |\sin 3x| + |\cos 3x| + |\tan 4x|$ is :
   (a) $\frac{\pi}{4}$ (b) $\frac{\pi}{2}$
   (c) π (d) 2π

6. Period of $f(x)=\sin\sqrt{2}x+\cos\sqrt{3}x$ is :

(a) π (b) $\frac{\pi}{\sqrt{6}}$

(c) 2π (d) None of these

7. If $f(x+2a)=f(x-2a)$, then $f(x)$ is :

(a) a periodic function with period $4a$
(b) a periodic function with period $2a$
(c) a periodic function with indeterminate period
(d) a non-periodic function

8. If $f(x+10)+f(x+4)=0$ then $f(x)$ is a periodic function with period :

(a) 2 (b) 4 (c) 6 (d) 12

9. The period of

$$f(x)=\frac{|\sin 4x|+|\cos 4x|}{|\sin 4x-\cos 4x|+|\sin 4x+\cos 4x|}\text{ is :}$$

(a) $\frac{\pi}{8}$ (b) $\frac{\pi}{4}$ (c) $\frac{\pi}{2}$ (d) π

10. The period of $f(x)=\sin x+\tan\frac{x}{2}+\sin\frac{x}{2^2}+\tan\frac{x}{2^3}$ $+\ldots+\sin\frac{x}{2^{n-1}}+\tan\frac{x}{2^n}$ is :

(a) π (b) 2π (c) $2^{n-1}\pi$ (d) $2^n\pi$

11. Consider a real valued function $f(x)$ satisfying $2f(x)=(f(x))^y+(f(y))^x\ \forall\, x, y\in R$ and $f(1)=a$, where $a\neq 1$, then $(a-1)\sum_{i=1}^{n} f(i)$ is :

(a) a^n-1 (b) $a^{n+1}+1$
(c) a^n+1 (d) $a^{n+1}-a$

12. If $f(x+y+1)=(\sqrt{f(x)}+\sqrt{f(y)})^2$ and $f(0)=1\ \forall\, x, y\in R$, then :

(a) $f(x)=x^2$ (b) $f(x)=(x+1)^2$
(c) $f(x)=(x-1)^2$ (d) None of these

13. If $F(n+1)=\frac{2F(n)+1}{2}$, $n\in N$ and $f(1)=2$, then $f(101)$ equals :

(a) 49 (b) 50 (c) 51 (d) 52

14. The function f satisfies the functional equation $3f(x)+2f\left(\frac{x+59}{x-1}\right)=10x+30$ for all real $x\neq 1$, then $f(7)$ is :

(a) – 8 (b) 4 (c) 8 (d) 11

15. If $f:R^+\to R$, $f(x)+3xf\left(\frac{1}{x}\right)=2(x+1)$, then $f(99)$ is equal to :

(a) 30 (b) 40 (c) 50 (d) 60

16. For the differentiable function $f:R-\{0\}\to R$, let $3f(x)+2f\left(\frac{1}{x}\right)=\frac{1}{x}-10$, then $\left|f(3)+f'\left(\frac{1}{4}\right)\right|$ is equal to **[JEE Main 13 Apr 2023 (MS)]**

(a) $\frac{33}{5}$ (b) 8 (c) $\frac{29}{5}$ (d) 13

17. Function $f(x)$ satisfies the functional equation $x^2f(x)+f(1-x)=2x-x^4$ for all real x, then $f(x)$ must be :

(a) x^2 (b) $1-x^2$
(c) $1+x^2$ (d) x^2+x+1

18. Let $f:R-\{0,1\}\to R$ be a function such that $f(x)+f\left(\frac{1}{1-x}\right)=1+x$. Then $f(2)$ is equal to **[JEE Main 1 Feb 2023 (ES)]**

(a) 9/2 (b) 9/4
(c) 7/4 (d) 7/3

19. Consider a function $f:N\to R$, satisfying $f(1)+2f(2)+3f(3)+\ldots+xf(x)=x(x+1)f(x)$; $x\geq 2$ with $f(1)=1$. Then $\frac{1}{f(2022)}+\frac{1}{f(2028)}$ is equal to **[JEE Main 29 Jan 2023 (ES)]**

(a) 8200 (b) 8000
(c) 8400 (d) 8100

20. If $f(x+1)-f(x)=2$, and $f(0)=1$, then $f(7)$ is equal to :

(a) 13 (b) 15 (c) 22 (d) 20

21. If $f(x+1)-2f(x)=1$ and $f(0)=1$, then $f(4)$ is equal to :

(a) 7 (b) 15 (c) 31 (d) 63

22. If $f(x+2)-4f(x+1)+4f(x)=0$ and $f(0)=0$, $f(1)=1$, then $f(5)$ is equal to :

(a) 20 (b) 40
(c) 80 (d) 160

23. If $f(x+2)+f(x+1)-2f(x)=0$ and $f(0)=1$, $f(1)=2$, then $f(5)$ is equal to :

(a) 12 (b) 16
(c) 20 (d) 24

Some Important Functions

(a) Homogeneous Function : A function is said to be homogenous function if each term of this function is of the same degree with respect to these variables.

For example : $ax^2+2hxy+by^2$ is homogeneous in x and y. Symbolically if $f(tx,ty)=t^n f(x,y)$, then $f(x,y)$ is homogeneous function of degree n. where t is any real number $(\neq 0)$.

(b) Bounded Function : A function $f(x)$ is said to be bounded if $|f(x)|\leq M$, where M is a finite quantity.

(c) Explicit Function : A function $y = f(x)$ is said to be an explicit function of x if the dependent variable y can be expressed in terms of independent variable x only.

Forexample: (i) $y = x - \sin x$ (ii) $y = x + 2x^2 - x\ln x$

(iii) $f(x) = \sin^{-1} x + 2\log_5 x$

(d) Implicit Function : A function $y = f(x)$ is said to be an implicit function of x if y can not be written in terms of x only.

For example :

(i) $ax^2 + 2hxy + by^2 + 2gx + 2fy + c = 0$

(ii) $xy = \sin(x+y)$ (iii) $x^3 + y^3 = 1$

Some Important Results (Fast Track Results)

If x and y are independent variables, then :

(i) $f(x+y) = f(x)\cdot f(y) \Rightarrow f(x) = a^{kx}$, where k is a constant.

(ii) $f(x+y) = f(x) + f(y) \Rightarrow f(x) = kx$, where k is a constant.

(iii) $f(xy) = f(x) + f(y) \Rightarrow f(x) = k\ln x$ or $f(x) = 0$, where k is a constant.

(iv) $f(xy) = f(x)\, f(y) \Rightarrow f(x) = x^n, n \in R$.

Illustration 156. Let f be a function satisfying $f(x+y) = f(x)f(y)\ \forall\ x, y \in R$ if $f(1) = 3$, then find the value of $\sum_{r=1}^{n} f(r)$.

Sol : $\because\ f(x+y) = f(x)\cdot f(y)$

$\therefore f(x) = a^{kx}$, where k is constant

and $f(1) = 3 \Rightarrow a^k = 3$ or $f(x) = (a^k)^x = 3^x$

$$\therefore \sum_{r=1}^{n} f(r) = \sum_{r=1}^{n} 3^r = 3^1 + 3^2 + 3^3 + \ldots + 3^n$$

$$= \frac{3(3^n - 1)}{(3-1)} = \frac{3}{2}(3^n - 1)$$

Important Results for Maximum and Minimum Values

(i) If $f(x)$ is an increasing function defined on $[a, b]$ i.e., $f'(x) > 0\ \forall\ x \in [a,b]$

Then, max. $\{f(t) \mid a \le t \le x,\ a \le x \le b\} = f(x)$ and

min. $\{f(t) \mid a \le t \le x, a \le x \le b\} = f(a)$

(ii) If $f(x)$ is decreasing function defined on $[a, b]$ i.e., $f'(x) < 0\ \forall\ x \in [a,b]$,

Then, max. $\{f(t) \mid a \le t \le x, a \le x \le b\} = f(a)$ and

min. $\{f(t) \mid a \le t \le x, a \le x \le b\} = f(x)$

Illustration 157. If $f(x) = \begin{cases} |x+1| - 1, & -2 \le x \le 0 \\ 1 - |x-1|, & 0 < x \le 2 \end{cases}$,

then find $g(x) = \begin{cases} \min\{f(t) \mid -2 \le t \le x, & -2 \le x \le 0\} \\ \max\{f(t) \mid 0 \le t \le x, & 0 < x \le 2\} \end{cases}$

Sol : The function $f(x)$ can be written as

$$f(x) = \begin{cases} -x-2, & -2 \le x \le -1 \\ x, & -1 \le x \le 0 \\ x, & 0 < x \le 1 \\ 2-x, & 1 \le x \le 2 \end{cases}$$

$$\therefore \quad f'(x) = \begin{cases} -1, & -2 \le x \le -1 \\ 1, & -1 \le x \le 0 \\ 1, & 0 < x \le 1 \\ -1, & 1 \le x \le 2 \end{cases}$$

$\Rightarrow$ $f(x)$ is decreasing on $x \in [-2, -1]$, $f(x)$ is increasing on $[-1, 1]$ and $f(x)$ is decreasing on $x \in [1, 2]$

$\therefore$ min $\{f(t) \mid -2 \le t \le x, -2 \le x \le 0\}$

$$= \begin{cases} f(x), & -2 \le x \le -1 \\ f(-1), & -1 \le x \le 0 \end{cases}$$

$$= \begin{cases} -x-2, & -2 \le x \le -1 \\ -1\ , & -1 \le x \le 0 \end{cases}$$

and max $\{f(t) \mid 0 \le t \le x, 0 < x \le 2\}$

$$= \begin{cases} f(x), & 0 < x \le 1 \\ f(1), & 1 \le x \le 2 \end{cases}$$

$$= \begin{cases} x, & 0 < x \le 1 \\ 1, & 1 \le x \le 2 \end{cases}$$

Hence, $$g(x) = \begin{cases} -x-2, & -2 \le x \le -1 \\ -1, & -1 \le x \le 0 \\ x, & 0 < x \le 1 \\ 1, & 1 \le x \le 2 \end{cases}$$

Short Cuts and Important Results to Remember

1.

	Functions	Domain
(i)	$f(x) = \sqrt{(a^2 - x^2)}$	$x \in [-a, a]$
(ii)	$f(x) = \dfrac{1}{\sqrt{(a^2 - x^2)}}$	$x \in (-a, a)$
(iii)	$f(x) = \sqrt{(x^2 - a^2)}$	$x \in (-\infty, -a] \cup [a, \infty)$

(iv)	$f(x)=\dfrac{1}{\sqrt{(x^2-a^2)}}$	$x\in(-\infty,-a)\cup(a,\infty)$
(v)	$f(x)=\sqrt{(x-a)(b-x)}, a<b$	$x\in[a,b]$
(vi)	$f(x)=\dfrac{1}{\sqrt{(x-a)(b-x)}}, a<b$	$x\in(a,b)$
(vii)	$f(x)=\sqrt{(x-a)(x-b)}, a<b$	$x\in(-\infty,a]\cup[b,\infty)$
(viii)	$f(x)=\dfrac{1}{\sqrt{(x-a)(x-b)}}, a<b$	$x\in(-\infty,a)\cup(b,\infty)$
(ix)	$f(x)=\sqrt{\left(\dfrac{x-a}{b-x}\right)}, a<b$	$x\in[a,b)$
(x)	$f(x)=\sqrt{\left(\dfrac{x-a}{b-x}\right)}, a>b$	$x\in(b,a]$
(xi)	$f(x)=\sqrt{\left(\dfrac{x-a}{x-b}\right)}, a<b$	$x\in(-\infty,a]\cup(b,\infty)$
(xii)	$f(x)=\sqrt{\left(\dfrac{x-a}{x-b}\right)}, a>b$	$x\in(-\infty,b)\cup[a,\infty)$
(xiii)	$f(x)=\ln(a^2-x^2)$	$x\in(-a,a)$
(xiv)	$f(x)=\ln(x^2-a^2)$	$x\in(-\infty,-a)\cup(a,\infty)$
(xv)	$f(x)=\ln\{(x-a)(b-x)\}, a<b$	$x\in(a,b)$
(xvi)	$f(x)=\ln\{(x-a)(b-x)\}, a>b$	$x\in(b,a)$
(xvii)	$f(x)=\ln\{(x-a)(x-b)\}, a<b$	$x\in(-\infty,a)\cup(b,\infty)$
(xviii)	$f(x)=\ln\{(x-a)(x-b)\}, a>b$	$x\in(-\infty,b)\cup(a,\infty)$
(xix)	$f(x)=\dfrac{\|x+a\|}{(x+a)}$ or $\dfrac{(x+a)}{\|x+a\|}$	$x\in R\sim\{-a\}$
(xx)	$f(x)=\sqrt{\cos(\sin x)}$	$x\in R$
(xxi)	$f(x)=\sqrt{\sin(\cos x)}$	$x\in\left[2n\pi-\dfrac{\pi}{2}, 2n\pi+\dfrac{\pi}{2}\right], n\in I$
(xxii)	$f(x)=\sin^{-1}\left(\dfrac{1+x^2}{2x}\right)$ or $\cos^{-1}\left(\dfrac{1+x^2}{2x}\right)$	$x=\{-1,1\}$

2. Range of $f(x)=\sqrt{(a^2-x^2)}$ is $[0,a]$

3. Range of $f(x)=\dfrac{|x+a|}{(x+a)}$ or $\dfrac{(x+a)}{|x+a|}$ is $\{-1,1\}$

4. Range of $f(x)=a\sin x+b\cos x+c$ is $[-c+\sqrt{(a^2+b^2)}, c+\sqrt{(a^2+b^2)}]$

5. Range of $f(x)=a|\sin x|-b|\cos x|$ or $a|\cos x|-b|\sin x|$ is $[-b,a]$ for $a,b>0$.

6. Range of $f(x)=m^x+m^{-x}$ is $[2,\infty)\ \forall\ m\in N$

7. Range of $f(x)=a\sin\sqrt{(b^2-x^2)}$ is $[0, a\sin b]$, where $a>0$ and $0<b<\pi$.

8. Range of $f(x)=a\cos\sqrt{(b^2-x^2)}$ is $[a\cos b, a]$, where $a>0$ and $0<b<\dfrac{\pi}{2}$.

9. If $f(x)\cdot f(1/x)=f(x)+f(1/x)\ \forall\ x\in R-\{0\}$ and $f(x)$ be a polynomial function of degree $n\in N$, then $f(x)=1\pm x^n$.

10. If $f(x)=\dfrac{a^x}{a^x+\sqrt{a}}(a>0)$, then $f(x)+f(1-x)=1$ and $\sum_{r=1}^{2n} f\left(\dfrac{r}{2n+1}\right)=n$, $\sum_{r=1}^{2n-1} f\left(\dfrac{r}{2n}\right)=\dfrac{1}{2}(2n-1)$.

11. If $f(x+1)+pf(x)=q$, then

$$f(x) = \begin{cases} A+qx, & p=-1 \\ (-p)^x A + \dfrac{q\{1-(-p)^x\}}{1+p}, & p \neq -1 \end{cases}$$

where $A = f(0)$.

12. If $f(x+2)+p\,f(x+1)+q\,f(x)=0$, then

$$f(x) = \begin{cases} A\cdot\alpha^x + B\cdot\beta^x, & \alpha \neq \beta \\ (A+Bx)\alpha^x, & \alpha=\beta \end{cases}$$

where α, β are the roots of the equation $x^2+px+q=0$ and A, B are determined if $f(0)$ and $f(1)$ are given.

13. If $f(x)=\dfrac{ax+b}{c}$; $a, c \neq 0$ is bijective functions, then $f^{-1}(x)=\dfrac{cx-b}{a}$.

14. The number of onto functions that can be defined from a finite set A containing n elements onto a finite set B containing 2 elements $= 2^n - 2$.

15. The number of onto functions that can be defined from a finite set A containing n elements onto a finite set B containing 3 elements $= 3^n - 3 \cdot 2^n + 3$.

Solved Examples

SECTION-I (SINGLE CORRECT CHOICE TYPE)

This section contains **15 Multiple Choice Examples.** Each example has four choices (a), (b), (c) and (d) out of which **Only One** is correct.

Example 1. If $f\left(\sqrt{\left(\dfrac{x-1}{x+1}\right)}\right)=3x$, then the value of $f(3)$ is :

(a) $-\dfrac{15}{4}$ (b) $\dfrac{\sqrt{2}}{2}$ (c) 9 (d) $\dfrac{\sqrt{3}}{3}$

Sol : (a) Put $\sqrt{\dfrac{x-1}{x+1}}=3 \Rightarrow \dfrac{x-1}{x+1}=9$

or $x=-5/4$

$\Rightarrow$ $f(3)=3\times-\dfrac{5}{4}=-\dfrac{15}{4}$

Example 2. Let $f(x)$ be a function such that $f(x)+3f(-x)=\cos x$ for even real number x, then the value of $f(\pi)$ is :

(a) $-\dfrac{1}{4}$ (b) $-\dfrac{1}{2}$

(c) $\dfrac{1}{4}$ (d) 1

Sol : (a) Given $f(x)+3f(-x)=\cos x$

Putting $x=\pi$ and $x=-\pi$, then

$$f(\pi)+3f(-\pi)=\cos\pi=-1$$

$\Rightarrow$ $f(\pi)+3f(-\pi)=-1$ ...(i)

and $f(-\pi)+3f(\pi)=\cos(-\pi)=\cos\pi=-1$

$\Rightarrow$ $3f(\pi)+f(-\pi)=-1$ ...(ii)

Now from (ii), $9f(\pi)+3f(-\pi)=-3$ ...(iii)

Subtracting Eq. (i) from Eq. (iii), we get

$$8f(\pi)=-2$$

$\therefore$ $f(\pi)=-\dfrac{1}{4}$

Example 3. Let f : (0, 1) $\to R$ be a function defined by $f(x)=\dfrac{1}{1-e^{-x}}$ and $g(x) = [f(-x) - f(x)]$. Consider two statements

(I) g is an increasing function in (0, 1)

(II) g is one-one in (0, 1),

Then : **[JEE Main 25 Jan 2023 (MS)]**

(a) only (I) is true

(b) only (II) is true

(c) Neither (I) nor (II) is true

(d) Both (I) and (II) are true

Sol : (d) $\because$ $f(x)=\dfrac{1}{1-e^{-x}}$

$\therefore$ $g(x)=[f(-x)-f(x)]=\dfrac{1}{1-e^x}-\dfrac{1}{1-e^{-x}}$

$$=\frac{1}{1-e^x}-\frac{e^x}{e^x-1}=\frac{1+e^x}{1-e^x}$$

$$g'(x)=\frac{2e^x}{(1-e^x)^2}>0$$

Hence, $g(x)$ is one-one is (0, 1) and $g(x)$ is increasing in (0 1).

Example 4. The domain of the function $f(x)=\ln\left(\ln\dfrac{x}{\{x\}}\right)$ is : (where $\{\cdot\}$ denotes the fractional part function).

(a) $(0, \infty) - I$ (b) $(1, \infty) - I$

(c) $R - I$ (d) $(2, \infty) - I$

Sol : (b) Function $f(x)$ is defined, when

$$\ln\frac{x}{\{x\}}>0 \text{ and } \frac{x}{\{x\}}>0$$

$\Rightarrow$ $\dfrac{x}{\{x\}}>1$ and $\dfrac{x}{\{x\}}>0$ ...(i)

Here $0<\{x\}<1 \Rightarrow x \notin I$ ...(ii)

From Eq. (i), $x > \{x\}$ and $x > 0$

$\Rightarrow$ $x-\{x\}>0$ and $x>0$

$\Rightarrow$ $[x]>0$ and $x>0$ (where $[\cdot]$ denotes greatest integer function) or $\quad x \geq 1$ and $x > 0$

or $\quad x \geq 1$

or $\quad x > 1 \qquad (\because x \notin I)$ ...(iii)

From Eqs. (ii) and (iii), $D_f = (1, \infty) - I$

Example 5. Let $f(x) = [9^x - 3^x + 1] \ \forall \ x \in (-\infty, 1)$, then range of f(x) is : ([·] denotes the greatest integer function)

(a) {2, 3, 4, 5, 6} (b) {0, 1, 2, 3, 4, 5, 6, 7}
(c) {0, 1, 2, 3, 4, 5, 6} (d) {1, 2, 3, 4, 5, 6, 7}

Sol : (c) $\because \ 9^x - 3^x + 1 = 3^{2x} - 3^x + 1$

$$= \left(3^x - \frac{1}{2}\right)^2 - \frac{1}{4} + 1$$

$$= \left(3^x - \frac{1}{2}\right)^2 + \frac{3}{4} \geq \frac{3}{4}$$

For $\quad x \in (-\infty, 1)$

$\Rightarrow \quad \frac{3}{4} \leq 9^x - 3^x + 1 < 7$

Now $[9^x - 3^x + 1] = 0, 1, 2, 3, 4, 5, 6$

$\therefore \quad R_f = \{0, 1, 2, 3, 4, 5, 6\}$

Example 6. Let $f : R - \{0\} \to R$ be any function such that $f(x) + 2f\left(\frac{1}{x}\right) = 3x$. The sum of the values of x for which f(x) = 1 is :

(a) 1 (b) 2 (c) – 1 (d) – 2

Sol : (c) Given $f(x) + 2f\left(\frac{1}{x}\right) = 3x$ ...(i)

Replace x by $1/x$ in Eq. (i), then

$$f\left(\frac{1}{x}\right) + 2f(x) = \frac{3}{x}$$

or $\quad 4f(x) + 2f\left(\frac{1}{x}\right) = \frac{6}{x}$ ...(ii)

Subtracting Eq. (i) from Eq. (ii), then we get

$$3f(x) = \frac{6}{x} - 3x$$

or $\quad f(x) = \frac{2}{x} - x = 1$ (given)

$\Rightarrow \quad 2 - x^2 = x$

or $\quad x^2 + x - 2 = 0$

or $\quad (x+2)(x-1) = 0$

$\Rightarrow \quad x = -2, 1$

Hence, sum of the values of $x = -2 + 1 = -1$

Example 7. Suppose $f(x) = ax + b$ and $g(x) = bx + a$, where $a, b \in N$. If $f(g(50)) - g(f(50)) = 28$, then number of all possible values of the product (ab) is :

(a) 1 (b) 2 (c) 3 (d) 4

Sol : (b) $\because \ f(g(50)) = f(50b + a) = a(50b + a) + b$

$$= 50ab + a^2 + b$$

and $\quad g(f(50)) = g(50a + b) = b(50a + b) + a$

$$= 50ab + b^2 + a$$

$\therefore \quad f(g(50)) - g(f(50))$

$$= (50ab + a^2 + b) - (50ab + b^2 + a)$$

$$= (a^2 - b^2) - (a - b)$$

$$= (a - b)(a + b - 1)$$

But given $f(g(50)) - g(f(50)) = 28$

or $\quad (a-b)(a+b-1) = 28 = 1 \times 28, 2 \times 14, 4 \times 7$

$\Rightarrow \quad a - b = 1, a + b - 1 = 28$

$\Rightarrow \quad a = 15, \ b = 14$

$\therefore \quad ab = 210$

and $\quad a - b = 2, \ a + b - 1 = 14$

$\Rightarrow \quad a = \frac{17}{2}, b = \frac{13}{2}$ (Impossible) $\quad (\because a, b \in N)$

and $\quad a - b = 4, a + b - 1 = 7$

$\Rightarrow \quad a = 6, b = 2$

$\therefore \quad ab = 12$

Hence, required number = 2.

Example 8. Let $f(x) = x^2, x \in R$. For any $A \subseteq R$, define $g(A) = \{x \in R : f(x) \in A\}$. If $S = [0, 4]$, then which one of the following statements is not true?

[JEE Main 10 Apr 2019 (MS)]

(a) $g(f(S)) \neq S$ (b) $f(g(S)) \neq f(S)$
(c) $f(g(S)) = S$ (d) $g(f(S)) = g(S)$

Sol : (d) From the given information

$g(S) = \{x \in R : x^2 \in S\} = [-2, 2]$. So,

$f(g(S)) = [0, 4] = S$

and $\quad f(S) = S^2, S \in R = [0, 16]$. So,

$g(f(S)) = [-4, 4] \neq g(S)$

So, $\quad g(f(S)) \neq S$

Hence, $g(f(S)) = g(S)$ is not true.

Example 9. Let α, β and γ be three positive real numbers. Let $f(x) = \alpha x^5 + \beta x^3 + \gamma x, x \in R$ and $g : R \to R$ be such that $g(f(x)) = x$ for all $x \in R$. If $a_1, a_2, a_3, \ldots, a_n$ be in arithmatic progression with mean zero, then the value of $f\left[g\left(\frac{1}{n}\sum_{i=1}^{n} f(a_i)\right)\right]$ is equal to

[JEE Main 28 Jul 2022 (MS)]

(a) 0 (b) 3 (c) 9 (d) 27

Sol : (a) Given, mean is zero, so

$$\frac{a_1 + a_2 + a_3 + \ldots + a_n}{n} = 0$$

$\Rightarrow \quad a_1 + a_2 + a_3 + \ldots + a_n = 0$ ...(i)

$\Rightarrow \quad a_1 + a_n = a_2 + a_{n-1} = a_3 + a_{n-2} = \ldots = 0$

$\Rightarrow \quad a_n = -a_1, a_{n-1} = -a_2, a_{n-2} = -a_3, \ldots$

$\therefore \quad a_1^3 + a_2^3 + a_3^3 + \ldots + a_n^3$

$$= a_1^5 + a_2^5 + a_3^5 + \ldots + a_n^5 = 0$$ ...(ii)

Now $\sum_{i=1}^{n} f(a_i) = \alpha\,(a_1^5 + a_2^5 + a_3^5 + \ldots + a_n^5)$

$$+ \beta\,(a_1^3 + a_2^3 + a_3^3 + \ldots + a_n^3)$$
$$+ \gamma\,(a_1 + a_2 + a_3 + \ldots + a_n)$$
$$= \alpha(0) + \beta(0) + \gamma(0) \text{ [from (i) and (ii)]}$$
$$= 0 \qquad \ldots\text{(iii)}$$

Now given $\quad g(f(x)) = x$

$\Rightarrow \quad f(g(f(x)) = f(x)$

So, $\quad f\left[g\left(\frac{1}{n}\sum_{i=1}^{n} f(a_i)\right)\right] = \frac{1}{n}\sum_{i=1}^{n} f(a_i) = 0 \qquad$ [from (iii)]

Example 10. If $f(x)$ is an even function and satisfies the relation $x^2 f(x) - 2f\left(\frac{1}{x}\right) = g(x)$, where $g(x)$ is an odd function, then the value of $f(5)$ is :

(a) 0 (b) $\frac{37}{75}$ (c) 4 (d) $\frac{51}{77}$

Sol : (a) Given $Nx^2 f(x) - 2f\left(\frac{1}{x}\right) = g(x) \qquad \ldots$(i)

Replacing x by $\frac{1}{x}$ in Eq. (i), then

$$\frac{1}{x^2} f\left(\frac{1}{x}\right) - 2f(x) = g\left(\frac{1}{x}\right)$$

or $\quad 2f\left(\frac{1}{x}\right) - 4x^2 f(x) = 2x^2 g\left(\frac{1}{x}\right) \qquad \ldots$(ii)

Adding Eqs. (i) and (ii) then

$$-3x^2 f(x) = g(x) + 2x^2 g\left(\frac{1}{x}\right)$$

or $\quad f(x) = -\left(\dfrac{g(x) + 2x^2 g\left(\frac{1}{x}\right)}{3x^2}\right) \qquad \ldots$(iii)

$\Rightarrow \quad f(-x) = -\left(\dfrac{g(-x) + 2(-x)^2 g\left(\frac{1}{-x}\right)}{3(-x)^2}\right)$

$$= -\left(\frac{-g(x) - 2x^2 g\left(\frac{1}{x}\right)}{3x^2}\right)$$

($\because g(x)$ is an odd function)

$$= \left(\frac{g(x) + 2x^2 g\left(\frac{1}{x}\right)}{3x^2}\right) = -f(x) \qquad \text{(from (iii))}$$

or $\quad f(x) = -f(-x) = -f(x)$

($\because f(x)$ is an even function)

$\Rightarrow \quad 2f(x) = 0$

or $\quad f(x) = 0$

$\therefore \quad f(5) = 0$

Example 11. Suppose $f(x) = \frac{2x+7}{9}$ and $g(x) = \frac{9x+5}{7}$, then we define $h(x) = f^{-1}(x) + g^{-1}(x) - 5x$. If $h^{-1}(x) = \frac{ax+b}{c}$, where c is prime, then $(a+b+c)$ is equal to :

(a) 96 (b) 106
(c) 86 (d) None of these

Sol : (a) Let $F(x) = \frac{px+q}{r}$,

then $\quad F^{-1}(x) = \frac{rx-q}{p}$

$\therefore \quad f^{-1}(x) = \frac{9x-7}{2}$, $g^{-1}(x) = \frac{7x-5}{9}$

and $\quad h(x) = \frac{cx-b}{a} \qquad \left(\because h^{-1}(x) = \frac{ax+b}{c}\right)$

Given $\quad h(x) = f^{-1}(x) + g^{-1}(x) - 5x$

$\Rightarrow \quad \frac{cx-b}{a} = \frac{9x-7}{2} + \frac{7x-5}{9} - 5x$

Comparing coefficient of x and constant term, then

$$\frac{c}{a} = \frac{9}{2} + \frac{7}{9} - 5 = \frac{5}{18}$$

and $\quad \frac{b}{a} = \frac{73}{18}$

$\Rightarrow \quad c = \frac{5a}{18}$, $b = \frac{73}{18}a \qquad \because c$ is prime

$\therefore \quad a = 18$, $b = 73$ and $c = 5$

$\Rightarrow \quad a+b+c = 18+73+5 = 96$

Example 12. If

$f(x) = \tan^2\left(\frac{\pi x}{n^2 - 5n + 8}\right) + \cot(n+m)\pi x$, $n \in N$ $m \in Q$,

is a periodic function with 2 as its fundamental period, then m can not belong to :

(a) $(-\infty, -2) \cup (-1, \infty)$

(b) $(-\infty, -3) \cup (-2, \infty)$

(c) $(-3, -2) \cup (-2, -1)$

(d) $(-3, -5/2) \cup (-5/2, -2)$

Sol : (c) $\because$ Period of $\tan^2\left(\frac{\pi x}{n^2 - 5n + 8}\right)$

$$= \frac{\pi}{\pi / (n^2 - 5n + 8)}$$
$$= (n^2 - 5n + 8) \text{ and period of}$$

$$\cot(n+m)\pi x = \frac{\pi}{(n+m)\pi} = \frac{1}{(n+m)}$$

$\therefore$ Period of $f(x)$ = LCM of $\left(n^2-5n+8, \frac{1}{n+m}\right)=2$ (given)

Case I. $n^2-5n+8=1$ or $n^2-5n+8=2$

$\Rightarrow \quad n^2-5n+7=0$ or $n^2-5n+6=0$

$\therefore \quad n=\frac{5\pm\sqrt{3}i}{2}$ or $n=2, 3$ (where $i=\sqrt{-1}$)

but $n\in N$

$\therefore \quad n=2, 3$

Case II. $\left(\frac{1}{n+m}\right)\lambda=2, \lambda\in I \Rightarrow \frac{1}{n+m}\not> 1$

For $n=2$, $\frac{1}{2+m}\not> 1 \Rightarrow \frac{1}{2+m}-1\not> 0$

$\Rightarrow \quad \frac{-1-m}{2+m}\not> 0 \Rightarrow \frac{m+1}{m+2}\not< 0$

$\therefore \quad m\notin(-2,-1)$ ...(i)

and for $n=3$, $\frac{1}{3+m}\not> 0 \Rightarrow \frac{1}{3+m}-1\not> 0$

$\Rightarrow \quad \frac{-2-m}{3+m}\not> 0 \Rightarrow \frac{m+2}{m+3}\not< 0$

$\therefore \quad m\notin(-3,-2)$ ...(ii)

Combing Eqs. (i) and (ii), we get

$$m\notin(-3,-2)\cup(-2,-1)$$

Example 13. If $2f(x-1)-f\left(\frac{1-x}{x}\right)=x$, then $f(x)$ is :

(a) $\frac{1}{3}\left\{2(1+x)+\frac{1}{(1+x)}\right\}$ (b) $2(x-1)-\frac{(1-x)}{x}$

(c) $x^2+\frac{1}{x^2}+4$ (d) $\frac{1}{4}\left\{(x+2)+\frac{1}{(x+2)}\right\}$

Sol : (a) Given, $2f(x-1)-f\left(\frac{1-x}{x}\right)=x$ ...(i)

Replacing x by $\frac{1}{x}$, then

$$2f\left(\frac{1}{x}-1\right)-f\left(\frac{1-\frac{1}{x}}{\frac{1}{x}}\right)=\frac{1}{x}$$

$\Rightarrow \quad 2f\left(\frac{1-x}{x}\right)-f(x-1)=\frac{1}{x}$ ...(ii)

Now, multiplying Eq. (i) by 2 and Eq. (ii) by 1, then adding, we get

$$3f(x-1)=2x+\frac{1}{x}$$

Now, replacing x by $x+1$, then

$$3f(x)=2(x+1)+\frac{1}{(x+1)}$$

or $\quad f(x)=\frac{1}{3}\left\{2(x+1)+\frac{1}{(x+1)}\right\}$

Example 14. Let the graph of $f(x)=-3\cos(4\pi x)$ have amplitude a, period b and let c be the number of zeros of the function on the interval $[0, 2\pi]$, then the value of c^b-a is equal to :

(a) 2 (b) $2\sqrt{6}-3$ (c) $3\sqrt{3}-3$ (d) $2\sqrt{2}-3$

Sol : (a) Let $y=f(x)=-3\cos(4\pi x)$

$\therefore \quad a$ = amplitude $=|y_{max}|=3$

b = period of $f(x)=\frac{2\pi}{4\pi}=1/2$

For c, $f(x)=0$

$\Rightarrow \quad \cos(4\pi x)=0$

$\Rightarrow \quad 4\pi x=n\pi+\frac{\pi}{2}, n\in I$

$\Rightarrow \quad x=\frac{(2n+1)}{8}$

but $\quad x\in[0, 2\pi]$

$\therefore$ for $n=0, 1, 2, 3, 4, 5, 6, 7, \ldots, 24$

$\Rightarrow \quad c$ = number of zeros of the function $f(x)$ in $[0, 2\pi]$ = 25

Hence, $\quad c^b-a=(25)^{1/2}-3=5-3=2$

Example 15. The number of functions

$f:\{1, 2, 3, 4\}\to\{a\in Z:|a|\le 8\}$ satisfying

$f(n)+\frac{1}{n}f(n+1)=1, \forall n\in\{1, 2, 3\}$ is

[JEE Main 25 Jan 2023 (ES)]

(a) 3 (b) 4 (c) 1 (d) 2

Sol : (d) Given, $f(n)+\frac{1}{n}f(n+1)=1, \forall n\in\{1, 2, 3\}$

$\Rightarrow \quad nf(n)+f(n+1)=n$ ...(i)

Putting $n=1$ in (i), we get

$f(1)+f(2)=1=f(2)=1-f(1)$ ...(ii)

Putting $n=2$ in (i), we get

$2f(2)+f(3)=2$

$\Rightarrow \quad 2(1-f(1))+f(3)=2$ [using (ii)]

$\Rightarrow \quad f(3)=2f(1)$ ...(iii)

Putting $n=3$ in (i), we get

$3f(3)+f(4)=3$

$\Rightarrow \quad 6f(1)+f(4)=3$ [Using (iii)]

$f(4)=3-6f(1)$ ...(iv)

$\because \quad f:\{1, 2, 3, 4\}\to\{a\in Z:|a|\le 8\}$

$\Rightarrow \quad -8\le f(4)\le 8$

$\Rightarrow \quad -8\le 3-6f(1)\le 8$ [using (iv)]

$\Rightarrow \quad -11\le -6f(1)\le 5$

$\Rightarrow \quad -\frac{5}{6}\le f(1)\le\frac{11}{6}$

$\Rightarrow \quad f(1)=0, 1$

Case I : $f(1)=0$

$\Rightarrow \quad f(2)=1, f(3)=0, f(4)=3$ [from (ii), (iii) & (iv)]

Case II : $f(1) = 1$

$\Rightarrow f(2) = 0, f(3) = 2, f(4) = -3$ [from (ii), (iii) & (iv)]

Hence, two such functions are possible.

SECTION-II (MULTIPLE CORRECT CHOICE TYPE)

This section contains **6 Multiple Choice Examples**. Each example has four choices (a), (b), (c) and (d) out of which **More Than One** may be correct.

Example 16. If $f_1(x) = x$, $f_2(x) = 1 - x$, $f_3(x) = \frac{1}{x}$, $f_4(x) = \frac{1}{1-x}$, $f_5(x) = \frac{x}{x-1}$, $f_6(x) = \frac{x-1}{x}$ suppose that $f_6(f_m(x)) = f_4(x)$ and $f_n(f_4(x)) = f_3(x)$, then :

(a) $m = 5$ (b) $n = 5$

(c) $m = 6$ (d) $n = 6$

Sol : (b, c) $\because \; f_6(f_m(x)) = f_4(x)$

$$\Rightarrow \quad \frac{f_m(x) - 1}{f_m(x)} = \frac{1}{1-x}$$

$$\Rightarrow \quad 1 - \frac{1}{f_m(x)} = \frac{1}{1-x}$$

$$\Rightarrow \quad \frac{1}{f_m(x)} = 1 - \frac{1}{1-x} = \frac{-x}{1-x}$$

or $\quad f_m(x) = \frac{1-x}{-x} = \frac{x-1}{x} = f_6(x)$ (given)

$\therefore \quad m = 6(c)$

and $\quad f_n(f_4(x)) = f_3(x)$

$$\Rightarrow \quad f_n\left(\frac{1}{1-x}\right) = \frac{1}{x}$$

replace $\frac{1}{1-x}$ by x, then x by $\left(1 - \frac{1}{x}\right)$

or $\quad f_n(x) = \frac{1}{\left(1 - \frac{1}{x}\right)} = \frac{x}{x-1} = f_5(x)$ (given)

$\therefore \quad n = 5(b)$

Example 17. Which of the following denotes an identity function or R?

(a) $f(x) = |x| \operatorname{sgn} x$

(b) $g(x) = \tan(\tan^{-1} x)$

(c) $h(x) = \ln e^x$

(d) $l(x) = \lim_{x\to\infty} \frac{2|x|}{\pi} \tan^{-1}(nx)$

Sol : (a, b, c, d) The function $f : R \to R$ is called an identity function if $f(x) = x \; \forall \; x \in R$

Alternate (a) :

$$\because \quad \operatorname{sgn} x = \begin{cases} \frac{|x|}{x} \text{ or } \frac{x}{|x|}, & x \neq 0 \\ 0 \quad , & x = 0 \end{cases}$$

$$\therefore \quad |x| \operatorname{sgn} x = \begin{cases} x, & x \neq 0 \\ 0, & x = 0 \end{cases}$$

$$\Rightarrow \quad f(x) = x \; \forall \; x \in R$$

Alternate (b) : $g(x) = \tan(\tan^{-1} x) = x \; \forall \; x \in R$

Alternate (c) : $h(x) = \ln e^x \; \forall \; x \in R$

$$= x \ln e \; \forall \; x \in R = x \; \forall \; x \in R$$

Alternate (d) : $l(x) = \lim_{x\to\infty} \frac{2|x|}{\pi} \tan^{-1}(nx)$

For $x > 0$

$l(x) = \lim_{x\to\infty} \tan^{-1}(0) = 0$ and for $x < 0$

$$l(x) = \lim_{x\to\infty} \frac{2x}{\pi} \tan^{-1}(nx) = \frac{2x}{\pi} \tan^{-1}(\infty)$$

$$= \frac{2x}{\pi} \cdot \frac{\pi}{2} = x$$

For $x = 0$

$$l(x) = \lim_{x\to\infty} \frac{2(-x)}{\pi} \tan^{-1}(nx)$$

$$= -\frac{2x}{\pi} \tan^{-1}(-\infty) = -\frac{2x}{\pi} \times \frac{-\pi}{2} = x$$

Hence, $\quad l(x) = x \; \forall \; x \in R$

Example 18. If domain of $f(x)$ be $(-1, 2)$ then,

(a) domain of $f(\sin x)$ will be $(-\infty, \infty)$

(b) domain of $f(\ln x)$ will be $(1/e, e^2)$

(c) domain of $f(2x - 3)$ will be $(1, 5/2)$

(d) domain of $f([x])$ will be $[0, 2)$, where $[x] \leq x$

Sol : (b, c, d)

$\because$ Domain of $f(x)$ is $(-1, 2)$

Alternate (a) : $f(\sin x) : -1 < \sin x < 2$

but $\quad -1 \leq \sin x \leq 1$

$\therefore \quad -1 < \sin x \leq 1$

Hence $\quad \sin x \neq -1$

$\therefore \quad x \neq 2n\pi - \frac{\pi}{2}, n \in I$

Hence, domain of $f(\sin x) \notin (-\infty, \infty)$

Alternate (b) : $f(\ln x) : -1 < \ln x < 2$

$\Rightarrow \quad e^{-1} < x < e^2$

$\therefore$ domain of $f(\ln x)$ is $(1/e, e^2)$

Alternate (c) : $f(2x - 3) : -1 < 2x - 3 < 2$

$\Rightarrow \quad 2 < 2x < 5$ or $1 < x < 5/2$

Hence domain of $f(2x - 3)$ is $(1, 5/2)$.

Alternate (d) : $f([x]) : -1 < [x] < 2$

$\Rightarrow \quad 0 \leq x < 2$

Hence domain of $f([x])$ is $[0, 2)$.

Example 19. Which of the following pairs of functions is/are identical?

(a) $f(x) = \tan(\tan^{-1} x)$ and $g(x) = \cot(\cot^{-1} x)$

(b) $f(x) = \text{sgn}(x)$ and $g(x) = \text{sgn}(\text{sgn}(x))$

(c) $f(x) = \cot^2 x \cdot \cos^2 x$ and $g(x) = \cot^2 x - \cos^2 x$

(d) $f(x) = e^{\ln \sec^{-1} x}$ and $g(x) = \sec^{-1} x$

Sol : (a, b, c)

Alternate (a) : $f(x) = \tan(\tan^{-1} x) = x \;\forall\; x \in R$

and $\quad g(x) = \cot(\cot^{-1} x) = x \;\forall\; x \in R$

Hence, this pair is identical.

Alternate (b) : $f(x) = \text{sgn}(x) = \begin{cases} 1, & x > 0 \\ 0, & x = 0 \\ -1, & x < 0 \end{cases}$

$\therefore \quad D_f = R,\; R_f = \{-1, 0, 1\}$

and $\quad g(x) = \text{sgn}(\text{sgn}(x)) = \begin{cases} 1, & \text{sgn}(x) > 0 \\ 0, & \text{sgn}(x) = 0 \\ -1, & \text{sgn}(x) < 0 \end{cases}$

$\Rightarrow \quad \begin{cases} 1, & x > 0 \\ 0, & x = 0 \\ -1, & x < 0 \end{cases}$

$\therefore \quad D_g = R,\; R_g = \{-1, 0, 1\}$

Also, $\quad f(x) = g(x) \;\forall\; x \in R$

Hence, this pair is identical.

Alternate (c) : $f(x) = \cot^2 x \cdot \cos^2 x = \dfrac{\cos^2 x}{\sin^2 x} \cdot \cos^2 x$

$\therefore \quad D_f = R \sim n\pi, n \in I$

and $\quad R_f = [0, \infty)$

$$g(x) = \cot^2 x - \cos^2 x$$

$$= \cos^2 x \left(\frac{1}{\sin^2 x} - 1\right)$$

$$= \frac{\cos^2 x\,(1 - \sin^2 x)}{\sin^2 x}$$

$$= \frac{\cos^2 x \cdot \cos^2 x}{\sin^2 x} = f(x)$$

$$D_g = R \sim n\pi,\; n \in I$$

and $\quad R_g = [0, \infty]$

it is clear that $D_f = D_g$ and $R_f = R_g$ and $f(x) = g(x)$

Hence, this pair is identical.

Alternate (d) : $f(x) = e^{\ln \sec^{-1} x}$

$f(x)$ is defined, when $\sec^{-1} x > 0$ and ($x \le -1$ or $x \ge 1$)

$\Rightarrow \quad x > 1$ and $x \in (-\infty, -1] \cup [1, \infty)$

$\therefore \quad D_f = (1, \infty)$

whereas $g(x) = \sec^{-1} x$ has domain $(-\infty, -1] \cup [1, \infty)$

Hence, this pair is not identical.

Example 20. If $f : [a, \infty) \to [a, \infty)$ be a function defined by $f(x) = x^2 - 2ax + a(a+1)$. If one of the solution of the equation $f(x) = f^{-1}(x)$ is 2012, then the other may be :

(a) 2014 $\qquad$ (b) 2011

(c) 2015 $\qquad$ (d) 2013

Sol : (b, d) It is clear that $f(x)$ and $f^{-1}(x)$ are symmetrical about the line

$$y = x$$

$\Rightarrow \quad f(x) = x \qquad (\because y = f(x))$

$\Rightarrow \quad x^2 - 2ax + a(a+1) = x$

$\Rightarrow \quad (x-a)^2 - (x-a) = 0$

or $\quad (x-a)(x-a-1) = 0$

$\therefore \quad x = a \quad$ or $\quad a+1$

If $a = 2012$, then $a+1 = 2013$

and if $\quad a+1 = 2012$, then $a = 2011$

Example 21. If the fundamental period of the function $f(x) = \sin x + \cos(\sqrt{(4-a^2)})x$ is 4π, then the value of a is/are :

(a) $\dfrac{\sqrt{15}}{2}$ $\qquad$ (b) $-\dfrac{\sqrt{7}}{2}$

(c) $\dfrac{\sqrt{7}}{2}$ $\qquad$ (d) $\dfrac{\sqrt{15}}{2}$

Sol : (a, b, c, d)

$\because$ Period of $\sin x$ is $2x$ and period of $\cos(\sqrt{(4-a^2)})x$ is $\dfrac{2\pi}{\sqrt{(4-a^2)}}$

$\therefore$ Period of $f(x) = $ LCM of $\left(2\pi, \dfrac{2\pi}{\sqrt{(4-a^2)}}\right) = 4\pi$ (given)

i.e., $\quad \sqrt{(4-a^2)} = \dfrac{1}{2}, \dfrac{3}{2}, \dfrac{5}{2}, \dfrac{7}{2}, \ldots$

If $\quad \sqrt{(4-a^2)} = \dfrac{5}{2}, \dfrac{7}{2}, \ldots\ldots$

then $a^2 > 4$ (Impossible)

$\therefore \quad \sqrt{(4-a^2)} = \dfrac{1}{2}, \dfrac{3}{2}$

$\Rightarrow \quad 4 - a^2 = \dfrac{1}{4}, \dfrac{9}{4}$

or $\quad a^2 = \dfrac{15}{4}, \dfrac{7}{4}$

$\Rightarrow \quad a = \pm\dfrac{\sqrt{15}}{2}, \pm\dfrac{\sqrt{7}}{2}$

SECTION-III (PARAGRAPH TYPE)

This section contains three solved paragraph with **two questions, three questions and three questions** respectively. Each question has four choices (a), (b), (c) and (d) out of which **Only One is correct**.

Paragraph for Question No. 22 and 23

Let $P(x)$ be a quadratic polynomial with real coefficients such that for all real x the relation $2(1+P(x)) = P(x-1)+P(x+1)$ holds.

If $P(0) = 8$ and $P(2) = 32$, then

22. Sum of all the coefficients of $P(x)$ is :

(a) 15 (b) 17 (c) 19 (d) 21

23. If the range of $P(x)$ is $[m, \infty)$, then the value of m is :

(a) – 5 (b) – 12 (c) – 15 (d) – 17

Sol : Let $P(x) = ax^2 + bx + c$

and given $2(1+P(x)) = P(x-1)+P(x+1)$

$$\Rightarrow \quad 2(1+ax^2+bx+c)$$

$$= a(x-1)^2 + b(x-1) + c + a(x+1)^2 + b(x+1) + c$$

$$\Rightarrow \quad 2+2ax^2+2bx+2c = 2ax^2+2bx+2a+2c$$

$$\Rightarrow \quad 2a = 2 \Rightarrow a = 1$$

also, $\quad P(0) = 8 \Rightarrow 0 + 0 + c = 8$

$\therefore \quad c = 8$

and $\quad P(2) = 32 \Rightarrow 4a + 2b + c = 32$

$\Rightarrow \quad 4+2b+8 = 32 \qquad (\because a = 1, c = 8)$

$\Rightarrow \quad b = 10$

$\therefore \quad P(x) = x^2+10x+8$

22. (c) Sum of all coefficients of $P(x) = 1+10+8 = 19$

23. (d) $\because \quad P(x) = x^2+10x+8$

$$= (x+5)^2 - 25 + 8 = (x+5)^2 - 17$$

or $\quad P(x)+17 = (x+5)^2 \geq 0$

$\Rightarrow \quad P(x)+17 \geq 0 \Rightarrow P(x) \geq -17$

or $\quad P(x) \in [-17, \infty)$

$\therefore \quad m = -17$

Paragraph for Question Nos. 24 to 26

If $(f(x))^2 \times f\left(\frac{1-x}{1+x}\right) = 64x$, $\forall\, x \in D_f$ then

24. If $f(x) = (g(x))^{1/3}$, then $g(1/2)$ is equal to :

(a) 24 (b) 48

(c) 96 (d) 192

25. If $m f\left(\frac{1}{2}\right) = n f\left(\frac{1}{3}\right)$, then the value of n^m is (where m and n are prime numbers) :

(a) 8 (b) 9 (c) 32 (d) 125

26. The domain of $f(x)$ is :

(a) $[0, \infty)$ (b) $R \sim \{1\}$

(c) $(-\infty, \infty)$ (d) None of these

Sol : $\because \quad (f(x))^2 \times f\left(\frac{1-x}{1+x}\right) = 64x \qquad \text{...(i)}$

Replacing x by $\left(\frac{1-x}{1+x}\right)$ in Eq. (i), then

$$\left(f\left(\frac{1-x}{1+x}\right)\right)^2 \times f(x) = 64\left(\frac{1-x}{1+x}\right) \qquad \text{...(ii)}$$

From Eq. (i),

$$(f(x))^4 \times \left(f\left(\frac{1-x}{1+x}\right)\right)^2 = (64)^2 x^2 \qquad \text{...(iii)}$$

Now dividing Eq. (iii) by Eq. (ii), then

$$(f(x))^3 = 64x^2\left(\frac{1+x}{1-x}\right)$$

or $$f(x) = \left(64x^2\left(\frac{1+x}{1-x}\right)\right)^{1/3}$$

24. (b) Here $g(x) = 64x^2\left(\frac{1+x}{1-x}\right)$

$$\therefore \quad g\left(\frac{1}{2}\right) = 64\left(\frac{1}{2}\right)^2\left(\frac{1+\frac{1}{2}}{1-\frac{1}{2}}\right) = 16(3) = 48$$

25. (b) $\because \quad f(x) = \left(64x^2\left(\frac{1+x}{1-x}\right)\right)^{1/3}$

$$\therefore \quad f\left(\frac{1}{2}\right) = \left(64\left(\frac{1}{4}\right)\left(\frac{1+1/2}{1-1/2}\right)\right)^{1/3} = (48)^{1/3} \qquad \text{...(iv)}$$

and $$f\left(\frac{1}{3}\right) = \left(64\left(\frac{1}{9}\right)\left(\frac{1+1/3}{1-1/3}\right)\right)^{1/3}$$

$$= \left(\frac{64}{9}\times 2\right)^{1/3} = \left(\frac{128}{9}\right)^{1/3} \qquad \text{...(v)}$$

From Eqs. (iv) and (v), we get

$$\frac{f\left(\frac{1}{2}\right)}{f\left(\frac{1}{3}\right)} = \left(\frac{48\times 9}{128}\right)^{1/3} = \left(\frac{27}{8}\right)^{1/3} = \frac{3}{2}$$

$$\Rightarrow \quad 2f\left(\frac{1}{2}\right) = 3f\left(\frac{1}{3}\right)$$

$\therefore \quad m = 2$ and $n = 3$

$\Rightarrow \quad n^m = 3^2 = 9$

26. (b) $\because \quad f(x) = \left(64x^2\left(\frac{1+x}{1-x}\right)\right)^{1/3}$

$f(x)$ is to be defined, when $x \neq 1$

Hence $\quad D_f = R \sim \{1\}$

Paragraph for Question Nos. 27 to 29

Let we describe two function $f(x) = \log_{[x]}\{x\}$ and $g(x) = \log_{\{x\}}[x]$, where $[x]$ and $\{x\}$ denote respectively the integral and fractional part of real x. Let D_f and D_g are the sets of domain of $f(x)$ and $g(x)$ respectively. R_f and R_g are the sets of range of $f(x)$ and $g(x)$ respectively.

27. Which statement is correct?

(a) D_f is the subset of D_g

(b) D_g is the subset of D_f

(c) D_f and D_g are the same sets

(d) D_f and D_g are disjoint sets

28. The union of R_f and R_g is :

(a) $(-\infty, 0)$ (b) $(-\infty, \infty)$

(c) $(-\infty, 0]$ (d) $[0, \infty)$

29. The sum of all the non-negative integral values of the set R_g is :

(a) – 1 (b) 0

(c) 1 (d) 2

Sol : $\because\ f(x) = \log_{[x]}\{x\}$

$f(x)$ is to be defined, when

$$0 < \{x\} < 1 \text{ and } [x] > 0,\ [x] \neq 1$$

$$\Rightarrow\quad x \in R - I \text{ and } x \geq 1,\ x \notin [1, 2)$$

$$\Rightarrow\quad x \in R - I \text{ and } x \geq 2 \Rightarrow x \in (2, \infty) - I$$

$$\therefore\quad D_f = (2, \infty) - I$$

Also, $\quad 0 < \{x\} < 1$

$$\Rightarrow\quad \log_{[x]} 0 < \log_{[x]}\{x\} < \log_{\{x\}} 1$$

$$\Rightarrow\quad -\infty < f(x) < 0 \quad (\because x \in (2, \infty) - I)$$

$$\therefore\quad R_f = (-\infty, 0)$$

Again $\quad g(x) = \log_{\{x\}}[x]$

$g(x)$ is to be defined, when

$$0 < \{x\} < 1 \text{ and } [x] > 0$$

$$\Rightarrow\quad x \in R - I \text{ and } x \geq 1$$

$$\Rightarrow\quad x \in (1, \infty) - I$$

$$\therefore\quad D_g = (1, \infty) - I$$

Also, for $\quad x \in (1, 2)$

$$g(x) = 0$$

and for $x \in (2, \infty) - I$

$$g(x) \in (-\infty, 0)$$

$$\therefore\quad R_g = (-\infty, 0]$$

27. (a) $\because\ D_f = (2, \infty) - I$ and $D_g = (1, \infty) - I$

Hence $\quad D_g \subset D_f$

28. (c) $R_f \cup R_g = (-\infty, 0) \cup (-\infty, 0] = (-\infty, 0]$

29. (b) $\because\ R_g = (-\infty, 0]$

Non-negative integral values in R_g is 0 (only)

$\therefore\quad$ Required sum = 0

SECTION-IV (INTEGER TYPE)

This section contains **7 examples**. The answer to each example is a positive integer.

Example 30. If fractional parts of $1/x$ and x^2 for some $x \in (\sqrt{2}, \sqrt{3})$ are equal, then the value of $x^4 - \dfrac{3}{x}$ is equal to .....

Sol : (5) $\because\ x \in (\sqrt{2}, \sqrt{3})$

$$\Rightarrow\quad \sqrt{2} < x < \sqrt{3} \quad\text{or}\quad \frac{1}{\sqrt{3}} < \frac{1}{x} < \frac{1}{\sqrt{2}}$$

$$\therefore\quad \left\{\frac{1}{x}\right\} = \frac{1}{x} \quad\text{and}\quad 2 < x^2 < 3$$

$$\text{or}\quad 0 < x^2 - 2 < 1$$

$$\therefore\quad \{x^2\} = x^2 - 2$$

$$\text{Given}\quad \left\{\frac{1}{x}\right\} = \{x^2\} \Rightarrow \frac{1}{x} = x^2 - 2 \quad \text{...(i)}$$

$$\Rightarrow\quad x^3 - 2x - 1 = 0$$

$$\Rightarrow\quad (x+1)(x^2 - x - 1) = 0$$

$$\Rightarrow\quad x \neq -1 \quad (\because -1 \notin (\sqrt{2}, \sqrt{3}))$$

$$\therefore\quad x^2 - x - 1 = 0 \Rightarrow x^2 = x + 1$$

$$\Rightarrow\quad x^4 = x^2 + 2x + 1$$

$$= x + 1 + 2x + 1 \quad (\because x^2 = x + 1)$$

$$\Rightarrow\quad x^4 = 3x + 2 \quad \text{...(ii)}$$

$$\Rightarrow\quad x^4 - \frac{3}{x} = (3x + 2) - 3(x^2 - 2)$$

(from eqs. (i) and (ii))

$$= (3x + 2) - 3(x + 1 - 2) \quad (\because x^2 = x + 1)$$

$$= 5$$

Example 31. A non-zero function $f(x)$ is symmetrical about the line $y = x$, then the value of λ (constant) such that $f^2(x) = (f^{-1}(x))^2 - \lambda x f(x) f^{-1}(x) + 3x^2 f(x)\ \forall\ x \in R^+$ is equal to ....

Sol : (3) Given $f(x) \neq 0$

$\because f(x)$ is symmetrical about the line $y = x$, then

$$f^{-1}(x) = f(x) = y = x$$

$$\Rightarrow\quad f^2(x) = (f^{-1}(x))^2 - \lambda x f(x) f^{-1}(x) + 3x^2 f(x)$$

$$\Rightarrow\quad x^2 = x^2 - \lambda x \cdot x \cdot x + 3x^2 \cdot x \Rightarrow \lambda x^3 = 3x^3$$

$$\therefore\quad \lambda = 3,\ x \neq 0 \quad (\because x \in R^+)$$

Example 32. If $x^2 + 4y^2 = 4$ and range of $f(x, y)$ is $[L, M]$, where $f(x, y) = x^2 + y^2 - xy$, then LM is equal to .....

Sol : (3) $\because \quad x^2+4y^2=4$

$\Rightarrow \quad \left(\frac{x}{2}\right)^2+\left(\frac{y}{1}\right)^2=1$

Let $x=2\cos\theta$, $y=\sin\theta$

$$\Rightarrow \quad f(\theta)=(2\cos\theta)^2+(\sin\theta)^2-(2\cos\theta)(\sin\theta)$$
$$=4\cos^2\theta+\sin^2\theta-\sin 2\theta$$
$$=3\cos^2\theta+(\cos^2\theta+\sin^2\theta)-\sin 2\theta$$
$$=\frac{3}{2}(1+\cos 2\theta)+1-\sin 2\theta$$
$$=\frac{3}{2}\cos 2\theta-\sin 2\theta+\frac{5}{2}$$

$$\Rightarrow \quad \frac{5}{2}-\sqrt{\left\{\left(\frac{3}{2}\right)^2+(-1)^2\right\}}\le\frac{3}{2}\cos 2\theta-\sin 2\theta+\frac{5}{2}\le\frac{5}{2}+\sqrt{\left\{\left(\frac{3}{2}\right)^2+(-1)^2\right\}}$$

$$\Rightarrow \quad \frac{5-\sqrt{13}}{2}\le f(\theta)\le\frac{5+\sqrt{13}}{2}$$

Hence $\quad L=\frac{5-\sqrt{13}}{2}$ and $M=\frac{5+\sqrt{13}}{2}$

$$\therefore \quad LM=\left(\frac{5-\sqrt{13}}{2}\right)\left(\frac{5+\sqrt{13}}{2}\right)=\frac{25-13}{4}=\frac{12}{4}=3$$

Example 33. If the domain of the function $\log_e\left(\frac{6x^2+5x+1}{2x-1}\right)+\cos^{-1}\left(\frac{2x^2-3x+4}{3x-5}\right)$ is $(\alpha,\beta)\cup(\gamma,\delta)$, then $18(\alpha^2+\beta^2+\gamma^2+\delta^2)$ is equal to ...... **[JEE Main, 8 Apr 2023 (ES)]**

Sol : (20)

Let $f(x)=\log_e\left(\frac{6x^2+5x+1}{2x-1}\right)+\cos^{-1}\left(\frac{2x^2-3x+4}{3x-5}\right)$

$f(x)$ is defined, when

$$\frac{6x^2+5x+1}{2x-1}>0 \quad \text{...(i)}$$

and $$-1\le\frac{2x^2-3x+4}{3x-5}\le 1 \quad \text{...(ii)}$$

From (i),

$$\frac{6x^2+5x+1}{2x-1}>0$$

$$\Rightarrow \quad \frac{(3x+1)(2x+1)}{(2x-1)}>0$$

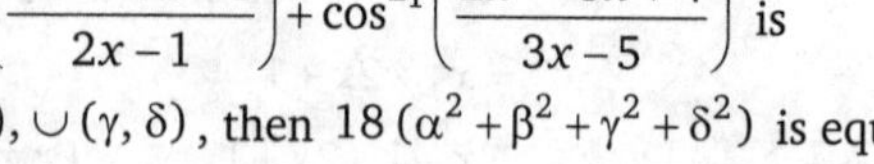

$$\therefore \quad x\in\left(-\frac{1}{2},-\frac{1}{3}\right)\cup\left(\frac{1}{2},\infty\right) \quad \text{...(a)}$$

From (ii),

$$\frac{2x^2-3x+4}{3x-5}\ge -1 \quad \text{...(iii)}$$

and $$\frac{2x^2-3x+4}{3x-5}\le 1 \quad \text{...(iv)}$$

From (iii), we get

$$\frac{2x^2-3x+4}{3x-5}+1\ge 0 \Rightarrow \frac{2x^2-1}{3x-5}\ge 0$$

$$\Rightarrow \quad \frac{(\sqrt{2}x+1)(\sqrt{2}x-1)}{(3x-5)}\ge 0$$

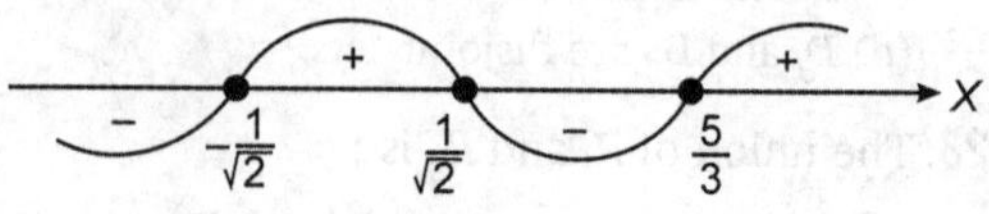

$$\therefore \quad x\in\left[-\frac{1}{\sqrt{2}},\frac{1}{\sqrt{2}}\right]\cup\left(\frac{5}{3},\infty\right) \quad \text{... (b)}$$

From (iv), we get

$$\frac{2x^2-3x+4}{3x-5}-1\le 0$$

$$\Rightarrow \quad \frac{2x^2-6x+9}{3x-5}\le 0$$

$$\Rightarrow \quad 3x-5<0 \quad \left(\begin{array}{l}\because D=(-6)^2-4.2.9<0\\ \therefore 2x^2-6x+9>0 \text{ for all } x\in R\end{array}\right)$$

$$\Rightarrow \quad x\in\left(-\infty,\frac{5}{3}\right) \quad \text{...(c)}$$

Domain of $f(x)=(a)\cap(b)\cap(c)$

$$=\left(-\frac{1}{2},-\frac{1}{3}\right)\cup\left(\frac{1}{2},\frac{1}{\sqrt{2}}\right]$$

On comparing, we get

$$\alpha=-\frac{1}{2},\ \beta=-\frac{1}{3},\ \gamma=\frac{1}{2},\ \delta=\frac{1}{\sqrt{2}}$$

$$\Rightarrow \quad \alpha^2+\beta^2+\gamma^2+\delta^2=\frac{10}{9}$$

$$\therefore \quad 18(\alpha^2+\beta^2+\gamma^2+\delta^2)=20$$

Example 34. Let $f:R\to R$ be a function defined by $f(x)=\left(2\left(1-\frac{x^{25}}{2}\right)(2+x^{25})\right)^{1/50}$. If the function $g(x)=f(f(f(x)))+f(f(x))$, then the greatest integer less than or equal to $g(1)$ is .... **[JEE Main 25 Jan 2022 (MS)]**

Sol : (2)

Given $f(x)=\left(2\left(1-\frac{x^{25}}{2}\right)(2+x^{25})\right)^{1/50}$

$$=((2-x^{25})(2+x^{25}))^{1/50}=(4-x^{50})^{1/50}$$

$\Rightarrow f(f(x))=(4-(f(x))^{50})^{1/50}=x$

Now, $g(x)=f(f(f(x)))+f(f(x))=f(x)+x$

So, $g(1) = f(1)+1 = 3^{1/50}+1 = 1.0222+1 = 2.0222$

Hence, $[g(1)] = [2.0222] = 2$

Example 35. Let ü, ∈ . If $f(x) = (c+1)x^2+(1-c^2)x+2k$ and $f(x+y) = f(x)+f(y)-xy$, for all ü, ∈ , then the value of $|2(f(1)+f(2)+f(3)+....+f(20))|$ is equal to

[JEE Main 29 Jun 2022]

Sol : (3395)

Given, $f(x) = (c+1)x^2+(1-c^2)\,x+2k$ ...(i)

and $f(x+y) = f(x)+f(y)-xy$ ...(ii)

Putting $x =$ ü $y =$ in (ii), we get

$$f(0) = f(0)+f(0)-0 \Rightarrow f(0) = 0$$

From (i), $f(0) = 0+0+2k = 0$

$\therefore$ $k = 0$

$\Rightarrow$ $f(x) = (c+1)x^2+(1-c^2)x$ ...(iii)

From (ii) and (iii),

$$(c+1)(x+y)^2+(1-c^2)(x+y)$$
$$= (c+1)\,x^2+(1-c^2)x + (c+1)y^2+(1-c^2)y-xy$$

$\Rightarrow$ $2xy\,(c+1) = -xy$

$\Rightarrow$ $2(c+1) = -1 \Rightarrow c = -\dfrac{3}{2}$

From (ii), $f(x) = -\dfrac{x^2}{2}-\dfrac{5}{4}x$

Now, $|2(f(1)+f(2)+f(3)+...+f(20))|$

$$= \left|2\sum_{r=1}^{20} f(r)\right| = \left|2\sum_{r=1}^{20}\left(-\frac{r^2}{2}-\frac{5}{4}r\right)\right|$$

$$= \sum_{r=1}^{20} r^2+\frac{5}{2}\sum_{r=1}^{20} r$$

$$= \frac{20\times21\times41}{6}+\frac{5}{2}\left(\frac{20\times21}{2}\right)$$

$$= 2870+525 = 3395$$

Example 36. Let $S = \{1,2,3,4,5,6\}$. Then the number of one-one functions $f: S \to P(S)$, where $P(S)$ denote the power set of S, such that $f(n) \subset f(m)$, where $n < m$ is ......

[JEE Main 30 Jan 2023 (MS)]

Sol : (3240)

$\because$ $S = \{1, 2, 3, 4, 5, 6\}$

$\therefore$ $P(S) = \{\phi, \{1\}, \{2\}, ..., \{6\}, \{1, 2\}, ..., \{5, 6\}, ... \{1,2,3,4,5,6\}\}$

Given, $f: S \to P(S)$

$\Rightarrow$ $n(S) = 6$ and $n(P(S)) = 2^6 = 64$

For number of one-one functions some cases arise (such that $f(n) \subset f(m)$, where $n < m$)

Case I : $f(5) =$ any 5 element subset A of S
$= {}^6C_5 = {}^6C_1 = 6$

$f(4) =$ any 4 element subset B of $A = {}^5C_4 = {}^5C_1 = 5$

$f(3) =$ any 3 element subset C of $B = {}^4C_3 = {}^4C_1 = 4$

$f(2) =$ any 2 element subset D of $C = {}^3C_2 = {}^3C_1 = 3$

$f(1) =$ any 1 element subset E of $D = {}^2C_1+1 = 2+1 = 3$ or empty subset

Thus, number of one-one functions
$= 6\times5\times4\times3\times3 = 1080$

Case II : $f(6) =$ any 5 element subset A of $S = {}^6C_5 = {}^6C_1 = 6$

$f(5) =$ any 4 element subset B of $A = {}^5C_4 = {}^5C_1 = 5$

$f(4) =$ any 3 element subset C of $B = {}^4C_3 = {}^4C_1 = 4$

$f(3) =$ any 2 element subset D of $C = {}^3C_2 = {}^3C_1 = 3$

$f(2) =$ any 1 element subset E of $D = {}^2C_1 = 2$

$f(1) =$ empty subset $= 1$

Thus, number of one-one functions
$= 6 \times 5 \times 4 \times 3 \times 2 \times 1 = 720$

Case III : $f(6) = S$

$f(5) =$ any 4 element subset A of $S = {}^6C_4 = {}^6C_2 = 15$

$f(4) =$ any 3 element subset B of $A = {}^4C_3 = {}^4C_1 = 4$

$f(3) =$ any 2 element subset C of $B = {}^3C_2 = {}^3C_1 = 3$

$f(2) =$ any 1 element subset D of $C = {}^2C_1 = 2$

$f(1) =$ empty subset $= 1$

Thus, number of one-one functions
$= 15\times4\times3\times2\times1 = 360$

Case IV : $f(6) = S$

$f(5) =$ any 5 element subset A of $S = {}^6C_5 = {}^6C_1 = 6$

$f(4) =$ any 3 element subset B of $A = {}^5C_3 = {}^5C_2 = 10$

$f(3) =$ any 2 element subset C of $B = {}^3C_2 = {}^3C_1 = 3$

$f(2) =$ any 1 element subset D of $C = {}^2C_1 = 2$

$f(1) =$ empty subset $= 1$

Thus, number of one-one functions
$= 6\times10\times3\times2\times1 = 360$

Case V : $f(6) = S$

$f(5) =$ any 5 element subset A of $S = {}^6C_5 = {}^6C_1 = 6$

$f(4) =$ any 4 element subset B of $A = {}^5C_4 = {}^5C_1 = 5$

$f(3) =$ any 2 element subset C of $B = {}^4C_2 = 6$

$f(2) =$ any 1 element subset D of $C = {}^2C_1 = 2$

$f(1) =$ empty subset $= 1$

Thus, number of one-one functions
$= 6\times5\times6\times2\times1 = 360$

Case VI : $f(6) = S$

$f(5) =$ any 5 element subset A of $S = {}^6C_5 = {}^6C_1 = 6$

$f(4) =$ any 4 element subset B of $A = {}^5C_4 = {}^5C_1 = 5$

$f(3) =$ any 3 element subset C of $B = {}^4C_3 = {}^4C_1 = 4$

$f(2) =$ any 1 element subset D of $C = {}^3C_1 = 3$

$f(1)$ = empty subset = 1

Thus, number of one-one functions

$$= 6 \times 5 \times 4 \times 3 \times 2 \times 1 = 360$$

Hence, total number of one-one functions

$$= 1080 + 720 + 360 + 360 + 360 + 360 = 3240$$

SECTION-V (MATRIX TYPE)

This section contains **2 examples**. Example 37 and Example 38 each has four statements (A, B, C and D) given in **Column I** and four statements (*p, q, r* and *s*) in **Column II**. Any given statements in **Column I** can have correct matching with one or more statement(s) given in **Column II**.

Example 37. Here [·] and { · } denotes the greatest integer and fractional part function.

	Column I		Column II
(A)	$f(x) = e^{\ln[1+\{x\}]}$	(p)	Range consists of only one natural
(B)	$g(x) = \sin^{-1}(\sin x)$	(q)	Domain is R
(C)	$h(x) = 2e^{-\lvert x\rvert}$	(r)	Periodic
(D)	$k(x) = \tan^{-1}$ $\sqrt{[x]+[-]} + \sqrt{3-\lvert x\rvert} + \frac{1}{x^2}$	(s)	Symmetric about y-axis

Sol : (A) → (p, q, r, s); **(B)** → (p, q, r); **(C)** → (q, s); **(D)** → (p, s).

(A) ∵ $f(x) = e^{\ln[1+\{x\}]} = [1+\{x\}]$ $(\because e^{\ln x} = x)$

$= 1$ $(\because 0 < \{x\} < 1)$

Here $D_f = R$, $R_f = \{1\}$

and $f(x)$ is periodic and symmetrical about y-axis.

(B) ∵ $g(x) = \sin^{-1}(\sin x)$

Here $D_g = R$, Range consist of only one natural and $g(x)$ is periodic and not symmetric about y-axis.

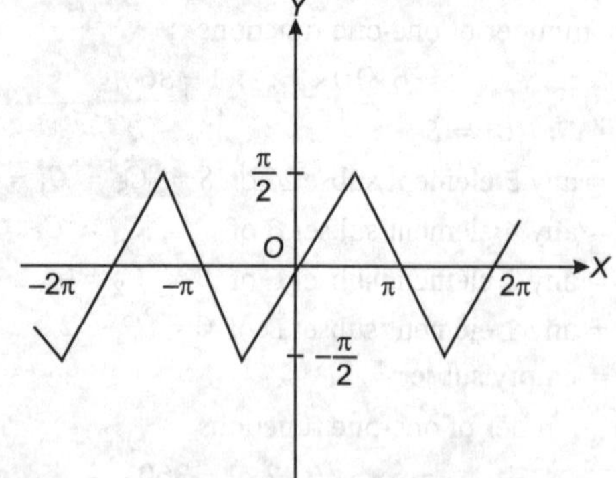

(C) $h(x) = 2e^{-|x|}$

Here $D_h = R$, Range consists of two (1 and 2) natural and $g(x)$ is non-periodic and symmetrical about y-axis

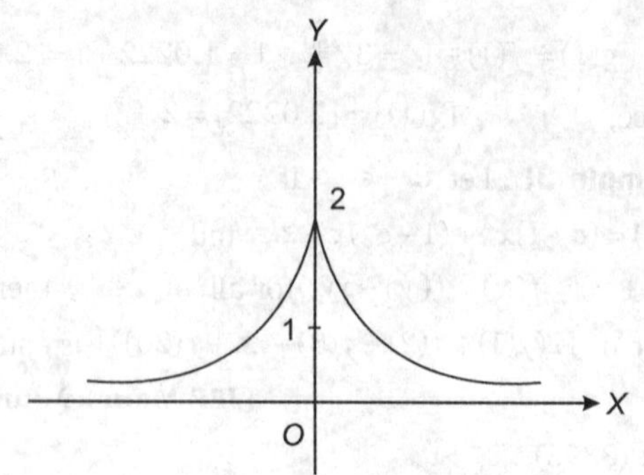

(D) $k(x) = \tan^{-1}\sqrt{[x]+[-x]} + \sqrt{3-|x|} + \frac{1}{x^2}$

∴ $k(-x) = \tan^{-1}\sqrt{[-x]+[x]} + \sqrt{3-|-x|} + \frac{1}{(-x)^2}$

$= \tan^{-1}\sqrt{[x]+[-x]} + \sqrt{3-|x|} + \frac{1}{x^2} = k(x)$

∴ $k(x)$ is symmetrical about y-axis.

and $k(x)$ is non-periodic.

The function $k(x)$ is defined only when

$$[x]+[-x] \geq 0, \; 3-|x| \geq 0$$

and $x \neq 0$

⇒ $x \in I$, $-3 \leq x \leq 3$

and $x \neq 0$

∴ $D_k = \{-3, -2, -1, 1, 2, 3\} \neq R$

at one value of x, $k(x)$ consists of only one natural.

Example 38. Column I contains two functions $f(x)$ and $g(x)$ and column II contains values of x for which both the functions are defined.

	Column I		Column II
(A)	$f(x) = \tan^{-1}\left(\frac{2x}{1-x^2}\right)$, $g(x) = 2\tan^{-1} x$	(p)	$x \in \{-1, 1\}$
(B)	$f(x) = \sin^{-1}(\sin x)$, $g(x) = \sin(\sin^{-1} x)$	(q)	$x \in (0,1)$
(C)	$f(x) = \log_{x^2} 25$, $g(x) = \log_x 5$	(r)	$x \in (-1, 1)$
(D)	$f(x) = \sin^{-1} x + \cos^{-1} x$, $g(x) = \sec^{-1} x + \text{cosec}^{-1} x$	(s)	$x \in [-1, 1]$

Sol : (A) → (q, r); **(B)** → (p, q, r, s); **(C)** → (q); **(D)** → (p)

(A) ∵ $\tan^{-1}\left(\frac{2x}{1-x^2}\right) = \begin{cases} 2\tan^{-1} x, & -1 < x < 1 \\ \pi + 2\tan^{-1} x, & x < -1 \\ -\pi + 2\tan^{-1} x, & x > 1 \end{cases}$

Hence, both the functions $f(x)$ and $g(x)$ are identical for $x \in (-1, 1)$.

(B) $\because \quad f(x) = \sin^{-1}(\sin x)$

$\therefore \quad D_f = R$ and $g(x) = \sin(\sin^{-1} x)$

$\therefore \quad D_g = [-1, 1]$

Hence, both the functions $f(x)$ and $g(x)$ are identical for $x \in [-1, 1]$

(C) $\because \quad f(x) = \log_{x^2} 25$

$f(x)$ is defined when $x^2 > 0$ and $\neq 1$

$\therefore \quad D_f = R \sim \{-1, 1\}$ and $g(x) = \log_x 5$

$g(x)$ is defined when $x > 0$ and $\neq 1$

$\therefore \quad D_g = (0, \infty) \sim \{1\}$

Hence, both the functions $f(x)$ and $g(x)$ are identical for $x \in (0, \infty) \sim \{1\}$.

(D) $\because \quad f(x) = \sin^{-1} x + \cos^{-1} x = \frac{\pi}{2}$ for $x \in [-1, 1]$

and $\quad g(x) = \sec^{-1} x + \operatorname{cosec}^{-1} x$

$$= \frac{\pi}{2} \text{ for } x \in (-\infty, -1] \cup [1, \infty)$$

Hence, both the functions are identical only if

$$x = \{-1, 1\}$$

SECTION-VI (ASSERTION-REASON TYPE)

Directions : Example number 39 to 42 are Assertion-Reason type questions. Each of these questions contains two statements.

Statement-1. (Assertion) and

Statement-2. (Reason)

Each of these examples also has four alternative choices, only one of which is the correct answer. You have to select the correct choice as given below.

(a) Statement-1. is true, Statement-2 is true and Statement-2 is correct explanation for Statement-1.
(b) Statement-1 is true, Statement-2 is true and Statement-2 is not the correct explanation for Statement-1.
(c) Statement-1 is true, Statement-2 is false.
(d) Statement-1 is false, Statement-2 is true.

Example 39. Consider the function

$$f(x) = \sin^{-1}\left(\frac{2x}{1+x^2}\right) + 2\tan^{-1} x$$

Statement 1. $f(10) = \pi$ because

Statement 2. $f(x) = \pi$ for all $x > 1$

Sol : (a)

$$\because \sin^{-1}\left(\frac{2x}{1+x^2}\right) = \begin{cases} 2\tan^{-1} x, & |x| \leq 1 \\ \pi - 2\tan^{-1} x, & x > 1 \\ -\pi - 2\tan^{-1} x, & x < -1 \end{cases} \quad \therefore \text{ for } x > 1$$

$$\sin^{-1}\left(\frac{2x}{1+x^2}\right) = \pi - 2\tan^{-1} x$$

$$\Rightarrow \quad \sin^{-1}\left(\frac{2x}{1+x^2}\right) + 2\tan^{-1} x = \pi$$

or $\quad f(x) = \pi$

$\therefore \quad f(10) = \pi$

Hence, both statements are true and Statement 2 is correct explanation of Statement 1.

Example 40. Statement-1. $f(x) = \log_e x$ can not be expressed as a sum of odd and even function.

Statement-2. $f(x) = \log_e x$ is neither odd nor even function.

Sol : (b) Any function can be expressed as a sum of odd and even function, no matter it is odd or even.

But $f(x) = \log_e x$ is not defined for $x < 0$, hence statement-2. is true but not correct explanation of statement-1.

Example 41. Statement-1. Fundamental period of the function

$$f(x) = |\sin 3x| + |\cos 3x| + |\tan 4x| \text{ is } \pi.$$

Statement-2. If period of $f_1(x)$ is T_1, period of $f_2(x)$ is T_2 and period of $f_3(x)$ is T_3, then the fundamental period of $f_1(x) + f_2(x) + f_3(x)$ is the LCM of T_1, T_2 and T_3.

Sol : (d) Note that period of $|\sin x| + |\cos x| = \frac{\pi}{2}$

$\therefore \quad$ Period of $|\sin 3x| + |\cos 3x| = \frac{\pi}{6}$

and period of $\quad |\tan 4x| = \frac{\pi}{4}$

Hence period of $|\sin 3x| + |\cos 3x| + |\tan 4x| =$ LCM of

$$\left\{\frac{\pi}{6} \text{ and } \frac{\pi}{4}\right\} = \frac{\text{LCM of } \pi \text{ and } \pi}{\text{HCF of } 6 \text{ and } 4}$$

$$= \frac{\pi}{2}$$

Hence, statement 1 is false but Statement 2 is true.

Example 42. Consider the function $f(x) = \cos^{-1} x$ and $g(x) = \tan^{-1}\left(\frac{\pi x}{2\sqrt{3}}\right)$, for $x \in$ domain of $f\left(\frac{\pi x}{2}\right)$

Statement-1. The range of $g(x)$ is $\left[-\frac{\pi}{6}, \frac{\pi}{6}\right]$

because

Statement-2. Domain of $g(x)$ is the subset of that of $f(x)$.

Sol : (b)

$\because \quad f(x) = \cos^{-1} x$

$\therefore \quad$ Domain of $f(x)$ is $[-1, 1]$ *i.e.*, $-1 \leq x \leq 1$

For domain of $f\left(\frac{\pi x}{2}\right) \Rightarrow -1 \le \frac{\pi x}{2} \le 1$

or $\quad -\frac{2}{\pi} \le x \le \frac{2}{\pi}$

$\therefore\quad$ Domain of $f\left(\frac{\pi x}{2}\right)$ is $\left[-\frac{2}{\pi}, \frac{2}{\pi}\right]$

Now, $\quad -\frac{2}{\pi} \times \frac{\pi}{2\sqrt{3}} \le \frac{\pi x}{2\sqrt{3}} \le \frac{\pi}{2\sqrt{3}} \times \frac{2}{\pi}$

$\Rightarrow \quad -\frac{1}{\sqrt{3}} \le \frac{\pi x}{2\sqrt{3}} \le \frac{1}{\sqrt{3}}$

$\Rightarrow \quad \tan^{-1}\left(-\frac{1}{\sqrt{3}}\right) \le \tan^{-1}\left(\frac{\pi x}{2\sqrt{3}}\right) \le \tan^{-1}\left(\frac{1}{\sqrt{3}}\right)$

$\Rightarrow \quad -\frac{\pi}{6} \le g(x) \le \frac{\pi}{6}$

$\therefore$ Range of $g(x)$ is $\left[-\frac{\pi}{6}, \frac{\pi}{6}\right]$

$\because$ Domain of $g(x)$ is $\left[-\frac{2}{\pi}, \frac{2}{\pi}\right]$

and Domain of $f(x)$ is $[-1, 1]$

It is clear that $\left[-\frac{2}{\pi}, \frac{2}{\pi}\right] \subset [-1, 1]$

Hence, both statements are true and Statement-2 is not correct explanation of Statement-1.

SECTION-VII (SUBJECTIVE TYPE)

In this section there are **18** subjective solved examples.

Example 43. Find the domain of definition of the function $f(x) = \frac{1}{[|x-1|]+[|7-x|]-6}$, where $[\cdot]$ denotes the greatest integer function.

Sol : Since $f(x) = \frac{1}{[|x-1|]+[|7-x|]-6}$

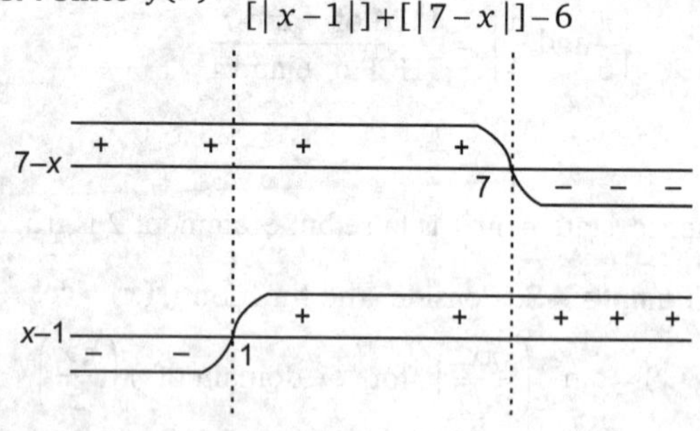

$f(x)$ is defined when

$$[|x-1|]+[|7-x|]-6 \ne 0$$

$\Rightarrow \quad [|x-1|]+[|7-x|] \ne 6$

For $x \le 1$:

$$[1-x] + [7-x] \ne 6$$

$\Rightarrow \quad 1 + [-x] + 7 + [-x] \ne 6$

$\Rightarrow \quad 2[-x] \ne 2$

or $\quad [-x] = -1$

$\Rightarrow \quad x \notin (0, 1) \quad$ ...(i)

For $1 < x \le 7$:

$$[x-1]+[7-x] \ne 6$$

$\Rightarrow \quad 1+[-x]+7+[-x] \ne 6$

$\Rightarrow \quad [x]+[-x] \ne 0$

$\Rightarrow \quad x \notin I$

$\Rightarrow \quad x \notin \{2, 3, 4, 5, 6, 7\} \quad$ ...(ii)

For $x > 7$:

$$[x-1]+[x-7] \ne 6$$

$\Rightarrow \quad [x]-1+[x]-7 \ne 6$

$\Rightarrow \quad 2[x] \ne 14 \Rightarrow [x] \ne 7$

$\Rightarrow \quad x \notin (7, 8) \quad$ ...(iii)

Combing (i), (ii) and (iii), we get

Domain of $f(x)$ is

$$x \in R \sim (0, 1] \cup \{2, 3, 4, 5, 6, 7\} \cup (7, 8)$$

Example 44. If $f(x) = \log_{100x}\left(\frac{2\log_{10} x + 2}{-x}\right)$, where $\{x\}$ denotes the fractional part of x. If the function $(fog)(x)$ exists, then find the range of $g(x)$.

Sol : Given $f(x) = \log_{100x}\left(\frac{2\log_{10} x + 2}{-x}\right)$ and $g(x) = \{x\}$

$\therefore \quad (fog)(x) = f(g(x))$

$$= f(\{x\}) = \log_{100\{x\}}\left(\frac{2\log_{10}\{x\}+2}{-\{x\}}\right)$$

$\because \quad (fog)(x)$ exists.

$\therefore \quad 0 < 100\{x\} < 1, 100\{x\} > 1 \quad$ ...(1)

and $\quad \frac{2\log_{10}\{x\}+2}{-\{x\}} > 0$

$\Rightarrow \quad \frac{\log_{10}\{x\}+1}{\{x\}} < 0$

$\therefore \quad \log_{10}\{x\}+1 < 0 \quad \{\because \{x\} > 0 \text{ from (i)}\}$

$\Rightarrow \quad \{x\} < 10^{-1} \quad$ ...(ii)

From (i), $0 < \{x\} < 10^{-2}$ and $\{x\} > 10^{-2}$ ...(iii)

From (ii) and (iii), we get $\{x\} \in (0, 10^{-2}) \cup (10^{-2}, 10^{-1})$

Hence, range of $g(x)$ = range of $\{x\}$

$$= (0, 10^{-2}) \cup (10^{-2}, 10^{-1})$$

Example 45. If the function $f(x) = \sin^{-1}\left(\frac{2x}{1+x^2}\right)$ and $g(x) = \cos^{-1}\left(\frac{1-x^2}{1+x^2}\right)$ are identical functions, then compute their domain and range.

Sol : Since $f(x) = \sin^{-1}\dfrac{2x}{(1+x^2)}$

$$= \begin{cases} 2\tan^{-1}x, & -1 \le x \le 1 \\ \pi - 2\tan^{-1}x, & x > 1 \\ -(\pi + 2\tan^{-1}x), & x < -1 \end{cases}$$

and $g(x) = \cos^{1}\left(\dfrac{1-x^2}{1+x^2}\right) = \begin{cases} 2\tan^{-1}x, & x \ge 0 \\ -2\tan^{-1}x, & x < 0 \end{cases}$

But given f(x) and g(x) are identical functions, then domain of f(x) = domain of g(x) and range of f(x) = range of g(x)

$\therefore \quad f(x) = g(x) = 2\tan^{-1}x \quad \forall\, 0 \le x \le 1$

$\therefore$ Domain = [0, 1]

$\Rightarrow \quad 0 \le x \le 1 \quad \Rightarrow \quad 0 \le \tan^{-1}x \le \tan^{-1}1$

or $\quad 0 \le 2\tan^{-1}x \le \dfrac{\pi}{2}$

$\therefore \quad$ Range $= \left[0, \dfrac{\pi}{2}\right]$

Example 46. Let f(x) = (2 cos x – 1) (2 cos 2x – 1) (2 cos 2^2 x – 1) .... (2 cos 2^{n-1} x – 1) (where $n \ge 1$). Then prove that $f\left(\dfrac{2\pi k}{2^n \pm 1}\right) = 1 \;\forall\, k \in I$

Sol : Since f(x) = (2 cos x – 1) (2 cos 2x – 1) (2 cos 2^2 x – 1) ...... (2 cos 2^{n-1} x – 1)

$$= \prod_{r=1}^{n}(2\cos 2^{r-1}x - 1)$$

$$= \prod_{r=1}^{n}\frac{(2\cos 2^{r-1}x - 1)(2\cos 2^{r-1}x + 1)}{(2\cos 2^{r-1}x + 1)}$$

$$= \prod_{r=1}^{n}\frac{(4\cos^2 2^{r-1}x - 1)}{(2\cos 2^{r-1}x + 1)}$$

$$= \prod_{r=1}^{n}\frac{(2\cos 2^{r}x + 1)}{(2\cos 2^{r-1}x + 1)}$$

$$= \frac{(2\cos 2^{n}x + 1)}{(2\cos x + 1)}$$

For $x = \dfrac{2\pi k}{2^n \pm 1} \Rightarrow 2^n x = 2\pi k \pm x$

$\therefore \quad \cos 2^n x = \cos(2\pi k \pm x) = \cos x$

Hence $f\left(\dfrac{2\pi k}{2^n \pm 1}\right) = 1$

Example 47. If $f(x) = ax^2 + bx + c$ is such that; $|f(0)| \le 1, |f(1)| \le 1$ and $|f(-1)| \le 1$. Prove that

$$|f(x)| \le \frac{5}{4} \;\forall\, x \in [-1, 1].$$

Sol : Given $\quad f(x) = ax^2 + bx + c \quad$ ...(i)

$\therefore \quad f(0) = c \quad$ ...(ii)

$f(1) = a + b + c \quad$ ...(iii)

and $\quad f(-1) = a - b + c \quad$ ...(iv)

Solving (ii), (iii) and (iv), we get

$$a = \frac{f(-1) + f(1) - 2f(0)}{2},\; b = \frac{f(1) - f(-1)}{2},\; c = f(0)$$

Substituting the values a, b and c in (i), then

$$f(x) = \left\{\frac{f(-1) + f(1) - 2f(0)}{2}\right\}x^2 + \left\{\frac{f(1) - f(-1)}{2}\right\}x + f(0)$$

$\Rightarrow 2f(x) = (x^2 - x)f(-1) + (x^2 + x)f(1) + 2(1 - x^2)f(0)$

$\therefore \; |2f(x)|$

$= |(x^2 - x)f(-1) + (x^2 + x)f(1) + 2(1 - x^2)f(0)|$

$\le |x^2 - x||f(-1)| + |x^2 + x||f(1)| + 2|(1 - x^2)||f(0)|$

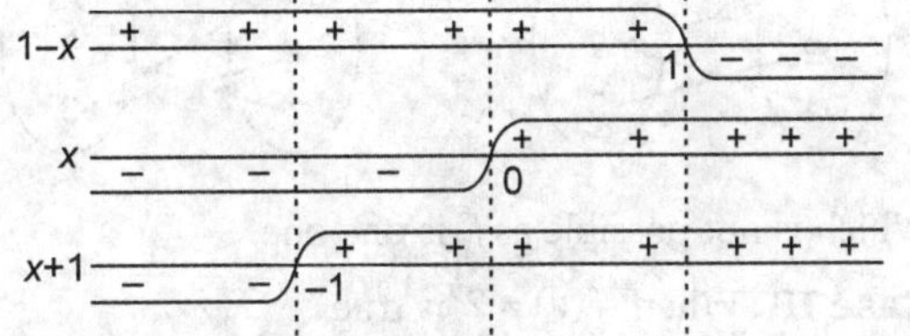

$\le |x^2 - x| + |x^2 + x| + 2|1 - x^2|$

$\{\because |f(-1)| \le 1, |f(0)| \le 1$ and $|f(1)| \le 1\}$

$= |x||x - 1| + |x||x + 1| + 2|1 - x||1 + x|$

$\because \; x \in [-1, 1] \qquad \{\therefore \; 0 \le 1 + x \le 2, 0 \le 1 - x \le 2\}$

and also $0 \le 1 - x^2 \le 4$

Then

$2|f(x)| \le -|x|(x - 1) + |x|(x + 1) + 2(1 - x^2)$

$= 2|x| + 2 - 2x^2$

or $\quad |f(x)| \le |x| + 1 - x^2 = -(x^2 - |x| - 1)$

$= -(|x|^2 - |x| - 1)$

$= -\left\{\left(|x| - \dfrac{1}{2}\right)^2 - \dfrac{1}{4} - 1\right\}$

$= \dfrac{5}{4} - \left(|x| - \dfrac{1}{2}\right)^2 \le \dfrac{5}{4}$

Hence $\quad |f(x)| \le \dfrac{5}{4}$

Example 48. Let f be a one-one function with domain (x, y, z) and range {1, 2, 3}. It is given that exactly one of the following statements is true and the remaining two are false.

$$f(x) = 1;\; f(y) \ne 1;\; f(z) \ne 2.\text{ Determine } f^{-1}(1)$$

Sol : Case I. When $f(x) = 1$ is true.

and $f(y) \ne 1, f(z) \ne 2$ are both false

Therefore $f(x)=1, f(y)=1, f(z)=2$

Which is not possible as f is one-one.

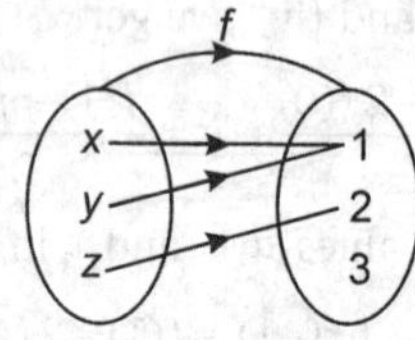

Case II. When $f(y)\neq 1$ is true and

$f(x)=1, f(z)\neq 2$ are both false.

Therefore $f(x)\neq 1$, $f(z)=2$, $f(y)=3$

i.e., $f(x)=2$ or 3

If $f(x)=2$ If $f(x)=3$

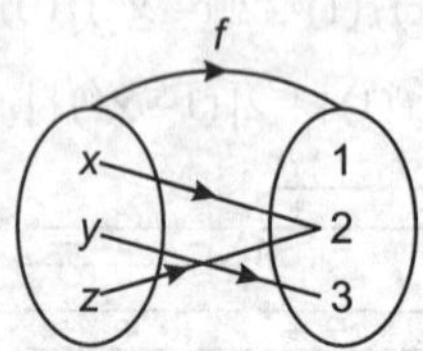

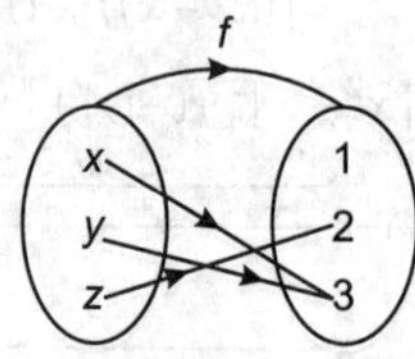

Which is not possible as f is one-one.

Case III. When $f(z)\neq 2$ is true

and $f(y)\neq 1, f(x)=1$ are both false

Therefore $f(y)=1$, $f(z)=3$, $f(x)\neq 1$ *i.e.,* $f(x)=2$

which is possible as f is one-one.

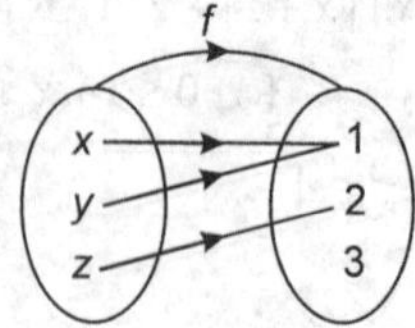

Thus $f(x)=2$, $f(y)=1$, $f(z)=3$

$\Rightarrow f^{-1}(1)=y$

Example 49. If f and g be two linear functions from $[-1, 1]$ onto $[0, 2]$ and $\phi : R^{+}-\{-1,1\}\to R$ be defined by $\phi(x)=\dfrac{f(x)}{g(x)}$, then show that

$$\left|\phi(\phi(x))+\phi\left(\phi\left(\frac{1}{x}\right)\right)\right|\geq 2.$$

Sol : Let $F(x)$ be a linear function from $[-1, 1]$ onto $[0, 2]$.

Let $F(x)=ax+b$

and $F(-1)=0$ and $F(1)=2$,

then $-a+b=0$ and $a+b=2$

$\Rightarrow$ $a=1$, $b=1$

$\therefore$ $F(x)=x+1\,(f(x)$ or $g(x))$

and also, $F(-1)=2$

and $F(1)=0$

then $-a+b=2$ and $a+b=0$

$\Rightarrow$ $a=-1, b=1$ $\therefore$ $F(x)=-x+1(f(x)$ or $g(x))$

Hence $\phi(x)=\dfrac{1+x}{1-x}$ or $\dfrac{1-x}{1+x}$

Case I. If $\phi(x)=\dfrac{1+x}{1-x}, x\neq 1$

then $$\phi(\phi(x))=\frac{1+\phi(x)}{1-\phi(x)}=\frac{1+\dfrac{1+x}{1-x}}{1-\dfrac{1+x}{1-x}}=\frac{1}{x}$$

and $$\phi\left(\phi\left(\frac{1}{x}\right)\right)=-x$$

Case II. If $\phi(x)=\dfrac{1-x}{1+x}, x\neq -1$

then $$\phi(\phi(x))=\frac{1-\phi(x)}{1+\phi(x)}=\frac{1-\dfrac{1-x}{1+x}}{1+\dfrac{1-x}{1+x}}=x$$

and $$\phi\left(\phi\left(\frac{1}{x}\right)\right)=\frac{1}{x}$$

From both cases $\left|\phi(\phi(x)+\phi\left(\phi\left(\dfrac{1}{x}\right)\right)\right|=\left|x+\dfrac{1}{x}\right|$

(given $x>0$ and $n\neq \pm 1$)

≥ 2 $(\because$ A.M. $\geq$ G.M.)

Hence $$\left|\phi(\phi(x))+\phi\left(\phi\left(\frac{1}{x}\right)\right)\right|\geq 2$$

Example 50. Let $f:[-2, 2]\to R$ be a function if; for $x\in[0, 2]$

$$f(x)=\begin{cases} x\tan x, & 0<x\leq \dfrac{\pi}{2} \\ \dfrac{\pi}{2}[x], & \dfrac{\pi}{2}<x\leq 2 \end{cases}$$

define f for $x\in[-2, 0]$, then

(i) f is an odd function (ii) f is an even function

(where $[\cdot]$ denotes the greatest integer function)

Sol : Since $$f(x)=\begin{cases} x\tan x, & 0<x\leq \dfrac{\pi}{2} \\ \dfrac{\pi}{2}[x], & \dfrac{\pi}{2}<x\leq 2 \end{cases}$$

$$\therefore\quad f(-x)=\begin{cases} (-x)\tan(-x), & 0<-x\leq \dfrac{\pi}{2} \\ \dfrac{\pi}{2}[-x], & \dfrac{\pi}{2}<-x\leq 2 \end{cases}$$

or $$f(-x)=\begin{cases} x\tan x, & -\dfrac{\pi}{2}\leq x<0 \\ \dfrac{\pi}{2}[-x], & -2\leq x<-\dfrac{\pi}{2} \end{cases}$$

(i) If f is an odd function, then $f(x) = -f(-x)$

$$= \begin{cases} -x\tan x, & -\frac{\pi}{2} \le x < 0 \\ -\frac{\pi}{2}[-x], & -2 \le x < -\frac{\pi}{2} \end{cases}$$

(ii) If f is an even function, then

$$f(x) = f(-x) = \begin{cases} x\tan x, & -\frac{\pi}{2} \le x < 0 \\ \frac{\pi}{2}[-x], & -2 \le x < -\frac{\pi}{2} \end{cases}$$

Example 51. Let $g : R \to R$ be given by $g(x) = 3 + 4x$. If $g^n(x) = \underbrace{gogo \ldots og(x)}_{n \text{ times}}$, show that $g^n(x) = (4^n - 1) + 4^n x$ if $g^{-n}(x)$ denotes the inverse of $g^n(x)$.

Sol : Since $g(x) = 3 + 4x$

$\therefore \quad g^2(x) = (gog)(x) = g\{g(x)\} = g(3+4x)$

$= 3 + 4(3+4x)$

or $\quad g^2(x) = 15 + 4^2 x = (4^2 - 1) + 4^2 x$

Now $\quad g^3(x) = (gogog)(x) = g\{g^2(x)\}$

$= g(15 + 4^2 x)$

$= 3 + 4(15 + 4^2 x) = 63 + 4^3 x$

$= (4^3 - 1) + 4^3 x$

Similarly, we get $g^n(n) = (4^n - 1) + 4^n x$

Now let $g^n(x) = y \Rightarrow x = g^{-n}(y)$ ...(i)

$\therefore \quad y = (4^n - 1) + 4^n x$ or $x = (y + 1 - 4^n)4^{-n}$ ...(ii)

From (i) and (iii), we get

$g^{-n}(y) = (y + 1 - 4^n)4^{-n}$

Hence $\quad g^{-n}(x) = (x + 1 - 4^n)4^{-n}$

Example 52. Find the inverse of the function,

$$f(x) = \begin{cases} x, & x < 1 \\ x^2, & 1 \le x \le 4 \\ 8\sqrt{x}, & x > 4 \end{cases}$$

Sol : Given $f(x) = \begin{cases} x, & x < 1 \\ x^2, & 1 \le x \le 4 \\ 8\sqrt{x}, & x > 4 \end{cases}$

Let $\quad f(x) = y \Rightarrow x = f^{-1}(y)$ ...(i)

$$\therefore \quad x = \begin{cases} y, & y < 1 \\ \sqrt{y}, & 1 \le \sqrt{y} \le 4 \\ y^2/64, & y^2/64 > 4 \end{cases}$$

$$= \begin{cases} y, & y < 1 \\ \sqrt{y}, & 1 \le y \le 16 \\ y^2/64, & y > 16 \end{cases}$$

$$\Rightarrow \quad f^{-1}(y) = \begin{cases} y, & y < 1 \\ \sqrt{y}, & 1 \le y \le 16 \\ y^2/64, & y > 16 \end{cases} \quad \text{[From (i)]}$$

Hence $f^{-1}(x) = \begin{cases} x, & x < 1 \\ \sqrt{x}, & 1 \le x \le 16 \\ x^2/64, & x > 16 \end{cases}$

Example 53. For what integral value of n, is 3π period of the function $\cos(nx)\sin\left(\frac{5x}{n}\right)$?

Sol : Let $f(x) = \cos nx \sin\left(\frac{5x}{n}\right)$

$\because$ $f(x)$ is periodic $\Rightarrow f(x+\lambda) = f(x)$, where λ is period.

$\Rightarrow \quad \cos(nx + n\lambda)\sin\left(\frac{5x+5\lambda}{n}\right) = \cos nx \sin\left(\frac{5x}{n}\right)$

at $\quad x = 0$, $\cos n\lambda \sin\left(\frac{5\lambda}{n}\right) = 0$

If $\cos n\lambda = 0 \Rightarrow n\lambda = r\pi + \frac{\pi}{2}, r \in I$

$\Rightarrow \quad n(3\pi) = r\pi + \frac{\pi}{2}$ $\quad (\because \lambda = 3\pi)$

$\Rightarrow \quad 3n - r = \frac{1}{2}$ $\quad$ (Impossible)

Again let $\sin\left(\frac{5\lambda}{n}\right) = 0$

$\therefore \quad \frac{5\lambda}{n} = p\pi, p \in I \Rightarrow \frac{5(3\pi)}{n} = p\pi$ $\quad (\because \lambda = 3\pi)$

$\Rightarrow \quad n = \frac{15}{p}$

For $\quad p = \pm 1, \pm 3, \pm 5, \pm 15$

$\therefore \quad n = \pm 15, \pm 5, \pm 3, \pm 1$ $\quad (\because n \in I)$

Hence $\quad n = \pm 1, \pm 3, \pm 5, \pm 15$

Example 54. Let $f(x, y)$ be a periodic function satisfying $f(x,y) = f(2x+2y, 2y-2x)$ for all x, y; Define $g(x) = f(2^x, 0)$. Show that $g(x)$ is a periodic function with period 12.

Sol : Since $f(x,y) = f(2x+2y, 2y-2x)$ ...(i)

or $\quad f(I, II) = f(2I + 2II, 2II - 2I)$

$\therefore f(2x+2y, 2y-2x)$

$= f(2(2x+2y) + 2(2y-2x), 2(2y-2x) - 2(2x+2y))$

$= f(8y, -8x)$ ...(ii)

From (i) and (ii), we get

$f(x,y) = f(8y, -8x)$ ...(iii)

or $\quad f(I, II) = f(8II, -8I)$

$\therefore \quad f(8y, -8x) = f(8(-8x), -8(8y))$

$= f(-64x, -64y)$ ...(iv)

From (iii) and (iv), we get

or $\quad f(x,y) = f(-64,-64y) \quad \text{...(v)}$

$$f(I,II) = f(-64I,-64II)$$

$$\therefore\ f(-64x,-64y) = f(-64(-64x),-64(-64y))$$

$$= f(2^{12}x, 2^{12}y) \quad \text{...(vi)}$$

From (v) and (vi), we get

$$f(x,y) = f(2^{12}x, 2^{12}y)$$

$$\Rightarrow \quad f(x,0) = f(2^{12}x, 0)$$

Replace x by 2^x, then

$$f(2^x,0) = f(2^{x+12},0)$$

$$\Rightarrow \quad g(x) = g(x+12) \quad \{\because g(x) = f(2^x,0)\}$$

Hence $g(x)$ is periodic with period 12.

Example 55. If the periodic function $f(x)$ satisfies the equation $f(x+1)+f(x-1)=\sqrt{3}f(x) \quad \forall\, x\in R$, then find the period of $f(x)$.

Sol : We have

$$f(x+1)+f(x-1) = \sqrt{3}f(x),\ \forall\, x\in R \quad \text{...(i)}$$

Replacing x by $x-1$ and $x+1$ in (i), then

$$f(x)+f(x-2) = \sqrt{3}\,f(x-1) \quad \text{...(ii)}$$

and $\quad f(x+2)+f(x) = \sqrt{3}\,f(x+1) \quad \text{...(iii)}$

Adding (ii) and (iii), we get

$$2f(x)+f(x-2)+f(x+2) = \sqrt{3}(f(x-1)+f(x+1))$$

$$\Rightarrow 2f(x)+f(x-2)+f(x+2) = \sqrt{3}\cdot\sqrt{3}f(x) \quad \text{[From (i)]}$$

$$\therefore \quad f(x+2)+f(x-2) = f(x) \quad \text{...(iv)}$$

Replacing x by $x+2$ in equation (iv), then

$$f(x+4)+f(x) = f(x+2) \quad \text{...(v)}$$

Adding equation (iv) and (v), we get

$$f(x+4)+f(x-2) = 0 \quad \text{...(vi)}$$

Again replacing x by $x+6$ in (vi), then

$$f(x+10)+f(x+4) = 0 \quad \text{...(vii)}$$

Subtracting (vi) from (vii), we get

$$f(x+10)-f(x-2) = 0 \quad \text{...(viii)}$$

Replacing x by $x+2$ in (viii), then $f(x+12)-f(x)=0$

or $\quad f(x+12) = f(x)$

Hence $f(x)$ is periodic function with period 12.

Example 56. Let f and g be real valued functions such that $f(x+y)+f(x-y)=2f(x)\cdot g(y) \quad \forall\, x,y\in R$. Prove that if f is not identically zero and $|f(x)|\le 1\ \forall\, x\in R$, then $|g(y)|\le 1\ \forall\, y\in R$.

Sol : Given $0<|f(x)|\le 1$

Now let max $|f(x)| = M\ \forall\ 0<M\le 1 \quad \text{...(i)}$

$$\therefore \quad |f(x+y)|\le M \text{ and } |f(x-y)|\le M$$

Then $\quad |f(x+y)|+|f(x-y)|\le 2M \quad \text{...(ii)}$

$$2f(x)g(y) = f(x+y)+f(x-y)$$

$$\Rightarrow \quad |2f(x)g(y)|$$

$$= |f(x+y)+f(x-y)| \le |f(x+y)|+|f(x-y)| \le 2M$$

[From (ii)]

$$\Rightarrow \quad 2|f(x)||g(y)|\le 2M$$

$$\Rightarrow \quad |f(x)||g(y)|\le M\le 1 \quad \text{[From (i)]}$$

$$\Rightarrow \quad |f(x)||g(y)|\le 1$$

Given $\quad |f(x)|\le 1 \quad \therefore\ |g(y)|\le 1\ \forall\ y\in R$

Example 57. If $f(x) = x+\int_0^1 (xy^2+x^2y)\,f(y)\,dy$, then find $f(x)$.

Sol : Given $f(x) = x+\int_0^1 (xy^2+x^2y)\,f(y)\,dy$

or $\quad f(x) = x + x\int_0^1 y^2 f(y)\,dy + x^2\int_0^1 y\,f(y)\,dy$

Let $\quad A = \int_0^1 y^2 f(y)\,dy$ and $B = \int_0^1 yf(y)\,dy$

Then $\quad f(x) = x+Ax+Bx^2 \quad \text{...(i)}$

$$\therefore \quad A = \int_0^1 y^2 f(y)dy = \int_0^1 y^2(y+Ay+By^2)\,dy$$

[From (i)]

$$\Rightarrow \quad A = \frac{1}{4}+\frac{A}{4}+\frac{B}{5} \text{ or } \frac{3A}{4}-\frac{B}{5}=\frac{1}{4} \quad \text{...(ii)}$$

and $\quad B = \int_0^1 yf(y)\,dy = \int_0^1 y(y+Ay+By^2)\,dy \quad$ [From (i)]

$$\Rightarrow \quad B = \frac{1}{3}+\frac{A}{3}+\frac{B}{4} \text{ or } -\frac{A}{3}+\frac{3B}{4}=\frac{1}{3} \quad \text{...(iii)}$$

Solving (ii) and (iii), we get $A = \frac{61}{119}$, $B = \frac{80}{119}$

Now from (i), $f(x) = x+\frac{61}{119}x+\frac{80}{119}x^2$

Example 58. If f be a function defined on the set of non-negative integers and taking values in the same set. Given that

(i) $x-f(x) = 19\left[\frac{x}{19}\right]-90\left[\frac{f(x)}{90}\right]$ for all non-negative integers.

(ii) $1900 < f(1990) < 2000$

Find the possible values of $f(1990)$ can take.

(**Notation** : $[z]$ refers to the largest integer that is $\le z$, e.g., $[3.1415] = 3$).

Sol : Since $1900 < f(1990) < 2000$

Therefore $\quad \frac{1900}{90} < \frac{f(1990)}{90} < \frac{2000}{90}$

$$\Rightarrow \quad \left[\frac{1900}{90}\right] < \left[\frac{f(1990)}{90}\right] < \left[\frac{2000}{90}\right]$$

$$\Rightarrow \quad 21 < \left[\frac{f(1990)}{90}\right] < 22$$

{∵ [21.111] = 21 and [22.22] = 22}

Case I. Suppose $\left[\frac{f(1990)}{90}\right] = 21$, then

Substituting $x = 1990$ in (i)

$$1990 - f(1990) = 19\left[\frac{1990}{19}\right] - 90\left[\frac{f(1990)}{90}\right]$$

$$\Rightarrow \quad 1990 - f(1990) = 19.104 - 90.21$$

$$\Rightarrow \quad f(1990) = 1904$$

Case II. Suppose $\left[\frac{f(1990)}{90}\right] = 22$, then

Substituting $x = 1990$ in (i)

$$1990 - f(1990) = 19\left[\frac{1990}{19}\right] - 90\left[\frac{f(1990)}{90}\right]$$

$$\Rightarrow \quad 1990 - f(1990) = 19 \times 104 - 90 \times 22$$

$$\Rightarrow \quad f(1990) = 1994$$

Both the values 1904 and 1994 satisfy the given conditions (i) and (ii) and therefore $f(1990)$ can take the values 1904 and 1994.

Example 59. Determine all functions f satisfying the functional relation

$$f(x) + f\left(\frac{1}{1-x}\right) = \frac{2(1-2x)}{x(1-x)}, \text{ where } x \text{ is a real}$$

number, $x \neq 0, x \neq 1$.

Sol : Given $f(x) + f\left(\frac{1}{1-x}\right) = \frac{2(1-2x)}{x(1-x)} = \frac{2}{x} - \frac{2}{1-x}$ ...(i)

Replacing x by $\frac{1}{(1-x)}$, we obtain

$$f\left(\frac{1}{1-x}\right) + f\left(\frac{1}{1-\frac{1}{1-x}}\right) = 2(1-x) - \frac{2}{1-\frac{1}{1-x}}$$

$$\Rightarrow \quad f\left(\frac{1}{1-x}\right) + f\left(1-\frac{1}{x}\right) = 2(1-x) - 2\left(1-\frac{1}{x}\right)$$

$$= -2x + \frac{2}{x} \qquad \text{...(ii)}$$

Again replacing x by $\left(1-\frac{1}{x}\right)$ in (i), we obtain

$$f\left(1-\frac{1}{x}\right) + f\left(\frac{1}{1-\left(1-\frac{1}{x}\right)}\right) = \frac{2}{\left(1-\frac{1}{x}\right)} - \frac{2}{1-\left(1-\frac{1}{x}\right)}$$

$$\Rightarrow \quad f\left(1-\frac{1}{x}\right) + f(x) = \frac{2x}{(x-1)} - 2x \qquad \text{...(iii)}$$

Adding (i) and (iii) and subtracting (ii), we get

$$2f(x) = \frac{2x}{x-1} - \frac{2}{1-x}$$

$$\Rightarrow \quad f(x) = \left(\frac{x+1}{x-1}\right)$$

Alternative Method :

Given $f(x) + f\left(\frac{1}{1-x}\right) = \frac{2(1-2x)}{x(1-x)} = \frac{2}{x} - \frac{2}{1-x}$ ...(i)

Let $\quad y = \frac{1}{1-x} \Rightarrow x = 1 - \frac{1}{y}$

From (i), we have $f(x) + f(y) = \frac{2}{x} - 2y$ ...(ii)

Again from (i), $f(y) + f\left(\frac{1}{1-y}\right) = \frac{2}{y} - \frac{2}{1-y}$ ...(iii)

Introduce another variable $z = \frac{1}{1-y}$

$$\Rightarrow \quad y = 1 - \frac{1}{z}$$

From (iii), we have $f(y) + f(z) = \frac{2}{y} - 2z$ ...(iv)

whenever $y, z \neq 0$ or 1

the relations $\quad y = \frac{1}{1-x}$ and $z = \frac{1}{1-y}$,

then $\quad z = 1 - \frac{1}{x}$

Similarly, $\quad f(x) + f(z) = \frac{2}{z} - 2x$ ...(v)

Adding corresponding sides of (ii), (iv) and (v) and dividing throughout by 2 we have

$$\Rightarrow \quad f(x) + f(y) + f(z) = \left(\frac{1}{x} + \frac{1}{y} + \frac{1}{z}\right) - (x + y + z)$$

$$\Rightarrow \quad f(x) + \frac{2}{y} - 2z = \left(\frac{1}{x} + \frac{1}{y} + \frac{1}{z}\right) - (x + y + z)$$

$$\Rightarrow \quad f(x) = \left(\frac{1}{x} - \frac{1}{y} + \frac{1}{z}\right) - (x + y - z)$$

$$= \left(\frac{1}{x} - x\right) - \left(y + \frac{1}{y}\right) + \left(z + \frac{1}{z}\right)$$

$$= \left(\frac{1}{x} - x\right) - \left(\frac{1}{1-x} + 1 - x\right) + \left(1 - \frac{1}{x} + \frac{x}{x-1}\right)$$

$$\left(\because y = \frac{1}{1-x} \text{ and } z = 1 - \frac{1}{x}\right)$$

$$= -\frac{1}{1-x} + \frac{x}{x-1}$$

$= \left(\frac{x+1}{x-1}\right)$ for real values of x except $x = 0$ or 1.

Example 60. Let I denote the set of integers. Consider a function $f: I \to I$ with the properties

$$f(92+x) = f(92-x)$$
$$f(19.12+x) = f(19.92-x) \quad (19.92 = 1748)$$
$$f(1992+x) = f(1992-x)$$

for all $x \in I$. It is possible that all positive divisors of 92 occurs as value of f?

Sol : We have

$$f(92+x) = f(92-x) \quad \text{...(i)}$$
$$f(19.12+x) = f(19.92-x) \quad \text{...(ii)}$$
$$f(1992+x) = f(1992-x) \quad \text{...(iii)}$$

from Eq. (ii). $f(228+x) = f(1748-x)$ ...(iv)

and from Eq. (iii), $f(1992+x) = f(1992-x)$

$= f(1748+244-x)$

$= f(1748-(x-244)) = f(228+x-244)$ [from Eq. (iv)]

$= f(x-16)$ ...(v)

Replacing x by $(x+16)$ in Eq. (v), we get

$$f(x+2008) = f(x)$$

$\therefore$ $f(x)$ is periodic with period 2008

From Eq. (iv),

$$f(228+x) = f(1748-x) = f(1748-x-2008)$$
$$= f(-x-260) = f(92-x-260-92)$$
$$= f(92-(x+352))$$
$$= f(92+(x+352)) \text{ [from eq. (i)]}$$
$$= f(x+444)$$

$\therefore$ $f(x+228) = f(x+444)$

Replacing x by $x-228$, we get

$$f(x) = f(x-228+444) = f(x+216)$$

$\therefore$ $f(x)$ is periodic with period 216.

Now, $f(x) = f(x+2008) = f(x+216\times 9+64)$

$= f(x+64)$

$\therefore$ $f(x)$ is periodic with period 64.

Now, $f(x) = f(x+216) = f(x+64\times 3+24)$

$= f(x+24)$

$\therefore$ $f(x)$ is periodic with period 24.

Again, $f(x) = f(64+x) = f(x+24\times 3-8) = f(x-8)$

Replacing x by $x+8$, we get $f(x+8) = f(x)$

$\therefore$ $f(x)$ is periodic with period 8.

Since, 8 completely divides 24. So 8 is the principal period.

Since 8 is the period, function can assume, $f(1), f(2), f(3), f(4), f(5), f(6), f(7), f(8)$ values.

($\because f(9) = f(8+1) = f(1), f(10) = f(8+2) = f(2)$ and so on)

So, the function can assume 6 values, *i.e.*, 1, 2, 4, 23, 46, 92

Exercises

SECTION-I (SINGLE CORRECT CHOICE TYPE)

This section contain **45 *multiple choice questions.*** Each question has four choices (a), (b), (c) and (d) out of which ***ONLY ONE*** is correct.

1. The domain of definition of the function $f(x) = \ln\{x\} + \sqrt{(x-2\{x\})}$ is :
(where $\{\cdot\}$ denotes fractional part function)

(a) $(1, \infty)$ (b) $(0, \infty)$
(c) $(1, \infty) - I^+$ (d) None of these

2. The domain of definition of the function

$$f9x) = (x+0.5)^{\log_{(x+0.5)}\left(\frac{x^2+2x-3}{4x^2-4x-3}\right)} \text{ is :}$$

(a) $\left(-\frac{1}{2}, \infty\right)$
(b) $[1, 3]$
(c) $\left(\frac{1}{2}, 1\right) \cup \left(\frac{3}{2}, \infty\right)$
(d) $\left(-\frac{1}{2}, \frac{1}{2}\right) \cup \left(\frac{1}{2}, 1\right) \cup \left(\frac{3}{2}, \infty\right)$

3. If $f(x) = 2x + |x|$, $g(x) = \frac{1}{3}(2x - |x|)$ and $h(x) = f(g(x))$, then domain of

$$\sin^{-1}\underbrace{(h(h(h(h \ldots\ldots h(x))\ldots\ldots))))}_{n\text{-times}} \text{ is :}$$

(a) $[-1, 1]$ (b) $\left[-1, -\frac{1}{2}\right] \cup \left[\frac{1}{2}, 1\right]$
(c) $\left[-1, -\frac{1}{2}\right]$ (d) $\left[\frac{1}{2}, 1\right]$

4. The domain of definition of the function $f(x) = \sqrt{\log_{(\sin x + \cos x)}(|\cos x| + \cos x)}, 0 \le x \le \pi$ is :

(a) $(0, \pi)$ (b) $\left(0, \frac{\pi}{2}\right)$
(c) $\left(0, \frac{\pi}{3}\right)$ (d) None of these

5. The domain of definition of the function

$$f(x) = \frac{1}{\sqrt{\{\sin x\} + \{-\sin x\}}} \text{ is :}$$

(where $\{\cdot\}$ denotes the fractional part function)

(a) $R \sim n\pi, n \in I$ (b) $R \sim \frac{n\pi}{2}, n \in I$

(c) $R \sim I$ (d) None of these

6. Range of the function $f(x) = \frac{\{x\}}{1+\{x\}}$ is :

(where $\{\cdot\}$ denotes the fractional part function)

(a) [0, 1) (b) $\left[0, \frac{1}{2}\right]$

(c) $\left[0, \frac{1}{2}\right)$ (d) $\left(0, \frac{1}{2}\right)$

7. The rage of the function

$f(x) = \tan^{-1}\sqrt{[x]+[-x]} + \sqrt{2-|x|} + \frac{1}{x^2}$ is :

(where $[\cdot]$ denotes the greatest integer function)

(a) $\left[\frac{1}{4}, \infty\right)$ (b) $\left\{\frac{1}{4}\right\} \cup [2, \infty)$

(c) $\left\{\frac{1}{4}, 2\right\}$ (d) $\left[\frac{1}{4}, 2\right]$

8. Therangeofthefunction $f(x) = \sqrt{(x-4)} + 2\sqrt{(6-x)}$ is :

(a) $[\sqrt{2}, 2\sqrt{2}]$ (b) $[\sqrt{2}, \sqrt{10}]$

(c) $[2\sqrt{2}, \sqrt{10}]$ (d) [4, 6]

9. If $f(x) = \log_{e^2 x}\left(\frac{2\ln x + 2}{-x}\right)$ and $g(x) = \{x\}$, then range of $g(x)$ for the existence of $f(g(x))$ is :
(where $\{\cdot\}$ denotes the fractional part function)

(a) $(0, e^{-1}) - \{e^{-2}\}$ (b) $(0, 2e^{-1}) - \{2e^{-2}\}$

(c) $(0, 3e^{-1}) - \{3e^{-2}\}$ (d) $(0, 4e^{-1}) - \{4e^{-2}\}$

10. The range of $f(x) = \sum_{r=1}^{n}\left[r + \sin\left(\frac{x}{r}\right)\right] \forall x \in [0, \pi]$ is :

(where $[\cdot]$ denotes the greatest integer function)

(a) $\left\{\frac{n^2+n-2}{2}, \frac{n(n+1)}{2}\right\}$

(b) $\left\{\frac{n(n+1)}{2}\right\}$

(c) $\left\{\frac{n^2+n-2}{2}, \frac{n(n+1)}{2}, \frac{n^2+n+2}{2}\right\}$

(d) $\left\{\frac{n(n+1)}{2}, \frac{n^2+n+2}{2}\right\}$

11. Let D be the domain of the function $f(x) = \sin^{-1}\left(\log_{3x}\left(\frac{6+2\log_3 x}{-5x}\right)\right)$. If the range of the function $g: D \to R$ defined by $g(x) = x - [x]$, ($[x]$ is the greatest integer function) is (α, β), then $\alpha^2 + \frac{5}{\beta}$ is approximately equal to

[JEE Main 12 Apr 2023 (MS)]

(a) 135 (b) 45

(c) 46 (d) 136

12. The function $f(x)$ is defined as

$f(x) = \begin{cases} \frac{1}{2^x}, & x \geq 5 \\ f(x+1), & x < 5 \end{cases}$, then the value of

$f\left(2 + \log_2\left(\frac{3}{2}\right)\right)$ is :

(a) $\frac{1}{12}$ (b) $\frac{1}{24}$

(c) $\frac{1}{32}$ (d) $\frac{1}{48}$

13. Let us consider a function

$f(x) = x^2 + \ln\left(\frac{\pi+x}{\pi-x}\right) \cdot \cos x + \ln\left(\frac{\pi-x}{\pi+x}\right)$ and if

$f(10) = 100$, then the value of $f(-10)$ is :

(a) 99 (b) 100

(c) 101 (d) 1000

14. If $f(x) = \frac{2^{2x}}{2^{2x}+2}, x \in R$, then

$f\left(\frac{1}{2023}\right) + f\left(\frac{2}{2023}\right) + f\left(\frac{3}{2023}\right) + \ldots + f\left(\frac{2022}{2023}\right)$ is equal to

[JEE Main 24 Jan 2023 (ES)]

(a) 2011 (b) 1010 (c) 2010 (d) 1011

15. If the two roots of the equation $(\lambda-1)(x^2+x+1)^2 - (\lambda+1)(x^4+x^2+1) = 0$ are real and distinct and $f(x) = \frac{1-x}{1+x}$, then $f(f(x)) + f\left(f\left(\frac{1}{x}\right)\right)$ is :

(a) -2λ (b) $-\lambda$

(c) λ (d) 2λ

16. The sum of the maximum and minimum values of the function $f(x) = \frac{1}{2+(3\cos x - 4\sin x)^2}$ is :

(a) $\frac{54}{29}$ (b) $\frac{29}{54}$ (c) $\frac{53}{28}$ (d) $\frac{28}{53}$

17. If $f(x) \cdot f(y) = f(x) + f(y) + f(xy) - 2 \forall x, y \in R$ and $f(x)$ is not a constant function, then the value of $f(1)$ is :

(a) – 1 (b) 0

(c) 1 (d) 2

18. Suppose $f: R \to (0, \infty)$ be a differentiable function such that $5f(x+y) = f(x) \cdot f(y),\ \forall\ x, y \in R$.
If $f(3) = 320$, then $\sum_{n=0}^{5} f(n)$ is equal to

[JEE Main 30 Jan 2023 (MS)]

(a) 6875 (b) 6575
(c) 6825 (d) 6528

19. Given $f(x) = \dfrac{x^4 - 7x^2 + 9}{x - (3/x) + 1}$. Its zeros are of the form $\dfrac{a \pm \sqrt{b}}{c}$, where a, b and c are positive integers. Then the value of $b - a - c$ is :
(a) 8 (b) 10
(c) 12 (d) 14

20. Given $f(x) = \sqrt{\left(\dfrac{8}{1-x} + \dfrac{8}{1-x}\right)}$ and $g(x) = \dfrac{4}{f(\sin x)} + \dfrac{4}{f(\cos x)}$, then $g(x)$ is :
(a) periodic with period $\pi/2$
(b) periodic with period π
(c) periodic with period 2π
(d) aperiodic

21. If $f(x)$ and $g(x)$ are periodic functions with periods 7 and 11 respectively. Then the period of $F(x) = f(x) \cdot g\left(\dfrac{x}{5}\right) - g(x) \cdot f\left(\dfrac{x}{3}\right)$ is :
(a) 177 (b) 222
(c) 433 (d) 1155

22. If $f\left(x + \dfrac{1}{2}\right) + f\left(x - \dfrac{1}{2}\right) = f(x)$ for all $x \in R$, then the period of $f(x)$ is :
(a) 1 (b) 2 (c) 3 (d) 4

23. If $f(x) = \dfrac{x}{\ln x}$ and $g(x) = \dfrac{\ln x}{x}$, then :
(a) $\dfrac{1}{f(x)}$ and $g(x)$ are identical functions
(b) $\dfrac{1}{g(x)}$ and $f(x)$ are identical functions
(c) $f(x) \cdot g(x) = 1\ \forall\ x > 0$
(d) $\dfrac{1}{f(x)g(x)} = 1\ \forall\ x > 0$

24. If $f: [-2, 2] \to R$, where $f(x) = x^3 + \tan x + \left[\dfrac{x^2+1}{\lambda}\right]$ is an odd function, then the value of parameter λ is : (where $[\cdot]$ denotes the greatest integer function)
(a) $|\lambda| < 5$ (b) $\lambda < 5$
(c) $\lambda > 5$ (d) None of these

25. If $f: \left[-\dfrac{\pi}{2}, \dfrac{\pi}{2}\right] \to R$ given by $f(x) = \cos x + \sin\left[\dfrac{x+1}{\lambda}\right]$ is an even function. Then the set of values of $\lambda\ (\lambda > 0)$ is : (where $[\cdot]$ denotes the greatest integer function)
(a) $\left(-\dfrac{\pi}{2}, \dfrac{\pi}{2}\right) \sim \{0\}$
(b) $\left(\dfrac{\pi+2}{2}, \infty\right)$
(c) $\left(0, \dfrac{\pi+2}{2}\right) \cup \left(\dfrac{\pi+2}{2}, \infty\right)$
(d) $\left(-\dfrac{\pi}{2}, \dfrac{\pi+2}{2}\right) - \{0\}$

26. Let $f: [-10, 10] \to R$, where $f(x) = \sin x + \left[\dfrac{x^2}{\lambda}\right]$ be an odd function. Then the set of values of λ is : (where $[\cdot]$ denotes the greatest integer function)
(a) $(-10, 10) \sim \{0\}$
(b) $(0, 10)$
(c) $[100, \infty)$
(d) $(100, \infty)$

27. If the function $f(x) = [7/2 + a \sin x]$ is an even function (where $[\cdot]$ denotes the greatest integer function), then complete set of values of 'a' is :
(a) $\left(-\dfrac{1}{2}, \dfrac{1}{2}\right)$ (b) $\left[-\dfrac{1}{2}, \dfrac{1}{2}\right]$
(c) $(0, 1)$ (d) $[-1, 1]$

28. A real valued function $f(x)$ satisfies the function $f(x-y) = f(x)f(y) - f(a-x)f(a+y)$, where a is a given constant and $f(0) = 1$, then the graph of the function is symmetrical about a :
(a) point $(2a, 0)$ (b) point $(a, 0)$
(c) line $x = 2a$ (d) line $x = a$

29. If the graph of $y = f(x)$ is symmetrical about lines $x = 2$ and $x = 3$, then which of the following is true?
(a) $f(x+1) = f(x)$ (b) $f(x+2) = f(x)$
(c) $f(x+3) = f(x)$ (d) None of these

30. If $f(x) = x^3 + 3x^2 + 5x + p \sin x + q \cos x$, $\forall\ x \in R$ is a one-one function, then the value of $p^2 + q^2$ is :
(a) > 4 (b) ≥ 8
(c) < 4 (d) ≤ 8

31. The number of one-one function from $A\ \{1, 2, 3, 4\} \to B\ \{1, 2, 3, 4, 5, 6, 7\}$ such that $f(i) \neq i$ are :
(a) 840 (b) 675
(c) 465 (d) 255

32. Let $f: R \rightarrow \left[0, \frac{\pi}{2}\right)$ defined by

$f(x) = \tan^{-1}(x^2 + x + a)$, then the set of values of a for which f is onto is :

(a) $[0, \infty)$ (b) $[1, 2]$

(c) $\left[\frac{1}{4}, \infty\right)$ (d) None of these

33. If $f: R \rightarrow \left[\frac{\pi}{6}, \frac{\pi}{2}\right)$, $f(x) = \sin^{-1}\left(\frac{x^2 - a}{x^2 + 1}\right)$ is an onto function, then set of values of a is :

(a) $\left\{-\frac{1}{2}\right\}$

(b) $\left[-\frac{1}{2}, -1\right)$

(c) $(-1, \infty)$

(d) None of the above

34. Let a function f defined from $R \rightarrow R$ as :

$$f(x) = \begin{cases} x + a & \text{for } x \leq 1 \\ 2ax - 1 & \text{for } x > 1 \end{cases}$$

If the function is surjective, then a must lie in the interval :

(a) $(0, 2]$ (b) $(-\infty, 0]$

(c) $(-\infty, 0)$ (d) $(0, \infty)$

35. The number of bijective function $f(1, 3, 5, 7, ..., 99) \rightarrow (2, 4, 6, 8, ..., 100)$ if $f(3) > f(5) > f(7) ... > f(99)$ is

[JEE Main 25 Jul 2022 (ES)]

(a) ${}^{50}C_1$ (b) ${}^{50}C_2$

(c) $\frac{50!}{2}$ (d) ${}^{50}C_3 \times 3!$

36. The inverse of the function $f(x) = \frac{-x|x|}{1+x^2}$ is :

(a) $-\sqrt{\left(\frac{|x|}{1+|x|}\right)}$

(b) $\sqrt{\left(\frac{|x|}{1-|x|}\right)}$

(c) $-(\text{sgn}(x))\sqrt{\left(\frac{|x|}{1-|x|}\right)}$

(d) $(\text{sgn}(x))\sqrt{\left(\frac{|x|}{1-|x|}\right)}$

37. Suppose $f(x) = \frac{2x+5}{7}$ and $g(x) = \frac{7x+3}{5}$, then we define $h(x) = 2f^{-1}(x) - 7g^{-1}(x) + 3x$, then find the value of h (11) is :

(a) 33 (b) 43 (c) 53 (d) 63

38. Let $f: R - \{3\} \rightarrow R - \{1\}$ be defined by $f(x) = \frac{x-2}{x-3}$. Let $g: R \rightarrow R$ be given as $g(x) = 2x - 3$. Then, the sum of all the values of x for which $f^{-1}(x) + g^{-1}(x) = \frac{13}{2}$ is equal to

[JEE Main 18 Mar 2021 (ES)]

(a) 7 (b) 2 (c) 5 (d) 3

39. Let $f_o(x) = \frac{x}{2x+1}$ and $f_{n+1} = f_o o f_n$ for $n = 0, 1, 2, 3, ...$, then the value of $f_2(x)$ is :

(a) $\frac{x}{2x+1}$ (b) $\frac{2x}{4x+1}$

(c) $\frac{x}{6x+1}$ (d) $\frac{2x}{8x+1}$

40. Let $f_1(x) = \begin{cases} x, & 0 \leq x \leq 1 \\ 1, & x > 1 \\ 0, & \text{otherwise} \end{cases}$

and $f_2(x) = f_1(-x)$ for all x

$f_3(x) = f_2(-x)$ for all x

$f_4(x) = f_3(-x)$ for all x

Which of the following in necessarily true?

(a) $f_4(x) = f_1(x)$ for all x

(b) $f_1(x) = -f_3(-x)$ for all x

(c) $f_2(-x) = f_4(x)$ for all x

(d) $f_1(x) + f_3(x) = 0$ for all x

41. Let S be the set of all triangle and R^+ be the set of positive real numbers. Then the function $f: S \rightarrow R^+$, $f(\Delta) =$ area of the Δ, where $\Delta \in S$ as :

(a) injective but not surjective

(b) surjective but not injective

(c) injective as well as surjective

(d) neither injective nor serjective

42. Let $f: R \rightarrow R$ and $g: R \rightarrow R$ be two one-one and onto functions such that they are the mirror images of each other about the line $y = a$. If $h(x) = f(x) + g(x)$, then $h(x)$ is :

(a) one-one and onto

(b) only one-one not onto

(c) only onto but not one-one

(d) neither one-one nor onto

43. Let $f(x) = e^{\{e^{|x|} \text{sgn } x\}}$ and $g(x) = e^{[e^{|x|} \text{sgn } x]}$, $x \in R$, where $\{\cdot\}$ and $[\cdot]$ denotes the fractional part and integral part of the functions respectively. Also $h(x) = \ln(f(x)) + \ln(g(x))$, then for all real x, $h(x)$ is :

(a) an odd function

(b) an even function

(c) neither an odd nor an even function

(d) both odd as well as even functions

44. The domain of the function $f(x)$ given by $3^x + 3^{f(x)} = \min.\{2t^3 - 15t^2 + 36t - 25, 2 + |\sin t|;$ $2 \le t \le 4\}$ is :

(a) $(-\infty, 1)$

(b) $(-\infty, \log_3 e)$

(c) $(0, \log_3 2)$

(d) $(-\infty, \log_3 2)$

45. Let $f(x) = \max.\{\sin t : 0 \le t \le x\}$

$g(x) = \min.\{\sin t : 0 \le t \le x\}$

and $h(x) = [f(x) - g(x)]$

(where [·] denotes the greatest integer function), then the range of $h(x)$ is :

(a) $\{-3, -2, -1, 0, 1, 2, 3\}$

(b) $\{0, 1\}$

(c) $\{1, 2\}$

(d) $\{0, 1, 2\}$

SECTION-II (MULTIPLE CORRECT CHOICE TYPE)

This section contains **17 multiple choice questions**. Each question has four choices (a), (b), (c) and (d) out of which **MORE THAN ONE** may be correct.

46. If $f(x) = \log_{[x-1]}\left(\frac{|x|}{x}\right)$, where [·] denotes greatest integer function, then :

(a) $D_f = (2, \infty)$ (b) $R_f = \{0, 1\}$

(c) $D_f = [3, \infty)$ (d) $R_f = \{0\}$

47. For the function $f(x) = \frac{\sin^2 x}{\sin x - \cos x} - \frac{\sin x + \cos x}{\tan^2 x - 1}$, $x \ne n\pi \pm \frac{\pi}{4}, n \in I$ which of the following does not hold good?

(a) f is even

(b) f is odd

(c) f is neither even nor odd but periodic with period 2π

(d) f is neither even nor odd and is periodic

48. Let $f : \left(-\frac{\pi}{2}, \frac{\pi}{2}\right) \to R$ be given by $f(x) = [\ln(\sec x + \tan x)]^3$. Then :

[JEE Advanced 2014]

(a) $f(x)$ is an odd function

(b) $f(x)$ is an one-one function

(c) $f(x)$ is an onto function

(d) $f(x)$ is an even function

49. If $f : R \to R$ is a polynomial function satisfying the functional equation $f(f(x)) + f(x) = 12x$, then $f(-7)$ is equal to :

(a) – 7 (b) 14

(c) – 21 (d) 28

50. If $f(x)$ is a polynomial function satisfying the condition $f(x) \cdot f\left(\frac{1}{x}\right) = f(x) + f\left(\frac{1}{x}\right)$ and $f(3) = -26$, if $\prod_{r=3}^{8}\left(\frac{1 - f(r)}{1 - f(r-1)}\right) = 2^\lambda$, then λ is :

(a) prime number

(b) composite number

(c) perfect number

(d) product of co-prime numbers

51. For $x \in R - \{1\}$, If $f(x)$ is a function satisfies $f(x) + 2f\left(\frac{1}{1-x}\right) = x$, then :

(a) $f(-1) = f(2)$ (b) $f(-2) = f(3)$

(c) $2f(-2) = f(3)$ (d) $f(-2) = 2f(3)$

52. Let $f(x) = \ln(2x - x^2) + \sin\left(\frac{\pi x}{2}\right)$, then :

(a) graph of f is symmetrical about the line $x = 1$

(b) graph of f is symmetrical about the line $x = 2$

(c) maximum value of f is 1

(d) minimum value of f does not exist

53. Identify the statement(s) which is/are incorrect?

(a) The function $f(x) = \cos(\cos^{-1} x)$ is neither odd nor even

(b) The fundamental period of $f(x) = \cos(\sin x) + \cos(\cos x)$ is π

(c) The range of the function $f(x) = \cos(3 \sin x)$ is $[-1, 1]$

(d) The function $f(x, y) = \sqrt{(y^2 - x^2)} + x$ is a homogeneous function of degree 1

54. Which of the following is not a function (where [·] and { · } denotes the greatest integer and fractional part functions respectively)?

(a) $\frac{x!}{\{x\}}$ (b) $\{x\}\, x!$

(c) $\frac{1}{\ln[1 - |x|]}$ (d) $\frac{\ln(x-1)}{\sqrt{(1 - x^2)}}$

55. Which of the following function is periodic?

(a) $\text{sgn}(e^{-x})$

(b) $|\sin x| + \sin x$

(c) min. $(\sin x, |x|)$

(d) $\left[x + \frac{1}{2}\right] + \left[x - \frac{1}{2}\right] + 2[-x], [x] \le x$

56. Let $f(x) = \min.(|e - x|, \pi - |x|)$. Then which of the following statement(s) is (are) correct?

(a) $R_f = \left(-\infty, \frac{\pi - e}{2}\right]$

(b) $f(x)$ is many one

(c) $f(x)$ is not a periodic function

(d) $f(x)$ is neither even nor odd

57. If L and M are the minimum and maximum values of the function

$$f(x)\frac{\sin x}{\sqrt{(1-\cos^2 x)}}+\frac{\cos x}{\sqrt{(1-\sin^2 x)}}+\frac{\tan x}{\sqrt{(\sec^2 x-1)}}+\frac{\cot x}{\sqrt{(\text{cosec}^2 x-1)}}$$

respectively as x varies over all numbers in the largest possible domain of $f(x)$, then :

(a) $M + L = 2$ (b) $M + L = 4$

(c) $M - L = 4$ (d) $M - L = 6$

58. Which of the following pairs of functions is/are identical?

(a) $f(x)=\cot^2 x\cdot\cos^2 x$ and $g(x)=\cot^2 x-\cos^2 x$

(b) $f(x)=\sqrt{(x^2-1)}$ and $g(x)=\sqrt{(x+1)}\cdot\sqrt{(x-1)}$

(c) $f(x)=\sqrt{\left(\frac{1-\cos 2x}{2}\right)}$ and $g(x)=\sin x$

(d) $f(x)=[\{x\}]$ and $g(x)=\{[x]\}$

(where [·] and { · } are denotes the greatest integer and fractional part functions respectively)

59. Let $f(x)=\begin{cases}|2x-1|, & -2\le x\le 1\\ x^2-4, & 1<x\le 8\end{cases}$,

then possible integers in the domain of $y = f(f(x))$ is/are :

(a) –2 (b) 2 (c) 4 (d) 6

60. Let $f(x)=\sin\left[\frac{\pi}{6}\sin\left(\frac{\pi}{2}\sin x\right)\right]$ for all $x\in R$ and $g(x)=\frac{\pi}{2}\sin x$ for all $x\in R$. Let $(fog)(x)$ denotes $f(g(x))$ and $(gof)(x)$ denotes $g(f(x))$. Then, which of the following is/are true?

[JEE Advanced 2015]

(a) Range of f is $\left[-\frac{1}{2},\frac{1}{2}\right]$

(b) Range of fog is $\left[-\frac{1}{2},\frac{1}{2}\right]$

(c) $\lim_{x\to 0}\frac{f(x)}{g(x)}=\frac{\pi}{6}$

(d) There is an $x\in R$ such that $(gof)(x)=1$

61. Which of the following functions defined below are one-one functions(s)?

(a) $f(x)=(x+1), x\ge -1$

(b) $g(x)=x+\frac{1}{x}, x>0$

(c) $h(x)=x^2+4x-5, x>0$

(d) $k(x)=e^{-x}, x\ge 0$

62. Let $f(x)=\begin{cases}x-[x], & \lambda\le x<\lambda+0.5\\ [x], & \lambda+0.5\le x<\lambda+1\end{cases}$,

(where λ is an integer and [·] denotes greatest integer function) and $g(x) = \sin^4 x + \cos^4 x$, then which of the following are true.

(a) $f(g(x))$ is an odd function

(b) $f(g(x))$ is an even function

(c) $f(g(x))$ is a periodic function

(d) $f(g(x))$ is a one-one function

SECTION-III (PARAGRAPH TYPE)

This section contains 8 paragraphs. Based upon each of the paragraph 3 **multiple choice questions** have to be answered. Each of these question has four choices (a), (b), (c) and (d) out of which **ONLY ONE** is correct.

Paragraph for Question Nos. 63 to 65

Let $\frac{1}{4}f(x)=x+3x^2\int_0^1 y\,f(y)dy+5\int_0^1 xy^2 f(y)dy$

63. The maximum value of $f(x)$ is :

(a) $\frac{1}{16}$ (b) $\frac{1}{8}$

(c) 8 (d) 16

64. The number of solutions of the equation $|f(|x|)|=e^{-|x|}$ are :

(a) 2 (b) 4

(c) 6 (d) 8

65. If range of $f(-\cos\theta)$ is $[a,b]$, then the value of $\left|\frac{a}{b}\right|$ is :

(a) 8 (b) 16

(c) 24 (d) 32

Paragraph for Question Nos. 66 to 68

Let $f(x)=\frac{\cos 2x-(a+2)\cos x+a+1}{x}$, where a is constant such that $a > 0$ and period of $g(x)=\cos\left(\frac{\pi x}{a}\right)$ is 6.

66. The value of the constant a is :

(a) 2 (b) 3

(c) 5 (d) 6

67. Difference between the maximum and minimum values of $xf(x)$ is :

(a) 81/8 (b) 5

(c) 10 (d) 2

68. The value of $\pi f\left(\frac{\pi}{3}\right)+2g(2)$ is :

(a) 1 (b) 2

(c) 3 (d) $\frac{2}{\sqrt{3}}$

Paragraph for Question Nos. 69 to 71

Let $f(x) = -x^2 + 3x - 2$, $g(x) = f(|x|)$, $h(x) = |g(x)|$ and $I(x) = h(x) - [x]$ are four functions, where $[x]$ is the integral part of real x.

69. The value of 'a' such that the equation $g(x) - a = 0$ has exactly 3 real and distinct roots :
(a) –3 (b) –2 (c) 0 (d) 1/4

70. The set of values of all x, such that equation $h(x) - g(x) = 0$ is satisfied :
(a) $x \in [-2, -1]$ (b) $x \in [-2, -1] \cup [1, 2]$
(c) $x \in [1, 2]$ (d) none of these

71. Which statement is true for $I(x) = 0$?
(a) two values of x is satisfied for $I(x) = 0$
(b) one value of x is satisfied for $I(x) = 0$ and that x lie between 1 and 2
(c) one value of x is satisfied for $I(x) = 0$ and that x lie between 3 and 4
(d) None of the above

Paragraph for Question Nos. 72 to 74

Let $f(x) = \log_{[x]} \{x\}$ and $g(x) = \log_{\{x\}} [x]$ are two functions, where $[x]$ and $\{x\}$ denotes respectively the integral and fractional part of real x. Let D_f and D_g are the set of domain of $f(x)$ and $g(x)$ respectively. R_f and R_g are the set of range of $f(x)$ and $g(x)$ respectively.

72. Which statement is correct?
(a) D_f is the subset of D_g
(b) D_g is the subset of D_f
(c) D_f and D_g are the same set
(d) None of the above

73. The union of R_f and R_g is :
(a) $(-\infty, 0)$ (b) $(-\infty, \infty)$
(c) $(-\infty, 0]$ (d) None of these

74. The sum of all the non-negative integral values of the set R_g is :
(a) –1 (b) 0
(c) 1 (d) None of these

Paragraph for Question Nos. 75 to 77

Let $2f(x) + xf\left(\frac{1}{x}\right) - 2f\left(\left|\sqrt{2}\sin\left(\pi\left(x + \frac{1}{4}\right)\right)\right|\right)$

$$= 4\cos^2\left(\frac{\pi x}{2}\right) + x\cos\left(\frac{\pi}{x}\right)$$

75. The value of $f(2)$ is :
(a) 0 (b) 1 (c) 2 (d) 3

76. The value of $f(2) + f\left(\frac{1}{2}\right)$ is :
(a) –1 (b) 0
(c) 1 (d) 2

77. The value of $f(2) + f(1)$ is :
(a) 0 (b) 1 (c) 2 (d) 4

Paragraph for Question Nos. 78 to 80

Let $(x - y)\, f(x + y) - (x + y)\, f(x - y) = 2(x^2 y - y^3)\ \forall\, x, y \in R$ and $f(1) = 2$.

78. The set of value of 'a' such that equation $|f(x) - 2| - a = 0$ has exactly 4 real roots :
(a) $a \in \left(0, \frac{9}{4}\right]$ (b) $a \in \left[0, \frac{9}{4}\right)$
(c) $a \in \left(0, \frac{9}{4}\right)$ (d) $a \in \left[0, \frac{9}{4}\right]$

79. The number of solutions of the equation $f(x) = \sqrt{(1 - x^2)}$ is :
(a) 1 (b) 2 (c) 3 (d) 4

80. The range of $-f(2^x)$ is :
(a) $(-\infty, 0)$ (b) $\left(-\infty, \frac{1}{4}\right)$
(c) $(0, \infty)$ (d) $\left(\frac{1}{4}, \infty\right)$

Paragraph for Question Nos. 81 to 83

Consider the function $f(x)$ satisfying the identity

$$f(x) + f\left(\frac{x-1}{x}\right) = \frac{2x(2-x)}{(x-1)}\ \forall\, x \in R - \{0, 1\}$$

and $$g(x) = 2f(x) + x - 1$$

81. The domain of $y = \sqrt{g(x)}$ is :
(a) $(-\infty, -2)$ (b) $(1, \infty)$
(c) $(0, 1)$ (d) None of these

82. The range of $y = g(x)$ is :
(a) $(-2, 6)$ (b) $(-\infty, -2]$
(c) $[6, \infty)$ (d) $(-\infty, -2] \cup [6, \infty)$

83. The number of roots of the equation $g(x) = 2$ is :
(a) 3 (b) 2 (c) 1 (d) 0

Paragraph for Question Nos. 84 to 86

Let $f(x) = \begin{cases} x + 1, & x < 1 \\ 2x + 1, & 1 \le x < 2 \end{cases}$ and $g(x) = \begin{cases} x^2, & -1 \le x < 2 \\ x + 2, & 2 \le x < 3 \end{cases}$ are two functions.

84. The domain of the function $(gof)\,x$ is :
(a) $[-1, 2)$ (b) $[-2, 1)$
(c) $(-1, \sqrt{2})$ (d) $[\sqrt{2}, 2)$

85. The range of the function $(gof)\,x$ is ;
(a) $[0, 7]$ (b) $[0, 5]$ (c) $[0, 4)$ (d) $[0, 6)$

86. The number of solutions of the equation $(gof)x = 3$ is :
(a) 0 (b) 1 (c) 2 (d) 3

SECTION-IV (INTEGER TYPE)

This section contains **15 questions.** The answer to each question is a **positive integer.**

87. If $3f(x+1)+2f\left(\frac{1}{x+1}\right)=x$, $x \neq -1$, then the value of $5f(2)$ is .... .

88. If $a+\alpha=1, b+\beta=2$ and $af(x)+\alpha f\left(\frac{1}{x}\right)=bx+\frac{\beta}{x}$, $x \neq 0$, then the value of expression $\frac{f(x)+f\left(\frac{1}{x}\right)}{x+\frac{1}{x}}$ is ..... **[JEE Main 24 Feb 2021 (ES)]**

89. A function $f(x)$ takes a domain D onto a range R if for each $y \in R$, there is some $x \in D$ for which $f(x) = y$. Number of functions that can be defined from the domain $D = \{1, 2, 3\}$ onto the range $R = \{4, 5\}$ is ......

90. Let $f(x)=\sin^{23} x-\cos^{22} x$ and $g(x)=1+\frac{1}{2}\tan^{-1}|x|$, then the number of non-positive values of x in the interval $[-10\pi, 20\pi]$ satisfying the equation $f(x)=\text{sgn}\,(g(x))$, is .....

91. Let $f(x)$ and $g(x)$ be two real polynomials of degree 2 and 1 respectively. If $f(g(x))=8x^2-2x$, and $g(f(x))=4x^2+6x+1$, then the value of $f(2)+g(2)$ is ..... **[JEE Main 29 June 2022 (ES)]**

92. If $f(x+1)+f(x-1)=\sqrt{2}\,f(x)$, then the period of $f(x)$ is ....

93. Let f be a real valued function such that $f(x)+2f\left(\frac{2002}{x}\right)=3x$ for all $x > 0$. Then the value of $f(1001)+1001$ is ....

94. A non-zero function $f(x)$ is symmetrical about $y = x$, then the value of k (constant) such that $f^2(x)=(f^{-1}(x))^2-k\,x\,f(x)f^{-1}(x)+2x^2f(x)$ $\forall\, x \in R^+$ is ....

95. For some $a, b, c \in N$, let $f(x)=ax-3$ and $g(x)=x^b+c, x \in R$. If $(fog)^{-1}(x)=\left(\frac{x-7}{2}\right)^{1/3}$, then $(fog)(ac)+(gof)(b)$ is equal to ..... **[JEE Main 25 Jan 2023 (MS)]**

96. Let X be a set of all positive integers greater than or equal to 8 *i.e.*, $X = \{8, 9, 10, 11,...\}$ and let $f: x \to y$ be a function such that $f(x+y)=f(xy)$ for all $x \geq 4$, $y \geq 4$. If $f(8)=9$, then the value of $f(9)$ is .....

97. If $f(x,y)=\max.(x,y)$ and $g(x,y)=\max.(x,y)-\min.(x,y)$ and the value of $f(g(5,-3), g(-1.25, 14.75))$ is λ^{λ^2}, then the value of λ is ....

98. If range of the function $f(x)=3|\sin x|-2|\cos x|$ is $[a, b]$, then the value of $|b-a|$ is .....

99. Let $f: R \to R$ be a function defined by $f(x)=\frac{2e^{2x}}{e^{2x}+e}$. Then $f\left(\frac{1}{100}\right)+f\left(\frac{2}{100}\right)+f\left(\frac{3}{100}\right)+...+f\left(\frac{99}{100}\right)$ is equal to .... **[JEE Main 27 June 2022 (MS)]**

100. If $f(x)=2x^3+7x-5$, then the value of $f^{-1}(4)$ is ....

101. Let $A=\{0, 1, 2, 3, 4, 5, 6, 7\}$. Then the number of bijective functions $f: A \to A$ such that $f(1)+f(2)=3-f(3)$ is equal to .... **[JEE Main 22 Jul 2021 (ES)]**

SECTION-V (MATRIX TYPE)

This section contains **5 questions**. Question 102, 103 has four statements (A, B, C and D), questions 104, 105 has three statements (A, B and C) and questions 106 has four statements (A, B, C and D) given in **Column-I** and questions 102, 103, 104, 105, 106 has four statements (p, q, r and s) in **Column-II**. Any given statement in **Column-I** can have correct matching with one or more statement(s) given in **Column-II**.

102.

	Column-I		Column-II
(A)	Constant function $f(x)=c, c \in R$	**(p)**	Periodic
(B)	The function $g(x)=\int_1^x \frac{dt}{t} (x>0)$ is	**(q)**	Bound
(C)	The function $h(x)=$ arc $\tan x$ is	**(r)**	Monotonic
(D)	The function $k(x)=$ arc $\cot x$ is	**(s)**	Neither even nor odd

103.

	Column-I		Column-II
(A)	Let f be a function satisfying of x, then $f(xy)=\frac{f(x)}{y}$ for a positive real numbers x and y if $f(30) = 20$, then the value of $f(50)$ is divisible by	(p)	2
(B)	If the period of $\frac{\cos(\sin nx)}{\tan\left(\frac{x}{n}\right)}$, $n \in N$ is 6π, then n is divisible by	(q)	3
(C)	The number of real solutions of the equation $\log_{0.5}\|x\| = 2\|x\|$ is divisible by	(r)	4
(D)	If $f(3x+2)+f(3x+29)=0\ \forall\ x \in R$, then the period of $f(x)$ is divisible by	(s)	6

104.

	Column-I		Column-II
(A)	Let $h(x)=\|\lambda x+5\|$, the domain of $f(x)$ is $[-5, 7]$, the domain of $f(h(x))$ is $[-6, 1]$ and the range of $h(x)$ is the same as the domain of $f(x)$, then the value of λ is a	(p)	Natural number
(B)	A function $f: I \to I$ defined by $f(x)=\begin{cases} n+3, & n \in \text{odd} \\ n/2, & n \in \text{even} \end{cases}$ suppose $\lambda \in$ odd and $f(f(f(\lambda)))=27$, then sum of the digits of λ is a	(q)	Composite number
(C)	Let $f: R \to R$ be a function defined by $f(x)=\{\|\cos x\|\}$, where $\{x\}$ represents fractional part of x. Let S be the set containing all real values x lying in the interval $[0, 2\pi]$ for which $f(x) \neq \|\cos x\|$. Then, the number of elements in the set S is a	(r)	Prime number
		(s)	Perfect number

105. Column-I contains rational algebraic expressions and column-II contains possible integers which lie in their range.

	Column-I		Column-II
(A)	$y=\frac{x^2-2x+4}{x^2+2x+4}, x \in R$	(p)	–9
(B)	$y=\frac{x^2-3x-2}{2x-3}, x \in R$	(q)	–3
(C)	$y=\frac{2x^2-2x+4}{x^2-4x+3}, x \in R$	(r)	1
		(s)	3

106. Let $f(x) = \ln x$ and $g(x) = x^2 - 1$ column-I contains composite functions and column-II contains their domain.

	Column-I		Column-II
(A)	*fog*	(p)	$(-\infty, \infty)$
(B)	*gof*	(q)	$(-\infty, -1) \cup (1, \infty)$
(C)	*fof*	(r)	$(0, \infty)$
(D)	*gog*	(s)	$(1, \infty)$

SECTION-VI (ASSERTION-REASON TYPE)

Directions : Questions number 107 to 118 are Assertion-Reason type questions. Each of these questions contains two statements:

Statement-1 : (Assertion) and

Statement-2 : (Reason)

Each of these questions also has four alternative choices, only one of which is the correct answer. You have to select the correct choice, as given below.

(a) Statement 1 is true, statement-2 is true; Statement-2 is a correct explanation for Statement-1.

(b) Statement 1 is true, Statement-2 is true; Statement-2 is not a correct explanation for Statement-1.

(c) Statement 1 is true, Statement-2 is false.

(d) Statement 1 is false, Statement-2 is true.

107. Statement-1 : If $f(x) = x$ and $g(x) = \frac{x^2}{x}$, then $f(x) = g(x)$ always

Statement-2 : At $x = 0$, $g(x)$ is not defined.

108. Consider the function $f(x) = \frac{1}{2\{-x\}} - \{x\}$, where $\{x\}$ denotes the fractional part of x and x is not an integer.

Statement-1 : The minimum value of $f(x)$ is $\sqrt{2}-1$.

because

Statement-2 : If the product of two positive numbers is a constant, then the minimum value of their sum is 2 times the square root of their product.

109. Statement-1 : Fundamental period of $\cos x + \tan x$ is 2π

because

Statement-2 : If the period of $f(x)$ is T_1 and the period of $g(x)$ is T_2, then the fundamental period of $f(x) + g(x)$ is LCM of T_1 and T_2.

110. Let $f : R \to R$; $f(x) = x^4 + 2x^3 - x^2 + 1$

Statement-1 : $f(x)$ is an into.

because

Statement-2 : A polynomial of degree even defined from $R \to R$ is always into.

111. Statement-1 : The modulus function $f(x) = |x|$ is NOT one-one.

Statement-2 : The negative real number are NOT the images of any real numbers.

112. Consider the function

$$f(x) = \cos^{-1}\left(\frac{1+x^2}{2x}\right)$$

Statement-1 : Range of the function $f(x)$ has exactly two elements.

because

Statement-2 : Domain of the function $f(x)$ has exactly two elements.

113. Statement-1 : If $f(1 + x) = f(1 - x)$ and $f(4 + x) = f(4 - x)$, then $f(x)$ is periodic with period 6.

Statement-2 : 6 is not necessarily fundamental period of $f(x)$.

114. Let f : defined from $\left(\frac{1}{4}, \infty\right) \to R^+$ as,

$$f(x) = \log_{1/4}\left(x - \frac{1}{4}\right) + \frac{1}{2}\log_4(16x^2 - 8x + 1)$$

Statement-1 : $f(x)$ is neither injective nor surjective because

Statement-2 : $f(x)$ is a constant function.

115. Statement-1 : If $g(x) = f(x) - 1$ and $f(x) + f(1-x) = 2\ \forall\ x \in R$, then $g(x)$ is symmetrical about the point $\left(\frac{1}{2}, 0\right)$.

Statement-2 : If $g(a-x) = -g(a+x)\ \ \forall\ x \in R$, then $g(x)$ is symmetrical about the point $(a, 0)$.

116. Consider the function $f(x) = |\sin(kx)| + \{x\}$, where (x) denotes the fractional part fraction of x.

Statement-1 : $f(x)$ is periodic for $k = m\pi$, where m is a rational number. because

Statement-2 : The sum of two periodic function is always periodic.

117. Statement-1 : Let $f : R - \{1, 2, 3\} \to R$ be a function defined by

$f(x) = \frac{1}{x-1} + \frac{2}{x-2} + \frac{3}{x-3}$, then f is many one function.

Statement-2 : If either $f'(x) > 0$ or $f'(x) < 0$, $\forall\ x \in$ domain of f, then $y = f(x)$ is one-one function.

118. Statement-1 : The function $f(x) = x^4 + 2x + 3$ defined from R to R is not injective.

Statement-2 : Every polynomial function of even degree defined from R to R is always not injective.

SECTION-VII (SUBJECTIVE TYPE)

In this section there are 18 subjective questions.

119. Let $f(x + y + 1) = (\sqrt{f(x)} + \sqrt{f(y)})^2$ and $f(0) = 1$ $\forall\ x, y \in R$. Determine $f(x)$.

120. Determine all functions $f : R \to R$ such that $f(x - f(y)) = f(f(y)) + xf(y) + f(x) - 1\ \forall\ x, y \in R$.

121. A function f, defined for all $x, y \in R$ is such that $f(1) = 2$; $f(2) = 8$ and $f(x+y) - kxy = f(x) + 2y^2$, where k is some constant. Find $f(x)$ and show that :

$$f(x+y)\, f\left(\frac{1}{x+y}\right) = k \text{ for } x+y \neq 0$$

122. Consider a real valued function $f(x)$ satisfying $2f(xy)=(f(x))^y+(f(y))^x$ for all $x, y \in R$ and $f(1)=a$, where $a \neq 1$. Prove that

$$(a-1)\sum_{i=1}^{n} f(i)=a^{n+1}-a$$

123. Let a function $f: R \to R$ such that $f(1)=2$ and $f(x+y)=2^x f(y)+4^y f(x)\ \forall x, y \in R$. If $f'(2)=\lambda \ln 2$, then find the value of λ.

124. If the function f satisfies the relation $f(x+y)+f(x-y)=2f(x)\cdot f(y)\ \ \forall x, y \in R$ and $f(0)\neq 0$, then prove that $f(x)$ is an even function.

125. If $f(a-x)=f(a+x)$ and $f(b-x)=f(b+x)$ for all real x, where a, b $(a > b)$ are constants, then prove that $f(x)$ is a periodic function.

126. If p and q are positive integers, f is a function defined for positive numbers and attains only positive values, such that $f(xf(y))=x^p y^q$, then prove that $q=p^2$.

127. Let $f_1(x)=\dfrac{x}{3}+10\ \forall x \in R$ and $f_n(x)=f_1(f_{n-1}(x))$ for $n \geq 2$, then find $f_n(x)$.

128. Let $f: R \to R$ be defined as $f(x+y)+f(x-y)=2f(x)f(y),\ f\left(\dfrac{1}{2}\right)=-1.$ Then find the value of $\displaystyle\sum_{k=1}^{30}\frac{1}{\sin k \sin(k+f(k))}$.

129. Let f be a function from the set of positive integers to the set of real number *i.e.*, $f: N \to R$, such that

(i) $f(1)=1$
(ii) $f(1)+2f(2)+3f(3)+\ldots+nf(n)$ for $n \geq 2$, then find the value of f(2013).

130. Prove that : $f(x)=\dfrac{2x(\sin x+\tan x)}{2\left[\dfrac{x+2\pi}{\pi}\right]-3}$ is an odd function, where $[\cdot]$ denotes the greatest integer function.

131. Let n be a positive integer and define $f(n)=1!+2!+3!+\ldots+n!$, where $n!=n(n-1)(n-2)\ldots 3\cdot 2\cdot 1$.
Find polynomials $P(x)$ and $Q(x)$ such that $f(n+2)=P(n)f(n+1)+Q(n)f(n)$, for all $n \geq 1$

132. Let f be a real valued function with domain R. Now if for some positive constant a, the equation, $f(x+a)=1+\{2-3f(x)+3(f(x))^2-(f(x))^3\}^{1/3}$ holds good for $x \in R$, then prove that $f(x)$ is a periodic function.

133. Find the number of functions, $f:\{1,2,3\}\to\{1,2,3,4,5\}$, such that $f(i)\leq f(j)$ for $i<j$.

134. If $P(x)$ be a polynomial with integer coefficients such that there exists four distinct integers a, b, c and d with $P(a)=P(b)=P(c)=P(d)=1991$, then show that there exist no integer α such that $P(\alpha)=1993$.

135. Find total number of functions, $f:\{1,2,3,4,5\}\to\{1,2,3,4,5,6\}$ such that $f(1)+f(2)+1=f(4)$.

136. Find the range of the function $f(x)=\dfrac{\sin^2 x+\sin x-1}{\sin^2 x-\sin x+2}$ (Use differentiation, maxima and minimas is not allowed).

SECTION VIII (Selected Questions Asked (JEE (Main and Advanced) in Previous Years)

137. Let $5f(x)+4f\left(\dfrac{1}{x}\right)=\dfrac{1}{x}+3,\ x>0$. Then $18\int_1^2 f(x)dx$ is equal to

(a) $5\log_e 2+3$ (b) $10\log_e 2+6$
(c) $10\log_e 2-6$ (d) $5\log_e 2-3$

[JEE Main 6 Apr 2023 (MS)]

138. If $f(x)=\dfrac{(\tan 1°)x+\log_e(123)}{x\log_e(1234)-(\tan 1°)}$, $x>0$, then the least value of $f(f(x))+f\left(f\left(\dfrac{4}{x}\right)\right)$ is

(a) 0 (b) 8 (c) 2 (d) 4

[JEE Main 10 Apr 2023 (MS)]

139. If the domain of the function $f(x)=\log_e(4x^2+11x+6)+\sin^{-1}(4x+3)+\cos^{-1}\left(\dfrac{10x+6}{3}\right)$ is $(\alpha, \beta]$, then $36|\alpha+\beta|$ is equal to

(a) 54 (b) 72
(c) 63 (d) 45

[JEE Main 15 Apr 2023 (MS)]

140. Let $f^1(x)=\dfrac{3x+2}{2x+3}$, $x \in R-\left\{-\dfrac{3}{2}\right\}$. For $n \geq 2$, define $f^n(x)=f^1 o f^{n-1}(x)$. If $f^5(x)=\dfrac{ax+b}{bx+a}$, $gcd(a,b)=1$, then $a+b$ is equal to ....

[JEE Main 30 Jan 2023 (MS)]

141. Let $R=\{a,b,c,d,e\}$ and $S=\{1,2,3,4\}$. Total number of onto functions $f: R\to S$ such that $f(a)\neq 1$, is equal to .... **[JEE Main 8 Apr 2023 (ES)]**

142. Let $S=\{1,2,3,4,5,6,7,8,9,10\}$. Define $f: S\to S$ as

$$f(n)=\begin{cases}2n, \text{ if } n=1,2,3,4,5\\ 2n-11, \text{ if } n=6,7,8,9,10\end{cases}$$

Let $g : S \to S$ be a function such that

$fog(n) = \begin{cases} n+1, \text{ if } n \text{ is odd} \\ n-1, \text{ if } n \text{ is even} \end{cases}$, then

$g(10)(g(1)+g(2)+g(3)+ g(4)+g(5))$ is equal to ....

[JEE Main 27 Jun 2022 (ES)]

143. The range of the function

$f(x) = \log_{\sqrt{5}}\left(3+\cos\left(\frac{3\pi}{4}+x\right)+\cos\left(\frac{\pi}{4}+x\right)+\cos\left(\frac{\pi}{4}-x\right) - \cos\left(\frac{3\pi}{4}-x\right)\right)$ is

(a) $\left[\frac{1}{\sqrt{5}}, \sqrt{5}\right]$ (b) [0, 2]

(c) $(0, \sqrt{5})$ (d) [−2, 2]

[JEE Main 1 Sep 2021 (ES)]

144. Consider function $f : A \to B$ and $g : B \to C$ $(A, B, C \subseteq R)$ such that $(gof)^{-1}$ exists, then

(a) *f* and *g* both are one-one

(b) *f* and *g* both are onto

(c) *f* is one-one and *g* is onto

(d) *f* is onto and *g* is one-one

[JEE Main 25 Jul 2021 (ES)]

145. If the function $f : R \to R$ is defined by $f(x) = |x|(x - \sin x)$, then which of the following statements is true? **[JEE Advanced 2020]**

(a) *f* is one-one, but not onto

(b) *f* is onto, but not one-one

(c) *f* is both one-one and onto

(d) *f* is neither one-one nor onto

146. Let the function $f : [0, 2] \to R$ be defined by

$f(x) = \frac{4^x}{4^x + 2}$. Then, the value of

$f\left(\frac{1}{40}\right)+f\left(\frac{2}{40}\right)+f\left(\frac{3}{40}\right)+\ldots+f\left(\frac{39}{40}\right)-f\left(\frac{1}{2}\right)$ is ...

[JEE Advanced 2020]

147. Let $A = \{a, b, c\}$ and $B = \{1, 2, 3, 4\}$. Then the number of elements in the set $C = \{f : A \to B \mid 2 \in f(A)$ and f is not one-one$\}$ is .... **[JEE Main 5 Sep 2020 (ES)]**

148. For $x \in R - \{0, 1\}$, let $f_1(x) = \frac{1}{x}$, $f_2(x) = 1 - x$ and $f_3(x) = \frac{1}{1-x}$ be three given functions. If a function, $J(x)$ satisfies $(f_2 \, o \, J \, o \, f_1)(x) = f_3(x)$, then $J(x)$ is equal to **[JEE Main 9 Jan 2019 (MS)]**

(a) $f_3(x)$ (b) $\frac{1}{x} f_3(x)$ (c) $f_1(x)$ (d) $f_2(x)$

149. Let X be a set with exactly 5 elements and Y be a set with exactly 7 elements. If α is the number of one-one functions from X to Y and β is the number of onto functions from Y to X, then the value of $\frac{1}{5!}(\beta - \alpha)$ is .... **[JEE Advanced 2018]**

150. Let $E_1 = \{x \in R : x \neq 1$ and $\frac{x}{x-1} > 0\}$ and

$E_2 = \{x \in E_1 : \sin^{-1}\left(\log_e\left(\frac{x}{x-1}\right)\right)$ is a real number$\}$.

(Here, the inverse trigonometric function $\sin^{-1} x$ assumes values in $\left[-\frac{\pi}{2}, \frac{\pi}{2}\right]$).

Let $f : E_1 \to R$ be the function defined by $f(x) = \log_e\left(\frac{x}{x-1}\right)$ and $g : E_2 \to R$ be the function defined by the $g(x) = \sin^{-1}\left(\log_e\left(\frac{x}{x-1}\right)\right)$.

	List-I		List-II
P.	The range of *f* is	**1.**	$\left(-\infty, \frac{1}{1-e}\right] \cup \left[\frac{e}{e-1}, \infty\right)$
Q.	The range of *g* contains	**2.**	(0, 1)
R.	The domain of *f* contains	**3.**	$\left[-\frac{1}{2}, \frac{1}{2}\right]$
S.	The domain of *g* is	**4.**	$(-\infty, 0) \cup (0, \infty)$
		5.	$\left(-\infty, \frac{e}{e-1}\right]$
		6.	$(-\infty, 0) \cup \left(\frac{1}{2}, \frac{e}{e-1}\right]$

The correct option is

(a) P → 4; Q → 2; R → 1; S → 1

(b) P → 3; Q → 3; R → 6; S → 5

(c) P → 4; Q → 2; R → 1; S → 6

(d) P → 4; Q → 3; R → 6; S → 5

[JEE Advanced 2018]

151. If $f(x) + 2f\left(\frac{1}{x}\right) = 3x$, $x \neq 0$ and $S = \{x \in R : f(x) = f(-x)\}$; then S :

(a) Contain exactly two elements

(b) Contains move than two elements

(c) is an empty set

(d) contains exactly one element

[JEE Advanced 2016]

Answers

Introductory Exercise 2.1

1.	(b)	2.	(b)	3.	(c)	4.	(c)	5.	(c)										

Introductory Exercise 2.2

1.	(b)	2.	(a)	3.	(a)	4.	(b)	5.	(c)	6.	(a)	7.	(b)	8.	(b)	9.	(c)	10.	(d)

Introductory Exercise 2.3

1.	(c)	2.	(b)	3.	(b)	4.	(c)	5.	(c)	6.	(b)	7.	(b)	8.	(b)	9.	(c)	10.	(b)
11.	(a)	12.	(b)	13.	(a)	14.	(b)	15.	(c)	16.	(d)								

Introductory Exercise 2.4

1.	(d)	2.	(c)	3.	(d)	4.	(a)	5.	(d)	6.	(b)	7.	(d)	8.	(c)	9.	(c)	10.	(a)
11.	(d)	12.	(c)	13.	(b)	14.	(c)	15.	(b)	16.	(b)	17.	(b)	18.	(d)				

Introductory Exercise 2.5

1.	(a)	2.	(c)	3.	(c)	4.	(a)	5.	(a)	6.	(c)	7.	(b)	8.	(a)	9.	(d)	10.	(a)
11.	(c)	12.	(a)	13.	(b)	14.	(b)	15.	(b)	16.	(a)	17.	(a)	18.	(d)	19.	(c)	20.	(d)
21.	(c)	22.	(b)	23.	(c)	24.	(d)	25.	(b)	26.	(b)	27.	(c)						

Introductory Exercise 2.6

1.	(d)	2.	(d)	3.	(c)	4.	(a)	5.	(c)	6.	(b)	7.	(b)	8.	(c)	9.	(c)	10.	(b)
11.	(d)	12.	(b)	13.	(c)	14.	(d)	15.	(b)	16.	(c)	17.	(b)						

Introductory Exercise 2.7

1.	(a)	2.	(d)	3.	(d)	4.	(c)	5.	(a)	6.	(b)	7.	(a)	8.	(a)	9.	(c)	10.	(d)
11.	(c)	12.	(b)	13.	(d)	14.	(b)	15.	(b)	16.	(b)	17.	(a)	18.	(b)	19.	(b)	20.	(b)
21.	(c)	22.	(c)	23.	(d)	24.	(a)	25.	(d)	26.	(b)	27.	(d)						

Introductory Exercise 2.8

1.	(b)	2.	(a)	3.	(a)	4.	(b)	5.	(b)	6.	(d)	7.	(a)	8.	(a)	9.	(b)	10.	(c)
11.	(b)	12.	(a)																

Introductory Exercise 2.9

1.	(c)	2.	(b)	3.	(a)	4.	(a)	5.	(b)	6.	(d)	7.	(a)	8.	(d)	9.	(b)	10.	(d)
11.	(d)	12.	(b)	13.	(d)	14.	(b)	15.	(c)	16.	(d)	17.	(b)	18.	(b)	19.	(d)	20.	(b)
21.	(c)	22.	(c)	23.	(a)														

Section-I (Single Correct Choice Type)

1.	(c)	2.	(d)	3.	(a)	4.	(d)	5.	(b)	6.	(c)	7.	(c)	8.	(b)	9.	(a)	10.	(d)
11.	(a)	12.	(d)	13.	(b)	14.	(d)	15.	(c)	16.	(b)	17.	(d)	18.	(c)	19.	(b)	20.	(a)
21.	(d)	22.	(c)	23.	(b)	24.	(c)	25.	(b)	26.	(d)	27.	(a)	28.	(c)	29.	(b)	30.	(c)
31.	(c)	32.	(c)	33.	(c)	34.	(a)	35.	(a)	36.	(c)	37.	(c)	38.	(c)	39.	(c)	40.	(b)
41.	(b)	42.	(d)	43.	(a)	44.	(d)	45.	(d)										

Section-II (Multiple Correct Choice Type)

46.	(c, d)	47.	(a, b, d)	48.	(a, b, c)	49.	(c, d)	50.	(b, c, d)	51.	(a, c)
52.	(a, c, d)	53.	(a, b, c)	54.	(a, c, d)	55.	(a, b, c, d)	56.	(b, c, d)	57.	(a, d)
58.	(a, d)	59.	(a, b)	60.	(a, b, c)	61.	(a, c, d)	62.	(a, b, c)		

Section-III (Paragraph Type)

63.	(b)	**64.**	(a)	**65.**	(c)	**66.**	(b)	**67.**	(c)	**68.**	(b)	**69.**	(b)	**70.**	(b)	**71.**	(c)	**72.**	(a)
73.	(c)	**74.**	(b)	**75.**	(b)	**76.**	(c)	**77.**	(a)	**78.**	(c)	**79.**	(b)	**80.**	(a)	**81.**	(b)	**82.**	(d)
83.	(d)	**84.**	(b)	**85.**	(c)	**86.**	(c)												

Section-IV (Integer Type)

87.	(4)	**88.**	(2)	**89.**	(6)	**90.**	(5)	**91.**	(18)	**92.**	(8)
93.	(4)	**94.**	(2)	**95.**	(2039)	**96.**	(9)	**97.**	(2)	**98.**	(5)
99.	(99)	**100.**	(1)	**101.**	(720)						

Section-V (Matrix Type)

102. (A) → (p, q); (B) → (r, s); (C) → (q, r); (D) → (q, r, s)

103. (A) → (p, q, r, s); (B) → (p, q, s); (C) → (p); (D) → (p, q, s)

104. (A) → (p, r); (B) → (p, q, s); (C) → (p, r)

105. (A) → (r, s); (B) → (p, q, r, s); (C) → (p, r, s)

106. (A) → (q); (B) → (r); (C) → (s); (D) → (p)

Section-VI (Assertion-Reason Type)

107.	(d)	**108.**	(a)	**109.**	(c)	**110.**	(a)	**111.**	(c)	**112.**	(a)	**113.**	(b)	**114.**	(a)
115.	(a)	**116.**	(c)	**117.**	(c)	**118.**	(a)								

Section-VII (Subjective Type)

119. $f(x)=(x+1)^2$

120. $f(x)=1-\dfrac{x^2}{2}$

121. $f(x)=2x^2$

123. $\lambda=28$

127. $f_n(x)=\left(\dfrac{x-15}{3^n}\right)+15$

128. $\operatorname{cosec}^2(1)\cdot\operatorname{cosec}(31)\cdot\sin(30)$

129. $f(2013)=\dfrac{1}{4026}$

131. $P(x)=x+3$ and $Q(x)=-x-2$

133. (35)

135. (360)

136. $R_f=\left[\dfrac{3-2\sqrt{11}}{7},\dfrac{1}{2}\right]\cup\left[3,\dfrac{3+2\sqrt{11}}{7}\right]\cup\{1\}$

Section-VIII (Selected Questions Asked) (JEE (Main & Advanced) in Previous Years)

137.	(c)	**138.**	(d)	**139.**	(d)	**140.**	(3125)	**141.**	(180)	**142.**	(190)	**143.**	(b)
144.	(c)	**145.**	(c)	**146.**	(19)	**147.**	(19)	**148.**	(a)	**149.**	(119)	**150.**	(a)
151.	(a)												

SOLUTIONS

1. We have $f(x)=\ln\{x\}+\sqrt{x-2\{x\}}$

Clearly, $f(x)$ is defined, if

$$0<\{x\}<1 \text{ and } x-2\{x\}\geq 0$$

$$\Rightarrow \quad 0<\{x\}<1 \text{ and } x\geq 2\{x\}$$

$$\Rightarrow \quad x\in R-I \text{ and } x\geq 1 \Rightarrow x\in[1,\infty)-I$$

Hence, $D_f=(1,\infty)-I^+$

2. We have

$$f(x)=(x+0.5)^{\log_{(x+0.5)}\left(\frac{x^2+2x-3}{4x^2-4x-3}\right)}$$

Clearly, $f(x)$ is defined, if

$$x+0.5>0,\ x+0.5\neq 1 \text{ and } \left(\frac{x^2+2x-3}{4x^2-4x-3}\right)>0$$

$$\Rightarrow x>-0.5,\ x\neq 0.5 \text{ and } \frac{(x+3)(x-1)}{(2x+1)(2x-3)}>0 \quad \text{...(i)}$$

Now, for $\dfrac{(x+3)(x-1)}{(2x+1)(2x-3)}>0$

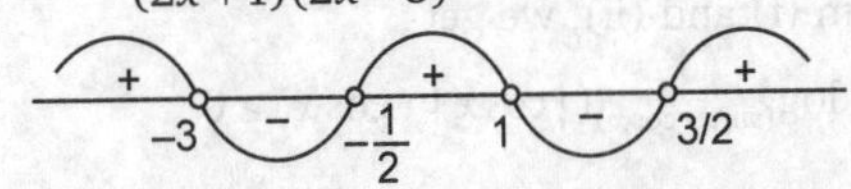

From wavy curve method,

$$x \in (-\infty, -3) \cup \left(-\frac{1}{2}, 1\right) \cup \left(\frac{3}{2}, \infty\right) \quad \ldots\text{(ii)}$$

From (i) and (ii), we get

$$\Rightarrow \; x > -\frac{1}{2}, x \neq \frac{1}{2} \text{ and } x \in (-\infty, -3) \cup \left(-\frac{1}{2}, 1\right) \cup \left(\frac{3}{2}, \infty\right)$$

$$\Rightarrow \quad x \in \left(-\frac{1}{2}, 1\right) \cup \left(\frac{3}{2}, \infty\right) - \left\{\frac{1}{2}\right\}$$

Hence, $D_f = \left(-\frac{1}{2}, \frac{1}{2}\right) \cup \left(\frac{1}{2}, 1\right) \cup \left(\frac{3}{2}, \infty\right)$

3. Given $f(x) = 2x + |x|$ and $g(x) = \frac{1}{3}(2x - |x|)$

For $x \geq 0$:

$$f(x) = 2x + x = 3x \text{ and } g(x) = \frac{1}{3}(2x - x) = \frac{x}{3}$$

$$\therefore \; h(x) = f(g(x)) = f\left(\frac{x}{3}\right) = 3\left(\frac{x}{3}\right) = x \quad \ldots\text{(i)}$$

For $x < 0$:

$$f(x) = 2x - x = x \text{ and } g(x) = \frac{1}{3}(2x + x) = x$$

$$\therefore \; h(x) = f(g(x)) = f(x) = x \quad \ldots\text{(ii)}$$

From (i) and (ii), we get

$$h(x) = x \; \forall \; x \in R$$

Now let $F(x) = \sin^{-1}\underbrace{(h(h(h(h\ldots\ldots h(x))\ldots\ldots))))}_{n\text{-times}}$

$$= \sin^{-1} x$$

Hence, $D_F = [-1, 1]$

4. We have

$$f(x) = \sqrt{\log_{(\sin x + \cos x)}(|\cos x| + \cos x)}$$

Clearly, $f(x)$ is defined, if

$\log_{(\sin x + \cos x)}(|\cos x| + \cos x) \geq 0,$

$\sin x + \cos x > 0, \; \sin x + \cos x \neq 1$

and $\quad |\cos x| + \cos x > 0 \quad \ldots\text{(i)}$

$\because \; |\cos x| + \cos x > 0 \quad \forall \; x \in \left[0, \frac{\pi}{2}\right)$

and $\quad 1 \leq \sin x + \cos x \leq \sqrt{2} \quad \forall \; x \in \left[0, \frac{\pi}{2}\right)$

but $\sin x + \cos x \neq 1$

$\therefore \quad 1 < \sin x + \cos x \leq \sqrt{2} \quad \forall \; x \in \left(0, \frac{\pi}{2}\right) \quad \ldots\text{(ii)}$

From (i) and (ii), we get

$\log_{(\sin x + \cos x)}(|\cos x| + \cos x) \geq 0$

and $\quad x \in \left(0, \frac{\pi}{2}\right)$

$$\Rightarrow \; \log_{(\sin x + \cos x)}(2\cos x) \geq 0 \text{ and } x \in \left(0, \frac{\pi}{2}\right)$$

$$\Rightarrow \quad 2\cos x \geq 1 \text{ and } x \in \left(0, \frac{\pi}{2}\right)$$

$$\Rightarrow \quad \cos x \geq \frac{1}{2} \text{ and } x \in \left(0, \frac{\pi}{2}\right)$$

$$\Rightarrow \quad x \leq \frac{x}{3} \text{ and } x \in \left(0, \frac{\pi}{2}\right)$$

$$\therefore \quad x \in \left(0, \frac{\pi}{3}\right]$$

Hence, $D_f = \left(0, \frac{\pi}{3}\right]$

5. We have

$$f(x) = \frac{1}{\sqrt{\{\sin x\} + \{-\sin x\}}}$$

Clearly, $f(x)$ is defined, if $\{\sin x\} + \{-\sin x\} > 0$

$$\Rightarrow \quad \sin \notin I$$

$$\therefore \quad \sin x \neq -1, 0, 1 \quad (\because -1 \leq \sin x \leq 1)$$

$$\Rightarrow \quad x \neq \frac{n\pi}{2}, n \in I$$

Hence, $D_f = R \sim \frac{n\pi}{2}, n \in I$

6. Let $y = f(x) = \frac{\{x\}}{1 + \{x\}} = 1 - \frac{1}{1 + \{x\}} \quad \ldots\text{(i)}$

$$\because \quad 0 \leq \{x\} < 1 \Rightarrow 1 \leq 1 + \{x\} < 2$$

$$\Rightarrow \quad 1 \geq \frac{1}{1 + \{x\}} > \frac{1}{2}$$

$$\Rightarrow \quad -1 \leq -\frac{1}{1 + \{x\}} < -\frac{1}{2}$$

$$\Rightarrow \quad 1 - 1 \leq 1 - \frac{1}{1 + \{x\}} < 1 - \frac{1}{2}$$

$$\Rightarrow \quad 0 \leq y < \frac{1}{2} \quad \{\text{from (i)}\}$$

$$\Rightarrow \quad y \in \left[0, \frac{1}{2}\right)$$

Hence, $R_f = \left[0, \frac{1}{2}\right)$

7. We have

$$f(x) = \tan^{-1}\sqrt{[x] + [-x]} + \sqrt{2 - |x|} + \frac{1}{x^2}$$

Clearly, $f(x)$ is defined, if

$$[x] + [-x] \geq 0, \; 2 - |x| \geq 0 \text{ and } x \neq 0$$

$$\Rightarrow [x] + [-x] = 0, \; |x| \leq 2$$

and $\quad x \neq 0 \quad (\because [x] + [-x] \leq 0 \; \forall \; x \in R)$

$$\Rightarrow \quad x \in I, \; -2 \leq x \leq 2 \text{ and } x \neq 0$$

$$\Rightarrow \quad x = -2, -1, 1, 2$$

$\Rightarrow \quad D_f = \{-2,-1,1,2\}$

Now, $\quad f(\pm 2) = 0+0+\frac{1}{4} = \frac{1}{4}$

and $\quad f(\pm 1) = 0+1+\frac{1}{1} = 2$

Hence, $\quad R_f = \left\{\frac{1}{4}, 2\right\}$

8. We have

$$f(x) = \sqrt{(x-4)} + 2\sqrt{(6-x)}$$

Clearly, $f(x)$ is defined, if

$x-4 \geq 0$ and $6-x \geq 0$

$\Rightarrow \quad x \geq 4$ and $x \leq 6$

Hence, $\quad D_f = [4,6]$

Now, $\quad f'(x) = \frac{1}{2\sqrt{(x-4)}} - \frac{1}{\sqrt{(6-x)}} = 0$

$\Rightarrow \quad \sqrt{(6-x)} = 2\sqrt{(x-4)}$

$\Rightarrow \quad 6-x = 4x-16 \Rightarrow x = \frac{22}{5} \in (4,6)$

$\therefore$ Critical points are $4, \frac{22}{5}$ and 6.

$\Rightarrow \quad f(4) = 2\sqrt{2},\ f\left(\frac{22}{5}\right) = \sqrt{10}$ and $f(6) = \sqrt{2}$

Hence, $\quad R_f = [\sqrt{2}, \sqrt{10}]$

9. We have

$$f(x) = \log_{e^2 x}\left(\frac{2\ln x + 2}{-x}\right) \text{ and } g(x) = \{x\}$$

$\therefore \quad f(g(x)) = \log_{e^2 g(x)}\left(\frac{2\ln g(x)+2}{-g(x)}\right)$

$= \log_{e^2\{x\}}\left(\frac{2\ln\{x\}+2}{-\{x\}}\right)$

Clearly, $f(g(x))$ is defined, if

$e^2\{x\} > 0,\ e^2\{x\} \neq 1$ and $\left(\frac{2\ln\{x\}+2}{-\{x\}}\right) > 0$

$\Rightarrow \{x\} > 0,\ \{x\} \neq e^{-2}$ and $\left(\frac{\ln\{x\}+1}{\{x\}}\right) < 0$

$\Rightarrow \{x\} > 0,\ \{x\} \neq e^{-2}$ and $\ln\{x\} + 1 < 0$

$\Rightarrow \{x\} > 0,\ \{x\} \neq e^2$ and $\{x\} < e^{-1}$

$\therefore \quad \{x\} \in (0, e^{-2}) \cup (e^{-2}, e^{-1})$

or $\quad g(x) \in (0, e^{-1}) - \{e^{-2}\}$

Hence, $\quad R_g = (0, e^{-1}) - \{e^{-2}\}$

10. We have

$$f(x) = \sum_{r=1}^{n}\left[r + \sin\left(\frac{x}{r}\right)\right] \forall\, x \in [0, \pi]$$

$$= \sum_{r=1}^{n}\left(r + \left[\sin\left(\frac{x}{r}\right)\right]\right)$$

$$= \sum_{r=1}^{n} r + \sum_{r=1}^{n}\left[\sin\left(\frac{x}{r}\right)\right]$$

$$= \frac{n(n+1)}{2} + \sum_{r=1}^{n}\left[\sin\left(\frac{x}{r}\right)\right]$$

$\because \quad \sum_{r=1}^{n}\left[\sin\left(\frac{x}{r}\right)\right] = [\sin x] + \left[\sin\left(\frac{\pi}{2}\right)\right] + \left[\sin\left(\frac{x}{3}\right)\right] + \ldots + \left[\sin\left(\frac{x}{n}\right)\right]$

$= 0, 1\ \forall\, x \in [0, \pi]$

$\therefore \quad f(x) = \frac{n(n+1)}{2}, \frac{n(n+1)}{2} + 1$

Hence, $\quad R_f = \left\{\frac{n(n+1)}{2}, \frac{n^2+n+2}{2}\right\}$

11. $\frac{6+2\log_3 x}{-5x} > 0$ and $x > 0$ and $x \neq \frac{1}{3}$

$\Rightarrow \quad 6 + 2\log_3 x < 0$

$\Rightarrow \quad \log_3 x < -3 \Rightarrow x < 3^{-3}$

$\therefore \quad x \in \left(0, \frac{1}{27}\right)$ ...(i)

and $\quad -1 \leq \log_{3x}\left(\frac{6+2\log_3 x}{-5x}\right) \leq 1$

$\Rightarrow \quad (3x)^{-1} \geq \frac{6+2\log_3 x}{-5x} \geq 3x$

$\Rightarrow \quad -\frac{5}{3} \leq 6 + 2\log_3 x \leq -15x^2$

$\Rightarrow \quad 15x^2 + 6 + 2\log_3 x \leq 0$

and $\quad 6 + 2\log_3 x + \frac{5}{3} \geq 0$

$\Rightarrow \quad x \in \left(0, \frac{1}{27}\right)$ ...(ii)

and $\quad x \geq 3^{\frac{-23}{6}}$ ...(iii)

From (i), (ii) and (iii), we get $x \in \left[3^{-\frac{23}{6}}, \frac{1}{27}\right)$

$\therefore$ α is small positive quantity and $\beta = \frac{1}{27}$

Hence, $\alpha^2 + \frac{5}{\beta}$ is just greater than 135

$\therefore$ $\alpha^2 + \frac{5}{\beta}$ is approximately equal to 135.

12. We have

$$f(x) = \begin{cases} \frac{1}{2^x}, & x \geq 5 \\ f(x+1), & x < 5 \end{cases}$$

$$\therefore \quad f\left(2+\log_2\left(\frac{3}{2}\right)\right) = f(2+\log_2 3-1)$$
$$= f(1+\log_2 3) \quad (\because 1+\log_2 3<5)$$
$$= f(2+\log_2 3) \quad (\because 1+\log_2 3<5)$$
$$= f(3+\log_2 3) \quad (\because 3+\log_2 3<5)$$
$$= f(4+\log_2 3) \quad (\because 4+\log_2 3>5)$$
$$= \frac{1}{2^{4+\log_2 3}}$$
$$= \frac{1}{2^4\cdot 3} = \frac{1}{48}$$

13. We have

$$f(x) = x^2 + \ln\left(\frac{\pi+x}{\pi-x}\right)\cdot\cos x + \ln\left(\frac{\pi-x}{\pi+x}\right) \quad \text{...(i)}$$
$$\therefore \; f(-x) = (-x)^2 + \ln\left(\frac{\pi-x}{\pi+x}\right)\cdot\cos(-x) + \ln\left(\frac{\pi+x}{\pi-x}\right)$$
$$\Rightarrow f(-x) = x^2 - \ln\left(\frac{\pi+x}{\pi-x}\right)\cdot\cos x - \ln\left(\frac{\pi-x}{\pi+x}\right) \quad \text{...(ii)}$$

Adding (i) and (ii), we get

$$f(x)+f(-x) = 2x^2$$

Put $x = 10$, then

$$f(10)+f(-10) = 2(10)^2$$
$$\Rightarrow \quad 100+f(-10) = 200 \quad [\because f(10)=100]$$
$$\therefore \quad f(-10) = 100$$

14. Given, $f(x) = \dfrac{2^{2x}}{2^{2x}+2} = \dfrac{4^x}{4^x+2}$,

So, $$f(1-x) = \frac{4^{1-x}}{4^{1-x}+2}$$
$$= \frac{4}{4+2.4^x} = \frac{2}{2+4^x}$$

Now, $$f(x)+f(1-x) = \frac{4^x}{4^x+2} + \frac{2}{2+4^x}$$
$$= \frac{4^x+2}{4^x+2} = 1 \quad \text{...(i)}$$

Putting $x = \dfrac{1}{2023}, \dfrac{2}{2023}, \dfrac{3}{2023}, \ldots, \dfrac{1011}{2023}$ in (i), we get

$$f\left(\frac{1}{2023}\right) + f\left(\frac{2022}{2023}\right) = 1$$
$$f\left(\frac{2}{2023}\right) + f\left(\frac{2021}{2023}\right) = 1$$
$$f\left(\frac{3}{2023}\right) + f\left(\frac{2020}{2023}\right) = 1$$

....

$$f\left(\frac{1011}{2023}\right) + f\left(\frac{1012}{2023}\right) = 1$$

Adding all, we get

$$f\left(\frac{1}{2023}\right) + f\left(\frac{1}{2023}\right) + f\left(\frac{3}{2023}\right) + \ldots + f\left(\frac{2022}{2023}\right)$$
$$= \underbrace{1+1+1+\ldots+1}_{1011 \text{ times}} = 1011$$

15. We have

$$(\lambda-1)(x^2+x+1)^2 - (\lambda+1)\;(x^4+x^2+1) = 0$$
$$\Rightarrow \quad \frac{\lambda+1}{\lambda-1} = \frac{(x^2+x+1)^2}{(x^4+x^2+1)}$$
$$= \frac{(x^2+x+1)^2}{(x^2+x+1)(x^2-x+1)}$$
$$= \frac{x^2+x+1}{x^2-x+1}$$
$$\Rightarrow \quad \frac{(\lambda+1)+(\lambda-1)}{(\lambda+1)-(\lambda-1)} = \frac{(x^2+x+1)+(x^2-x+1)}{(x^2+x+1)-(x^2-x+1)}$$
$$\Rightarrow \quad \lambda = \frac{x^2+1}{x} = x+\frac{1}{x} \quad \text{...(i)}$$

Also, $$f(x) = \frac{1-x}{1+x}$$
$$\therefore \quad f(f(x)) = \frac{1-f(x)}{1+f(x)} = \frac{1-\dfrac{1-x}{1+x}}{1+\dfrac{1-x}{1+x}} = x$$

then $$f\left(f\left(\frac{1}{x}\right)\right) = \frac{1}{x}$$
$$\Rightarrow \quad f(f(x)) + f\left(f\left(\frac{1}{x}\right)\right) = x+\frac{1}{x} \quad \text{...(ii)}$$

From (i) and (ii), we get

$$f(f(x)) + f\left(f\left(\frac{1}{x}\right)\right) = \lambda$$

16. We have

$$f(x) = 2x^n + a$$
$$\therefore \quad f(2) = 2(2)^n + a = 26 \quad \text{...(i)} \; (\because f(2)=26)$$
and $$f(4) = 2(4)^n + a = 138 \quad \text{...(ii)} \; (\because f(4)=138)$$

From (i) and (ii), we get

$$(2(4)^n+a) - (2(2)^n+a) = 138-26$$
$$\Rightarrow \quad 2(4)^n - 2(2)^n = 112$$
$$\Rightarrow \quad 4^n - 2^n = 56 = 4^3 - 2^3$$
$$\therefore \quad n = 3$$

Now, from (i),

$$2(2)^3 + a = 26 \Rightarrow a = 10$$
$$\therefore \quad f(x) = 2x^3+10$$
$$\Rightarrow \quad f(3) = 2\cdot(3)^3+10 = 64$$

17. We have

$$f(x)\cdot f(y)=f(x)+f(y)+f(xy)-2$$

Put $x=y=1$, then

$$(f(1))^2-3f(1)+2=0$$

$$\Rightarrow \qquad f(1)=1 \text{ or } 2$$

Let $f(1)=1$, then put $y=1$

$$f(x)\cdot f(1)=f(x)+f(1)+f(x)-2$$

$$\Rightarrow \qquad f(x)=f(x)+1+f(x)-2$$

$\Rightarrow$ $f(x)=1$ constant function

$\Rightarrow$ $f(x)\neq 1$, Hence, $f(1)=2$

18. Given, $5f(x+y)=f(x)\cdot f(y)$ ...(i)

At $x=0$, $y=0$, we get

$$5f(0)=(f(0))^2 \Rightarrow f(0)=5 \qquad (\because f(0)\neq 0)$$

Differentiating (i) w.r.t. x keeping y as constant, we get

$$5f'(x+y)=f'(x)\cdot f(y)$$

At $x=0$, we get

$$\frac{f'(y)}{f(y)}=\frac{f'(0)}{5}=\lambda \text{ (say)}$$

$$\Rightarrow \qquad \frac{f'(y)}{f(y)}dy=\int \lambda\, dy$$

$$\Rightarrow \qquad \ln f(y)=\lambda y+\ln c$$

$$\Rightarrow \qquad \ln\left(\frac{f(y)}{c}\right)=\lambda y$$

$$\Rightarrow \qquad \frac{f(y)}{c}=e^{\lambda y}$$

$$\Rightarrow \qquad f(y)=ce^{\lambda y} \Rightarrow f(0)=c=5$$

$$\therefore \qquad f(y)=5e^{\lambda y} \qquad \text{...(ii)}$$

and given, $\quad f(3)=320$

$$\therefore \qquad f(3)=5e^{3\lambda}=320$$

$$\Rightarrow \qquad e^{3\lambda}=64=4^3$$

$$\Rightarrow \qquad e^{\lambda}=4$$

From (ii), $\quad f(y)=5(e^{\lambda})^y=5\cdot 4^y$

Now, $\quad \displaystyle\sum_{n=0}^{5} f(n)=\sum_{n=0}^{5} 5\cdot 4^n$

$$=5\,(4^0+4^1+4^2+4^3+4^4+4^5)$$

$$=5\times\left(\frac{1\cdot(4^6-1)}{4-1}\right)$$

$$=\frac{5}{3}\times 4095=5\times 1365=6825$$

19. We have

$$f(x)=\frac{x^4-7x^2+9}{x-\left(\frac{3}{x}\right)+1}$$

$$=\frac{x^2\left(\frac{x^2+9}{x^2-7}\right)}{\left(\frac{x-3}{x+1}\right)}$$

$$=\frac{x^2\left\{\left(\frac{x-3}{x}\right)^2-1\right\}}{\left(\frac{x-3}{x+1}\right)}$$

$$=\frac{x^2\left(\frac{x-3}{x+1}\right)\left(\frac{x-3}{x-1}\right)}{\left(\frac{x-3}{x+1}\right)}$$

$$=x^2\left(x-\frac{3}{x}-1\right), \frac{x-3}{x+1}\neq 0$$

$$\Rightarrow \qquad f(x)=0$$

$$\therefore \qquad x^2\left(x-\frac{3}{x}-1\right)=0$$

$$\Rightarrow \qquad x-\frac{3}{x}-1=0, x\neq 0$$

$$\Rightarrow \qquad x^2-x-3=0,\ x\neq 0$$

$$\therefore \qquad x=\frac{1\pm\sqrt{1+12}}{2}$$

$$=\frac{1\pm\sqrt{13}}{2}=\frac{a\pm\sqrt{b}}{c} \text{ (given)}$$

$$\Rightarrow \qquad a=1, b=13, c=2$$

Now, $\quad b-a-c=13-1-2=10$

20. We have

$$f(x)=\sqrt{\left(\frac{8}{1-x}+\frac{8}{1+x}\right)}=\sqrt{\frac{16}{1-x^2}}=\frac{4}{\sqrt{1-x^2}}$$

$$\therefore \qquad f(\sin x)=\frac{4}{\sqrt{1-\sin^2 x}}=\frac{4}{|\cos x|}$$

and $\quad f(\cos x)=\dfrac{4}{\sqrt{(1-\cos^2 x)}}=\dfrac{4}{|\sin x|}$

Now, $\quad g(x)=\dfrac{4}{f(\sin x)}+\dfrac{4}{f(\cos x)}$

$$=|\cos x|+|\sin x|$$

Hence, $g(x)$ is periodic with period $\frac{\pi}{2}$.

21. Given period of $f(x)$ is $7\Rightarrow$ period of $f\left(\frac{x}{3}\right)$ is $\frac{7}{1/3}$

and given period of $g(x)$ is $11\Rightarrow$ period of $g\left(\frac{x}{5}\right)$ is

$\frac{11}{1/5}$ *i.e.*, 55

$\Rightarrow T_1=$ Period of $f(x)\cdot g\left(\frac{x}{5}\right)=$ LCM of $\{7, 55\}=385$

and $T_2 =$ Period of $g(x)\cdot f\left(\frac{x}{3}\right) =$ LCM of {11, 21}

$= 231$

Hence, Period of $F(x) =$ LCM of $\{T_1, T_2\}$

$=$ LCM of {385, 231}

$=$ LCM of $\{7\times11\times5, 11\times7\times3\}$

$= 7\times11\times15$

$= 1155$

22. We have

$$f\left(x+\frac{1}{x}\right)+f\left(x-\frac{1}{2}\right)=f(x) \quad \text{...(i)}$$

Replacing x by $x+\frac{1}{2}$ in (i), then

$$f(x+1)+f(x)=f\left(x+\frac{1}{2}\right) \quad \text{...(i)}$$

Adding (i) and (ii), we get

$$f(x+1)+f\left(x-\frac{1}{2}\right)=0 \quad \text{...(iii)}$$

Now replacing x by $x+\frac{1}{2}$ in (iii), we get

$$f\left(x+\frac{3}{2}\right)+f(x)=0 \quad \text{...(iv)}$$

And replacing x by $x+\frac{3}{2}$ in (iv), we get

$$f(x+3)+f\left(x+\frac{3}{2}\right)=0 \quad \text{...(v)}$$

From (iv) and (v), we get

$$f(x+3)=f(x)$$

$\therefore$ Period of $f(x)$ is 3.

23. We have

$$f(x)=\frac{x}{\ln x} \text{ and } g(x)=\frac{\ln x}{x}$$

Clearly, $f(x)$ is defined, if $x>0$ and $\ln x \neq 0$

$\Rightarrow \quad x>0$ and $x\neq 1$

$\therefore \quad D_f=(0,\infty)-\{1\}$

Now in $x\in(0,\infty)-\{1\}$, $f(x)\neq 0$

$\Rightarrow \quad D_{1/f}=(0,\infty)\sim\{1\}$

and $g(x)$ is defined, if $x>0$

$\therefore \quad D_1=(0,\infty)$

and $g(x)=0$ at $x=1$

$\therefore \quad D_{1/g}=(0,\infty)-\{1\}$

Hence, $\frac{1}{g(x)}$ and $f(x)$ are identical function

Also, $\quad D_f \neq D_g$

$\therefore \quad f(x)\cdot g(x)\neq 1$

and $D_{1/f}\neq D_{1/g}$ for $x>0$

$\therefore \quad \frac{1}{f(x)g(x)}\neq 1$

24. We have

$$f(x)=x^3+\tan x+\left[\frac{x^2+1}{\lambda}\right]$$

$\because$ $f(x)$ is an odd function.

$\therefore \quad f(x)+f(-x)=0$

$\Rightarrow \quad x^3+\tan x+\left[\frac{x^2+1}{\lambda}\right]-x^3-\tan x+\left[\frac{x^2+1}{\lambda}\right]=0$

$\Rightarrow \quad 2\left[\frac{x^2+1}{\lambda}\right]=0 \Rightarrow \left[\frac{x^2+1}{\lambda}\right]=0$

$\therefore \quad 0\leq\frac{x^2+1}{\lambda}<1$ (because $x\in[-2,2]$)

$\Rightarrow \quad 0\leq\frac{5}{\lambda}<1$

$\therefore \quad \lambda>5$

25. We have

$$f(x)=\cos x+\sin\left[\frac{x+1}{\lambda}\right]$$

$\because$ $f(x)$ is an even function.

$\therefore \quad f(x)=f(-x)$

$\Rightarrow \quad \cos x+\sin\left[\frac{x+1}{\lambda}\right]=\cos x+\sin\left[\frac{-x+1}{\lambda}\right]$

$\Rightarrow \quad \sin\left[\frac{x+1}{\lambda}\right]=\sin\left[\frac{-x+1}{\lambda}\right] \ \forall x\in\left[-\frac{\pi}{2},\frac{\pi}{2}\right]$

$\therefore \quad \left[\frac{x+1}{\lambda}\right]=0 \ \forall x\in\left[-\frac{\pi}{2},\frac{\pi}{2}\right]$

and $\quad -x\in\left[-\frac{\pi}{2},\frac{\pi}{2}\right]$

$\Rightarrow \quad 0\leq\frac{x+1}{\lambda}<1 \ \forall x\in\left[-\frac{\pi}{2},\frac{\pi}{2}\right]$

$\Rightarrow \quad 0\leq\frac{\frac{\pi}{2}+1}{\lambda}<1$

$\therefore \quad \lambda>\frac{\pi}{2}+1$

or $\quad \lambda\in\left(\frac{\pi+2}{2},\infty\right)$

26. We have

$$f(x)=\sin x+\left[\frac{x^2}{\lambda}\right]$$

$f(x)$ is an odd function.

$\therefore \quad f(x)+f(-x)=0$

$\Rightarrow \sin x+\left[\frac{x^2}{\lambda}\right]-\sin x+\left[\frac{x^2}{\lambda}\right]=0$

$\Rightarrow \quad 2\left[\frac{x^2}{\lambda}\right]=0$

$\therefore \left[\frac{x^2}{\lambda}\right]=0$ for all $x\in[-10,10]$

$\Rightarrow \quad 0\le\frac{x^2}{\lambda}<1$ for all $x\in[-10,10]$

$\Rightarrow \quad 0\le\frac{100}{\lambda}<1$

$\therefore \quad \lambda>100$

$\Rightarrow \quad \lambda\in(100,\infty)$

27. We have

$$f(x)=\left[\frac{7}{2}+a\sin x\right]$$

$\because f(x)$ is an even function.

$\therefore \quad f(x)=f(-x)\ \forall\ x\in R$

$\Rightarrow \left[\frac{7}{2}+a\sin x\right]=\left[\frac{7}{2}-a\sin x\right] \quad \begin{pmatrix}\because 1\le\sin x\le 1\\ \therefore -1\le-\sin x\le 1\end{pmatrix}$

$\therefore \quad 3<\frac{7}{2}+a\sin x<4 \quad \forall\ x\in R$

$\Rightarrow \quad -\frac{1}{2}<a\sin x<\frac{1}{2}\ \forall x\in R$

$\therefore \quad -\frac{1}{2}<a<\frac{1}{2}$

$\Rightarrow \quad a\in\left(-\frac{1}{2},\frac{1}{2}\right)$

28. We have

$f(x-y)=f(x)f(y)-f(a-x)f(a+y)$ ...(i)

Replace x be a and y by $x-a$ in (i), we get

$f(a-(x-a)=f(a)\cdot f(x-a)\ -f(0)\cdot f(a+x-a)$

$\Rightarrow \quad f(2a-x)=f(a)\cdot f(x-a)-f(0)f(x)$ ...(ii)

and put $x=y=0$ in (i), we get

$f(0)=f(0)\cdot f(0)-f(a)\cdot f(a)$

$\Rightarrow \quad 1=1-(f(a))^2 \quad (\because f(0)=1)$

$\therefore \quad f(a)=0$

Put $f(a)=0$ and $f(0)=1$ in (ii),

$f(2a-x)=0-f(x)$ ...(iii)

Also put x = 0 in (i), then

$f(-y)=f(0)f(y)-f(a)f(a+y)$

$=f(y)-0$

$\Rightarrow \quad f(-y)=f(y)$

$\therefore f$ is an even function.

$\therefore \quad f(-x)=f(x)$

Replace x by $-x$ in (iii), then

$f(2a+x)=-f(-x)$

$=-f(x)$ ...(iv) $(\because f(-x)=f(x))$

Finally, from (iii) and (iv), we get

$f(2a-x)=f(2a+x)$

Hence, $f(x)$ is symmetrical about the line $x=2a$

29. Given $y=f(x)$ is symmetrical about the lines x = 2 and x = 3

$\therefore \quad f(2-x)=f(2+x)$ ...(i)

and $\quad f(3-x)=f(3+x)$ ...(ii)

Replacing x by x + 1 in (ii), then

$f(2-x)=f(4+x)$ ...(iii)

From (i) and (ii), we get

$f(4+x)=f(2+x)$ ...(iv)

Now, replacing x by x – 2 in (iv), then we get

$f(x+2)=f(x)$

30. We have

$f(x)=x^3+3x^2+5x+p\sin x\ +q\cos x\ \forall\ x\in R$

$\therefore \quad f'(x)=3x^2+6x+5+p\cos x-q\sin x$

For $f(x)$ is one-one

$f'(x)>0\ \forall x\in R$

$\Rightarrow 3x^2+6x+5+(p\cos x-q\sin x)>0 \quad \forall\ x\in R$

$\Rightarrow 3x^2+6x+5-\sqrt{(p^2+q^2)}>0 \quad \forall\ x\in R$

$\therefore \quad D<0$

$\Rightarrow 36-4\cdot 3\cdot(5-\sqrt{(p^2-q^2)})<0$

$\Rightarrow \quad 3-(5-\sqrt{(p^2+q^2)})<0$

$\Rightarrow \quad \sqrt{(p^2+q^2)}<2$

$\therefore \quad (p^2+q^2)<4$

31. The number of one-one function from $A\{1,2,3,4\}\to B\{1,2,3,4,5,6\}$ such that $f(i)\ne i$ are

${}^7P_4-\{{}^4C_1\times 6\times 5\times 4-{}^4C_2\times 5\times 4+{}^4C_3\times 4-{}^4C_4\times 1\}$

$=7\times 6\times 5\times 4-\{4\times 6\times 5\times 4\ -6\times 5\times 4+4\times 4-1\}$

$=840-\{480-120+16-1\}$

= 465

32. (c) Since Co-domain $=\left[0,\frac{\pi}{2}\right)$

$\because f$ is onto, the range = Co-domain $=\left[0,\frac{\pi}{2}\right)$

Let $y = f(x) = \tan^{-1}(x^2 + x + a)$

$\therefore \quad x^2 + x + a = \tan y \geq 0 \quad \forall\, y \in \left[0, \frac{\pi}{2}\right)$

$\Rightarrow \quad x^2 + x + a \geq 0 \quad \forall\, x \in R$

$\therefore \quad (1)^2 - 4 \cdot 1 \cdot a \leq 0$

$\Rightarrow \quad a \geq \frac{1}{4}$

$\therefore \quad a \in \left[\frac{1}{4}, \infty\right)$

33. Since Co-domain $= \left[\frac{\pi}{6}, \frac{\pi}{2}\right)$

$\because$ f is onto, $\therefore$ The range = Co-domain $= \left[\frac{\pi}{6}, \frac{\pi}{2}\right)$

Let $\quad y = f(x) = \sin^{-1}\left(\frac{x^2 - a}{x^2 + 1}\right)$

$\Rightarrow \quad \frac{\pi}{6} \leq \sin^{-1}\left(\frac{x^2 - a}{x^2 + 1}\right) < \frac{\pi}{2}$

$\Rightarrow \quad \frac{1}{2} \leq \frac{x^2 - a}{x^2 + 1} < 1$

$\Rightarrow \quad \frac{x^2}{2} + \frac{1}{2} \leq x^2 - a < x^2 + 1$

From first two relations, we get $x^2 \geq 2a + 1$ ...(i)

and last two relations, we get $a > -1$ ...(ii)

From (i) and (ii), we get

Finally $a > -1$ or $a \in (-1, \infty)$

34. Since, Co-domain $= R$

$\because$ $f(x)$ is onto, $\therefore$ Range = Co-domain $= R$

$\Rightarrow$ (value of $(x + a)$ at $x = 1$)

$\geq$ (Value of $(2ax - 1)$ at $x = 1$)

$\Rightarrow \quad 1 + a \geq 2a - 1$

$\therefore \quad a \leq 2$ ...(i)

and for $a \leq 0$, Range $(\leq 1) \neq R$

and for $a > 2$, Range $\neq R$

Hence, $a \in (0, 2]$ for $f(x)$ is surjective.

35. Given, $f(1, 3, 5, 7, ..., 99) \to (2, 4, 6, 8, ..., 100)$

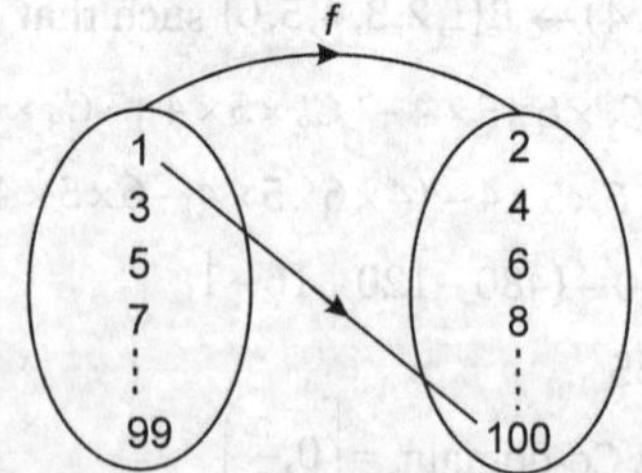

Let us assume $f(1) = 100$, we have only 1 way for arranging

$f(3) > f(5) > f(7) ... > f(99)$

Similarly if we choose $f(1) = 98$, then again we have only 1 way for arranging

$f(3) > f(5) > f(7) ... > f(99)$

So we can see the arrangement

$f(3) > f(5) > f(7) > ... > f(99)$

depends upon f(1)

So number of ways for choosing $f(1)$ is ${}^{50}C_1$

36. Let $\quad y = f(x) = \begin{cases} -\dfrac{x^2}{1+x^2}, & x \geq 0 \\ \dfrac{x^2}{1+x^2}; & x < 0 \end{cases}$

$\Rightarrow \quad f^{-1}(y) = \begin{cases} \sqrt{\dfrac{-y}{1+y}}, & y < 0 \\ -\sqrt{\dfrac{y}{1-y}}, & y > 0 \end{cases}$

$\therefore \quad f^{-1}(x) = \begin{cases} \sqrt{\dfrac{-x}{1+x}}, & x < 0 \\ -\sqrt{\dfrac{x}{1-x}}, & x > 0 \end{cases}$

$= -(\text{sgn})(x))\sqrt{\left(\dfrac{|x|}{1 - |x|}\right)}$

37. Let $\quad F(x) = \frac{ax + b}{c}$

$\therefore \quad F^{-1}(x) = \frac{cx - b}{a}$

Hence, $f(x) = \frac{2x + 5}{7} \Rightarrow f^{-1}(x) = \frac{7x - 5}{2}$

and $\quad g(x) = \frac{7x + 3}{5}$

$\Rightarrow \quad g^{-1}(x) = \frac{5x - 3}{7}$

Also given, $h(x) = 2f^{-1}(x) - 7g^{-1}(x) + 3x$

$= (7x - 5) - (5x - 3) + 3x$

$= (5x - 2)$

$\therefore \quad h(11) = 5 \times 11 - 2 = 53$

38. Let $y = f(x) = \frac{x - 2}{x - 3} \Rightarrow x = \frac{3y - 2}{y - 1}$

$\Rightarrow \quad f^{-1}(y) = \frac{3y - 2}{y - 1},$

$\therefore \quad f^{-1}(x) = \frac{3x - 2}{x - 1}$ ...(i)

and let $\quad y = g(x) = 2x - 3$

$\Rightarrow \quad x = \frac{y + 3}{2}$

$$\Rightarrow \qquad g^{-1}(y) = \frac{y+3}{2}$$

$$\therefore \qquad g^{-1}(x) = \frac{x+3}{2} \qquad \ldots\text{(ii)}$$

Given, $f^{-1}(x) + g^{-1}(x) = \frac{13}{2}$

$$\Rightarrow \qquad \frac{3x-2}{x-1} + \frac{x+3}{2} = \frac{13}{2}$$

$$\Rightarrow 2(3x-2) + (x-1)(x+3) = 13(x-1) \qquad (\because x \neq 1)$$

$$\Rightarrow \qquad x^2 - 5x + 6 = 0$$

$$\Rightarrow \qquad (x-3)(x-2) = 0$$

$$\Rightarrow \qquad x = 3, 2$$

$\therefore$ Sum of all values of $x = 3 + 2 = 5$

39. We have $f_0(x) = \frac{x}{2x+1}$

and $\qquad f_{n+1} = f_0 o f_n$

For $n = 0$,

$$f_1 = f_0 \, o \, f_0$$

$$\Rightarrow \qquad f_1(x) = f_0(f_0(x)) = f_0\left(\frac{x}{2x+1}\right)$$

$$= \frac{\left(\frac{x}{2x+1}\right)}{2\left(\frac{x}{2x+1}\right)+1} = \frac{x}{4x+1}$$

For $n = 1$,

$$f_2(x) = f_0(f_1(x)) = f_0\left(\frac{x}{4x+1}\right)$$

$$= \frac{\frac{x}{4x+1}}{2\left(\frac{x}{4x+1}\right)+1} = \frac{x}{6x+1}$$

40.

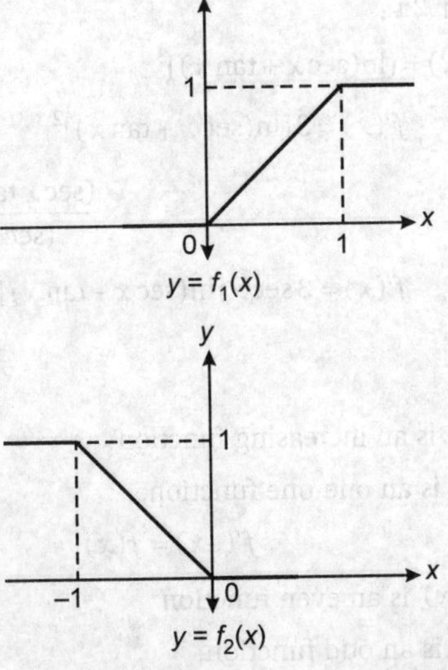

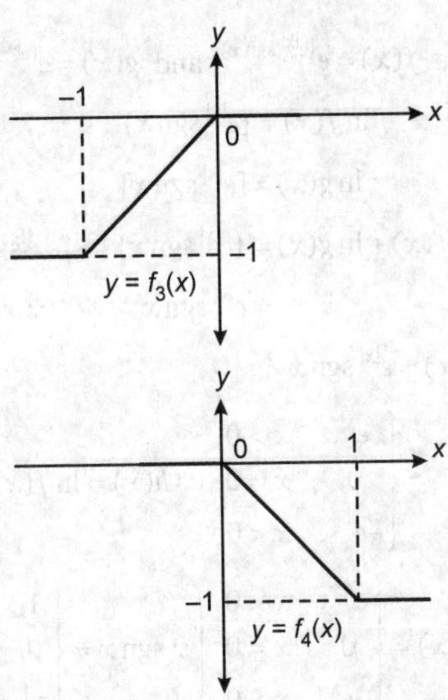

It is clear from the graph

$$f_1(x) = -f_3(-x) \text{ for all } x.$$

41. Let $\Delta_1, \Delta_2 \in S$

$f(\Delta_1) = f(\Delta_2) \Rightarrow \Delta_1 \neq \Delta_2$ (Two triangle may have equal axis)

$\therefore f$ is not injective

Since each positive real number can represent area of a triangle.

$$\therefore \qquad \text{Range} = R^+$$

$$\Rightarrow \qquad \text{Co-domain} = \text{Range}$$

$\therefore f$ is surjective

Hence, f is surjective but not injective.

42. Since $y = f(x)$ and $y = g(x)$ are mirror images of each other about the line $y = a$ at x.

$$f(x) - a = a - g(x)$$

$$\Rightarrow \qquad f(x) + g(x) = 2a \Rightarrow \quad h(x) = 2a$$

Hence, $h(x)$ is constant function. Thus it is neither one-one nor onto.

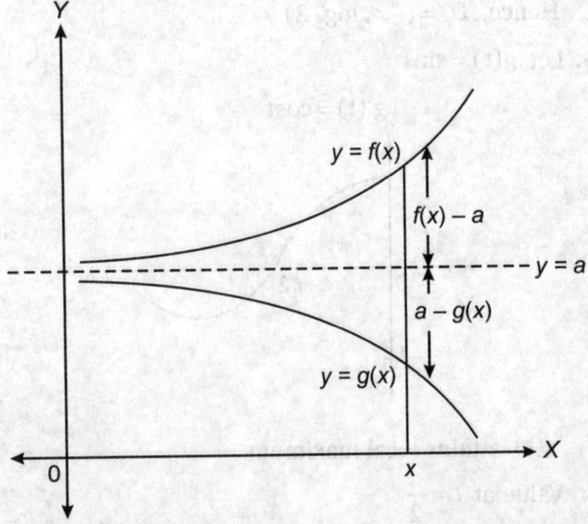

43. We have $f(x)=e^{\{e^{|x|}\operatorname{sgn}x\}}$ and $g(x)=e^{[e^{|x|}\operatorname{sgn}x]}$

$\therefore \quad \ln f(x)=\{e^{|x|}\operatorname{sgn}x\}$

and $\quad \ln g(x)=[e^{|x|}\operatorname{sgn}x]$

$\Rightarrow \ln f(x)+\ln g(x)=\{e^{|x|}\operatorname{sgn}x\}+[e^{|x|}\operatorname{sgn}x]$

$=e^{|x|}\operatorname{sgn}x \qquad (\because x=[x]+\{x\})$

$\Rightarrow h(x)=e^{|x|}\operatorname{sgn}x$

$$=\begin{cases} e^{x}, & x>0 \\ 0, & x=0 \quad (\because h(x)=\ln f(x)+\ln g(x)) \\ -e^{-x}, & x<0 \end{cases}$$

$$\Rightarrow h(-x)=\begin{cases} e^{-x}, & x<0 \\ 0, & x=0 \\ -e^{x}, & x>0 \end{cases} \left(\because \operatorname{sgn}x=\begin{cases} 1, & x>0 \\ 0, & x=0 \\ -1, & x<0 \end{cases}\right)$$

$\Rightarrow h(x)+h(-x)=0$ for all x

Hence $h(x)$ is an odd function.

44. Let $\quad g(t)=2t^3-15t^2+36t-25$

$\therefore \quad g'(t)=6t^2-30t+36$

$=6(t-2)(t-3)$

g attains local minimum value at $t=3$

$\therefore \{g(t)\}_{\min}=g(3)=54-135+108-25=2$

also $\quad 2\le 2+|\sin t|\le 3 \qquad (\because 0\le|\sin t|\le 1)$

Hence, $\min.\{2t^3-15t^2+36t-25,\ 2+|\sin t|;\ 2\le t\le 4\}=2$

$\therefore \quad 3^x+3^{f(x)}=2 \Rightarrow 3^{f(x)}=2-3^x$

$\Rightarrow \quad f(x)=\log_3(2-3^x)$

Clearly, $f(x)$ is defined, if

$2-3^x>0 \Rightarrow 3^x<2 \Rightarrow x<\log_3 2$

Hence, $D_t=(-\infty,\log_3 2)$

45. Let $g(t)=\sin t$

$\therefore \quad g'(t)=\cos t$

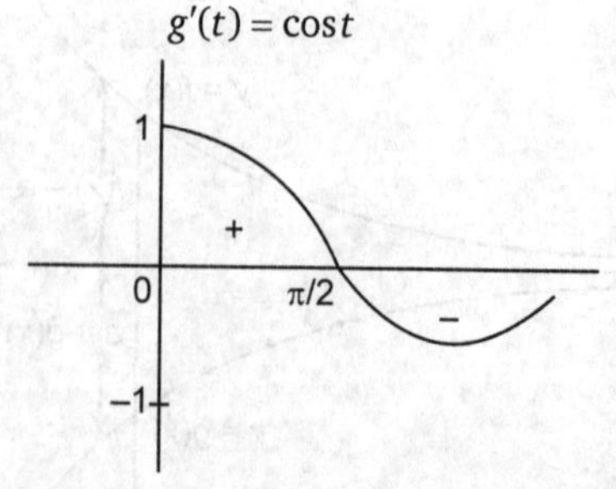

$g(t)$ attains local maximum

Value at $t=\dfrac{\pi}{2}$

$\therefore \quad \{g(t)\}_{\max}=\sin\dfrac{\pi}{2}=1$

and $\quad \{g(t)\}_{\min}=\sin x$

$\Rightarrow \quad h(x)=[f(x)-g(x)]=[1-\sin x]$

$\therefore \quad h(x)=0,1,2 \qquad (\because -1\le\sin x\le 1)$

Hence, $R_h=\{0,1,2\}$

46. Given, $f(x)=\log_{[x-1]}\left(\dfrac{|x|}{x}\right)$

$f(x)$ is defined, when

$\dfrac{|x|}{x}>0$ and $[x-1]>0$ and $[x-1]\ne 1$

$\Rightarrow x>0$ and $x-1\ge 1$

and $x-1\notin[1,2)$

$\Rightarrow x>0$ and $x\ge 2$ and $x\notin[2,3)$

$\therefore \quad x\in[3,\infty)$

and $f(x)=\log_{[x-1]}(1)\ \left(\dfrac{|x|}{x}=1 \text{ for all } x\ge 3\right)=0$

Hence, $D_f=[3,\infty)$ and $R_f=\{0\}$

47. We have,

$$f(x)=\frac{\sin^2 x}{\sin x-\cos x}-\frac{\sin x+\cos x}{\tan^2 x-1},\ x\ne n\pi\pm\frac{\pi}{4}, n\in I$$

$$=\frac{\sin^2 x}{\sin x-\cos x}-\frac{\cos^2 x(\sin x+\cos x)}{\sin^2 x-\cos^2 x},$$

$$x\ne n\pi\pm\frac{\pi}{4}, n\in I$$

$$=\frac{\sin^2 x}{\sin x-\cos x}-\frac{\cos^2 x}{\sin x-\cos x},\ x\ne n\pi\pm\frac{\pi}{4}, n\in I$$

$\Rightarrow f(x)=\sin x+\cos x,\ x\ne n\pi\pm\dfrac{\pi}{4}, n\in I$

$f(x)$ is periodic with period 2π and $f(-x)\ne\pm f(x)$

Hence, $f(x)$ is neither even nor odd but periodic with period 2π.

48. $\because \ f(x)=[\ln(\sec x+\tan x)]^3$

$\therefore \quad f'(x)=3[\ln(\sec x+\tan x)]^2 \times\dfrac{(\sec x\tan x+\sec^2 x)}{(\sec x+\tan x)}$

$\Rightarrow \quad f'(x)=3\sec x[\ln(\sec x+\tan x)]^2>0$

$\forall x\in\left(-\dfrac{\pi}{2},\dfrac{\pi}{2}\right)$

$\Rightarrow f(x)$ is an increasing function

$\therefore f(x)$ is an one-one function.

and $\quad f'(-x)=f(x)$

$\Rightarrow f'(x)$ is an even function

$\therefore f(x)$ is an odd function.

Also, $\sec x+\tan x=\tan\left(\frac{\pi}{4}+\frac{x}{2}\right)>0$, $\forall x\in\left(-\frac{\pi}{2},\frac{\pi}{2}\right)$

$\Rightarrow \quad 0<\sec x+\tan x<\infty$

$\Rightarrow \quad -\infty<[\ln(\sec x+\tan x)]<\infty$

$\Rightarrow \quad -\infty<[\ln(\sec x+\tan x)]^3<\infty$

$\Rightarrow \quad -\infty<f(x)<\infty$

$\therefore \quad R_f=R=C_f$

Hence, $f(x)$ is an onto function.

49. We have,

$$f(f(x))+f(x)=12x$$

$f(x)$ must be a linear function. Let $f(x)=ax+b$ then,

$$f(ax+b)+ax+b=12x$$

$\Rightarrow \quad a(ax+b)+b+ax+b=12x$

$\Rightarrow \quad a^2+a=12$ or $ab+2b=0$

$\Rightarrow \quad a^2+a-12=0$ or $(a+2)b=0$

$\Rightarrow \quad a=-4,3;\ b=0$

$\therefore \quad f(x)=-4x$ or $3x$

$\Rightarrow \quad f(-7)=28$ or -21

50. Given, $f(x)\cdot f\left(\frac{1}{x}\right)=f(x)+f\left(\frac{1}{x}\right)$ and $f(3)=-26$

$\therefore \quad f(x)=1\pm x^n$

$\Rightarrow \quad f(3)=1\pm 3^n=-26$

$\Rightarrow \quad \pm 3^n=-3^3$

Takinig (-) sign, then $n=3$

$\therefore \quad f(x)=1-x^3$

Now, $$\prod_{r=3}^{8}\frac{1-f(r)}{1-f(r-1)}=\prod_{r=3}^{8}\frac{r^3}{(r-1)^3}$$

$$=\frac{8^3}{2^3}=4^3=2^6=2^\lambda \text{ (given)}$$

$\therefore \quad \lambda=6$

which is composite number, perfect number and product of co-prime numbers.

51. We have,

$$f(x)+2f\left(\frac{1}{1-x}\right)=x \qquad \text{...(i)}$$

replacing x by $\frac{1}{1-x}$ in (i), we get

$$f\left(\frac{1}{1-x}\right)+2f\left(1-\frac{1}{x}\right)=\frac{1}{1-x} \qquad \text{...(ii)}$$

and replacing x by $1-\frac{1}{x}$ in (i), we get

$$f\left(1-\frac{1}{x}\right)+2f(x)=1-\frac{1}{x} \qquad \text{...(iii)}$$

Now $\frac{1}{2}$(i) + 2(iii) - (ii), we get

$$\frac{1}{2}f(x)+4f(x)=\frac{x}{2}+2-\frac{2}{x}-\frac{1}{1-x}$$

$\Rightarrow \quad f(x)=\frac{2}{9}\left(\frac{x}{2}+2-\frac{2}{x}-\frac{1}{1-x}\right)$

$\therefore \quad f(-1)=\frac{2}{3},\ f(-2)=\frac{10}{27},\ f(2)=\frac{2}{3}$

and $\quad f(3)=\frac{20}{27}$

$\Rightarrow f(-1)=f(2)$ and $2f(-2)=f(3)$

52. We have,

$$f(x)=\ln(2x-x^2)+\sin\left(\frac{\pi x}{2}\right)$$

$\because \quad f(1+x)=f(1-x)$ and $f(2+x)\neq f(2-x)$

$\therefore$ f is symmetrical about the line $x=1$

Also, $\quad f'(x)=\frac{2(1-x)}{(2x-x^2)}+\frac{\pi}{2}\cos\left(\frac{\pi x}{2}\right)$

$\therefore \quad f'(1)=0$ and $f''(1)<0$

$\Rightarrow f(x)$ is maximum at $x=1$

$\therefore$ maximum value of f is 1

and minimum value of f does not exist.

53. (a) We have,

$$f(x)=\cos(\cos^{-1}x)$$

$\therefore \quad f(-x)=\cos(\cos^{-1}(-x))=\cos(\pi-\cos^{-1}x)$

$$=-\cos(\cos^{-1}x)=-f(x)$$

$\Rightarrow f(x)$ is an odd function.

(b) We have,

$$f(x)=\cos(\sin x)+\cos(\cos x)$$

$\therefore \quad f\left(\frac{\pi}{2}+x\right)=\cos(\cos x)+\cos(-\sin x)$

$$=\cos(\cos x)+\cos(\sin x)=f(x)$$

$\Rightarrow$ Period of $f(x)$ is $\frac{\pi}{2}$.

(c) $\because -3\le 3\sin x\le 3$

$\therefore \cos 3\le\cos(3\sin x)\le 1\Rightarrow \cos 3\le f(x)\le 1$

$\Rightarrow$ Range of $f(x)$ is [cos 3, 1]

(d) $\because \quad f(\lambda x,\lambda y)=\lambda' f(x,y)$

$\therefore \quad f(x,y)$ is a homogeneous function of degree 1.

54. (a) $\because$ $x\,!$ is defined only when $x\in N$. but $\{x\}$ become zero for these values of x. Hence, $\frac{x!}{\{x\}}$ is not defined.

(b) $\{x\}x!$ is defined for $x \in N$. Hence, $\{x\}x!$ is a function for all $x \in N$

(c) $\dfrac{1}{\ln[1-|x|]}$ is defined, when

$$[1-|x|] > 0 \text{ and } [1-|x|] \neq 1$$

$$\Rightarrow \quad 1-|x| \geq 1 \text{ and } 1-|x| \notin [1, 2)$$

$$\Rightarrow \quad 1-|x| \geq 2$$

$\Rightarrow |x| \leq -1$. Which is not possible.

Hence, $\dfrac{1}{\ln[1-|x|]}$ is not defined.

(d) $\dfrac{\ln(x-1)}{\sqrt{(1-x^2)}}$ is defined, when

$$x-1>0 \text{ and } 1-x^2>0$$

$$\Rightarrow \quad x>1 \text{ and } x^2<1$$

$$\Rightarrow \quad x>1 \text{ and } -1<x<1$$

$$\Rightarrow \quad x \in \phi$$

Hence, $\dfrac{\ln(x-1)}{\sqrt{(1-x^2)}}$ is not defined.

55. (a) $\because \ \text{sgn}(x) = \begin{cases} 1, & x>0 \\ 0, & x=0 \\ -1, & x<0 \end{cases}$

$\therefore \quad \text{sgn}(e^{-x}) = 1 \quad (\because e^{-x} > 0)$

Hence, it is periodic.

(b) Period of $|\sin x|$ is π

and period of $\sin x$ is 2π

$\therefore$ Period of $|\sin x| + \sin x$ is L.C.M. of $(2\pi, \pi) = 2\pi$

(c) $\because$ min. $(\sin x, |x|) = \sin x \quad (\because \sin x < |x|)$

$\therefore$ Period of $\sin x$ is 2π

Hence, period of min. $(\sin x, |x|)$ is 2π

(d) $\because \left[x+\dfrac{1}{2}\right]+\left[x-\dfrac{1}{2}\right]+2[-x]$

$$= \left(x+\frac{1}{2}\right)-\left\{x+\frac{1}{2}\right\}+\left(x-\frac{1}{2}\right) - \left\{x-\frac{1}{2}\right\}+2((-x)-\{-x\})$$

$$= -\left\{x+\frac{1}{2}\right\}-\left\{x-\frac{1}{2}\right\}-2\{-x\}$$

Hence, it is periodic.

56. $\because \ |e-x| = \begin{cases} x-e, & x \geq e \\ e-x, & x<e \end{cases}$

and $\pi-|x| = \begin{cases} \pi-x, & x \geq 0 \\ \pi+x, & x<0 \end{cases}$

$\therefore$ Graph of $f(x) = \min.(|e-x|, \pi-|x|)$

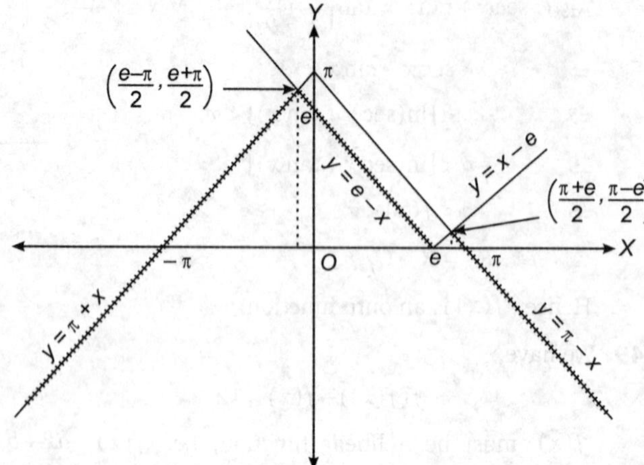

$\therefore \ R_f = \left(-\infty, \dfrac{\pi+e}{2}\right]$, $f(x)$ is not a periodic function but $f(x)$ is many one.

Also $f(x)$ is neither even nor odd.

57. We have,

$$f(x) = \frac{\sin x}{\sqrt{(1-\cos^2 x)}} + \frac{\cos x}{\sqrt{(1-\sin^2 x)}} + \frac{\tan x}{\sqrt{(\sec^2 x-1)}} + \frac{\cot x}{\sqrt{(\text{cosec}^2 x-1)}}$$

$$= \frac{\sin x}{|\sin x|} + \frac{\cos x}{|\cos x|} + \frac{\tan x}{|\tan x|} + \frac{\cot x}{|\cot x|}$$

$\because$ Periods of $\sin x$ and $\cos x$ is 2π,

periods of $\tan x$ and $\cot x$ is π

and periods of $|\sin x|$, $|\cos x|$, $|\tan x|$, $|\cot x|$ is π

Hence, period of $f(x)$ is 2π and $x \neq 0, \dfrac{\pi}{2}, \pi, \dfrac{3\pi}{2}, 2\pi$

$$\therefore \quad f(x) = \begin{cases} 4, & 0<x<\dfrac{\pi}{2} \\ -2, & \dfrac{\pi}{2}<x<\pi \\ 0, & \pi<x<\dfrac{3\pi}{2} \\ -2, & \dfrac{3\pi}{2}<x<2\pi \end{cases}$$

$\Rightarrow L=-2, M=4$, then $M+L=2, M-L=6$

58. (a) $\because f(x) = \cot^2 x \cdot \cos^2 x = \dfrac{\cos^4 x}{\sin^2 x}$

$\therefore \ f(x)$ is defined, when $\sin^2 x \neq 0$

$$\Rightarrow \quad x \neq n\pi, n \in I$$

$$\Rightarrow \quad D_f = R - n\pi, n \in I$$

and $\quad g(x) = \cot^2 x - \cos^2 x$

$$= \cos^2 x\left(\frac{1}{\sin^2 x}-1\right) = \frac{\cos^4 x}{\sin^2 x}$$

Similarly, $D_g = R - n\pi, n \in I$

Hence, $f(x)$ and $g(x)$ are identical $(\because D_f = D_g)$

(b) $\because f(x) = \sqrt{(x^2-1)}$ is defined, when $x^2 - 1 \geq 0$

$\Rightarrow \quad x \leq -1$ or $x \geq 1$

$D_f = (-\infty, -1] \cup [1, \infty)$

and $g(x) = \sqrt{(x+1)} \cdot \sqrt{(x-1)}$ is defined, when

$x+1 \geq 0$ and $x - 1 \geq 0 \Rightarrow x \geq 1$

$\therefore \quad D_g = [1, \infty)$

Hence, $f(x)$ and $g(x)$ are not identical $(\because D_f \neq D_g)$.

(c) $\because f(x) = \sqrt{\left(\frac{1-\cos 2x}{2}\right)} = \sqrt{\sin^2 x} = |\sin x|$

and $g(x) = \sin x$

$\therefore \quad R_f = [0, 1]$ and $R_g = [-1, 1]$

but $D_f = D_g = R$

Hence, $f(x)$ and $g(x)$ are not identical $(\because R_f \neq R_g)$.

(d) $\because \quad f(x) = [\{x\}] = 0 \quad (\because 0 \leq \{x\} < 1)$

and $\quad g(x) = \{[x]\} = 0$

Hence, $f(x)$ and $g(x)$ are identical $(\because D_f = D_g = R$ and $R_f = R_g = \{0\})$

59. We have,

$$f(x) = \begin{cases} |2x-1|, & -2 \leq x \leq 1 \\ x^2 - 4, & 1 < x \leq 8 \end{cases}$$

$$\therefore \quad y = f(f(x)) = \begin{cases} |2f(x)-1|, & -2 \leq f(x) \leq 1 \\ (f(x))^2 - 4, & 1 < f(x) \leq 8 \end{cases}$$

$$= \begin{cases} |2|2x-1|-1|, & -2 \leq |2x-1| \leq 1 \text{ and } -2 \leq x \leq 1 \\ (2x-1)^2 - 4, & 1 < |2x-1| \leq 8 \text{ and } -2 \leq x \leq 1 \\ |2x^2-9|, & -2 \leq x^2 - 4 \leq 1 \text{ and } 1 < x \leq 8 \\ (x^2-4)^2 - 4, & 1 < x^2 - 4 \leq 8 \text{ and } 1 < x \leq 8 \end{cases}$$

$$= \begin{cases} |2|2x-1|-1|, & 0 \leq x \leq 1 \\ 4x^2 - 4x - 3, & -2 \leq x < 0 \\ |2x^2-9|, & \sqrt{2} \leq x \leq \sqrt{5} \\ x^4 - 8x^2 + 12, & \sqrt{5} < x \leq 2\sqrt{3} \end{cases}$$

Integers in the domain of $y = f(f(x))$ are

$-2, -1, 0, 1, 2, 3$

60. (a) $\because -\frac{\pi}{2} \leq \frac{\pi}{2} \sin x \leq \frac{\pi}{2}$

$\Rightarrow \quad -1 \leq \sin\left(\frac{\pi}{2} \sin x\right) \leq 1$

$\Rightarrow \quad -\frac{\pi}{6} \leq \frac{\pi}{6} \sin\left(\frac{\pi}{2} \sin x\right) \leq \frac{\pi}{6}$

$\Rightarrow \quad -\frac{1}{2} \leq \sin\left[\frac{\pi}{6} \sin\left(\frac{\pi}{2} \sin x\right)\right] \leq \frac{1}{2}$

$\Rightarrow \quad -\frac{1}{2} \leq f(x) \leq \frac{1}{2}$

$\therefore$ Range of f is $\left[-\frac{1}{2}, \frac{1}{2}\right]$

(b) $(fog)(x) = f(g(x)) = f(t), \; t \in \left[-\frac{\pi}{2}, \frac{\pi}{2}\right]$

$\Rightarrow \quad f(t) \in \left[-\frac{1}{2}, \frac{1}{2}\right]$

Hence, range of fog is $\left[-\frac{1}{2}, \frac{1}{2}\right]$

(c) $$\lim_{x \to 0} \frac{f(x)}{g(x)} = \lim_{x \to 0} \frac{\sin\left[\frac{\pi}{6} \sin\left(\frac{\pi}{2} \sin x\right)\right]}{\frac{\pi}{2} \sin x} \times \frac{\frac{\pi}{6} \sin\left(\frac{\pi}{2} \sin x\right)}{\frac{\pi}{6} \sin\left(\frac{\pi}{2} \sin x\right)}$$

$$= \lim_{x \to 0} \frac{\sin\left[\frac{\pi}{6} \sin\left(\frac{\pi}{2} \sin x\right)\right]}{\frac{\pi}{6} \sin\left(\frac{\pi}{2} \sin x\right)} \times \frac{\pi}{6} \lim_{x \to 0} \frac{\sin\left(\frac{\pi}{2} \sin x\right)}{\left(\frac{\pi}{2} \sin x\right)}$$

$$= 1 \times \frac{\pi}{6} \times 1 = \frac{\pi}{6}$$

(d) $\because (gof)(x) = 1$

$\Rightarrow \quad g(f(x)) = 1$

$\Rightarrow \quad \frac{\pi}{2} \sin f(x) = 1$

$\Rightarrow \quad \sin f(x) = \frac{2}{\pi} \quad$ ...(i)

But $f(x) \in \left[-\frac{1}{2}, \frac{1}{2}\right] \subset \left[-\frac{\pi}{6}, \frac{\pi}{6}\right]$

$\therefore \quad \sin f(x) \in \left[-\frac{1}{2}, \frac{1}{2}\right] \quad$ ...(ii)

From (i) and (ii), $\sin f(x) \neq \frac{2}{\pi}$

i.e., No solution.

61. (a) $\because f(x) = (x+1), x \geq -1$

$\therefore \quad f'(x) = 1 > 0, x \geq -1$

Hence, $f(x)$ is one-one function.

(b) $\because g(x) = x + \frac{1}{x}, x > 0$

$\therefore \quad g'(x) = 1 - \frac{1}{x^2} = \frac{(x+1)(x-1)}{x^2}$

$\Rightarrow g'(x) < 0$ for $x \in (0, 1)$

and $g'(x) > 0$ for $x \in (1, \infty)$

In whole domain $g(x)$ is decreasing and increasing both.

Hence, $g(x)$ is not one-one function.

(c) $\because h(x) = x^2 + 4x - 5, x > 0$

$\therefore \quad h'(x) = 2x + 4 > 0, x > 0$

Hence, $h(x)$ is one-one function.

(d) $\because k(x) = e^{-x}, x \geq 0$

$\therefore k'(x) = -e^{-x} < 0, x \geq 0$

Hence, $k(x)$ is one-one function.

62. $\because f(x) = \begin{cases} x - [x], \lambda \leq x < \lambda + 0.5 \\ [x], \lambda + 0.5 \leq x < \lambda + 1 \end{cases}$

$= \begin{cases} x - \lambda, \lambda \leq x < \lambda + 0.5 \\ \lambda, \lambda + 0.5 \leq x < \lambda + 1 \end{cases}$

and $\quad g(x) = \sin^4 x + \cos^4 x$

$f(x)$ is both even and odd functions

and $g(x)$ is always even function. Then $f(g(x))$ is an even and an odd function.

and $g(x)$ is periodic function with period $\pi/2$, then $f(g(x))$ is also periodic.

Also, $g(x)$ is many one function

$\therefore f(g(x))$ is not a one-one function.

Paragraph for Question Nos. 63 to 65

Given $\frac{1}{4} f(x) = x + 3x^2 \int_0^1 y f(y) dy + 5 \int_0^1 x y^2 f(y) dy$

$\Rightarrow f(x) = 4x + 12x^2 \int_0^1 y f(y) dy + 20x \int_0^1 y^2 f(y) dy$

$= 12x^2 \int_0^1 y f(y) dy + x\left(4 + 20 \int_0^1 y^2 f(y) dy\right)$

Now let $12 \int_0^1 y f(y) dy = A$ and $4 + 20 \int_0^1 y^2 f(y) dy = B$

then $f(x) = Ax^2 + Bx \Rightarrow f(y) = Ay^2 + By$

$\because \quad A = 12 \int_0^1 y f(y) dy$

$= 12 \int_0^1 y (Ay^2 + By) dy$

$= 12\left(\frac{A}{4} + \frac{B}{3}\right)$

$\Rightarrow \quad A = 3A + 4B$

or $\quad A = -2B \quad$ ...(i)

and $\quad B = 4 + 20 \int_0^1 y^2 f(y) dy$

$= 4 + 20 \int_0^1 y^2 (Ay^2 + By) dy$

$= 4 + 20\left(\frac{A}{5} + \frac{B}{4}\right) = 4 + 4A + 5B$

$\Rightarrow \quad 4 + 4A + 4B = 0 \Rightarrow A + B = -1 \quad$ ...(ii)

From Eqs. (i) and (ii), we get

$A = -2$ and $B = 1$

$\therefore \quad f(x) = -2x^2 + x$

63. $f(x) = -2\left(x^2 - \frac{1}{2}x\right) = -2\left(\left(x - \frac{1}{4}\right)^2 - \frac{1}{16}\right)$

$= \frac{1}{8} - 2\left(x - \frac{1}{4}\right)^2$

$\therefore$ Maximum value of $f(x) = \frac{1}{8}$ (at $x = \frac{1}{4}$)

64. Let $y = |f(|x|)| = e^{-|x|}$

It is clear that graphs of $y = |f(|x|)|$ and $y = e^{-|x|}$

Cut at two points

Hence number of solutions are two.

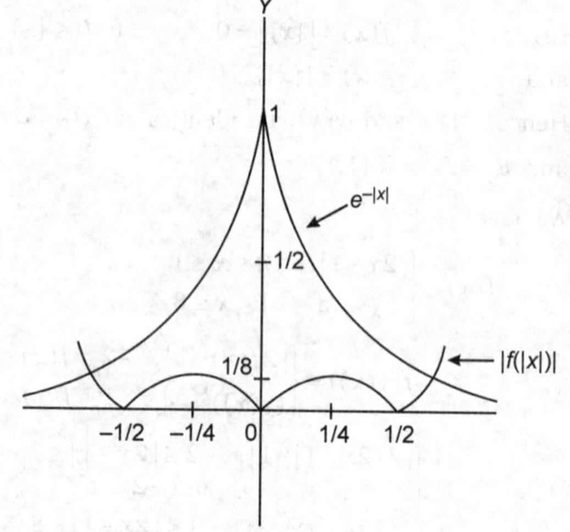

65. $\because f(-\cos\theta) = -2\cos^2\theta - \cos\theta$

$= -2\left(\cos^2\theta + \frac{1}{2}\cos\theta\right)$

$= -2\left(\left(\cos\theta + \frac{1}{4}\right)^2 - \frac{1}{16}\right)$

$= \frac{1}{8} - 2\left(\cos\theta + \frac{1}{4}\right)^2$

Maximum value of $f(-\cos\theta) = \frac{1}{8}$

and minimum value of $f(-\cos\theta) = \frac{1}{8} - 2\left(1 + \frac{1}{4}\right)^2 = -3$

Hence, Range of $f(-\cos\theta)$ is $\left[-3, \frac{1}{8}\right]$.

$\therefore \quad a = -3, b = \frac{1}{8} \Rightarrow \frac{a}{b} = -24$

or $\quad \left|\frac{a}{b}\right| = 24$

Paragraph for Question Nos. 66 to 68

66. $\because \quad g(x)=\cos\left(\frac{\pi x}{a}\right)$

$\therefore$ Period of $g(x)=\dfrac{2\pi}{\left|\frac{\pi}{a}\right|}=2|a|=6$ (given)

but given $a>0$

$\therefore \quad 2a=6$

$\Rightarrow \quad a=3$

67. Now $f(x)=\dfrac{\cos 2x-5\cos x+4}{x}$

$$\Rightarrow \quad xf(x)=2\cos^2 x-1-5\cos x+4$$
$$=2\cos^2 x-5\cos x+3$$
$$=2\left\{\cos^2 x-\frac{5}{2}\cos x+\frac{3}{2}\right\}$$
$$=2\left\{\left(\cos x-\frac{5}{4}\right)^2-\frac{25}{16}+\frac{3}{2}\right\}$$
$$=2\left(\cos x-\frac{5}{4}\right)^2-\frac{1}{8}$$

For minimum value of $xf(x)$, put $\cos x=1$

$\left(\because \cos x\neq\frac{5}{4}\right)$

$\therefore$ Minimum value of $xf(x)=0$

and for maximum value of $xf(x)$ put $\cos x=-1$

$\therefore$ Maximum value of $xf(x)=10$

$\Rightarrow$ Required difference $=10-0=10$

68. $\because \frac{\pi}{3}f\left(\frac{\pi}{3}\right)=\cos\frac{2\pi}{3}-5\cos\frac{\pi}{3}+4$

$$=-\frac{1}{2}-\frac{5}{2}+4$$
$$=1$$

$\therefore \quad \pi f\left(\frac{\pi}{3}\right)=3$

and $2g(2)=2\cos\left(\frac{2\pi}{a}\right)=2\cos\left(\frac{2\pi}{3}\right)$

$$=2\times-\frac{1}{2}=-1$$

$\Rightarrow \quad \pi f\left(\frac{\pi}{3}\right)+2g(2)=3-1=2$

Paragraph for Question Nos. 69 to 71

69. $\because f(x)=-x^2+3x-2$

$$=-(x^2-3x+2)=-(x-1)(x-2)$$

Now, $g(x)=f(|x|)$

Graph of $g(x)=$ Delete left portion of y-axis of $f(x)$ and then mirror images of right portion

Let $\quad y=g(x)=a$

for 3 real and distinct roots of $g(x)-a=0$

$a=-2$

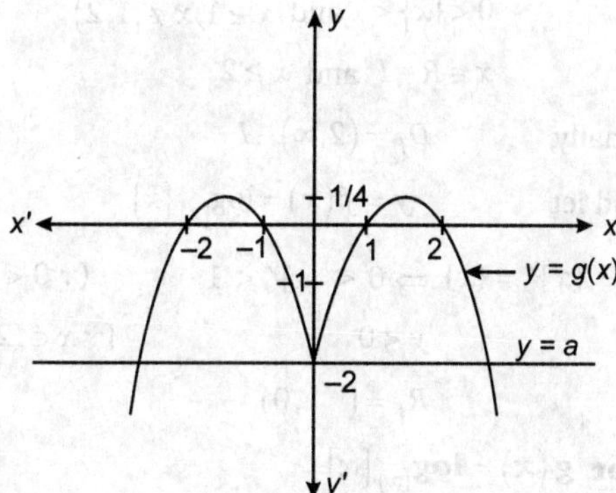

70. $\because h(x)-g(x)=0$

$\Rightarrow \quad |g(x)|-g(x)=0$

$\therefore \quad g(x)\geq 0$

It is clear from above graph

$x\in[-2,-1]\cup[1,2]$

71. $\because I(x)=0$

$\Rightarrow \quad h(x)-[x]=0$

$\Rightarrow \quad |g(x)|-[x]=0$

only when $3\leq x<4$

then $\quad |g(x)|=3$

$\Rightarrow \quad |-x^2+3x-2|=3$

$\Rightarrow \quad -x^2+3x-2=\pm 3$

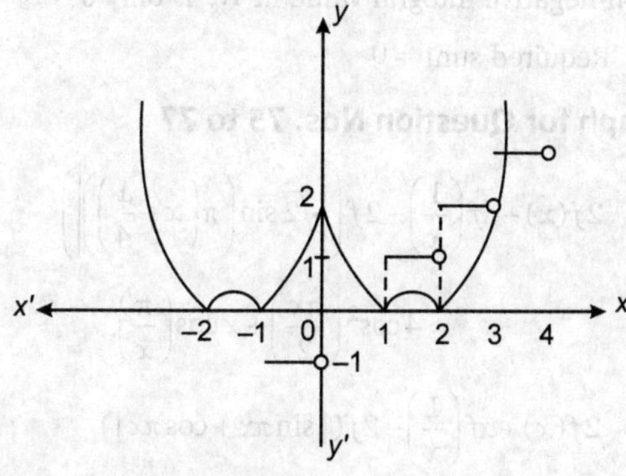

$\Rightarrow \quad x^2-3x+2\pm 3=0$

$\Rightarrow \quad x^2-3x+5=0$

$\Rightarrow \quad D<0$ (fail)

and $\quad x^2-3x-1=0$

$\therefore \quad x=\dfrac{3\pm\sqrt{9+4}}{2}=\dfrac{3+\sqrt{13}}{2}$ which is lies between 3 and 4 .

Paragraph for Question Nos. 72 to 74

For $f(x) = \log_{[x]}\{x\}$

$f(x)$ is defined when

$$\{x\} > 0 \text{ and } [x] > 0 \text{ \& } [x] \neq 1$$

$$\Rightarrow \quad 0 < \{x\} < 1 \text{ and } x \geq 1, x \notin [1,2)$$

$$\Rightarrow \quad x \in R - I \text{ and } x \geq 2$$

Finally, $D_f = (2, \infty) - I$

and let $y = f(x) = \log_{[x]}\{x\}$

$$\Rightarrow [x]^y = \{x\} \Rightarrow 0 < [x]^y < 1 \qquad (\because 0 < \{x\} < 1)$$

$$\therefore \quad y < 0 \qquad [\because x \in (2, \infty) - I]$$

$$\therefore \quad R_f = (\quad ,0)$$

For $g(x) = \log_{(x)}[x]$

$g(x)$ is defined when

$$0 < \{x\} < 1 \text{ and } [x] > 0$$

$$\Rightarrow \quad x \in R - I \text{ and } x \geq 1$$

Finally, $D_g = (1, \infty) - I$

and let $y = g(x) = \log_{(x)}[x]$

$$\Rightarrow \quad \{x\}^y = [x]$$

which is possible only when $y \leq 0$.

$$\therefore \quad R_g = (\quad ,0]$$

72. $\because D_f = (2, \infty) - I$ and $D_g = (1, \infty) - I$

$$\therefore \quad D_f \subset D_g$$

73. $R_f \cup R_g = (-\infty, 0) \cup (-\infty, 0] = (\quad ,0]$

74. Non-negative integral value in R_g is only 0

$\therefore$ Required sum $= 0$

Paragraph for Question Nos. 75 to 77

$$\because \quad 2f(x) + xf\left(\frac{1}{x}\right) - 2f\left(\left|\sqrt{2}\sin\left(\pi\left(x + \frac{1}{4}\right)\right)\right|\right)$$

$$= 4\cos^2\left(\frac{\pi x}{2}\right) + x\cos\left(\frac{\pi}{x}\right)$$

$$\Rightarrow 2f(x) + xf\left(\frac{1}{x}\right) - 2f(|\sin \pi x + \cos \pi x|)$$

$$= 4\cos^2\left(\frac{\pi x}{2}\right) + x\cos\left(\frac{\pi}{x}\right) \qquad \text{...(i)}$$

Put $x = 1$ in (i), then

$$2f(1) + 1f(1) - 2f(|\cos \pi|) = 4\cos^2\frac{\pi}{2} + \cos \pi$$

$$\Rightarrow \quad 3f(1) - 2f(1) = 0 - 1$$

$$\therefore \quad f(1) = -1 \qquad \text{...(ii)}$$

Now put $x = \frac{1}{2}$ in (i), then

$$2f\left(\frac{1}{2}\right) + \frac{1}{2}f(2) - 2f(1) = 2 + \frac{1}{2}$$

$$\Rightarrow \quad 2f\left(\frac{1}{2}\right) + \frac{1}{2}f(2) + 2 = 2 + \frac{1}{2} \qquad [\because f(1) = -1]$$

$$\Rightarrow \quad 4f\left(\frac{1}{2}\right) + f(2) = 1 \qquad \text{...(iii)}$$

and put $x = 2$ in (i), then

$$2f(2) + 2f\left(\frac{1}{2}\right) - 2f(1) = 4 + 0$$

$$\Rightarrow \quad 2f(2) + 2f\left(\frac{1}{2}\right) + 2 = 4 \qquad [\because f(1) = -1]$$

$$\therefore \quad f(2) + f\left(\frac{1}{2}\right) = 1 \qquad \text{[Ans. 76] ...(iv)}$$

From Eqs. (iii) and (iv), we get

$$4(1 - f(2)) + f(2) = 1$$

$$\Rightarrow \quad 3f(2) = 3$$

$$\therefore \quad f(2) = 1 \qquad \text{[Ans 75] ...(v)}$$

77. (a) From eqs, (ii) and (v), we get

$$f(2) + f(1) = 0$$

Paragraph for Question Nos. 78 to 80

$$\because (x - y)f(x + y) - (x + y)f(x - y) = 2(x^2y - y^3)$$

$$= 2y(x^2 - y^2)$$

$$= 2y(x + y)(x - y)$$

Let $x + y = u$ and $x - y = v$

$$\Rightarrow \quad vf(u) - uf(v) = (u - v)uv$$

$$\Rightarrow \quad \frac{f(u)}{u} - \frac{f(v)}{v} = u - v$$

$$\Rightarrow \quad \left(\frac{f(u)}{u} - u\right) = \left(\frac{f(v)}{v} - v\right) = k(\text{say}) \qquad \text{...(i)}$$

For $u = v = 1$

$$\Rightarrow \quad \left(\frac{f(1)}{1} - 1\right) = \left(\frac{f(1)}{1} - 1\right) = k$$

$$\Rightarrow \quad (2 - 1) = (2 - 1) = k \qquad (\because f(1) = 2)$$

$$\therefore \quad k = 1$$

From (i),

$$\left(\frac{f(u)}{u} - u\right) = \left(\frac{f(v)}{v} - v\right) = 1$$

$$\therefore \quad \frac{f(x)}{x} - x = 1$$

or

$$f(x) = x^2 + x$$

78. $\because\ f(x)-2=x^2+x-2\ =(x+2)(x-1)$

$\Rightarrow\qquad |f(x)-2|=|x^2+x-2|=|(x+2)(x-1)|$

and let $\qquad y=|f(x)-2|=a$

It is clear from graphs for exactly and real roots

$$0<a<\frac{9}{4}$$

$$\therefore\qquad a\in\left(0,\frac{9}{4}\right)$$

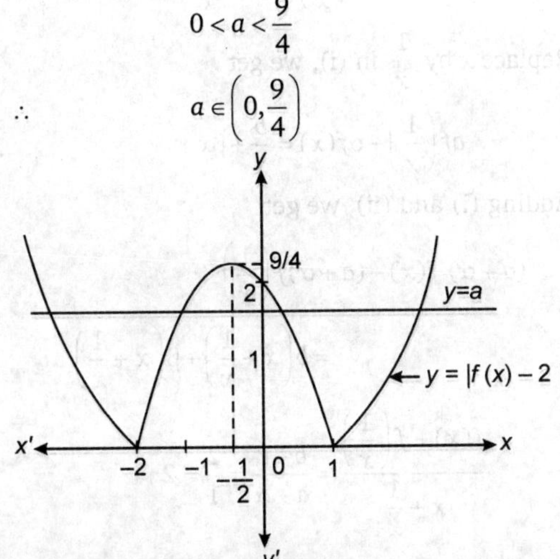

79. Let $\quad y=f(x)=\sqrt{(1-x^2)}$

$$y=x^2+x=\sqrt{(1-x^2)}$$

It is clear from graphs

No. of solution = 2

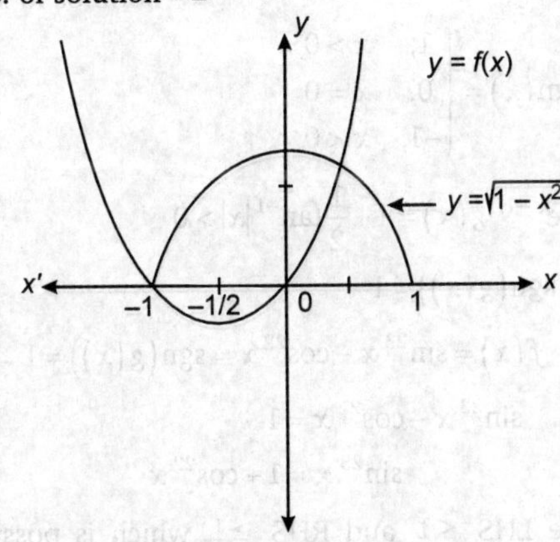

80. Let $\quad y=-f\left(2^x\right)=-\left(\left(2^x\right)^2+2^x\right)$

$$=-\left\{\left(2^x+\frac{1}{2}\right)^2-\frac{1}{4}\right\}$$

$$=\frac{1}{4}-\left(2^x+\frac{1}{2}\right)^2<0\ \forall x\in R$$

$\therefore$ Range = (,0)

Paragraph for Question Nos. 81 to 83

Given $\quad f(x)+f\left(\frac{x-1}{x}\right)=\frac{2x(2-x)}{(x-1)}$

$$\Rightarrow\qquad f(x)+f\left(1-\frac{1}{x}\right)=\frac{2x(2-x)}{(x-1)}\qquad ...(i)$$

Replacing x by $\left(1-\frac{1}{x}\right)$ in (i), we obtain

$$f\left(1-\frac{1}{x}\right)+f\left(\frac{1}{1-x}\right)=\frac{2\left(1-x^2\right)}{x}\qquad ...(ii)$$

Again replacing x by $\frac{1}{(1-x)}$ in (i), we obtain

$$f\left(\frac{1}{1-x}\right)+f(x)=\frac{2(2x-1)}{x(x-1)}\qquad ...(iii)$$

Adding (i) and (iii) and then subtracting (ii), we get

$$2f(x)=\frac{2(x+1)}{(x-1)}$$

or $\qquad f(x)=\frac{(x+1)}{(x-1)}$

and $\qquad g(x)=2f(x)+(x-1)$

$$=\frac{2(x+1)}{(x-1)}+(x-1)$$

$$=\frac{2(x+1)+(x-1)^2}{(x-1)}=\frac{(x^2+3)}{(x-1)}$$

$$\therefore\qquad g(x)=\frac{(x^2+3)}{(x-1)}$$

81. $\because\ y=\sqrt{g(x)}=\sqrt{\frac{(x^2+3)}{(x-1)}}$

$$\therefore\qquad \frac{(x^2+3)}{(x-1)}\geq 0\ \Rightarrow\ x-1>0$$

$$\therefore\qquad x\in(1,\infty)$$

82. $\because\ y=g(x)=\frac{(x^2+3)}{(x-1)}$

$\Rightarrow\ xy-y=x^2+3$

$\Rightarrow\ x^2-xy+(y+3)=0$

$\Rightarrow\ B^2-4AC\geq 0$

$\Rightarrow\ y^2-4\cdot(y+3)\geq 0$

$\Rightarrow\ y^2-4y-12\geq 0$

$\Rightarrow\ (y-6)(y+2)\geq 0$

$\therefore\ \ y\in(-\infty,-2]\cup[6,\infty)$

83. $g(x)=2$

but $g(x)\in(-\infty,-2]\cup[6,\infty)$

$\therefore g(x)\notin(-2,6)$

Hence, no solution.

$\therefore$ Number of solutions is 0.

Paragraph for Question Nos. 84 to 86

$$\because f(x)=\begin{cases}x+1, & x<1\\ 2x+1, & 1\le x<2\end{cases}$$

and $g(x)=\begin{cases}x^2, & -1\le x<2\\ x+2, & 2\le x<3\end{cases}$

$$\therefore\ (gof)x=g(f(x))=\begin{cases}(f(x))^2, & -1\le f(x)<2\\ f(x)+2, & 2\le f(x)<3\end{cases}$$

$$=\begin{cases}(x+1)^2, & -1\le x+1<2 \text{ and } x<1\\ x+1+2, & 2\le x+1<3 \text{ and } x<1\\ (2x+1)^2, & -1\le 2x+1<2 \text{ and } 1\le x<2\\ 2x+1+2, & 2\le 2x+1<3 \text{ and } 1\le x<2\end{cases}$$

$$=\begin{cases}(x+1)^2, & -2\le x<1 \text{ and } x<1\\ x+3, & 1\le x<2 \text{ and } x<1\\ (2x+1)^2, & -1\le x<\frac{1}{2} \text{ and } 1\le x<2\\ 2x+3, & \frac{1}{2}\le x<1 \text{ and } 1\le x<2\end{cases}$$

$$=\begin{cases}(x+1)^2, & -2\le x<1\\ x+3, & x\in\phi\\ (2x+1)^2, & x\in\phi\\ 2x+3, & x\in\phi\end{cases}$$

$\therefore\ (gof)x=(x+1)^2, -2\le x<1$

84. The domain of $(gof)\ x$ is $[-2,1)$.

85. $\because -2\le x<1$

$\Rightarrow\quad -1\le(x+1)<2$

$\therefore\quad 0\le(x+1)^2<4$

$\therefore$ Range of $(gof)x$ is $[0,4)$.

86. $(gof)x=3$

$$(x+1)^2=3$$

$$x+1=\pm\sqrt{3}$$

$\therefore\quad x=-1\pm\sqrt{3}$

Hence number of solutions is 2 .

87. $\because 3f(x+1)+2f\left(\frac{1}{x+1}\right)=x$ ...(i)

Replacing $(x+1)$ by $\frac{1}{(x+1)}$, we get

$$3f\left(\frac{1}{x+1}\right)+2f(x+1)=\frac{1}{x+1}-1 \quad ...(ii)$$

Multiplying (i) by 3 and (ii) by 2 and subtracting, we get

$$5f(x+1)=3x-\frac{2}{x+1}+2$$

Put $x=1$, we get

$$5f(2)=3-1+2=4$$

88. We have,

$$af(x)+\alpha f\left(\frac{1}{x}\right)=bx+\frac{\beta}{x} \quad ...(i)$$

Replace x by $\frac{1}{x}$ in (i), we get

$$af\left(\frac{1}{x}\right)+\alpha f(x)=\frac{b}{x}+\beta x \quad ...(ii)$$

Adding (i) and (ii), we get

$$(a+\alpha)\ f(x)+(a+\alpha)f\left(\frac{1}{x}\right)$$
$$=b\left(x+\frac{1}{x}\right)+\beta\left(x+\frac{1}{x}\right)$$

$$\Rightarrow\quad \frac{f(x)+f\left(\frac{1}{x}\right)}{x+\frac{1}{x}}=\frac{b+\beta}{a+\alpha}=\frac{2}{1}=2$$

$(\because a+\alpha=1, b+\beta=2)$

89. Number of functions

$$={}^2C_0(2-0)^3-{}^2C_1(2-1)^3+{}^2C_2(2-2)^3$$
$$=8-2+0$$
$$=6$$

90. $\because \text{sgn}(x)=\begin{cases}1, & x>0\\ 0, & x=0\\ -1, & x<0\end{cases}$

Here $\quad g(x)=1+\frac{1}{2}\tan^{-1}|x|>0$

$\therefore\ \text{sgn}(g(x))=1$

and $f(x)=\sin^{23}x-\cos^{22}x=\text{sgn}(g(x))=1$ (given)

$\therefore\quad \sin^{23}x-\cos^{22}x=1$

$\Rightarrow\quad \sin^{23}x=1+\cos^{22}x$

Here LHS ≤ 1 and RHS ≥ 1 which is possible only when

$$\sin x=1 \text{ and } \cos x=0$$

$\therefore\quad x=2n\pi+\frac{\pi}{2}, n\in I$

Also, $\quad -10\pi\le x\le 20\pi$

$\Rightarrow\quad -10\pi\le(4n+1)\frac{\pi}{2}\le 20\pi$

$\Rightarrow\quad -20\le(4n+1)\le 40$

$\Rightarrow\quad -21\le 4n\le 39$

$\Rightarrow\quad -5.25\le n\le 9.75$

But $n\in I$

$\therefore\ n = -5,-4,-3,-2,-1,0,1,2,3,4,5,6,7,8,9$

For non-positive value of $x, n = -5,-4,-3,-2,-1$

Hence, required number of solutions $= 5$

91. Given, $f(g(x)) = 8x^2 - 2x$

and $g(f(x)) = 4x^2 + 6x + 1$

Let $f(x) = ax^2 + bx + c$ and $g(x) = dx + e$

So, $f(g(x)) = a(dx+e)^2 + b(dx+e) + c$

$= 8x^2 - 2x$

and $g(f(x)) = d(ax^2 + bx + c) + e$

$= 4x^2 + 6x + 1$

On comparing both sides, we get

$ad^2 = 8,\ ad = 4$

$\Rightarrow\ a = d = 2$

$bd = 6 \Rightarrow b = 3$

$2ade + bd = -2 \Rightarrow e = -1$

and $cd + e = 1 \Rightarrow c = 1$

So, $g(x) = dx + e = 2x - 1$

$\Rightarrow\ g(2) = 3$

and $f(x) = ax^2 + bx + c$

$= 2x^2 + 3x + 1$

$\Rightarrow\ f(2) = 15$

So, $f(2) + g(2) = 18$

92. $\because f(x+1) + f(x-1) = \sqrt{2} f(x)$ ...(i)

Replacing x by $x+1$ and x by $x-1$ in (i), then

$f(x+2) + f(x) = \sqrt{2} f(x+1)$ ...(ii)

and $f(x) + f(x-2) = \sqrt{2} f(x-1)$ ...(iii)

Adding Eqs. (ii) and (iii), we get

$2f(x) + f(x+2) + f(x-2) = \sqrt{2}[f(x+1) + f(x-1)]$

$= \sqrt{2}\cdot\sqrt{2} f(x)$ [from (i)]

$\Rightarrow 2f(x) + f(x+2) + f(x-2) = 2f(x)$

$\Rightarrow\ f(x+2) + f(x-2) = 0$ ...(iv)

Now replacing x by $x+4$ in Eq. (iv), we get

$f(x+6) + f(x+2) = 0$ ...(v)

Subtracting (iv) from (v), we get

$f(x+6) - f(x-2) = 0$

or $f(x+6) = f(x-2)$

Finally, replacing x by $x+2$, then we get

$f(x+8) = f(x)$

$\therefore$ Period of $f(x) = 8$

93. (4) Given $f(x) + 2f\left(\frac{2002}{x}\right) = 3x$ ...(i)

Put $x = 2$ in Eq. (i), then

$f(2) + 2f(1001) = 6$ ...(ii)

and put $x = 1001$ in Eq. (i), then

$f(1001) + 2f(2) = 3003$ ...(iii)

From Eqs. (ii) and (iii), we get

$f(1001) + 2(6 - 2f(1001)) = 3003$

$\Rightarrow\ 3f(1001) = 12 - 3003$

$\therefore\ f(1001) = 4 - 1001$

$\Rightarrow\ f(1001) + 1001 = 4$

94. $\because f(x)$ is symmetrical about the line $y = x$

$\therefore\ f^{-1}(x) = f(x)$

$\Rightarrow\ f^2(x) = f^2(x) - kxf^2(x) + 2x^2 f(x)$

$\Rightarrow\ f(x) = \frac{2x}{k} = x$ $[\because y = f(x) = x]$

$\therefore\ k = 2$

95. Given, $f(x) = ax - 3$ and $g(x) = x^b + c,\ x \in R$

Let $y = (fog)^{-1}(x) = \left(\frac{x-7}{2}\right)^{1/3}$

$\Rightarrow\ (fog)(y) = x$ ...(i)

and $y = \left(\frac{x-7}{2}\right)^{1/3}$

$\Rightarrow\ x = 2y^3 + 7$ ...(ii)

From (i) and (ii), we get

$(fog)(y) = 2y^3 + 7$

$\Rightarrow\ (fog)(x) = 2x^3 + 7$

$\Rightarrow\ f(g(x)) = 2x^3 + 7$

$\Rightarrow\ f(x^b + c) = 2x^3 + 7$

$\Rightarrow\ a(x^b + c) - 3 = 2x^3 + 7$

On comparing, we get

$a = 2,\ b = 3,\ c = 5$

So, $f(x) = 2x - 3$

and $g(x) = x^3 + 5$

Now, $(fog)(ac) = (fog)(10) = 2(10)^3 + 7 = 2007$

and $(gof)(b) = (gof)(3) = g(f(3)) = g(3) = 32$

Therefore, $(fog)(ac) + (gof)(b) = 2007 + 32 = 2039$

96. $f(9) = f(4+5) = f(4\times 5)$

$= f(20) = f(16+4)$

$= f(16\times 4)$

$= f(64) = f(8\times 8)$

$= f(8+8)$

$= f(16) = f(4\times 4)$

$= f(4+4) \ = f(8) = 9$

97. $\because \ g(5,-3) = 5-(-3) = 8$

and $g(-1.25, 14.75) = 14.75-(-1.25) = 16$

$\therefore \quad f(g(5,-3), g(-1.25, 14.75) = f(8,16)$

$$= 16 = 2^4 = 2^{2^2}$$

$\Rightarrow \qquad \lambda = 2$

98. $\because 0 \le |\sin x| \le 1$ and $0 \le |\cos x| \le 1$

Minimum value of $f(x) = 0-2 = -2$

and maximum value of $f(x) = 3-0 = 3$

$\therefore$ Range of $f(x)$ is $[-2, 3]$

Here $a=-2, b=3 \Rightarrow b-a=5$ or $|b-a|=5$

99. $\because \qquad f(x) = \dfrac{2e^{2x}}{e^{2x}+e}$ ...(i)

$\therefore \qquad f(1-x) = \dfrac{2e^{2(1-x)}}{e^{2(1-x)}+e}$

$$= \frac{2e^2}{e^2+e\cdot e^{2x}} = \frac{2e}{e+e^{2x}} \quad \text{...(ii)}$$

Adding (i) and (ii), we get

$$f(x)+f(1-x) = 2 \quad \text{...(iii)}$$

Putting $x = \dfrac{1}{100}, \dfrac{2}{100}, \dfrac{3}{100}, \ldots, \dfrac{49}{100}, \dfrac{50}{100}$ in (iii), then

$$f\left(\frac{1}{100}\right)+f\left(\frac{99}{100}\right) = 2$$

$$f\left(\frac{2}{100}\right)+f\left(\frac{98}{100}\right) = 2$$

$$f\left(\frac{3}{100}\right)+f\left(\frac{97}{100}\right) = 2$$

....

$$f\left(\frac{49}{100}\right)+f\left(\frac{51}{100}\right) = 2$$

$$f\left(\frac{50}{100}\right)+f\left(\frac{50}{100}\right) = 2$$

$$\Rightarrow \quad f\left(\frac{50}{100}\right) = 1$$

Adding all, we get

$$f\left(\frac{1}{100}\right)+f\left(\frac{2}{100}\right)+f\left(\frac{3}{100}\right)+\ldots + f\left(\frac{99}{100}\right) = 2\times 49+1 = 99$$

100. Here $\qquad f(x) = 2x^3+7x-5$

$\therefore \qquad f'(x) = 6x^2+7 > 0$

$\therefore$ $f(x)$ is strictly increasing $\forall x \in R$

$\Rightarrow f(x)$ is one-one

and range $= R =$ co-domain

$\Rightarrow f(x)$ is onto

Hence, $f(x)$ is one-one onto.

$\therefore$ $f^{-1}(x)$ exists.

Now, let $\qquad f^{-1}(4) = \lambda$

$\Rightarrow \qquad f(\lambda) = 4$

$\Rightarrow \qquad 2\lambda^3+7\lambda-5 = 4$

$\Rightarrow \qquad 2\lambda^3+7\lambda-9 = 0$

$\Rightarrow \qquad (\lambda-1)(2\lambda^2+2\lambda+9) = 0$

$\because 2\lambda^2+2\lambda+9 \ne 0$

$\therefore \qquad \lambda-1 = 0$

$\Rightarrow \qquad \lambda = 1$

or $\qquad f^{-1}(4) = 1$

101. Given, $A = \{0, 1, 2, 3, 4, 5, 6, 7\}$ and also a function $f: A \to A$ satisfying

$$f(1)+f(2) = 3-f(3)$$

$\Rightarrow \qquad f(1)+f(2)+f(3) = 3$

Since, the range of the function is also A, hence the only possibility satisfying the given condition is :

$0+1+2=3$

Since the given function is bijective. Hence, the elements 1, 2, 3 in the domain can be mapped with only 0, 1, 2 in the co-domain in 3! ways and the remaining 5 elements 0, 4, 5, 6, 7 in the domain can be mapped with any of the remaining 5 elements 3, 4, 5, 6, 7 in 5! ways.

So, the number of bijective functions are

$$= 3!\times 5! = 6\times 120 = 720.$$

102. A. Constant function

$$f(x) = c$$

is periodic with no fundamental period.

$\because \qquad c \in R$

$\therefore \qquad f(x) \in R$ is bound

$\Rightarrow \qquad f'(x) = 0$ no monotonic

and $\quad f(-x) = c = f(x)$

Even ($\therefore$ constant function symmetric about y-axis)

B. $\because g(x) = \int_1^x \dfrac{dt}{t}$

$\therefore \qquad g'(x) = \dfrac{1}{x} > 0 \qquad (\because x > 0)$

$\Rightarrow g(x)$ is monotonic

Also, $g(-x) \neq \pm g(x)$

$\Rightarrow g(x)$ is neither even nor odd.

and $g(x)$ is non-periodic and no bound.

C. $\because h(x) = \text{arc}\tan x = \tan^{-1} x$

$\therefore -\frac{\pi}{2} < h(x) < \frac{\pi}{2} \Rightarrow h(x) \in \left(-\frac{\pi}{2}, \frac{\pi}{2}\right)$

$\Rightarrow h(x)$ is non-periodic and $h(x)$ is bound

$\Rightarrow \quad h(-x) = \tan^{-1}(-x)$

$= -\tan^{-1} x = -h(x)$

$\therefore h(x)$ is an odd function.

Also, $h'(x) = \frac{1}{1+x^2} > 0$

$\therefore h(x)$ is monotonic. (Strictly increasing)

D. $\because k(x) = \text{arc}\cot x = \cot^{-1} x$

$\therefore 0 < k(x) < \pi \Rightarrow k(x) \in (0, \pi)$

$\Rightarrow k(x)$ is non-periodic and $k(x)$ is bound.

$\Rightarrow k(-x) = \cot^{-1}(-x) = \pi - \cot^{-1} x \neq \pm k(x)$

$\therefore k(x)$ is neither even nor odd.

Also, $k'(x) = -\frac{1}{(1+x^2)} < 0$

$\therefore k(x)$ is monotonic (Strictly decreasing)

103. A. $f(xy) = \frac{f(x)}{y}$

$\Rightarrow \quad f(y) = \frac{f(1)}{y}$ (putting $x = 1$)

$\Rightarrow \quad f(30) = \frac{f(1)}{30}$

or $f(1) = 30 f(30) = 30 \times 20 = 600$

Now, $f(50) = \frac{f(1)}{50} = \frac{600}{50} = 12$

B. The period of $\cos(\sin nx)$ is $\frac{\pi}{n}$ and period of $\tan\left(\frac{x}{n}\right)$ is $n\pi$.

Thus, $6\pi = \text{LCM}\left(\frac{\pi}{n}, n\pi\right)$

$\Rightarrow \quad 6\pi = \frac{\pi}{n} k_1 \Rightarrow n = \frac{k_1}{6}$

and $6\pi = n\pi k_2 \Rightarrow n = \frac{6}{k_2}$

and $k_1, k_2 \in I^+$

From $n = \frac{6}{k_2} \Rightarrow n = 6, 3, 2, 1$

Clearly for $n = 6$, we get the period of $f(x)$ is 6π.

$\therefore \quad n = 6$

C. Draw the graphs of $y = \log_{0.5}|x|$ and

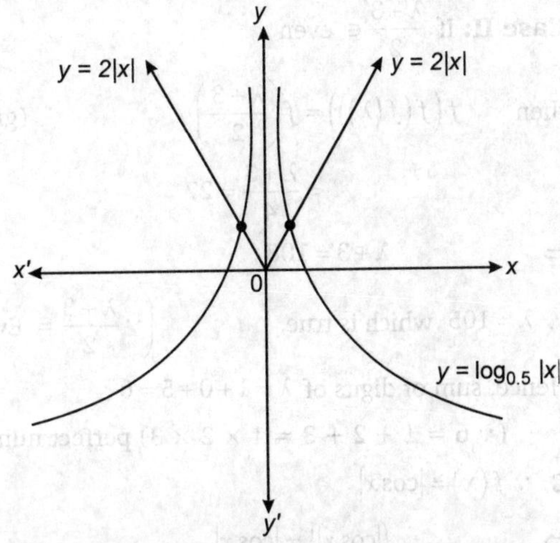

Clearly from the graph, there are two solutions.

D. $\because f(3x+2) + f(3x+29) = 0$ ...(i)

Replacing x by $x+9$, we get

$f(3(x+9)+2) + f(3(x+9)+29) = 0$

$\Rightarrow f(3x+29) + f(3x+56) = 0$ ...(i)

From Eqs. (i) and (ii), we get

$f(3x+2) = f(3x+56)$

$= f(3(x+18)+2)$

$\Rightarrow f(x)$ is periodic with period 18.

104. A. $\because$ Domain of $f(x)$ is $[-5, 7]$

$\therefore \quad -5 \le |\lambda x + 5| \le 7$, where $-6 \le x \le 1$

$\Rightarrow \quad 0 \le |\lambda x + 5| \le 7$, where $-6 \le x \le 1$

$\Rightarrow \quad -7 \le \lambda x + 5 \le 7$, where $-6 \le x \le 1$

$\Rightarrow \quad -12 \le \lambda x \le 2$, where $-6 \le x \le 1$

$\Rightarrow \quad -6 \le \left(\frac{\lambda}{2}\right) x \le 1$, where $-6 \le x \le 1$

$\therefore \quad \lambda = 2 \quad (\because R_h = D_f)$

B. $\because \lambda \in$ odd

$\therefore \quad f(\lambda) = (\lambda + 3)$

$\Rightarrow \quad f(f(\lambda)) = f(\lambda + 3)$

$= \frac{(\lambda+3)}{2}$ ($\because \lambda + 3$ is even)

Case I: If $\frac{\lambda+3}{2} \in$ odd

then $f(f(f(\lambda))) = f\left(\frac{\lambda+3}{2}\right) = \frac{\lambda+3}{2} + 3 = 27$ (given)

$\Rightarrow \quad \lambda + 3 = 48$

$\therefore \lambda = 45$ not possible $\left(\because \frac{\lambda+3}{2} = \text{Even}\right)$

Case II: If $\frac{\lambda+3}{2} \in$ even

then $f(f(f(\lambda))) = f\left(\frac{\lambda+3}{2}\right)$ (given)

$= \frac{\lambda+3}{4} = 27$

$\Rightarrow \quad \lambda + 3 = 108$

$\therefore \lambda = 105$ which is true $\left(\because \frac{\lambda+3}{2} = \text{Even}\right)$

Hence, sum of digits of $\lambda = 1 + 0 + 5 = 6$

($\because 6 = 1 + 2 + 3 = 1 \times 2 \times 3$) perfect number)

C. $\because f(x) \neq |\cos x|$

$\Rightarrow \quad \{|\cos x|\} \neq |\cos x|$

($\because 6 = 1+2+3 = 1\times2\times3 =$ Perfect number)

which is possible only when

$|\cos x| = 1$

$\Rightarrow \quad \cos x = \pm 1$

$\therefore \quad x = 0, \pi, 2\pi \quad \because x \in [0, 2\pi]$

$\therefore$ Required number of elements = 3

105. A. $\because y = \frac{x^2 - 2x + 4}{x^2 + 2x + 4}$

$\Rightarrow \quad x^2y + 2xy + 4y = x^2 - 2x + 4$

$\Rightarrow \quad x^2(y-1) + 2x(y+1) + 4y - 4 = 0 \quad \because x \in R,$

$\therefore \quad 4(y+1)^2 - 16(y-1)^2 \geq 0$

$\Rightarrow \quad 4\{(y+1)^2 - (2y-2)^2\} \geq 0$

$\Rightarrow \quad 4(3y-1)(3-y) \geq 0$

$\Rightarrow \quad (3y-1)(y-3) \leq 0$

$\therefore \quad \frac{1}{3} \leq y \leq 3 \quad \because y \in I$

$\therefore \quad y = 1, 2, 3$

B. $\because \quad y = \frac{x^2 - 3x - 2}{2x - 3}, x \neq \frac{3}{2}$

$\Rightarrow \quad 2xy - 3y = x^2 - 3x - 2$

$\Rightarrow \quad x^2 - x(2y+3) + 3y - 2 = 0$

$\because x \in R, \therefore (2y+3)^2 - 4(3y-2) \geq 0$

$\Rightarrow \quad 4y^2 + 17 \geq 0$

$\therefore \quad y \in R$

C. $\because \quad y = \frac{2x^2 - 2x + 4}{x^2 - 4x + 3}, \ x \neq 1, 3$

$\Rightarrow \quad x^2y - 4xy + 3y = 2x^2 - 2x + 4$

$\Rightarrow \quad x^2(y-2) - 2x(2y-1) + 3y - 4 = 0$

$\because x \in R, \therefore 4(2y-1)^2 - 4(y-2)(3y-4) \geq 0$

$\Rightarrow \quad (2y-1)^2 - (y-2)(3y-4) \geq 0$

$\Rightarrow \quad y^2 + 6y - 7 \geq 0$

$\Rightarrow \quad (y+7)(y-1) \geq 0$

$\therefore y \leq -7$ and $y \geq 1$

$\because f(x) = \ln x$ and $g(x) = x^2 - 1$

106. A. $(fog)(x) = f(g(x)) = f(x^2 - 1) = \ln(x^2 - 1)$

$(fog)(x)$ is defined, when $x^2 - 1 > 0$

$\Rightarrow \quad (x+1)(x-1) > 0$

$\therefore \quad x \in (-\infty, -1) \cup (1, \infty)$

B. $(gof)x = g(f(x)) = g(\ln x) = (\ln x)^2 - 1$

$(gof)x$ is defined, when $x > 0$

$\therefore \quad x \in (0, \infty)$

C. $(fof)\,x = f(f(x)) = f(\ln x) = \ln \ln x$

$(fof)(x)$ is defined, when $x > 0$ and $\ln x > 0$

$\Rightarrow x > 0$ and $x > 1$

$\therefore \quad x \in (1, \infty)$

D. $(gog)(x) = g(g(x)) = g(x^2 - 1)$

$= (x^2 - 1)^2 - 1$

$= x^4 - 2x^2$

$(gog)(x)$ is defined, when $x \in R$.

107. $\because D_f = R$ and $D_g = R \sim \{0\}$

$\therefore \quad f(x) \neq g(x)$

$\Rightarrow$ Statement-1 is false, statement-2 is true.

108. Let $x = n + h$, $n \in I$ and $0 < h < 1$ ($\because x$ is not an integer)

$\therefore \quad \{x\} = h$

and $\quad -x = -n - h = (-n-1) + (1-h)$

$\therefore \quad \{-x\} = (1-h)$

$\Rightarrow \quad f(x) = \frac{1}{2\{-x\}} - \{x\}$

$= \frac{1}{2(1-h)} - h$

$= \frac{1}{2(1-h)} + (1-h) - 1$

Now, $\frac{1}{2(1-h)} + (1-h) \geq 2\sqrt{\frac{1}{2(1-h)} \cdot (1-h)} = \sqrt{2}$

$$\Rightarrow \quad \frac{1}{2(1-h)}+(1-h)\geq\sqrt{2}$$

$$\Rightarrow \quad \frac{1}{2(1-h)}+(1-h)-1\geq\sqrt{2}-1$$

$$\Rightarrow \quad f(x)\geq\sqrt{2}-1$$

$\therefore$ Minimum value of $f(x)$ is $\sqrt{2}-1$.

Hence, both statements are true and statement-2 is correct explanation of statement-1.

109. Statement-1 : obvious

Statement-2 : is not true always

If $f(x)=|\sin x|+|\cos x|$

then fundamental period of $f(x)$ is $\frac{\pi}{2}\neq\pi$.

110. A polynomial of degree even will always be into.

111. The function $f(x)=|x|$ is many one

$\because \quad f(-x)=f(x)$

i.e., not one-one and the image of $-x$ is x.

112. $\because \; f(x)=\cos^{-1}\left(\frac{1+x^2}{2x}\right)$

$\therefore \; f(x)$ is defined, when

$$-1\leq\frac{1+x^2}{2x}\leq 1$$

$$\Rightarrow \quad \left|\frac{1+x^2}{2x}\right|\leq 1$$

$$\Rightarrow \quad \frac{|1+x^2|}{2|x|}\leq 1$$

$$\Rightarrow \quad \frac{1+x^2}{2|x|}\leq 1$$

$$\Rightarrow \quad \frac{1+|x|^2}{2|x|}-1\leq 0$$

$$\Rightarrow \quad \frac{(|x|-1)^2}{2|x|}\leq 0$$

Which is possible only when

$$\frac{(|x|-1)^2}{2|x|}=0$$

$$\Rightarrow \quad (|x|-1)^2=0 \Rightarrow |x|-1=0$$

$$\Rightarrow \quad |x|=1$$

$$\therefore \quad x=\pm 1 \Rightarrow D_f=\{-1,1\}$$

and $\quad R_f=\{f(-1),\, f(1)\}$

$\qquad =\{\pi,0\}$ or $\{0,\pi\}$

both statements are true and statement-2 is the correct explanation of statement-1.

113. $\because \quad f(1+x)=f(1-x)$ ...(i)

and $\quad f(4+x)=f(4-x)$ ...(ii)

Replacing x by $x-1$ in Eq. (i), then

$f(x)=f(2-x)$ ...(iii)

and replacing x by $x+2$ in Eq. (ii), then

$f(6+x)=f(2-x)$ ...(iv)

From Eqs. (iii) and (iv), we get

$$f(6+x)=f(x)$$

$\Rightarrow f(x)$ is periodic with period 6

Obviously 6 is not necessary the fundamental period.

114. $\because \quad f(x)=\log_{1/4}\left(x-\frac{1}{4}\right)+\frac{1}{2}\log_4(16x^2-8x+1)$

$$=\log_{4^{-1}}\left(x-\frac{1}{4}\right)+\frac{1}{2}\log_4\left(16\left(x^2-\frac{x}{2}+\frac{1}{16}\right)\right)$$

$$=-\log_4\left(x-\frac{1}{4}\right)+\frac{1}{2}\log_4\left(16\left(x-\frac{1}{4}\right)^2\right)$$

$$=-\log_4\left(x-\frac{1}{4}\right)+\frac{1}{2}\log_4 16+\frac{1}{2}\log_4\left(x-\frac{1}{4}\right)^2$$

$$=-\log_4\left(x-\frac{1}{4}\right)+\frac{1}{2}\log_4 4^2+\frac{1}{2}\log_4\left(x-\frac{1}{4}\right)^2$$

$$=-\log_4\left(x-\frac{1}{4}\right)+\frac{2}{2}+\frac{2}{2}\log_4\left|x-\frac{1}{4}\right|$$

$$=-\log_4\left(x-\frac{1}{4}\right)+1+\log_4\left(x-\frac{1}{4}\right)\quad\left(\because x>\frac{1}{4}\right)$$

$$=1$$

$\Rightarrow \quad f(x)=1 \qquad (\because f'(x)=0)$

which is constant

$\therefore f(x)$ is not injective.

Also, Range $=\{1\}\subset R^+$ (co-domain)

$\therefore f(x)$ is into.

Hence, $f(x)$ is neither injective nor surjective

115. $\because \quad f(x)+f(1-x)=2$

$$\Rightarrow \quad f(x)-1+f(1-x)-1=0$$

$$\Rightarrow \quad g(x)+g(1-x)=0 \quad (\because g(x)=f(x)-1)$$

Replacing x by $x+\frac{1}{2}$, then we get

$$g\left(x+\frac{1}{2}\right)+g\left(\frac{1}{2}-x\right)=0$$

$$\Rightarrow \quad g\left(\frac{1}{2}-x\right)=-g\left(x+\frac{1}{2}\right)$$

So, it is symmetrical about $\left(\frac{1}{2},0\right)$.

116. Since $|\sin kx|$ has period $\frac{\pi}{k}$ and period of $\{x\}$ is 1.

Now, LCM of $\frac{\pi}{k}$ and 1 exists only if k is a rational multiple of π (as LCM of rational and irrational number does not exist). Hence statement-1 is true but statement-2 is false as sum of two periodic function is not necessarily periodic. Consider $f(x) = \sin x + \{x\}$

117. From the graph of $f(x)$ it is clear that $f(x)$ is not one-one *i.e.*, many one.

$\therefore$ Statement-1 is true.

Also $\quad f'(x) < 0 \ \forall x \in D_f$

$\therefore$ $f(x)$ is one-one.

$\therefore$ Statement-2 is false.

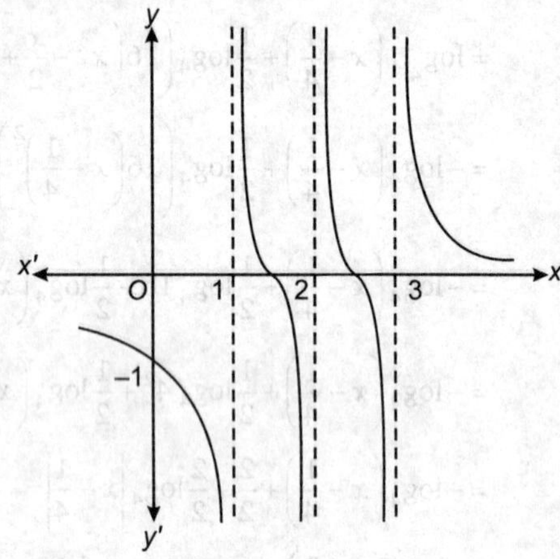

118. If $f(x)$ is an even degree polynomial then $f'(x)$ is an odd degree polynomial.

for $\quad f'(x) = 0$

has at least one real roots.

$\therefore$ $f(x)$ is maximum or minimum.

$\Rightarrow$ $f(x)$ is many one.

Hence, every polynomial function of even degree defined from R to R is always not injective. Hence, both statements are true., staetment-2 is correct explanation of statement-1.

119. Given, $f(x + y + 1) = (\sqrt{f(x)} + \sqrt{f(y)})^2$

Putting $x = y = 0$,

then $\quad f(1) = (\sqrt{f(0)} + \sqrt{f(0)})^2 = (1 + 1)^2 = 2^2$

Again putting $x = 0, y = 1$

Then $\quad f(2) = (\sqrt{f(0)} + \sqrt{f(1)})^2 = (1 + 2)^2 = 3^2$

and for $x = 1, y = 1$

$$f(3) = (\sqrt{f(1)} + \sqrt{f(1)})^2$$
$$= (2 + 2)^2 = 4^2$$

Similarly, $f(x) = (x + 1)^2$

120. Given $f(x - f(y)) = f(f(y)) + xf(y) + f(x) - 1 \quad$ ...(i)

Putting $x = f(y) = 0$, then $f(0) = f(0) + 0 + f(0) - 1$

$$f(0) = 1 \quad \text{...(ii)}$$

Again putting $x = f(y) = \lambda$ in (i)

Then $\quad f(0) = f(\lambda) + \lambda^2 + f(\lambda) - 1$

$\Rightarrow \quad 1 = 2f(\lambda) + \lambda^2 - 1 \quad$ {from (ii)}

$\therefore \quad f(\lambda) = \frac{2-\lambda^2}{2} = 1 - \frac{\lambda^2}{2}$

Hence, $f(x) = 1 - \frac{x^2}{2}$, is the unique function.

121. Given, $f(x+y) - kxy = f(x) + 2y^2$

Replacing y by $-x$, then we get

$$f(0) + kx^2 = f(x) + 2x^2$$

$\Rightarrow \quad f(x) = f(0) + kx^2 - 2x^2 \quad$ ...(i)

$\therefore \quad f(1) = f(0) + k - 2 = 2$ (given) $\quad$ ...(ii)

and $\quad f(2) = f(0) + 4k - 8 = 8$ (given) $\quad$ ...(iii)

Solving (ii) and (iii) we get

$$k = 4 \text{ and } f(0) = 0 \quad \text{...(iv)}$$

$\therefore$ From (i), we get $f(x) = 2x^2$

Also, $f(x+y) \cdot f\left(\frac{1}{x+y}\right) = 2(x+y)^2 \cdot 2\left(\frac{1}{x+y}\right)^2$

$= 4 = k \quad$ {From (iv)}

122. Given, $2f(xy) = (f(x))^y + (f(y))^x$

Replacing y by 1, then we get

$$2f(x) = f(x) + (f(1))^x$$

$\Rightarrow \quad f(x) = (f(1))^x = a^x \quad (\because f(1) = a)$

$\therefore \quad \text{LHS} = (a-1)\sum_{i=1}^{n} f(i) = (a-1)\sum_{i=1}^{n} a^i$

$$= (a-1)(a + a^2 + a^3 + \dots + a^n)$$
$$= (a-1)\left\{\frac{a(a^n - 1)}{(a-1)}\right\} \quad (\text{If } a > 1)$$
$$= a(a^n - 1) = (a^{n+1} - a) = \text{RHS}$$

123. Given, $f(x+y) = 2^x f(y) + 4^y f(x) \quad$ ...(i)

Also, $f(y+x) = 2^y f(x) + 4^x f(y) \quad$ ...(ii)

From (i) and (ii), we get

$$2^x f(y) + 4^y f(x) = 2^y f(x) + 4^x f(y)$$

$\Rightarrow \quad \frac{f(x)}{4^x - 2^x} = \frac{f(y)}{4^y - 2^y} = k$ (say)

$\Rightarrow \quad f(x) = k(4^x - 2^x) \quad$ ...(iii)

$\Rightarrow \quad f(1) = k(4 - 2) = 2 \quad$ (given)

$\therefore \quad k = 1$

From (iii), $\quad f(x) = 4^x - 2^x$

$\Rightarrow \quad f'(x) = 4^x \ln 4 - 2^x \ln 2$

$\Rightarrow \quad f'(2) = 4^2 \ln 4 - 2^2 \ln 2 = 28 \ln 2$

$\Rightarrow \quad \lambda \ln 2 = 28 \ln 2 \qquad (\because f'(2) = \lambda \ln 2)$

$\therefore \quad \lambda = 28$

124. Given, $f(x+y) + f(x-y) = 2f(x)f(y)$ ...(i)

Replacing x by y and y by x in (i)

then $f(x+y) + f(x-y) = 2f(x)f(y)$ ...(ii)

$\therefore$ From (i) and (ii) we get

$$f(y-x) = f(x-y)$$

Putting $y = 2x$, then

$$f(x) = f(-x)$$

Hence $f(x)$ is an even function.

125. Given, $f(b + x) = f(b - x)$

Replacing x by $x + b$

Then $f(x + 2b) = f(b - (x + b)) = f(-x)$

Again replacing x by $-x$, then

$f(x) = f(2b - x) = f(2b + x)$ $\{\because f(a - x) = f(a + x)$ and $a > b$, here $a = 2b\}$

$\Rightarrow \quad f(2b + x) = f(x)$

Hence $f(x)$ is periodic with period $2b$.

126. Given, $f(xf(y)) = x^p y^q$

or $[f(xf(y))]^{1/p} = xy^{q/p}$ or $x = \dfrac{[f(xf(y))]^{1/p}}{y^{q/p}}$ ...(i)

Now let $x f(y) = 1 \Rightarrow x = \dfrac{1}{f(y)}$ ...(ii)

Then from (i), $f(y) = \dfrac{y^{q/p}}{\{f(1)\}^{1/p}}$

$\therefore \quad f(1) = \dfrac{1}{\{f(1)\}^{1/p}} \Rightarrow f(1) = 1$

Then, $f(y) = y^{q/p}$ ...(iii)

Hence $f(xy^{q/p}) = x^p y^q$

put $y^{q/p} = z$

Then $f(xz) = (xz)^p$

$\Rightarrow \quad f(\lambda) = \lambda^p$ ...(iv)

From (iii) and (iv), $y^p = y^{q/p} \Rightarrow p = \dfrac{q}{p}$

Hence, $p^2 = q$

127. Given, $f_n(x) = f_1(f_{n-1}(x))$ for $n \geq 2$

$\therefore \quad f_2(x) = f_1(f_1(x)) = f_1\left(\dfrac{x}{3} + 10\right)$

$$= \frac{\left(\dfrac{x}{3}+10\right)}{3} + 10 = \frac{x}{3^2} + \frac{10}{3} + 10 \quad \text{...(i)}$$

$f_3(x) = f_1(f_2(x)) = f_1\left(\dfrac{x}{3^2} + \dfrac{10}{3} + 10\right)$ (from (i))

$$= \frac{x}{3^3} + \frac{10}{3^2} + \frac{10}{3} + 10 \quad \text{...(ii)}$$

and $f_4(x) = f_1(f_3(x)) = f_1\left(\dfrac{x}{3^3} + \dfrac{10}{3^2} + \dfrac{10}{3} + 10\right)$ (from (ii))

$$= \frac{x}{3^4} + \frac{10}{3^3} + \frac{10}{3^2} + \frac{10}{3} + 10$$

....

Similarly, $f_n(x)$

$$= \frac{x}{3^n} + \frac{10}{3^{n-1}} + \frac{10}{3^{n-2}} + \dots + \frac{10}{3^2} + \frac{10}{3} + 10$$

$$= \frac{x}{3^n} + 10\left(1 + \frac{1}{3} + \frac{1}{3^2} + \dots + \frac{1}{3^{n-1}}\right)$$

$$= \frac{x}{3^n} + 10\left\{\frac{1\cdot\left(1-\left(\frac{1}{3}\right)^n\right)}{1-\frac{1}{3}}\right\}$$

$$= \frac{x}{3^n} + 15\left(1 - \frac{1}{3^n}\right)$$

$$= \left(\frac{x-15}{3^n}\right) + 15$$

128. Let $f(x) = \cos px$ ...(i)

$\because \quad f\left(\dfrac{1}{2}\right) = -1$

$\therefore \quad \cos\left(\dfrac{p}{2}\right) = -1 = \cos \pi$

$\Rightarrow \quad \dfrac{p}{2} = \pi \Rightarrow p = 2\pi$

From (i), $f(x) = \cos 2\pi x$

$\therefore \quad f(k) = \cos 2\pi k = 1 \ \forall\ k \in N$

$$\therefore \sum_{k=1}^{30} \frac{1}{\sin k \cdot \sin(k + f(k))}$$

$$= \sum_{k=1}^{30} \frac{1}{\sin k \cdot \sin(k+1)}$$

$$= \frac{1}{\sin 1}\sum_{k=1}^{30} \frac{\sin((k+1)-k)}{\sin k \cdot \sin(k+1)}$$

$$= \frac{1}{\sin 1}\sum_{k=1}^{30} (\cot k - \cot(k+1))$$

$$= \frac{1}{\sin 1}(\cot 1 - \cot 31) = \frac{1}{\sin 1}\left(\frac{\sin(31-1)}{\sin 1 \sin 31}\right)$$

$$= \operatorname{cosec}^2(1)\cdot \operatorname{cosec}(31)\cdot \sin(30)$$

129. Given, $f(1) + 2f(2) + 3f(3) + \ldots + nf(n) = n(n+1)f(n)$ ...(i)

Replacing n by $(n+1)$, then

$$f(1)+2f(2)+3f(3)+\ldots+nf(n)+(n+1)f(n+1) = (n+1)(n+2)f(n+1) \quad \ldots\text{(ii)}$$

Subtracting (i) from (ii), then we get

$$(n+1)f(n+1) = (n+1)(n+2)f(n+1) - n(n+1)f(n)$$

$$\Rightarrow \quad nf(n) = (n+1)f(n+1)$$

From which we conclude that

$$2f(2) = 3f(3) = 4f(4) = \ldots = nf(n)$$

Substituting the above values of

$2f(2), 3f(3), \ldots$ in terms of $nf(n)$ in (i), we have

$$f(1) + (n-1)\,nf(n) = n(n+1)f(n)$$

$$\Rightarrow \quad f(1) = 2n\,f(n)$$

$$\therefore \quad f(n) = \frac{f(1)}{2n} = \frac{1}{2n} \quad (\because f(1) = 1)$$

$$\Rightarrow \quad f(2013) = \frac{1}{2\cdot 2013} = \frac{1}{4026}$$

130. Given $\quad f(x) = \dfrac{2x(\sin x + \tan x)}{2\left[\dfrac{x+2\pi}{\pi}\right] - 3}$

$$= \frac{2x(\sin x + \tan x)}{2\left[\dfrac{x}{\pi} + 2\right] - 3} = \frac{2x(\sin x + \tan x)}{2\left(\left[\dfrac{x}{\pi}\right] + 2\right) - 3}$$

$$= \frac{2x(\sin x + \tan x)}{2\left[\dfrac{x}{\pi}\right] + 1} = \frac{x(\sin x + \tan x)}{\left[\dfrac{x}{\pi}\right] + \dfrac{1}{2}}$$

Now $\quad f(-x) = \dfrac{(-x)(\sin(-x) + \tan(-x))}{\left[-\dfrac{x}{\pi}\right] + \dfrac{1}{2}}$

$$= \frac{x(\sin x + \tan x)}{\left[-\dfrac{x}{\pi}\right] + \dfrac{1}{2}}$$

Case I. If $x = n\pi,\ n \in I$,

then $\left[-\dfrac{x}{\pi}\right] = -\left[\dfrac{x}{\pi}\right]$ and numerator of $f(-x)$ is zero

$$\Rightarrow \quad f(-x) = 0$$

Case II. If $x \neq n\pi,\ n \in I$

$$\Rightarrow \quad \left[-\frac{x}{\pi}\right] = -\left[\frac{x}{\pi}\right] - 1$$

$$\therefore \quad f(-x) = \frac{x(\sin x + \tan x)}{-\left[\dfrac{x}{\pi}\right] - 1 + \dfrac{1}{2}} = -f(x)$$

Hence, $f(x)$ is an odd function.

131. Since $\quad f(n) = 1! + 2! + 3! + \ldots + n!$ ...(i)

$$\therefore \quad f(n+1) = 1! + 2! + 3! + \ldots + n! + (n+1)! \quad \ldots\text{(ii)}$$

Subtracting (i) from (ii), then

$$f(n+1) - f(n) = (n+1)! \quad \ldots\text{(iii)}$$

Now $f(n+2) = 1! + 2! + 3! + \ldots + (n+1)! + (n+2)!$

$$= f(n+1) + (n+2)(n+1)!$$

$$= f(n+1) + (n+2)\,\{f(n+1) - f(n)\} \quad \{\text{from (iii)}\}$$

$$\therefore \quad f(n+2) = (n+3)f(n+1) + (-n-2)f(n) \quad \ldots\text{(iv)}$$

But given $f(n+2) = P(n)f(n+1) + Q(n)f(n)$ ...(v)

Comparing (iv) and (v) we get

$$P(n) = n+3 \text{ and } Q(n) = -n-2$$

Hence $\quad P(x) = x+3$ and $Q(x) = -x-2$

132. Given,

$$f(x+a) = 1 + \{2 - 3f(x) + 3(f(x))^2 - (f(x))^3\}^{1/3}$$

$$= 1 + \{1 + (1 - f(x))^3\}^{1/3} \quad \ldots\text{(i)}$$

Replacing x by $(x+a)$, then

$$f(x+2a) = 1 + \{1 + (1 - f(x+a))^3\}^{1/3}$$

$$= 1 + (1 - (f(x+a) - 1)^3\}^{1/3}$$

$$= 1 + \{1 - (1 + (1 - f(x))^3\}^{1/3} \quad \{\text{from (i)}\}$$

$$= 1 - \{(1 - f(x))^3\}^{1/3}$$

$$= 1 - (1 - f(x)) = f(x)$$

Hence, $f(x)$ is periodic with period $2a$.

133.

f
A → B
A: 1, 2, 3
B: 1, 2, 3, 4, 5

Total function without restriction $= 5 \times 5 \times 5 = 125$

When 1 is associated with (5) :

(1) → (5) →[2] (5) →[3] (5) }1

When 1 is associated with (4) :

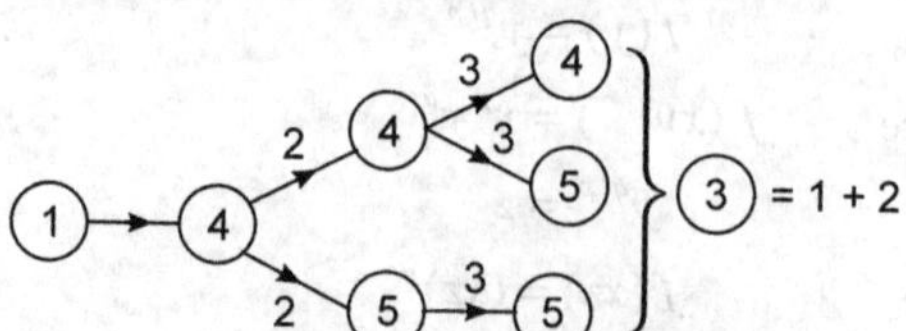

When 1 is associated with (3) :

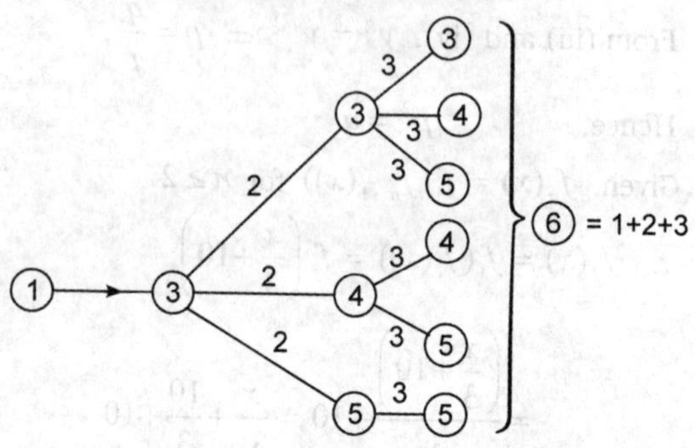

Similarly, when 1 is associated with (2), then number of such functions $=1+2+3+4=10$ and when 1 is associated with (1), then number of such functions $=1+2+3+4+5=15$.

Hence, total functions = 1 + 3 + 6 + 10 + 15 = 35

134. Let α is an integer, then $P(\alpha)=1993$

Again, let $F(x)=P(x)-1991$

Now, $F(a)=F(b)=F(c)=F(d)=0$

Thus, $F(x)=(x-a)(x-b)(x-c)(x-d)\times$ integer

$\therefore\quad F(\alpha)=P(\alpha)-1991=1993-1991=2$ ...(i)

Also, $F(\alpha)=(\alpha-a)(\alpha-b)(\alpha-c)(\alpha-d)$ Integer ...(ii)

From eqs. (i) and (ii), we get

$(\alpha-a)(\alpha-b)(\alpha-c)(\alpha-d)$ integer $=2$

Thus $(\alpha-a),(\alpha-b),(\alpha-c)$ and $(\alpha-d)$ are all divisors of 2 and are distinct.

$\Rightarrow\therefore(\alpha-a),(\alpha-b),(\alpha-c),(\alpha-d)$ are 1, – 1, 2, – 2 in same order

$\Rightarrow\quad F(\alpha)=4$ (Integer) $\neq 2$ (from (i))

($\because$ $\alpha-a$, $\alpha-b$, $\alpha-c$ and $\alpha-d$ are distinct integers)

Hence, such 'α' does not exist.

135. $\because\ f(1)+f(2)+1=f(4)\le 6\Rightarrow f(1)+f(2)\le 5$

Case I : $f(1)=1\Rightarrow f(2)=1,2,3,4$ *i.e.*, 4 mappings

Case II : $f(1)=2\Rightarrow f(2)=1,2,3$ *i.e.*, 3 mappings

Case III : $f(1)=3\Rightarrow f(2)=1,2$ *i.e.*, 2 mappings

Case IV : $f(1)=4\Rightarrow f(2)=1$ *i.e.*, 1 mapping

$f(5)$ and $f(6)$ both have 6 mappings each.

$\therefore$ Number of functions $=(4+3+2+1)\times 6\times 6$

$=10\times 36=360$

136. Let $y=f(x)=\dfrac{\sin^2 x+\sin x-1}{\sin^2 x-\sin x+2}$

and let $\sin x=z$, then

$$y=\frac{z^2+z-1}{z^2-z+2}\quad(\because -1\le z\le 1)$$

or $(y-1)z^2-(y+1)z+(2y+1)=0$ ...(i)

$$z=\frac{(y+1)\pm\sqrt{(y+1)^2-4\cdot(y-1)(2y+1)}}{2(y-1)}$$

$$z=\frac{(y+1)\pm\sqrt{(-7y^2+6y+5)}}{2(y-1)}$$

but $-7y^2+6y+5\ge 0$ or $7y^2-6y-5\le 0$

$\Rightarrow\quad \dfrac{3-2\sqrt{11}}{7}\le y\le\dfrac{3+2\sqrt{11}}{7}$ ...(ii)

Case I. If both roots of (i) are greater than 1, then $z_1>1$ and $z_2>1$

$\Rightarrow z_1+z_2>2$ and $(z_1-1)(z_2-1)>0$

$\Rightarrow\quad \dfrac{y+1}{y-1}>2$

and $\dfrac{2y+1}{y-1}-\dfrac{y+1}{y-1}+1>0$

$\Rightarrow\quad \dfrac{y-3}{y-1}<0$

and $\dfrac{2y-1}{y-1}>0$

$\Rightarrow 1<y<3$ and $y>1$ and $y<\dfrac{1}{2}$

Hence $y\in(1,3)$ ...(iii)

Case II. If both roots of (i) are less in – 1, then $z_1<-1$ and $z_2<-1$.

$\Rightarrow\quad z_1+z_2<-2$ and $(z_1+1)(z_2+1)>0$

$\Rightarrow\quad \dfrac{y+1}{y-1}+2<0$

and $\dfrac{2y+1}{y-1}+\dfrac{y+1}{y-1}+1>0$

$\Rightarrow\quad \dfrac{3y-1}{y-1}<0$ and $\dfrac{4y+1}{y-1}>0$

$\Rightarrow\quad \dfrac{1}{3}<y<1$ and $y>1$ and $y<-\dfrac{1}{4}$

Hence $y\in\phi$ ...(iv)

Case III. If one root < – 1 and other root > 1, then $z_1<-1$ and $z_2>1$.

$\Rightarrow\quad z_1<-1<z_2$ and $z_1<1<z_2$

$\Rightarrow\quad z_1<-1$ and $z_2>-1$ and $z_1<1$ and $z_2>1$

$\Rightarrow\quad (z_1+1)(z_2+1)<0$

and $(z_1-1)(z_2-1)<0$

$\Rightarrow\quad z_1z_2+(z_1+z_2)+1<0$

and $z_1z_2-(z_1+z_2)+1<0$

$\Rightarrow\quad \dfrac{2y+1}{y-1}+\dfrac{y+1}{y-1}+1<0$

and $\dfrac{2y+1}{y-1}-\dfrac{y+1}{y-1}+1<0$

$\Rightarrow\quad \dfrac{4y+1}{y-1}<0$ and $\dfrac{2y-1}{y-1}<0$

$\Rightarrow\quad -\dfrac{1}{4}<y<1$ and $\dfrac{1}{2}<y<1$

Hence $\dfrac{1}{2}<y<1$

$\therefore$ Range of $f(x)$

$$=\left[\frac{3-2\sqrt{11}}{7},\frac{3+2\sqrt{11}}{7}\right]\sim\left\{\left(\frac{1}{2},1\right)\cup(1,3)\right\}$$

$$=\left[\frac{3-2\sqrt{11}}{7},\frac{1}{2}\right]\cup\left[3,\frac{3+2\sqrt{11}}{7}\right]\cup\{1\}$$

137. We have,

$$5f(x)+4f\left(\frac{1}{x}\right)=\frac{1}{x}+3 \quad \text{...(i)}$$

Replace x by $\frac{1}{x}$, then we get

$$5f\left(\frac{1}{x}\right)+4f(x)=x+3 \quad \text{...(ii)}$$

Now, Eq. (i) × 5 – Eq. (ii) × 4, we get

$$f(x)=\frac{1}{9}\left(\frac{5}{x}-4x+3\right)$$

$$\therefore \ 18\int_1^2 f(x)\,dx = 2\int_1^2\left(\frac{5}{x}-4x+3\right)dx$$

$$=2\left[5\log_e x-2x^2+3\right]_1^2$$

$$=2(5\log_e 2-3)=10\log_e 2-6$$

138. Let $f(x)=\frac{Ax+B}{Cx-A}$

$$\therefore \quad f(f(x))=\frac{Af(x)+B}{Cf(x)-A}$$

$$=\frac{A\left(\frac{Ax+B}{Cx-A}\right)+B}{C\left(\frac{Ax+B}{Cx-A}\right)-A}=x$$

and $f\left(f\left(\frac{4}{x}\right)\right)=\frac{4}{x}$

Now, $f(f(x))+f\left(f\left(\frac{4}{x}\right)\right)=x+\frac{4}{x}\geq 4 \quad (\because AM \geq GM)$

Hence, least value of $f(f(x))+f\left(f\left(\frac{4}{x}\right)\right)$ is 4.

139. We have, $f(x)=\log_e(4x^2+11x+6)+\sin^{-1}(4x+3)$

$$+\cos^{-1}\left(\frac{10x+6}{3}\right)$$

$f(x)$ is defined, when

$4x^2+11x+6>0$ and $-1\leq 4x+3\leq 1$

and $-1\leq\frac{10x+6}{3}\leq 1$

$(x+2)(4x+3)>0$ and $-1\leq x\leq -\frac{1}{2}$

and $-\frac{9}{10}\leq x\leq -\frac{3}{10}$

$\Rightarrow \left(x<-2 \text{ or } x>-\frac{3}{4}\right)$ and $-1\leq x\leq -\frac{1}{2}$

and $-\frac{9}{10}\leq x\leq -\frac{3}{10}$

$\Rightarrow -\frac{3}{4}<x\leq -\frac{1}{2}$ i.e., $x\in\left(-\frac{3}{4},-\frac{1}{2}\right]$

$\therefore \ \alpha=-\frac{3}{4}, \beta=-\frac{1}{2}$

$\Rightarrow \alpha+\beta=-\frac{5}{4} \Rightarrow |\alpha+\beta|=\frac{5}{4}$

Hence, $36|\alpha+\beta|=36\times\frac{5}{4}=9\times 5=45$

140. Given, $f'(x)=\frac{3x+2}{2x+3}$

Now, $\quad f^2(x)=f'of'(x)$

$$=f'(f'(x))=\frac{3\left(\frac{3x+2}{2x+3}\right)+2}{2\left(\frac{3x+2}{2x+3}\right)+3}$$

$$=\frac{13x+12}{12x+13}$$

Similarly, $f^3(x)=\frac{63x+62}{62x+63}$,

$$f^4(x)=\frac{313x+312}{312x+313}$$

and $\quad f^5(x)=\frac{1563x+1562}{1562x+1563}=\frac{ax+b}{bx+a}$ (given)

$\therefore \quad a+b=1563+1562=3125$

141. Total number of onto functions

$$={}^4C_0(4-0)^5-{}^4C_1(4-1)^5$$

$$+{}^4C_2(4-2)^5-{}^4C_3(4-3)^5$$

$$=1024-4\times 243+6\times 32-4=240$$

Now taking $f(a)=1$, we get one of $f(b), f(c), f(d), f(e)$ $=1$, then total such cases $=4\times 3!=24$

Now, if $f(a)=1$, then we have distribute {2, 3, 4} amongst {b, c, d, e}. So, total cases

$$={}^3C_0(3-0)^4-{}^3C_1(3-1)^4+{}^3C_2(3-2)^4=36$$

So, number of onto function, when

$f(a)=1$ is $24+36=60$

Number of required functions, when $f(a)\neq 1$ will be

$=240-60$

$=180$

142. $\because f(1)=2, f(2)=4, f(3)=6, f(4)=8, f(5)=10,$

$f(6)=1, f(7)=3, f(8)=5, f(9)=7, f(10)=9$

and $\quad f(g(n))=\begin{cases} n+1, \text{ if } n \text{ is odd} \\ n-1, \text{ is } n \text{ is even}\end{cases}$

$\therefore \quad f(g(10))=9=f(10)\Rightarrow g(10)=10$

$f(g(1))=2=f(1)\Rightarrow g(1)=1$

$f(g(2))=1=f(6)\Rightarrow g(2)=6$

$f\left(\frac{g}{3}\right)=4=f(2)\Rightarrow g(3)=2$

$f(g(4))=3=f(7)\Rightarrow g(4)=7$

and $\quad f\left(\frac{g}{5}\right)=6=f(3)\Rightarrow g(5)=3$

So, $g(10)\,(g(1)+g(2)+g(3)+g(4)+g(5))$

$$= 10\,(1+6+2+7+3) = 10\times 19 = 190$$

143. $\because \cos\left(\frac{3\pi}{4}+x\right)-\cos\left(\frac{3\pi}{4}-x\right)$

$$= 2\sin\left(\frac{3\pi}{4}\right)\sin(-x) = -\sqrt{2}\sin x$$

and $\cos\left(\frac{\pi}{4}+x\right)+\cos\left(\frac{\pi}{4}-x\right)$

$$= 2\cos\left(\frac{\pi}{4}\right)\cos x = \sqrt{2}\cos x$$

$\therefore f(x) = \log_{\sqrt{5}}(3-\sqrt{2}\sin x+\sqrt{2}\cos x)$

Now, $1 \le 3-\sqrt{2}\sin x+\sqrt{2}\cos x \le 5$

$\Rightarrow \log_{\sqrt{5}^1} \le f(x) \le \log_{\sqrt{5}}(5)$

$\Rightarrow 0 \le f(x) \le 2$

$\Rightarrow R_f = [0, 2]$

144. $\because (gof)^{-1}(x)$ exits $\Rightarrow (gof)(x) = g(f(x))$ is bijective.

$\Rightarrow f$ must be one-one and g must be onto.

145. $\because f(x) = |x|(x-\sin x)$

$$= \begin{cases} x(x-\sin x), & x \ge 0 \\ -x(x-\sin x), & x < 0 \end{cases}$$

$$\therefore f'(x) = \begin{cases} (x-\sin x)+x(1-\cos x), & x \ge 0 \\ -(x-\sin x)-x(1-\cos x), & x < 0 \end{cases}$$

$\because 1-\cos x > 0, \forall x \in R$ and $x-\sin x > 0, \forall x > 0$

$\therefore \quad f'(x) > 0, \forall x \in (0, \infty)$

and $\quad f'(x) > 0, \forall x \in (-\infty, 0)$

$\Rightarrow f(x)$ is increasing in $R \Rightarrow f(x)$ is one-one.

Also f is odd and continuous function and as

$$\lim_{x\to\infty} f(x) = \infty \text{ and } \lim_{x\to-\infty} f(x) = -\infty$$

$\Rightarrow$ Range of $f(x)$ = R = Co-domain $\Rightarrow f(x)$ is a onto function

146. $\because f(x) = \frac{4^x}{4^x+2}$

$$\therefore f(x)+f(1-x) = \frac{4^x}{4^x+2}+\frac{4^{1-x}}{4^{1-x}+2}$$

$$= \frac{4^x}{4^x+2}+\frac{2}{2+4^x} = 2$$

$\Rightarrow \quad f(x)+f(1-x) = 1 \quad$ ...(i)

Putting $x = \frac{1}{40}, \frac{2}{40}, \frac{3}{40}, \ldots, \frac{19}{40}, \frac{20}{40}$ in (i), we get

$$f\left(\frac{1}{40}\right)+f\left(\frac{39}{40}\right) = 1$$

$$f\left(\frac{2}{40}\right)+f\left(\frac{38}{40}\right) = 1$$

$$f\left(\frac{3}{40}\right)+f\left(\frac{37}{40}\right) = 1$$

....

$$f\left(\frac{19}{40}\right)+f\left(\frac{21}{40}\right) = 1$$

$$f\left(\frac{20}{40}\right)+f\left(\frac{20}{40}\right) = 1$$

$$\Rightarrow \quad f\left(\frac{20}{40}\right) = \frac{1}{2}$$

Adding all, we get

$$f\left(\frac{1}{40}\right)+f\left(\frac{2}{40}\right)+f\left(\frac{3}{40}\right)+\ldots+f\left(\frac{39}{40}\right)$$

$$= 19+\frac{1}{2} = 19+f\left(\frac{1}{2}\right)$$

Hence, $f\left(\frac{1}{40}\right)+f\left(\frac{2}{40}\right)+f\left(\frac{3}{40}\right)+\ldots$

$$+f\left(\frac{39}{40}\right)-f\left(\frac{1}{2}\right) = 19$$

147. Only '2' in range $\Rightarrow$ 1 function.

Total number of many one functions without any restriction $= 4^3 - {}^4P_3 = 64-24 = 40$

Number of many one function with 2 is not in range.

$$= 3^3 - {}^3P_3$$

$$= 27-6 = 21$$

Hence, number of not one-one functions with 2 in range = 40 – 21 = 19

148. Given, $(f_2 \; o \; J \; o \; f_1)(x) = f_3(x)$

$\Rightarrow f_2(J(f_1(x)) = f_3(x)$

$$\Rightarrow \quad f_2\left(J\left(\frac{1}{x}\right)\right) = \frac{1}{1-x}$$

$$\Rightarrow \quad 1-J\left(\frac{1}{x}\right) = \frac{1}{1-x}$$

$$\Rightarrow \quad J\left(\frac{1}{x}\right) = 1-\frac{1}{1-x} = -\frac{x}{1-x}$$

$$\Rightarrow \quad J(x) = -\frac{\frac{1}{x}}{1-\frac{1}{x}} = \frac{1}{1-x} = f_3(x)$$

149. $\because \alpha = {}^7P_5 = 7\times 6\times 5\times 4\times 3$

$$\Rightarrow \quad \frac{\alpha}{5!} = 21$$

and $\beta = {}^5C_0(5-0)^7 - {}^5C_1(5-1)^7$

$$+\,{}^5C_2(5-2)^7 - {}^5C_3(5-3)^7 + {}^5C_4(5-4)^7$$

$$\Rightarrow \quad \frac{\beta}{5!} = 5\times 28 = 140$$

Hence, $\frac{1}{5!}(\beta-\alpha) = 140-21 = 119$

150. For E_1, $\frac{x}{x-1} > 0$ and $x \neq 1 \Rightarrow x \in (-\infty, 0) \cup (1, \infty)$

For E_2, $\quad -1 \le \log_e\left(\frac{x}{x-1}\right) \le 1 \Rightarrow \frac{1}{e} \le \frac{x}{x-1} \le e$

$$\Rightarrow \quad \frac{1}{e} - 1 \le \frac{1}{x-1} \le e - 1$$

$$\Rightarrow \quad (x-1) \in \left(-\infty, \frac{e}{1-e}\right] \cup \left[\frac{1}{e-1}, \infty\right)$$

$$\Rightarrow \quad x \in \left(-\infty, \frac{1}{1-e}\right] \cup \left[\frac{e}{e-1}, \infty\right)$$

For E_1, $\frac{x}{x-1} \in (0, \infty) - \{1\}$

$$\Rightarrow \log_e\left(\frac{x}{x-1}\right) \in (-\infty, \infty) - \{0\}$$

$$\Rightarrow f(x) \in (-\infty, 0) \cup (0, \infty)$$

and $g(x) = \sin^{-1}\left(\log\left(\frac{x}{x-1}\right)\right) \in \left[-\frac{\pi}{2}, \frac{\pi}{2}\right] - \{0\}$

$\Rightarrow$ P $\rightarrow$ 4; Q $\rightarrow$ 2; R $\rightarrow$ 1; S $\rightarrow$ 1

151. We have,

$$f(x) + 2f\left(\frac{1}{x}\right) = 3x \quad \text{...(i)}$$

Replace x by $\frac{1}{x}$, then

$$f\left(\frac{1}{x}\right) + 2f(x) = \frac{3}{x} \quad \text{...(ii)}$$

Subtract (i) from 2 × (ii), we get

$$3f(x) = \frac{6}{x} - 3x$$

$$\Rightarrow \quad f(x) = \frac{2}{x} - x$$

$$\because \quad f(x) = f(-x)$$

$$\Rightarrow \quad \frac{2}{x} - x = -\frac{2}{x} + x$$

$$\Rightarrow \quad x = \frac{2}{x}$$

$$\Rightarrow \quad x^1 = 2 \Rightarrow x = \sqrt{2}, -\sqrt{2}$$